De C

Hans-Otto Ge

Hans-Otto Georgii

Stochastics

Introduction to Probability and Statistics

2nd Revised and Extended Edition

Translated by
Marcel Ortgiese, Ellen Baake and the Author

De Gruyter

Prof. Dr. Hans-Otto Georgii
Mathematical Institute
Ludwig-Maximilians-Universität Munich
Theresienstr. 39
80333 Munich
Germany

Mathematics Subject Classification 2010: Primary: 60-01; Secondary: 62-01

ISBN 978-3-11-029254-1
e-ISBN 978-3-11-029360-9

Library of Congress Cataloging-in-Publication Data

A CIP catalog record for this book has been applied for at the Library of Congress.

Bibliographic information published by the Deutsche Nationalbibliothek

The Deutsche Nationalbibliothek lists this publication in the Deutsche Nationalbibliografie; detailed bibliographic data are available in the Internet at http://dnb.dnb.de.

© 2013 Walter de Gruyter GmbH, Berlin/Boston

Typesetting: Dimler & Albroscheit Partnerschaft, Müncheberg
Printing and binding: Hubert & Co. GmbH & Co. KG, Göttingen
∞ Printed on acid-free paper

Printed in Germany

www.degruyter.com

Preface

Chance – or what appears to us as such – is ubiquitous. Not only in the games of chance such as lottery or roulette where risk is played with, but also in substantial parts of everyday life. Every time an insurance company uses the claim frequencies to calculate the future premium, or a fund manager the most recent stock charts to rearrange his portfolio, every time cars are jamming at a traffic node or data packages at an internet router, every time an infection spreads out or a bacterium turns into a resistant mutant, every time pollutant concentrations are measured or political decisions are based on opinion polls – in all such cases a considerable amount of randomness comes into play, and there is a need to analyse the random situation and to reach at rational conclusions in spite of the inherent uncertainty. Precisely this is the objective of the field of stochastics, the 'mathematics of chance'. Stochastics is therefore a highly applied science, which tries to solve concrete demands of many disciplines. At the same time, it is genuine mathematics – with sound systematics, clear-cut concepts, deep theorems and sometimes surprising cross-connections. This interplay between applicability on the one side and mathematical precision and elegance on the other makes up the specific appeal of stochastics, and a variety of natural questions determines its lively and broad development.

This book offers an introduction to the typical way of thinking, as well as the basic methods and results of stochastics. It grew out of a two-semester course which I gave repeatedly at the University of Munich. It is addressed to students of mathematics in the second year, and also to scientists and computer scientists who intend not only to apply stochastics, but also to understand its mathematical side. The two parts of stochastics – probability theory and statistics – are presented in two separate parts of the book because of their own scopes and methods, but are united under the same cover on purpose. For, the statistics is built on the concepts and methods of probability theory, whereas the latter needs the former as a bridge to reality. In the choice of the material I confined myself deliberately to the central subjects belonging to the standard curriculum of the corresponding mathematical courses. (It is thus unavoidable that some readers will miss their favourite subjects, e.g., the resampling methods of statistics.) The standard themes, however, are discussed with all necessary details. Rather than starting with discrete models I preferred to present (and motivate) the general measure theoretic framework right from the beginning, and some theoretical issues are also treated later as the case arises. In general, however, the measure theoretic apparatus is confined to what is absolutely necessary, and the emphasis is on the development of a stochastic intuition.

This text comprises a little more material than can be presented in a four-hour course over two semesters. The reader may thus want to make a selection. Several possibilities present themselves. For a first overview, the reader may concentrate on concepts, theorems, and examples and skip all proofs. In particular, this is a practicable route for non-mathematicians. A deeper understanding, of course, requires the study of a representative selection of proofs. On the other hand, a reader mainly interested in the theory and familiar with some applications may skip a portion of examples. For a short route through Part I leading directly to statistics, one can restrict oneself to the essentials of the first chapters up to Section 3.4, as well as Sections 4.1 and 4.3, and 5.1 and 5.2. The core of Part II consists of Sections 7.1 – 7.5, 8.1 – 8.2, 9.2, Chapter 10, as well as 11.2 and 12.1. Depending on the specific interests, it will facilitate orientation to browse through some portions of the text and return to them later when needed. A list of notational conventions can be found on page 395.

At the end of each chapter there is a collection of problems offering applications, additions, or supplements to the text. Their difficulty varies, but is not indicated because the reader should follow only his or her interests. The main point is to try for oneself. Nevertheless, this second English edition now provides draft solutions of selected problems, marked with [S]. These should be used for self-testing, rather than lulling the reader's willingness to tackle the problems independently.

As every textbook, this one grew out of more sources than can possibly be identified. Much inspiration came from the classical German texts of Ulrich Krengel [38] and Klaus Krickeberg and Herbert Ziezold [39], which had strong influence on the introductory courses in stochastics all over Germany. I also got many impulses from my Munich stochastics colleagues Peter Gänßler and Helmut Pruscha as well as all those responsible for the problem classes of my lectures during more than two decades: Peter Imkeller, Andreas Schief, Franz Strobl, Karin Münch-Berndl, Klaus Ziegler, Bernhard Emmer, and Stefan Adams. I am very grateful to all of them.

The English translation of the German original would not have appeared without the assistance of two further colleagues: Marcel Ortgiese accepted to lay the foundation by preparing an initial English version, so that I could concentrate on details and cosmetic changes. Ellen Baake took pleasure in giving a final polish to the English and suggesting numerous clarifications. I gratefully acknowledge their help.

Munich, June 2012   *Hans-Otto Georgii*

Contents

Mathematics and Chance

What is stochastics[1]? In dictionaries of classical Greek one finds

στόχος	(stóchos)	goal, aim, guess, conjecture
στοχαστικός	(stochastikós)	skilful in aiming at, able to hit
στοχάζομαι	(stocházomai)	I aim at, guess at, infer, explore

Its current usage is captured by the sentence

Stochastics is the science of the rules of chance.

At first sight, this statement seems to be a contradiction in itself, since, in everyday life, one speaks of chance when no rules are apparent. A little thought, however, reveals that chance does indeed follow some rules. For example, if you flip a coin very often, you will have no doubt that heads will come up approximately half of the time. Obviously, this is a law of chance, and is generally accepted as such. Nevertheless, it is a widespread belief that such laws are too vague to be made precise, let alone to be formalised mathematically. Intriguingly, the opposite is true: Mathematics offers an exact language even for such seemingly disordered phenomena; a language that allows to state, and prove, a variety of laws obeyed by chance. The experience mentioned above, namely that heads shows up in about one half of a large number of coin flips, thus turns into a mathematical theorem, the law of large numbers. Stochastics is the part of mathematics that develops an appropriate formalism for describing the principles of randomness, for detecting rules where they seem to be absent, and for using them. This book presents the basic ideas and central results.

*

But what is chance? This is a philosophical question, and a satisfactory answer is still missing. Whether 'god plays dice with the universe' or, in fact, he does not (as was postulated by Albert Einstein in his famous dictum), whether randomness is only fictitious and merely a consequence of our incomplete knowledge or, on the contrary, is inherent to nature – these questions are still unresolved.

---

[1] So far, the noun 'stochastics' is not a well-established English word, in contrast to the adjective 'stochastic'. But since there is a need for a term comprising both probability theory and statistics, its use is spreading, and is expected to become standard, as it did in other languages. So we use it in this book.

As a matter of fact, however, we may refrain from trying to understand the nature of 'chance per se', by good reasons. We will never be able to investigate the universe as a whole. Rather, we will always have to focus on quite a special, restricted phenomenon, and we will restrict attention to the nature of this small part of reality. Even if the phenomenon considered could be explained in parts by its general circumstances (as we could, for instance, anticipate the number shown by a dice if we only knew in sufficient detail how it is thrown) – even then, it is much more practicable, and better adapted to our human perspective, if we adopt the viewpoint that the phenomenon is governed by chance. This kind of chance then comprises both, an indeterminacy possibly inherent to nature, and our (perhaps unavoidable) ignorance of the determining factors.

*

How does mathematics then come into play? As soon as a definite part of the real world is selected for investigation, one can try and collect all its relevant aspects into a mathematical model. Typically, this is done

▷ by *abstracting* from 'dirty' details that might distract from the essentials of the problem at hand, that is, by 'smoothing away' all features that seem irrelevant; and, on the other hand,

▷ by *mathematical idealisation*, i.e., by widening the scope using mental or formal limiting procedures that allow for a clear-cut description of the relevant phenomena.

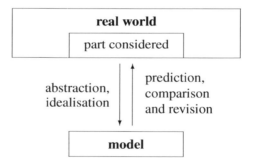

The model thus obtained can then be investigated mathematically, and the resulting predictions have to be checked against reality. If necessary, the model must be revised. In general, finding an appropriate model is a delicate process. It requires much flair and falls beyond mathematics. There are, however, some basic principles and mathematical structures; these will be discussed in this text.

*

Stochastics is composed of two equal parts: probability theory and statistics. The objective of probability theory is the description and investigation of specific random phenomena. Statistics seeks for methods of drawing rational conclusions from uncertain random observations. This, of course, requires and builds on the models of probability theory. The other way round, probability theory needs the validation obtained by comparing model and reality, which is made possible by statistics. Part I of this text offers an introduction to the basic concepts and results of probability theory. Part II then presents an introduction into theory and methods of mathematical statistics.

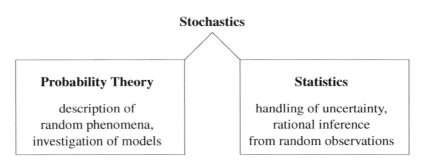

Stochastics

Probability Theory	**Statistics**
description of random phenomena, investigation of models	handling of uncertainty, rational inference from random observations

Part I

Probability Theory

Chapter 1

Principles of Modelling Chance

This chapter develops the fundamentals of modelling chance. The primary questions read: How can one describe a specific random phenomenon mathematically? What are the general properties of the stochastic model so obtained? How can one extract a particular aspect from a given model? In answering these questions, we will be led to the fundamental notions of 'probability space' and 'random variable'. We will also be faced with a few technical questions, which might be somewhat unpleasant to deal with at the beginning, but will enable us to concentrate on the principal ideas later on.

1.1 Probability Spaces

To build a mathematical model for a specific scenario of chance, one proceeds in three stages as follows.

1.1.1 Determining the Sample Space

If one aims at describing the effects of chance, the first question is: What can happen in the given situation? And which part of this is relevant? The possibilities that seem natural to distinguish are then collected into a set Ω. This is best understood by examples.

(1.1) Example. *Rolling a dice once.* If we roll a dice on a table, it can stop in infinitely many positions. We are not interested in its exact final position, and even less in the precise hand movement when throwing the dice, but only in the number that is showing. The interesting outcomes are thus captured by the set $\Omega = \{1, \ldots, 6\}$. With the restriction to this Ω, we fade out the irrelevant part of reality.

(1.2) Example. *Rolling a dice several times.* If the dice is thrown n times and we are interested in the sequence of numbers shown, then the relevant outcomes are those in the product space $\Omega = \{1, \ldots, 6\}^n$; for $\omega = (\omega_1, \ldots, \omega_n) \in \Omega$ and $1 \leq i \leq n$, ω_i represents the number showing at the ith throw.

On the other hand, if we are not interested in the exact sequence of numbers, but only in how often each number appears, it is more natural to choose

$$\widehat{\Omega} = \Big\{ (k_1, \ldots, k_6) \in \mathbb{Z}_+^6 : \sum_{a=1}^{6} k_a = n \Big\}$$

as the set of all possible outcomes. Here $\mathbb{Z}_+ = \{0, 1, 2, \dots\}$ is the set of all non-negative integers, and k_a stands for the number of throws the dice shows a.

(1.3) Example. *Tossing a coin infinitely often.* When we toss a coin n times, the appropriate set of outcomes is $\Omega = \{0, 1\}^n$, in analogy with the previous example (provided we are interested in the sequence of outcomes). If we decide to flip the coin once more, do we have to consider a new Ω? This would not be very practical; thus our model should not be limited to a fixed number of tosses. Moreover, we are especially interested in patterns that only appear for large n, that is in the limit as $n \to \infty$. Therefore it is often convenient to choose an idealised model, which admits an infinite number of tosses. (As an analogy, think of the mathematically natural transition from finite to infinite decimal fractions.) The set of all possible outcomes is then

$$\Omega = \{0, 1\}^{\mathbb{N}} = \{\omega = (\omega_i)_{i \in \mathbb{N}} : \omega_i \in \{0, 1\}\},$$

the set of all infinite sequences of zeros and ones.

As the examples demonstrate, the first step in the process of setting up a model for a random phenomenon is to decide which possibilities we would like to distinguish and observe, and which idealising assumptions might be useful. Considering this, one determines a set Ω of relevant outcomes. This Ω is called the *set of outcomes* or the *sample space*.

1.1.2 Determining a σ-Algebra of Events

In general we are not interested in the detailed outcome of a random experiment, but only in the occurrence of an *event* that consists of a certain selection of outcomes. Such events correspond to subsets of Ω.

(1.4) Example. *Events as sets of outcomes.* The event 'In n coin flips, heads shows at least k times' corresponds to the subset

$$A = \left\{\omega = (\omega_1, \dots, \omega_n) \in \Omega : \sum_{i=1}^n \omega_i \geq k\right\}$$

of the sample space $\Omega = \{0, 1\}^n$.

Our aim is to set up a system $\mathscr{F}$ of events, such that we can consistently assign to each event $A \in \mathscr{F}$ a probability $P(A)$ for A to occur.

Why so cautious? Why not assign a probability to *every* subset of Ω, in other words, why not simply take $\mathscr{F}$ to be the power set $\mathscr{P}(\Omega)$ (i.e., the set of *all* subsets of Ω)? Indeed, this is perfectly possible as long as Ω is countable. However, it is impossible in general, as the following 'no-go theorem' shows.

(1.5) Theorem. The power set is too large, Vitali 1905. *Let* $\Omega = \{0, 1\}^{\mathbb{N}}$ *be the sample space for infinite coin tossing. Then there is* no *mapping* $P : \mathscr{P}(\Omega) \to [0, 1]$ *with the properties*

(N) Normalisation. $P(\Omega) = 1$.

(A) σ-Additivity. *If* $A_1, A_2, \ldots \subset \Omega$ *are pairwise disjoint, then*

$$P\Big(\bigcup_{i \geq 1} A_i\Big) = \sum_{i \geq 1} P(A_i).$$

(The probabilities of countably many incompatible events can be added up.)

(I) Flip invariance. *For all* $A \subset \Omega$ *and* $n \geq 1$ *one has* $P(T_n A) = P(A)$; *here*

$$T_n : \omega = (\omega_1, \omega_2, \ldots) \to (\omega_1, \ldots, \omega_{n-1}, 1 - \omega_n, \omega_{n+1}, \ldots)$$

is the mapping from Ω *onto itself that inverts the result of the nth toss, and* $T_n A = \{T_n(\omega) : \omega \in A\}$ *is the image of* A *under* T_n. *(This expresses the fairness of the coin and the independence of the tosses.)*

At first reading, only the result is important, and its proof may be skipped.

Proof. We define an equivalence relation $\sim$ on Ω as follows. Say $\omega \sim \omega'$ if and only if $\omega_n = \omega'_n$ for all sufficiently large n. By the axiom of choice, there is a set $A \subset \Omega$ that contains exactly one element from each equivalence class.

Let $\mathscr{S} = \{S \subset \mathbb{N} : |S| < \infty\}$ be the set of all finite subsets of $\mathbb{N}$. As $\mathscr{S}$ is the union of the countably many finite sets $\{S \subset \mathbb{N} : \max S = m\}$ for $m \in \mathbb{N}$, it is countable. For $S = \{n_1, \ldots, n_k\} \in \mathscr{S}$ let $T_S := \prod_{n \in S} T_n = T_{n_1} \circ \cdots \circ T_{n_k}$ be the flip at all times contained in S. Then we have:

▷ $\Omega = \bigcup_{S \in \mathscr{S}} T_S A$, since for every $\omega \in \Omega$ there exists an $\omega' \in A$ such that $\omega \sim \omega'$, and thus an $S \in \mathscr{S}$ with $\omega = T_S \omega' \in T_S A$.

▷ The sets $(T_S A)_{S \in \mathscr{S}}$ are pairwise disjoint. For, suppose that $T_S A \cap T_{S'} A \neq \varnothing$ for some $S, S' \in \mathscr{S}$. Then there exist $\omega, \omega' \in A$ such that $T_S \omega = T_{S'} \omega'$ and so $\omega \sim T_S \omega = T_{S'} \omega' \sim \omega'$. By the choice of A, this means that $\omega = \omega'$ and thus $S = S'$.

Applying successively the properties (N), (A) and (I) of P we thus find

$$1 = P(\Omega) = \sum_{S \in \mathscr{S}} P(T_S A) = \sum_{S \in \mathscr{S}} P(A).$$

This is impossible, since the infinite sum of the same number is either 0 or ∞. ◇

How to proceed after this negative result? We have to insist on the properties (N), (A) and (I), since (N) and (A) are indispensable and elementary (just finite additivity is not sufficient, as we will see shortly), and (I) is characteristic of the coin tossing model. But the above proof has shown that the problems only arise for rather unusual, 'abnormal' sets $A \subset \Omega$. A natural way out is therefore to restrict the definition of probabilities to an appropriate *subsystem* $\mathscr{F} \subset \mathscr{P}(\Omega)$, which excludes the 'abnormal' sets. Fortunately, it turns out that this suffices both in theory and in practice. In particular, we will see in Example (3.29) that a function P satisfying (N), (A) and (I) can indeed be defined on a suitable $\mathscr{F}$ that is large enough for all practical purposes.

What are sensible properties we should ask of the system $\mathscr{F}$? The minimal requirements are apparently those in the following

Definition. Suppose $\Omega \neq \emptyset$. A system $\mathscr{F} \subset \mathscr{P}(\Omega)$ satisfying

(a) $\Omega \in \mathscr{F}$ ($\mathscr{F}$ contains the 'certain event')

(b) $A \in \mathscr{F} \Rightarrow A^c := \Omega \setminus A \in \mathscr{F}$ ($\mathscr{F}$ allows the logical negation)

(c) $A_1, A_2, \ldots \in \mathscr{F} \Rightarrow \bigcup_{i \geq 1} A_i \in \mathscr{F}$ ($\mathscr{F}$ allows the countable logical 'or')

is called a *σ-algebra* (or *σ-field*) on Ω. The pair $(\Omega, \mathscr{F})$ is then called an *event space* or a *measurable space*.

These three properties can be combined to obtain some further properties of a σ-algebra $\mathscr{F}$. By (a) and (b), the 'impossible event' $\emptyset$ belongs to $\mathscr{F}$. Together with (c) this gives, for $A, B \in \mathscr{F}$, that $A \cup B = A \cup B \cup \emptyset \cup \cdots \in \mathscr{F}$, $A \cap B = (A^c \cup B^c)^c \in \mathscr{F}$, and $A \setminus B = A \cap B^c \in \mathscr{F}$. Similarly, the countable intersection of sets in $\mathscr{F}$ also belongs to $\mathscr{F}$.

The σ in the name σ-algebra has become convention as a reminder of the fact that, in (c), we consider countably infinite (instead of only finite) unions (σ like 'sums'). Finite unions are not sufficient, because we also need to consider so-called tail events, such as 'coin falls heads for infinitely many tosses' or 'the relative frequency of heads tends to $1/2$, if the number of tosses tends to ∞'. Such events can not be captured by finite unions, but require countably infinite unions (and intersections).

At this point, let us pause for a moment to distinguish between the three set-theoretic levels we are moving on; see Figure 1.1. The base level consists of the set Ω containing all outcomes ω. Above this there is the event level $\mathscr{P}(\Omega)$; its *elements* are *subsets* of the base level Ω. This structure repeats itself once more: σ-algebras are *subsets* of $\mathscr{P}(\Omega)$, so they are *elements* of the top level $\mathscr{P}(\mathscr{P}(\Omega))$.

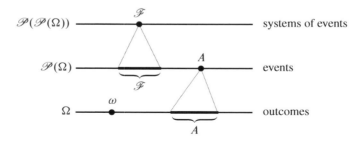

Figure 1.1. The three conceptual levels of stochastics.

How can one determine a σ-algebra in Ω? One starts from a system $\mathscr{G}$ of 'good', that is, especially simple or natural sets, whose probability one can easily guess or determine. Then this system is enlarged just as much as necessary to obtain a σ-algebra. More precisely, the following construction principle can be used.

(1.6) Remark and Definition. *Generating σ-algebras.* If $\Omega \neq \varnothing$ and $\mathscr{G} \subset \mathscr{P}(\Omega)$ is arbitrary, then there is a unique smallest σ-algebra $\mathscr{F} = \sigma(\mathscr{G})$ on Ω such that $\mathscr{F} \supset \mathscr{G}$. This $\mathscr{F}$ is called the *σ-algebra generated by* $\mathscr{G}$, and $\mathscr{G}$ is called a *generator* of $\mathscr{F}$.

Proof. Let Σ be the system of all σ-algebras $\mathscr{A}$ in Ω satisfying $\mathscr{A} \supset \mathscr{G}$. (So Σ is a subset of the top level in Figure 1.1.) Σ is non-empty, since $\mathscr{P}(\Omega) \in \Sigma$. Hence, we can set $\mathscr{F} := \bigcap_{\mathscr{A} \in \Sigma} \mathscr{A}$. As each $\mathscr{A} \in \Sigma$ is a σ-algebra, so is $\mathscr{F}$, as is easily checked by spelling out the defining properties (a) to (c). $\mathscr{F}$ thus belongs to Σ, and is obviously its unique smallest element. This proves the claim. $\diamond$

Here are three standard examples of this construction.

(1.7) Example. *The power set.* Suppose Ω is countable and $\mathscr{G} = \{\{\omega\} : \omega \in \Omega\}$ the system containing the singleton sets of Ω. Then, $\sigma(\mathscr{G}) = \mathscr{P}(\Omega)$. Indeed, since every $A \in \mathscr{P}(\Omega)$ is countable, it follows from axiom (c) that $A = \bigcup_{\omega \in A} \{\omega\} \in \sigma(\mathscr{G})$.

(1.8) Example and Definition. *The Borel σ-algebra.* Let $\Omega = \mathbb{R}^n$ and

$$\mathscr{G} = \left\{ \prod_{i=1}^n [a_i, b_i] : a_i < b_i, \ a_i, b_i \in \mathbb{Q} \right\}$$

be the system consisting of all compact rectangular boxes in $\mathbb{R}^n$ with rational vertices and edges parallel to the axes. In honour of Émile Borel (1871–1956), the system $\mathscr{B}^n := \sigma(\mathscr{G})$ is called the *Borel σ-algebra* on $\mathbb{R}^n$, and every $A \in \mathscr{B}^n$ a *Borel set*; in the case $n = 1$, we simply write $\mathscr{B}$ instead of $\mathscr{B}^1$. The Borel σ-algebra is much larger than one might expect at first sight. Namely, we have:

(a) Every open set $A \subset \mathbb{R}^n$ is Borel. To see this, it is sufficient to note that every $\omega \in A$ has a neighbourhood $Q \in \mathscr{G}$ with $Q \subset A$, so that $A = \bigcup_{Q \in \mathscr{G}, Q \subset A} Q$, a union of countably many sets. Our claim thus follows from property (c) of a σ-algebra.

(b) Every closed $A \subset \mathbb{R}^n$ is Borel, since A^c is open and hence by (a) Borel.

(c) It is impossible to describe $\mathscr{B}^n$ constructively. It does not only consist of count-able unions of boxes and their complements; rather, the procedure of taking complements and countable unions has to be repeated as many times as there are countable ordinal numbers, hence uncountably often; see [31, p. 139] or [6, pp. 24, 29]. But this does not cause problems. It suffices to know that $\mathscr{B}^n$ is large enough to contain all sets in $\mathbb{R}^n$ that may occur in practice, but still smaller than $\mathscr{P}(\mathbb{R})$. In fact, the existence of non-Borel sets follows from Theorem (1.5) and the proof of Theorem (3.12).

We will also need the following facts:

(d) Besides the system $\mathscr{G}$ of compact intervals, the Borel σ-algebra $\mathscr{B} = \mathscr{B}^1$ on $\mathbb{R}$ also admits the generator

$$\mathscr{G}' = \{]-\infty, c] : c \in \mathbb{R}\},$$

the system of all left-infinite closed half-lines. For, assertion (b) shows that $\mathscr{G}' \subset \mathscr{B}$ and thus, by the minimality of $\sigma(\mathscr{G}')$, $\sigma(\mathscr{G}') \subset \mathscr{B}$. Conversely, $\sigma(\mathscr{G}')$ contains all half-open intervals $]a, b] =]-\infty, b] \setminus]-\infty, a]$, and so all compact intervals $[a, b] = \bigcap_{n \geq 1}]a - \frac{1}{n}, b]$, hence also the σ-algebra $\mathscr{B}$ generated by them.

Likewise, $\mathscr{B}$ can also be generated by the left-infinite open half-lines, and also by the open or closed right-infinite half-lines.

(e) For $\varnothing \neq \Omega \subset \mathbb{R}^n$, the system $\mathscr{B}^n_\Omega = \{A \cap \Omega : A \in \mathscr{B}^n\}$ is a σ-algebra on Ω; it is called the *Borel σ-algebra on Ω.*

(1.9) Example and Definition. *Product σ-algebra.* Let Ω be the Cartesian product of arbitrary sets E_i, i.e., $\Omega = \prod_{i \in I} E_i$ for an index set $I \neq \varnothing$. Let $\mathscr{E}_i$ be a σ-algebra on E_i, $X_i : \Omega \to E_i$ the projection mapping onto the ith coordinate, and $\mathscr{G} = \{X_i^{-1} A_i : i \in I, A_i \in \mathscr{E}_i\}$ the system of all sets in Ω which are specified by an event in a single coordinate. Then, $\bigotimes_{i \in I} \mathscr{E}_i := \sigma(\mathscr{G})$ is called the *product σ-algebra* of the $\mathscr{E}_i$ on Ω. If $E_i = E$ and $\mathscr{E}_i = \mathscr{E}$ for all i, we write $\mathscr{E}^{\otimes I}$ instead of $\bigotimes_{i \in I} \mathscr{E}_i$. For example, the Borel σ-algebra $\mathscr{B}^n$ on $\mathbb{R}^n$ is exactly the n-fold product σ-algebra of the Borel σ-algebra $\mathscr{B} = \mathscr{B}^1$ on $\mathbb{R}$, meaning that $\mathscr{B}^n = \mathscr{B}^{\otimes n}$; cf. Problem 1.3.

The second step in the process of building a model can now be summarised as follows. Theorem (1.5) forces us to introduce a σ-algebra $\mathscr{F}$ of events in Ω. Fortunately, in most cases the choice of $\mathscr{F}$ is canonical. In this book, only the following three *standard cases* will appear.

▷ *Discrete case:* If Ω is at most countable, one can set $\mathscr{F} = \mathscr{P}(\Omega)$.

▷ *Real case:* For $\Omega \subset \mathbb{R}^n$, the natural choice is $\mathscr{F} = \mathscr{B}^n_\Omega$.

▷ *Product case:* If $\Omega = \prod_{i \in I} E_i$ and every E_i is equipped with a σ-algebra $\mathscr{E}_i$, one takes $\mathscr{F} = \bigotimes_{i \in I} \mathscr{E}_i$.

Once a σ-algebra $\mathscr{F}$ is fixed, every $A \in \mathscr{F}$ is called an *event* or a *measurable set*.

1.1.3   Assigning Probabilities to Events

The decisive point in the process of building a stochastic model is the next step: For each $A \in \mathscr{F}$ we need to define a value $P(A) \in [0, 1]$ that indicates the probability of A. Sensibly, this should be done so that the following holds.

(N) Normalisation: $P(\Omega) = 1$.

(A) σ-Additivity: For pairwise disjoint events $A_1, A_2, \ldots \in \mathscr{F}$ one has

$$P\Big(\bigcup_{i \geq 1} A_i\Big) = \sum_{i \geq 1} P(A_i).$$

(Pairwise disjoint means that $A_i \cap A_j = \varnothing$ for $i \neq j$.)

Definition. Let $(\Omega, \mathscr{F})$ be an event space. A function $P : \mathscr{F} \to [0, 1]$ satisfying the properties (N) and (A) is called a *probability measure* or a *probability distribution*, in short a *distribution* (or, a little old-fashioned, a *probability law*) on $(\Omega, \mathscr{F})$. Then, the triple $(\Omega, \mathscr{F}, P)$ is called a *probability space*.

> Properties (N) and (A), together with the non-negativity of the probability measure, are also known as *Kolmogorov's axioms*, since it was Andrej N. Kolmogorov (1903–1987) who in 1933 emphasised the significance of measures for the mathematical foundation of probability theory, and thus gave a decisive input for the development of modern probability theory.

Let us summarise: To describe a particular scenario of chance mathematically, one has to choose an appropriate probability space. Typically, the most delicate point is the choice of the probability measure P, since this contains all the relevant information about the kind of randomness. In Chapter 2, as well as in many examples later on, we will show how this can be done. At this point let us only mention the elementary, but degenerate, example of a probability measure that describes a situation without randomness.

(1.10) Example and Definition. *Deterministic case.* If $(\Omega, \mathscr{F})$ is an arbitrary event space and $\xi \in \Omega$, then

$$\delta_\xi(A) = \begin{cases} 1 & \text{if } \xi \in A, \\ 0 & \text{otherwise} \end{cases}$$

defines a probability measure δ_ξ on $(\Omega, \mathscr{F})$. It describes an experiment with the certain outcome ξ and is called the *Dirac distribution* or the *unit mass* at the point ξ.

We close this section with some remarks on the

Interpretation of probability measures. The concept of a probability space does not give an answer to the philosophical question what probability really is. The following are common answers:

(a) *The naive interpretation.* 'Nature' is uncertain about what it is doing, and $P(A)$ represents the degree of certainty of its decision to let A happen.

(b) *The frequency interpretation.* $P(A)$ is the relative frequency with which A occurs under some specified conditions.

(c) *The subjective interpretation.* $P(A)$ is the degree of certainty with which I would be willing to bet on the occurrence of A according to my personal evaluation of the situation.

(The interpretations (a) and (c) are dual concepts, the uncertainty moves from nature to the observer.)

In general, we cannot say which interpretation is to be preferred, since this depends on the nature of the problem at hand. If a random experiment can be repeated independently, the interpretations (a) and (b) may seem most natural. The probabilities of rain specified by weather forecasts are obviously based on (b), and so are the probabilities used in the insurance industry. The question that was asked before March 23 in 2001, namely 'What is the probability that humans will be injured by the crash of the space station *Mir*?', used the subjective interpretation (c), since it dealt with a singular event. A comprehensive and very stimulating historical and philosophical discussion of the notion of probability can be found in Gigerenzer et al. [23].

Fortunately, the validity of the mathematical statements about a probability model does not depend on its interpretation. The value of mathematics is not limited by the narrowness of human thought. This, however, should not to be misunderstood to mean that mathematics can take place in an 'ivory tower'. Stochastics thrives on the interaction with the real world.

1.2   Properties and Construction of Probability Measures

What are the implications of the assumption (A) of σ-additivity? We start with some elementary consequences.

(1.11) Theorem. Probability rules. *Every probability measure P on an event space $(\Omega, \mathscr{F})$ has the following properties, for arbitrary events $A, B, A_1, A_2, \ldots \in \mathscr{F}$.*

(a) $P(\emptyset) = 0$.

(b) Finite additivity. $P(A \cup B) + P(A \cap B) = P(A) + P(B)$, *and so in particular* $P(A) + P(A^c) = 1$.

(c) Monotonicity. *If $A \subset B$ then $P(A) \leq P(B)$.*

(d) σ-Subadditivity. $P\left(\bigcup_{i \geq 1} A_i\right) \leq \sum_{i \geq 1} P(A_i)$.

(e) σ-Continuity. *If either $A_n \uparrow A$ or $A_n \downarrow A$ (i.e., the A_n are either increasing with union A, or decreasing with intersection A), then $P(A_n) \xrightarrow[n \to \infty]{} P(A)$.*

Proof. (a) Since the empty set is disjoint from itself, the sequence $\varnothing, \varnothing, \ldots$ consists of pairwise disjoint events. In this extreme case, σ-additivity (A) thus gives

$$P(\varnothing) = P(\varnothing \cup \varnothing \cup \cdots) = \sum_{i=1}^{\infty} P(\varnothing).$$

But this is only possible when $P(\varnothing) = 0$.

(b) Suppose first that A and B are disjoint. Since property (A) requires an infinite sequence, we append the empty set infinitely often to the sets A and B. Hence we obtain from (A) and statement (a)

$$P(A \cup B) = P(A \cup B \cup \varnothing \cup \varnothing \cup \cdots) = P(A) + P(B) + 0 + 0 + \cdots.$$

So the probability is additive when an event is split into finitely many disjoint parts. In the general case, it thus follows that

$$P(A \cup B) + P(A \cap B) = P(A \setminus B) + P(B \setminus A) + 2P(A \cap B)$$
$$= P(A) + P(B).$$

The second assertion follows from the normalisation axiom (N) by taking $B = A^c$.

(c) For $B \supset A$ we conclude from (b) that $P(B) = P(A) + P(B \setminus A) \geq P(A)$ because probabilities are non-negative.

(d) Any union $\bigcup_{i \geq 1} A_i$ can actually be represented as a union of disjoint sets, by removing from A_i the part of A_i that is already contained in a 'previous' A_j. This procedure is known as the 'first entrance trick'. So we can write, using assumption (A) and statement (c),

$$P\left(\bigcup_{i \geq 1} A_i\right) = P\left(\bigcup_{i \geq 1} \left(A_i \setminus \bigcup_{j < i} A_j\right)\right) = \sum_{i \geq 1} P\left(A_i \setminus \bigcup_{j < i} A_j\right) \leq \sum_{i \geq 1} P(A_i).$$

(e) If $A_n \uparrow A$, the σ-additivity (A) and the finite additivity (b) give, with $A_0 := \varnothing$,

$$P(A) = P\left(\bigcup_{i \geq 1} (A_i \setminus A_{i-1})\right) = \sum_{i \geq 1} P(A_i \setminus A_{i-1})$$
$$= \lim_{n \to \infty} \sum_{i=1}^{n} P(A_i \setminus A_{i-1}) = \lim_{n \to \infty} P(A_n).$$

The case $A_n \downarrow A$ follows by taking complements and using (b). $\diamond$

A less obvious, but equally important consequence of σ-additivity is the fact that each probability measure is already determined by its restriction to a suitable generator of the σ-algebra.

(1.12) Theorem. Uniqueness theorem. *Let $(\Omega, \mathscr{F}, P)$ be a probability space, and suppose that $\mathscr{F} = \sigma(\mathscr{G})$ for a generator $\mathscr{G} \subset \mathscr{P}(\Omega)$. If $\mathscr{G}$ is intersection-stable, in the sense that $A, B \in \mathscr{G}$ implies $A \cap B \in \mathscr{G}$, then P is uniquely determined by its restriction $P|_{\mathscr{G}}$ to $\mathscr{G}$.*

Although we will apply the uniqueness theorem repeatedly, its proof should be skipped at first reading, since this kind of reasoning will not be used later on.

Proof. Let Q be an arbitrary probability measure on $(\Omega, \mathscr{F})$ such that $P|_{\mathscr{G}} = Q|_{\mathscr{G}}$. The system $\mathscr{D} = \{A \in \mathscr{F} : P(A) = Q(A)\}$ then exhibits the following properties.

(a) $\Omega \in \mathscr{D}$.

(b) If $A, B \in \mathscr{D}$ and $A \subset B$, then $B \setminus A \in \mathscr{D}$.

(c) If $A_1, A_2, \ldots \in \mathscr{D}$ are pairwise disjoint, then $\bigcup_{i \geq 1} A_i \in \mathscr{D}$.

Indeed, (a) follows from (N), (c) from (A), and (b) is immediate because $P(B \setminus A) = P(B) - P(A)$ for $A \subset B$. A system $\mathscr{D}$ satisfying (a) to (c) is called a *Dynkin system* (after the Russian mathematician Eugene B. Dynkin, *1924). By assumption we have $\mathscr{D} \supset \mathscr{G}$. Thus $\mathscr{D}$ also contains the Dynkin system $d(\mathscr{G})$ generated by $\mathscr{G}$. As in Remark (1.6), $d(\mathscr{G})$ is defined to be the smallest Dynkin system containing $\mathscr{G}$; the existence of such a smallest Dynkin system is proved in exactly the same way as indicated there. The following lemma will show that $d(\mathscr{G}) = \sigma(\mathscr{G}) = \mathscr{F}$. As a consequence, we have that $\mathscr{D} = \mathscr{F}$ and hence $P = Q$. $\diamond$

To complete the proof, we need the following lemma.

(1.13) Lemma. Generated Dynkin system. *For an intersection-stable system $\mathscr{G}$, the identity $d(\mathscr{G}) = \sigma(\mathscr{G})$ holds.*

Proof. Since $\sigma(\mathscr{G})$ is a σ-algebra, it is also a Dynkin system, and since $d(\mathscr{G})$ is minimal, we have $\sigma(\mathscr{G}) \supset d(\mathscr{G})$. Conversely, we will show that $d(\mathscr{G})$ is a σ-algebra. For, this implies that $\sigma(\mathscr{G}) \subset d(\mathscr{G})$ by the minimality of $\sigma(\mathscr{G})$.

Step 1. $d(\mathscr{G})$ is intersection-stable. Indeed, $\mathscr{D}_1 := \{A \subset \Omega : A \cap B \in d(\mathscr{G}) \text{ for all } B \in \mathscr{G}\}$ is obviously a Dynkin system, and since $\mathscr{G}$ is intersection-stable, we have $\mathscr{D}_1 \supset \mathscr{G}$. By the minimality of $d(\mathscr{G})$, it then follows that $\mathscr{D}_1 \supset d(\mathscr{G})$, i.e., we have $A \cap B \in d(\mathscr{G})$ for all $A \in d(\mathscr{G})$ and $B \in \mathscr{G}$.

Similarly, $\mathscr{D}_2 := \{A \subset \Omega : A \cap B \in d(\mathscr{G}) \text{ for all } B \in d(\mathscr{G})\}$ is also a Dynkin system, and, by the above, $\mathscr{D}_2 \supset \mathscr{G}$. Hence, we also have $\mathscr{D}_2 \supset d(\mathscr{G})$, i.e., $A \cap B \in d(\mathscr{G})$ for all $A, B \in d(\mathscr{G})$.

Step 2. $d(\mathscr{G})$ is a σ-algebra. For, let $A_1, A_2, \ldots \in d(\mathscr{G})$. By Step 1, the sets

$$B_i := A_i \setminus \bigcup_{j < i} A_j = A_i \cap \bigcap_{j < i} \Omega \setminus A_j$$

then also belong to $d(\mathscr{G})$ and are pairwise disjoint. Hence $\bigcup_{i \geq 1} A_i = \bigcup_{i \geq 1} B_i \in d(\mathscr{G})$. $\diamond$

Our next question is: How can one construct a probability measure on a σ-algebra? In view of the uniqueness theorem, we can reformulate this question as follows: Under which conditions can we extend a function P defined on a suitable system $\mathscr{G}$ to a probability measure defined on the generated σ-algebra $\sigma(\mathscr{G})$?

A satisfactory answer is given by a theorem from measure theory, namely Carathéodory's extension theorem, see for example [4, 6, 12, 15, 16]; here we will not discuss this further. However, to guarantee the existence of non-trivial probability measures on non-discrete sample spaces, we will have to take the existence of the Lebesgue integral for granted. We will make use of the following

(1.14) Fact. *Lebesgue integral.* For every function $f : \mathbb{R}^n \to [0, \infty]$ that satisfies the measurability criterion

$$(1.15) \qquad \{x \in \mathbb{R}^n : f(x) \leq c\} \in \mathscr{B}^n \quad \text{for all } c > 0$$

(which will be discussed further in Example (1.26)), one can define the *Lebesgue integral* $\int f(x)\, dx \in [0, \infty]$ in such a way that the following holds.

(a) For every Riemann-integrable function f, $\int f(x)\, dx$ coincides with the Riemann integral of f.

(b) For every sequence $f_1, f_2, \ldots$ of non-negative measurable functions as above, we have

$$\int \sum_{n \geq 1} f_n(x)\, dx = \sum_{n \geq 1} \int f_n(x)\, dx \,.$$

A proof of these statements can be found in numerous textbooks on analysis, such as [56, 57]. As property (a) shows, in concrete computations it often suffices to know the Riemann integral. However, the Riemann integral does not satisfy the σ-additivity property (b), which is essential for our purposes; it is equivalent to the monotone convergence theorem, see also Theorem (4.11c) later on.

In particular, the Lebesgue integral yields a sensible notion of volume for Borel sets in $\mathbb{R}^n$. Namely, let

$$(1.16) \qquad 1_A(x) = \begin{cases} 1 & \text{if } x \in A, \\ 0 & \text{otherwise} \end{cases}$$

denote the *indicator function* of a set A. The integral over $A \in \mathscr{B}^n$ is then defined as

$$\int_A f(x)\, dx := \int 1_A(x) f(x)\, dx \,.$$

In the special case $f \equiv 1$, property (1.14b) then gives us the following result.

(1.17) Remark and Definition. *Lebesgue measure.* The mapping $\lambda^n : \mathscr{B}^n \to [0, \infty]$ that assigns to each $A \in \mathscr{B}^n$ its n-dimensional volume

$$\lambda^n(A) := \int 1_A(x)\, dx$$

satisfies the σ-additivity property (A), and we have $\lambda^n(\varnothing) = 0$. Consequently, λ^n is a 'measure' on $(\mathbb{R}^n, \mathscr{B}^n)$. It is called the ($n$-dimensional) *Lebesgue measure on $\mathbb{R}^n$*. For $\Omega \in \mathscr{B}^n$, the restriction λ^n_Ω of λ^n to $\mathscr{B}^n_\Omega$ is called the *Lebesgue measure on Ω*.

We will see repeatedly that the existence of many interesting probability measures can be deduced from the existence of the Lebesgue measure. Here, we will use the Lebesgue measure for constructing probability measures on $\mathbb{R}^n$ (or subsets thereof) by means of density functions, a procedure which is obvious for discrete spaces. See the illustration in Figure 1.2.

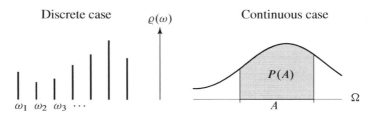

Figure 1.2. On the left: bar chart of a discrete density. On the right: Lebesgue density; its integral over an event A yields the probability $P(A)$.

(1.18) Theorem. Construction of probability measures via densities.

(a) Discrete case: *For countable Ω, the relations*

$$P(A) = \sum_{\omega \in A} \rho(\omega) \text{ for } A \in \mathscr{P}(\Omega), \quad \rho(\omega) = P(\{\omega\}) \text{ for } \omega \in \Omega$$

establish a one-to-one correspondence between the set of all probability measures P on $(\Omega, \mathscr{P}(\Omega))$ and the set of all sequences $\rho = (\rho(\omega))_{\omega \in \Omega}$ in $[0, 1]$ such that $\sum_{\omega \in \Omega} \rho(\omega) = 1$.

(b) Continuous case: *If $\Omega \subset \mathbb{R}^n$ is Borel, then every function $\rho : \Omega \to [0, \infty[$ satisfying the properties*

(i) $\{x \in \Omega : \rho(x) \le c\} \in \mathscr{B}^n_\Omega$ *for all $c > 0$ (cf. (1.15)),*

(ii) $\int_\Omega \rho(x)\, dx = 1$

determines a unique probability measure P on $(\Omega, \mathscr{B}_\Omega^n)$ via

$$P(A) = \int_A \rho(x)\, dx \ \text{ for } A \in \mathscr{B}_\Omega^n$$

(but not every probability measure on $(\Omega, \mathscr{B}_\Omega^n)$ is of this form).

Proof. The discrete case is obvious. In the continuous case, the claim follows immediately from Fact (1.14b), since $1_{\bigcup_{i \geq 1} A_i} = \sum_{i \geq 1} 1_{A_i}$ when the A_i are pairwise disjoint. $\diamond$

Definition. A sequence or function ρ as in Theorem (1.18) above is called a *density* (of P) or, more explicitly (to emphasise normalisation), a *probability density (function)*, often abbreviated as *pdf*. If a distinction between the discrete and continuous case is required, a sequence $\rho = (\rho(\omega))_{\omega \in \Omega}$ as in case (a) is called a *discrete density*, and a function $\rho : \Omega \to [0, \infty[$ as in case (b) a *Lebesgue density*.

A basic class of probability measures defined by densities is given by the uniform distributions, which are defined as follows and will be discussed in more detail in Section 2.1.

(1.19) Example and Definition. *The uniform distributions.* If Ω is finite, the probability measure having the constant discrete density $\rho(\omega) = 1/|\Omega|$ (so that all $\omega \in \Omega$ occur with the same probability) is called the (discrete) *uniform distribution* on Ω and is denoted by $\mathcal{U}_\Omega$.

Likewise, if $\Omega \subset \mathbb{R}^n$ is a Borel set with volume $0 < \lambda^n(\Omega) < \infty$, the probability measure on $(\Omega, \mathscr{B}_\Omega)$ with the constant Lebesgue density $\rho(x) = 1/\lambda^n(\Omega)$ is called the (continuous) *uniform distribution* on Ω; it is also denoted by $\mathcal{U}_\Omega$.

Let us conclude this section with some comments. In contrast to the discrete case, there is no one-to-one correspondence between probability measures and their densities in the continuous case. On the one hand, most probability measures on $\mathbb{R}^n$ fail to have a Lebesgue density, but many of the most common probability measures do. On the other hand, any two probability densities that differ only on a set of vanishing Lebesgue measure determine the same probability measure. For example, we have $\mathcal{U}_{]0,1[} = \mathcal{U}_{[0,1]}$.

Next we note that probability measures on Borel subsets of $\mathbb{R}^n$ can also be viewed as probability measures on all of $\mathbb{R}^n$. Specifically, let $\Omega \subset \mathbb{R}^n$ be a Borel set and P a probability measure on $(\Omega, \mathscr{B}_\Omega^n)$ with Lebesgue density ρ. The measure P can then be identified with the probability measure $\bar{P}$ on $(\mathbb{R}^n, \mathscr{B}^n)$ with density $\bar{\rho}$, where $\bar{\rho}(x) = \rho(x)$ for $x \in \Omega$ and $\bar{\rho}(x) = 0$ otherwise. Indeed, we have $\bar{P}(\mathbb{R}^n \setminus \Omega) = 0$, and $\bar{P}$ and P coincide on $\mathscr{B}_\Omega^n$. We will often carry out this identification without mentioning it explicitly. There is also an analogue in the discrete case: If $\Omega \subset \mathbb{R}^n$ is countable and P a probability measure on $(\Omega, \mathscr{P}(\Omega))$ with discrete density ρ, we can identify P with the probability measure $\sum_{\omega \in \Omega} \rho(\omega) \, \delta_\omega$, which is defined on $(\mathbb{R}^n, \mathscr{B}^n)$, or in fact even on $(\mathbb{R}^n, \mathscr{P}(\mathbb{R}^n))$; here δ_ω is the Dirac measure introduced in (1.10).

Finally, it is possible to combine discrete and continuous probability measures. For example,

$$(1.20) \qquad P(A) = \tfrac{1}{3} \delta_{-1/2}(A) + \tfrac{2}{3} \mathcal{U}_{]0,1/2[}(A) \,, \quad A \in \mathscr{B},$$

defines a probability measure on $(\mathbb{R}, \mathscr{B})$, which for two thirds is 'blurred uniformly' over the interval $]0, 1/2[$ and assigns the extra probability $1/3$ to the point $-1/2$.

1.3 Random Variables

Let us return for a moment to the first step of setting up a model, as described in Section 1.1.1. The choice of the sample space Ω is not unique, but depends on how many details of the random phenomenon should be included into the model, and is therefore a matter of the appropriate *observation depth*.

(1.21) Example. *Tossing a coin n times.* On the one hand, one can record the result of every single toss; then $\Omega = \{0, 1\}^n$ is the appropriate sample space. Alternatively, one may restrict attention to the number of tosses when heads is showing. Then, the natural sample space is $\Omega' = \{0, 1, \ldots, n\}$. The second case corresponds to a lower observation depth. The process of reducing the observation depth can be described by the mapping $X : \Omega \to \Omega'$ that assigns to each $\omega = (\omega_1, \ldots, \omega_n) \in \Omega$ the sum $\sum_{i=1}^{n} \omega_i \in \Omega'$, which indicates the 'number of successes'.

The example shows: The transition from a given event space $(\Omega, \mathscr{F})$ to a coarser model $(\Omega', \mathscr{F}')$ providing less information is captured by a mapping from the detailed to the coarser sample space, i.e., from Ω to Ω'. In the general case, such a mapping should satisfy the requirement

$$(1.22) \qquad A' \in \mathscr{F}' \; \Rightarrow \; X^{-1} A' \in \mathscr{F},$$

that is, all events on the coarse level can be traced back to events on the detailed level via the preimage mapping X^{-1}. The situation is visualised in Figure 1.3.

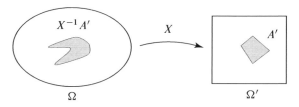

Figure 1.3. For a random variable, the preimage of an event in Ω' is an event in Ω.

Definition. Let $(\Omega, \mathscr{F})$ and $(\Omega', \mathscr{F}')$ be two event spaces. Then every mapping $X : \Omega \to \Omega'$ satisfying property (1.22) is called a *random variable from* $(\Omega, \mathscr{F})$ *to* $(\Omega', \mathscr{F}')$, or a *random element of* Ω'. Alternatively (in the terminology of measure theory), X is said to be *measurable* relative to $\mathscr{F}$ and $\mathscr{F}'$.

Due to (1.22), preimages will occur frequently in the following. In stochastics, it is common to use the suggestive notation

(1.23) $\{X \in A'\} := \{\omega \in \Omega : X(\omega) \in A'\} = X^{-1}A'$.

Let us note first that condition (1.22) holds automatically in the discrete case.

(1.24) Example. *Random variables on discrete spaces.* If $\mathscr{F} = \mathscr{P}(\Omega)$, then *every* mapping $X : \Omega \to \Omega'$ is a random variable.

In the general case, the following criterion is crucial.

(1.25) Remark. *Measurability criterion.* In the set-up of the previous definition, suppose that $\mathscr{F}'$ is generated by a system $\mathscr{G}'$, in that $\mathscr{F}' = \sigma(\mathscr{G}')$. Then $X : \Omega \to \Omega'$ is already a random variable when $X^{-1}A' \in \mathscr{F}$ for all $A' \in \mathscr{G}'$ only.

Proof. The system $\mathscr{A}' := \{A' \subset \Omega' : X^{-1}A' \in \mathscr{F}\}$ is a σ-algebra, which by assumption contains $\mathscr{G}'$. Since $\mathscr{F}'$ is by definition the smallest such σ-algebra, we also have $\mathscr{A}' \supset \mathscr{F}'$, which means that X satisfies condition (1.22). $\diamond$

(1.26) Example. *Real random variables.* Let $(\Omega', \mathscr{F}') = (\mathbb{R}, \mathscr{B})$. For a real function $X : \Omega \to \mathbb{R}$ to be a random variable, it is sufficient that all sets of the form

$$\{X \le c\} := X^{-1}\,]{-\infty}, c\,]$$

belong to $\mathscr{F}$. (Alternatively, one can replace '$\le$' by '$<$', '$\ge$' or '$>$'.) This follows immediately from Remark (1.25) and Fact (1.8d).

It is often convenient to consider so-called extended real functions taking values in $\bar{\mathbb{R}} = [-\infty, \infty]$. $\bar{\mathbb{R}}$ is equipped with the σ-algebra generated by the intervals $[-\infty, c]$, $c \in \mathbb{R}$. (Think about how this relates to the Borel σ-algebra on $\mathbb{R}$.) Consequently, an extended real function $X : \Omega \to \bar{\mathbb{R}}$ is a random variable if and only if $\{X \le c\} \in \mathscr{F}$ for all $c \in \mathbb{R}$.

(1.27) Example. *Continuous functions.* Let $\Omega \subset \mathbb{R}^n$ and $\mathscr{F} = \mathscr{B}_{\Omega}^n$. Then every continuous function $X : \Omega \to \mathbb{R}$ is a random variable. Indeed, for every $c \in \mathbb{R}$, $\{X \le c\}$ is closed in Ω, so by Example (1.8be) it belongs to $\mathscr{B}_{\Omega}^n$. Thus the claim follows from Example (1.26).

The next theorem describes an important principle for constructing new probability measures, which will be used repeatedly.

(1.28) Theorem. Distribution of a random variable. *If X is a random variable from a probability space $(\Omega, \mathscr{F}, P)$ to an event space $(\Omega', \mathscr{F}')$, then the prescription*

$$P'(A') := P(X^{-1}A') = P(\{X \in A'\}) \quad \textit{for } A' \in \mathscr{F}'$$

defines a probability measure P' on $(\Omega', \mathscr{F}')$.

To simplify the notation, we will omit the braces in expressions like $P(\{X \in A'\})$, and simply write $P(X \in A')$ in the future.

Proof. By (1.22) the definition of P' makes sense. Furthermore, P' satisfies the conditions (N) and (A). Indeed, on the one hand, $P'(\Omega') = P(X \in \Omega') = P(\Omega) = 1$. On the other hand, if $A_1', A_2', \ldots \in \mathscr{F}'$ are pairwise disjoint, so are their preimages $X^{-1}A_1', X^{-1}A_2', \ldots$, whence

$$P'\Big(\bigcup_{i \geq 1} A_i'\Big) = P\Big(X^{-1}\bigcup_{i \geq 1} A_i'\Big) = P\Big(\bigcup_{i \geq 1} X^{-1}A_i'\Big)$$

$$= \sum_{i \geq 1} P(X^{-1}A_i') = \sum_{i \geq 1} P'(A_i') \,.$$

Hence P' is a probability measure. $\diamond$

Definition.   (a) The probability measure P' in Theorem (1.28) is called the *distribution of X under P*, or the *image of P under X*, and is denoted by $P \circ X^{-1}$. (In the literature, one also finds the notations P_X or $\mathscr{L}(X; P)$. The letter $\mathscr{L}$ stands for the more traditional term *law*, or *loi* in French.)

(b) Two random variables are said to be *identically distributed* if they have the same distribution.

At this point, we need to emphasise that the term 'distribution' is used in an inflationary way in stochastics. Apart from the meaning that we have just introduced, it is also generally used as a synonym for probability measure. (In fact, every probability measure is the distribution of a random variable, namely the identity function of the underlying Ω.) This has to be distinguished from two further notions, namely 'distribution function' and 'distribution density', which refer to the real event space $(\mathbb{R}, \mathscr{B})$ and will be introduced now.

Each probability measure P on $(\mathbb{R}, \mathscr{B})$ is already uniquely determined by the function $F_P(c) := P(]-\infty, c])$ for $c \in \mathbb{R}$. Likewise, the distribution of a real-valued random variable X on a probability space $(\Omega, \mathscr{F}, P)$ is uniquely determined by the function $F_X : c \to P(X \leq c)$ on $\mathbb{R}$. This is because any two probability measures on $(\mathbb{R}, \mathscr{B})$ coincide if and only if they agree on all intervals of the form $]-\infty, c]$, by statement (1.8d) and the uniqueness theorem (1.12). This motivates the following concepts.

Definition. For a probability measure P on the real line $(\mathbb{R}, \mathscr{B})$, the function $F_P :$ $c \to P(]-\infty, c])$ from $\mathbb{R}$ to $[0, 1]$ is called the (cumulative) *distribution function of P*. Likewise, for a real random variable X on a probability space $(\Omega, \mathscr{F}, P)$, the distribution function $F_X(c) := F_{P \circ X^{-1}}(c) = P(X \le c)$ of its distribution is called the (cumulative) *distribution function of X*.

Every distribution function $F = F_X$ is increasing and right-continuous and has the asymptotic behaviour

(1.29) $$\lim_{c \to -\infty} F(c) = 0 \quad \text{and} \quad \lim_{c \to +\infty} F(c) = 1 .$$

This follows immediately from Theorem (1.11); see Problem 1.16. Figure 1.4 shows an example. Remarkably, *every* function with these properties is the distribution function of a random variable on the unit interval (equipped with the uniform distribution from Example (1.19)). The term 'quantile' occurring in the name of these random variables will play an important role in statistics, i.e., in Part II; see the definition on p. 231.

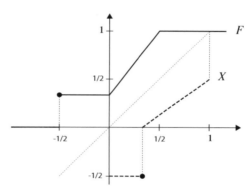

Figure 1.4. Distribution function F (bold) and quantile transformation X (dashed) of the probability measure $\frac{1}{3}\delta_{-1/2} + \frac{2}{3}\mathcal{U}_{]0,1/2[}$ from (1.20). The dotted lines illustrate that X is obtained from F by reflection at the diagonal. The values at the discontinuities are marked by bullets.

(1.30) Proposition. *Quantile transformation. For every increasing right-continuous function F on $\mathbb{R}$ with limit behaviour (1.29), there exists a real random variable X on the probability space $(]0, 1[, \mathscr{B}_{]0,1[}, \mathcal{U}_{]0,1[})$ such that $F_X = F$. This X is given explicitly by $X(u) = \inf\{c \in \mathbb{R} : F(c) \ge u\}$, $u \in]0, 1[$, and is called the 'quantile transformation'.*

Proof. By (1.29) we have $-\infty < X(u) < \infty$ for all $0 < u < 1$. In fact, X is a left-continuous inverse of F; compare Figure 1.4. Indeed, $X(u) \le c$ holds if and only if $u \le F(c)$; this is because, by the right-continuity of F, the infimum in the definition of X is in fact a minimum. In particular, $\{X \leq c\} =]0, F(c)] \cap]0, 1[\in \mathscr{B}_{]0,1[}$. Together with Example (1.26) this shows that X is a random variable. Furthermore, the set $\{X \leq c\}$ has Lebesgue measure $F(c)$. Hence F is the distribution function of X. $\Diamond$

Since every probability measure P on $(\mathbb{R}, \mathscr{B})$ is uniquely determined by its distribution function, we can rephrase the proposition as follows: Every P on $(\mathbb{R}, \mathscr{B})$ is the distribution of a random variable on the probability space $(]0, 1[, \mathscr{B}_{]0,1[}, \mathcal{U}_{]0,1[})$. This fact will repeatedly be useful.

The connection between distribution functions and probability densities is made by the notion of a distribution density.

(1.31) Remark and Definition. *Existence of a distribution density.* Let X be a real random variable on a probability space $(\Omega, \mathscr{F}, P)$. Its distribution $P \circ X^{-1}$ admits a Lebesgue density ρ if and only if

$$F_X(c) = \int_{-\infty}^{c} \rho(x)\, dx \quad \text{for all } c \in \mathbb{R}.$$

Such a ρ is called the *distribution density* of X. In particular, $P \circ X^{-1}$ admits a continuous density ρ if and only if F_X is continuously differentiable, and then $\rho = F_X'$. This follows directly from (1.8d) and the uniqueness theorem (1.12).

Problems

1.1 Let $(\Omega, \mathscr{F})$ be an event space, $A_1, A_2, \ldots \in \mathscr{F}$ and

$$A = \{\omega \in \Omega : \omega \in A_n \text{ for infinitely many } n\}.$$

Show that (a) $A = \bigcap_{N \geq 1} \bigcup_{n \geq N} A_n$, (b) $1_A = \limsup_{n \to \infty} 1_{A_n}$.

1.2 Let Ω be uncountable and $\mathscr{G} = \{\{\omega\} : \omega \in \Omega\}$ the system of the singleton subsets of Ω. Show that $\sigma(\mathscr{G}) = \{A \subset \Omega : A \text{ or } A^c \text{ is countable}\}$.

1.3[S] Show that the Borel σ-algebra $\mathscr{B}^n$ on $\mathbb{R}^n$ coincides with $\mathscr{B}^{\otimes n}$, the n-fold product of the Borel σ-algebra $\mathscr{B}$ on $\mathbb{R}$.

1.4 Let $\Omega \subset \mathbb{R}^n$ be at most countable. Show that $\mathscr{B}^n_\Omega = \mathscr{P}(\Omega)$.

1.5 Let E_i, $i \in \mathbb{N}$, be countable sets and $\Omega = \prod_{i \geq 1} E_i$ their Cartesian product. Denote by $X_i : \Omega \to E_i$ the projection onto the ith coordinate. Show that the system

$$\mathscr{G} = \{\{X_1 = x_1, \ldots, X_k = x_k\} : k \geq 1, x_i \in E_i\} \cup \{\varnothing\}$$

is an intersection-stable generator of the product σ-algebra $\bigotimes_{i \geq 1} \mathscr{P}(E_i)$.

1.6[S] Let $(\Omega_i, \mathscr{F}_i)$, $i = 1, 2$, be two event spaces and $\omega_1 \in \Omega_1$. Show the following. For every $A \in \mathscr{F}_1 \otimes \mathscr{F}_2$, the '$\omega_1$-section' $A_{\omega_1} := \{\omega_2 \in \Omega_2 : (\omega_1, \omega_2) \in A\}$ of A belongs to $\mathscr{F}_2$, and if f is a random variable on $(\Omega_1 \times \Omega_2, \mathscr{F}_1 \otimes \mathscr{F}_2)$ then the function $f(\omega_1, \cdot)$ is a random variable on $(\Omega_2, \mathscr{F}_2)$.

1.7 [S] *Inclusion–exclusion principle.* Let $(\Omega, \mathscr{F}, P)$ be a probability space and $A_i \in \mathscr{F}$, $i \in I = \{1, \ldots, n\}$. For $J \subset I$ let

$$B_J = \bigcap_{j \in J} A_j \cap \bigcap_{j \in I \setminus J} A_j^c \,;$$

by convention, an intersection over an empty index set is equal to Ω. Show the following:

(a) For all $K \subset I$,

$$P\Big(\bigcap_{k \in K} A_k\Big) = \sum_{K \subset J \subset I} P(B_J).$$

(b) For all $J \subset I$,

$$P(B_J) = \sum_{J \subset K \subset I} (-1)^{|K \setminus J|} P\Big(\bigcap_{k \in K} A_k\Big).$$

What does this imply for $J = \varnothing$?

1.8 *Bonferroni inequality.* Let $A_1, \ldots, A_n$ be any events in a probability space $(\Omega, \mathscr{F}, P)$. Show that

$$P\Big(\bigcup_{i=1}^n A_i\Big) \geq \sum_{i=1}^n P(A_i) - \sum_{1 \leq i < j \leq n} P(A_i \cap A_j).$$

1.9 A certain Chevalier de Méré, who has become famous in the history of probability theory for his gambling problems and their solutions by Pascal, once mentioned to Pascal how surprised he was that when throwing three dice he observed the total sum of 11 more often than the sum of 12, although 11 could be obtained by the combinations 6-4-1, 6-3-2, 5-5-1, 5-4-2, 5-3-3, 4-4-3, and the sum of 12 by as many combinations (which ones?). Can we consider his observation as caused by 'chance' or is there an error in his argument? To solve the problem, introduce a suitable probability space.

1.10 In a pack of six chocolate drinks every carton is supposed to have a straw, but it is missing with probability $1/3$, with probability $1/3$ it is broken and only with probability $1/3$ it is in perfect condition. Let A be the event 'at least one straw is missing and at least one is in perfect condition'. Exhibit a suitable probability space, formulate the event A set-theoretically, and determine its probability.

1.11 [S] Alice and Bob agree to play a fair game over 7 rounds. Each of them pays €5 as an initial stake, and the winner gets the total of €10. At the score of $2 : 3$ they have to stop the game. Alice suggests to split the winnings in this ratio. Should Bob accept the offer? Set up an appropriate model and calculate the probability of winning for Bob.

1.12 *The birthday paradox.* Let p_n be the probability that in a class of n children at least two have their birthday on the same day. For simplicity, we assume here that no birthday is on February 29th, and all other birthdays are equally likely. Show (using the inequality $1 - x \leq e^{-x}$) that

$$p_n \geq 1 - \exp\big(-n(n-1)/730\big),$$

and determine the smallest n such that $p_n \geq 1/2$.

1.13[S] *The rencontre problem.* Alice and Bob agree to play the following game: From two completely new, identical sets of playing cards, one is well shuffled. Both piles are put next to each other face down, and then revealed card by card simultaneously. Bob bets (for a stake of € 10) that in this procedure at least two identical cards will be revealed at the same time. Alice, however, is convinced that this is 'completely unlikely' and so bets the opposite way. Who do you think is more likely to win? Set up an appropriate model and calculate the probability of winning for Alice. *Hint:* Use Problem 1.7b; the sum that appears can be approximated by the corresponding infinite series.

1.14 Let $X, Y, X_1, X_2, \ldots$ be real random variables on an event space $(\Omega, \mathscr{F})$. Prove the following statements.

(a) $(X, Y) : \Omega \to \mathbb{R}^2$ is a random variable.

(b) $X + Y$ and XY are random variables.

(c) $\sup_{n \in \mathbb{N}} X_n$ and $\limsup_{n \to \infty} X_n$ are random variables (taking values in $\bar{\mathbb{R}}$).

(d) $\{X = Y\} \in \mathscr{F}$, $\{\lim_{n \to \infty} X_n \text{ exists}\} \in \mathscr{F}$, $\{X = \lim_{n \to \infty} X_n\} \in \mathscr{F}$.

1.15[S] Let $(\Omega, \mathscr{F}) = (\mathbb{R}, \mathscr{B})$ and $X : \Omega \to \mathbb{R}$ be an arbitrary real function. Verify the following:

(a) If X is piecewise monotone (i.e., $\mathbb{R}$ may be decomposed into at most countably many intervals, on each of which X is either increasing or decreasing), then X is a random variable.

(b) If X is differentiable with (not necessarily continuous) derivative X', then X' is a random variable.

1.16 *Properties of distribution functions.* Let P be a probability measure on $(\mathbb{R}, \mathscr{B})$ and $F(c) = P(]-\infty, c])$, for $c \in \mathbb{R}$, its distribution function. Show that F is increasing and right-continuous, and (1.29) holds.

1.17 Consider the two cases

(a) $\Omega = [0, \infty[$, $\rho(\omega) = e^{-\omega}$, $X(\omega) = (\omega/\alpha)^{1/\beta}$ for $\omega \in \Omega$ and $\alpha, \beta > 0$,

(b) $\Omega =]-\pi/2, \pi/2[$, $\rho(\omega) = 1/\pi$, $X(\omega) = \sin^2 \omega$ for $\omega \in \Omega$.

In each case, show that ρ is a probability density and X a random variable on $(\Omega, \mathscr{B}_\Omega)$, and calculate the distribution density of X with respect to the probability measure P with density ρ. (The distribution of X in case (a) is called the *Weibull distribution* with parameters α, β, in case (b) the *arcsine distribution*.)

1.18[S] *Transformation to uniformity.* Prove the following converse to Proposition (1.30): If X is a real random variable with a *continuous* distribution function $F_X = F$, then the random variable $F(X)$ is uniformly distributed on $[0, 1]$. Show further that the continuity of F is necessary for this to hold.

Chapter 2

Stochastic Standard Models

Having described the general structure of stochastic models, we will now discuss how to find suitable models for concrete random phenomena. In general, this can be quite delicate, and requires the right balance between being close to reality yet mathematically tractable. At this stage, however, we confine ourselves to several classical examples, for which the appropriate model is quite obvious. This gives us the opportunity to introduce some fundamental probability distributions along with typical applications. These distributions can be used as building blocks of more complex models, as we will see later on.

2.1 The Uniform Distributions

There are two different types of uniform distributions: the discrete ones on finite sets, and the continuous uniform distributions on Borel subsets of $\mathbb{R}^n$.

2.1.1 Discrete Uniform Distributions

Let us start with the simplest case of a random experiment with only finitely many possible outcomes, i.e., an experiment with a finite sample space Ω. For example, we can think of tossing a coin or rolling a dice several times. In these and many other examples, symmetry suggests the assumption that all single outcomes $\omega \in \Omega$ are equally likely. By Theorem (1.18a) this means that the probability measure P should have the *constant* density $\rho(\omega) = 1/|\Omega|$ (for $\omega \in \Omega$). This leads to the approach $P = \mathcal{U}_\Omega$, where

$$(2.1) \qquad \mathcal{U}_\Omega(A) = \frac{|A|}{|\Omega|} = \frac{\text{number of 'favourable' outcomes}}{\text{number of possible outcomes}} \quad \text{for all } A \subset \Omega.$$

Definition. For a finite set Ω, the probability measure $\mathcal{U}_\Omega$ on $(\Omega, \mathscr{P}(\Omega))$ defined by (2.1) is called the *(discrete) uniform distribution* on Ω. Sometimes $(\Omega, \mathscr{P}(\Omega), \mathcal{U}_\Omega)$ is also called a *Laplace space* (in honour of Pierre-Simon Laplace, 1749–1827).

Classical examples in which the uniform distribution shows up are tossing a coin or rolling a dice (once or several times), the lottery, playing cards, and many more. Several of these examples will be discussed soon, in particular in Sections 2.2 and 2.3. A less obvious example is the following.

(2.2) Example. *The Bose–Einstein distribution (1924).* Consider a system of n indistinguishable particles that are distributed over N different 'cells'; the cells are of the same type, but distinguishable. For example, one can imagine the seeds in the pits of the Syrian game Kalah, or – and this was Bose's and Einstein's motivation – physical particles, whose phase space is partitioned into finitely many cells. A (macro) state of the system is determined by specifying how many particles populate each cell. Hence,

$$\Omega = \left\{(k_1, \ldots, k_N) \in \mathbb{Z}_+^N : \sum_{j=1}^{N} k_j = n\right\}.$$

This sample space has cardinality $|\Omega| = \binom{n+N-1}{n}$, since each $(k_1, \ldots, k_N) \in \Omega$ is uniquely characterised by a sequence of the form

$$\underbrace{\bullet \cdots \bullet}_{k_1} | \underbrace{\bullet \cdots \bullet}_{k_2} | \cdots | \underbrace{\bullet \cdots \bullet}_{k_N},$$

where the blocks of $k_1, \ldots, k_N$ balls are separated from each other by a total of $N-1$ vertical bars. To determine a state, we only have to place the n balls (resp. the $N-1$ vertical bars) into the $n+N-1$ available positions. Hence, the uniform distribution $\mathcal{U}_\Omega$ on Ω is given by $\mathcal{U}_\Omega(\{\omega\}) = 1/\binom{n+N-1}{n}$, $\omega \in \Omega$. The assumption of a uniform distribution agrees with the experimental results in the case of so-called bosons (i.e., particles with integer spin, such as photons and mesons).

> Physicists often speak of Bose–Einstein 'statistics'. In this traditional terminology, statistics means the same as 'probability distribution' and has nothing in common with statistics in today's mathematical sense.

2.1.2 Continuous Uniform Distributions

Let us begin with a motivating example.

(2.3) Example. *Random choice of a direction.* Imagine we spin a roulette wheel. After the wheel has stopped, into which direction is the zero facing? The angle it forms relative to a fixed direction is a number in the interval $\Omega = [0, 2\pi[$, which is equipped with the Borel σ-algebra $\mathscr{F} := \mathscr{B}_\Omega$. Which probability measure P describes the situation? For every $n \geq 1$, Ω can be partitioned in the n disjoint intervals $[\frac{k}{n} 2\pi, \frac{k+1}{n} 2\pi[$ with $0 \leq k < n$. By symmetry, each of these should receive the same probability, which is then necessarily equal to $1/n$. That is, we should have

$$P\left(\left[\frac{k}{n} 2\pi, \frac{k+1}{n} 2\pi\right[\right) = \frac{1}{n} = \int_{\frac{k}{n} 2\pi}^{\frac{k+1}{n} 2\pi} \frac{1}{2\pi} \, dx$$

for $0 \leq k < n$ and, by additivity,

$$P\left(\left[\frac{k}{n} 2\pi, \frac{l}{n} 2\pi\right[\right) = \int_{\frac{k}{n} 2\pi}^{\frac{l}{n} 2\pi} \frac{1}{2\pi} \, dx$$

for $0 \le k < l \le n$. By Theorems (1.12) and (1.18), there exists only one probability measure P with this property, namely the one with the constant Lebesgue density $1/2\pi$ on $[0, 2\pi[$. This P reflects the idea that all possible directions are equally likely.

Definition. Let $\Omega \subset \mathbb{R}^n$ be a Borel set with n-dimensional volume $0 < \lambda^n(\Omega) < \infty$; cf. (1.17). The probability measure $\mathcal{U}_\Omega$ on $(\Omega, \mathscr{B}_\Omega^n)$ with constant Lebesgue density $\rho(x) = 1/\lambda^n(\Omega)$, which is given by

$$\mathcal{U}_\Omega(A) = \int_A \frac{1}{\lambda^n(\Omega)}\, dx = \frac{\lambda^n(A)}{\lambda^n(\Omega)}, \quad A \in \mathscr{B}_\Omega^n,$$

is called the *(continuous) uniform distribution* on Ω.

Note that, depending on the context, $\mathcal{U}_\Omega$ can either stand for a discrete or a continuous distribution. But both cases are completely analogous. The favourable resp. possible outcomes are simply counted in the discrete case (2.1), whereas in the continuous case they are measured with Lebesgue measure. The following application of continuous uniform distributions is an example of historical interest, and also a little taster from so-called stochastic geometry.

(2.4) Example. *Bertrand's paradox.* Given a circle with radius $r > 0$, a chord is chosen 'at random'. What is the probability that it is longer than the sides of an inscribed equilateral triangle? (This problem appeared in 1889 in a textbook of the French mathematician J. L. F. Bertrand, 1822–1900.)

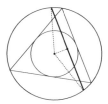

Figure 2.1. The geometry of Bertrand's paradox. The incircle of the inscribed equilateral triangle has half the radius.

The answer depends on what one considers a 'random choice' or, in other words, it depends on the method the chord is chosen.

First approach. The chord is uniquely determined by its midpoint (as long as it is not the centre of the circle; this case can be neglected). Hence, a possible sample space is $\Omega_1 = \{x \in \mathbb{R}^2 : |x| < r\}$, and it seems reasonable to interpret 'choosing at random' by taking the uniform distribution $\mathcal{U}_{\Omega_1}$ as the underlying probability measure. The event 'the chord is longer than the sides of the inscribed equilateral triangle' is then described by the set $A_1 = \{x \in \Omega_1 : |x| < r/2\}$, cf. Figure 2.1. Consequently,

$$\mathcal{U}_{\Omega_1}(A_1) = \frac{\pi(r/2)^2}{\pi r^2} = \frac{1}{4}.$$

Second approach. The chord is uniquely determined by the angle under which it is seen from the centre of the circle, and the direction of its perpendicular bisector; the latter is irrelevant by the rotational symmetry of the problem. The angle falls into $\Omega_2 =]0, \pi]$. The relevant event is $A_2 =]2\pi/3, \pi]$. If we again use the uniform distribution, it follows that

$$\mathcal{U}_{\Omega_2}(A_2) = \frac{\pi/3}{\pi} = \frac{1}{3}.$$

Third approach. The chord is also uniquely determined by its distance and direction from the centre; the latter can again be ignored. Hence, we can also take the sample space $\Omega_3 = [0, r[$. Then $A_3 = [0, r/2[$ is the event we are interested in, and we obtain $\mathcal{U}_{\Omega_3}(A_3) = 1/2$.

In Bertrand's times, this apparent paradox cast doubt on the legitimacy of non-discrete probability spaces. Today it is clear that the three versions describe different methods of choosing the chord 'at random', and it is all but surprising that the probability we are looking for depends on the choice of the method.

Some readers may consider this way of resolving the paradox as a cheap way out, because they think that there must be a unique 'natural' interpretation of 'choosing at random'. In fact the latter is true, but only if we reformulate the problem by requiring that the object chosen 'at random' is not a chord, but a straight line that intersects the circle. Such a random straight line is best described by the third approach, since one can show that this is the only case in which the probability that a random straight line hits a set A is invariant under rotations and translations of A.

This example demonstrates that the choice of a suitable model can be non-trivial, even in a simple case like this, which only involves uniform distributions. This is a main problem in all applications.

2.2   Urn Models with Replacement

The so-called urn models form a simple class of stochastic models with finite sample space. Their significance comes from the observation that many random experiments can be thought of as picking balls of different colours at random from a container, which is traditionally called an 'urn'. In this section we consider the case that the balls are replaced after being drawn. The case without replacement follows in the next section.

2.2.1   Ordered Samples

We begin with two examples.

(2.5) Example. *Investigation of a fish pond.* Consider a fish pond inhabited by several species of fish, and let E be the set of species. Suppose the pond contains N_a fish of species $a \in E$, so the total population size is $\sum_{a \in E} N_a = N$. We repeat the following procedure n times. A fish is caught, examined for parasites, say, and then thrown back into the pond. What is the probability that a random sample exhibits a specific sequence of species?

(2.6) Example. *A survey.* A local radio station interviews passers-by on the high street concerning a question of local interest, for instance the construction of a new football stadium. Let E be the set of viewpoints expressed (possible location, rejection on principle, etc.). The station interviews n people. What is the probability that a given sequence of views is observed?

Such problems, where samples are picked randomly from a given pool, are often formulated in an abstract way as *urn models*. An urn contains a certain number of coloured but otherwise identical balls. Let E be the set of colours, where $2 \le |E| < \infty$. A sample of size n is taken from the urn with replacement, meaning that n times a ball is drawn at random and returned immediately. We are interested in the colour of each ball we take out. Thus, the sample space is $\Omega = E^n$, equipped with the σ-algebra $\mathscr{F} = \mathscr{P}(\Omega)$. Which probability measure P describes the situation adequately?

To find out, we proceed as follows. We imagine that the balls are labelled with the numbers $1, \ldots, N$; the labels of the balls with colour $a \in E$ are collected in a class $C_a \subset \{1, \ldots, N\}$. Thus $|C_a| = N_a$. If we could observe the labels, we would describe our experiment by the sample space $\overline{\Omega} = \{1, \ldots, N\}^n$ (with the σ-algebra $\overline{\mathscr{F}} = \mathscr{P}(\overline{\Omega})$), and since all balls are identical (up to their colour), we would choose our probability measure to be the uniform distribution $\bar{P} = \mathcal{U}_{\overline{\Omega}}$. So, using the trick of increasing the observation depth artificially by labelling the balls, we arrive at a plausible stochastic model.

We now return to the true sample space $\Omega = E^n$. As we have seen in Section 1.3, this transition can be described by constructing a suitable random variable $X : \overline{\Omega} \to \Omega$. The colour of the ith draw is specified by the random variable

$$X_i : \overline{\Omega} \to E, \quad \bar{\omega} = (\bar{\omega}_1, \ldots, \bar{\omega}_n) \to a \ \text{ if } \ \bar{\omega}_i \in C_a \ .$$

The sequence of colours in our sample is then given by the n-step random variable $X = (X_1, \ldots, X_n) : \overline{\Omega} \to \Omega$.

What is the distribution of X? For every $\omega = (\omega_1, \ldots, \omega_n) \in E^n$ we have

$$\{X = \omega\} = C_{\omega_1} \times \cdots \times C_{\omega_n}$$

and thus

$$\bar{P} \circ X^{-1}(\{\omega\}) = \bar{P}(X = \omega) = \frac{|C_{\omega_1}| \dots |C_{\omega_n}|}{|\bar{\Omega}|} = \prod_{i=1}^{n} \rho(\omega_i) \,,$$

where $\rho(a) = |C_a|/N = N_a/N$ is the proportion of balls of colour a.

Definition. For every density ρ on E, the (discrete) density

$$\rho^{\otimes n}(\omega) = \prod_{i=1}^{n} \rho(\omega_i)$$

on E^n is called the *n-fold product density of* ρ, and the corresponding probability measure P on E^n is the *n-fold product measure of* ρ. (We will not introduce an extra notation for P, but instead we will use the notation $\rho^{\otimes n}$ for P as well.)

In the special case when $E = \{0, 1\}$ and $\rho(1) = p \in [0, 1]$, one obtains the product density

$$\rho^{\otimes n}(\omega) = p^{\sum_{i=1}^{n} \omega_i} (1 - p)^{\sum_{i=1}^{n}(1-\omega_i)}$$

on $\{0, 1\}^n$, and P is called the *Bernoulli measure* or the *Bernoulli distribution* for n trials with 'success probability' p. (The name refers to Jakob Bernoulli, 1654–1705.)

2.2.2   Unordered Samples

In the urn model, one is usually not interested in the (temporal) order in which the colours were selected, but only in how many balls of each colour were drawn. (In particular, this applies to the situations of Examples (2.5) and (2.6).) This corresponds to a lower observation depth and leads to the sample space

$$\widehat{\Omega} = \left\{ \vec{k} = (k_a)_{a \in E} \in \mathbb{Z}_+^E, \ \sum_{a \in E} k_a = n \right\},$$

which consists of the integral grid points in the simplex spanned by the $|E|$ vertices $(n\delta_{a,b})_{b \in E}$, $a \in E$; cf. Figure 2.3 on p. 37. The transition to $\widehat{\Omega}$ is described by the random variable

$$(2.7) \qquad\qquad S : \Omega \to \widehat{\Omega}, \quad \omega = (\omega_1, \dots, \omega_n) \to \big(S_a(\omega)\big)_{a \in E} \,;$$

where $S_a(\omega) = \sum_{i=1}^{n} 1_{\{a\}}(\omega_i)$ is the number of occurrences of colour a in sample point $\omega \in E^n$. $S(\omega)$ is called the *histogram* of ω. It can be visualised by plotting a rectangle of width 1 and height $S_a(\omega)$ for every $a \in E$. Notice that the total area of all rectangles is n.

Now, for $P = \rho^{\otimes n}$ and $\vec{k} = (k_a)_{a \in E} \in \widehat{\Omega}$, we find

$$P(S = \vec{k}) = \sum_{\omega \in \Omega: \, S(\omega) = \vec{k}} \prod_{i=1}^{n} \rho(\omega_i) = \binom{n}{\vec{k}} \prod_{a \in E} \rho(a)^{k_a} \,.$$

Here, we write

(2.8)
$$\binom{n}{\vec{k}} = \begin{cases} n! / \prod_{a \in E} k_a! & \text{if } \sum_{a \in E} k_a = n, \\ 0 & \text{otherwise,} \end{cases}$$

for the *multinomial coefficient*, which specifies the cardinality of the set $\{S = \vec{k}\}$; in the case $E = \{0, 1\}$ and $\vec{k} = (n - k, k)$, $\binom{n}{\vec{k}}$ coincides with the usual binomial coefficient $\binom{n}{k}$.

Definition. For every probability density ρ on E, the probability measure $\mathcal{M}_{n,\rho}$ on $(\widehat{\Omega}, \mathscr{P}(\widehat{\Omega}))$ with density

$$\mathcal{M}_{n,\rho}(\{\vec{k}\}) = \binom{n}{\vec{k}} \prod_{a \in E} \rho(a)^{k_a}, \quad \vec{k} \in \widehat{\Omega},$$

is called the *multinomial distribution* for n trials with success probabilities $\rho(a)$, $a \in E$.

For $|E| = 3$, the multinomial distribution is illustrated in Figure 2.3 on p. 37. The case $E = \{0, 1\}$ is particularly simple. This is because $\widehat{\Omega}$ can then be replaced by the sample space $\{0, \ldots, n\}$ by identifying every $k \in \{0, \ldots, n\}$ with the pair $(n - k, k) \in \widehat{\Omega}$. Setting $p = \rho(1) \in [0, 1]$, we find that the multinomial distribution $\mathcal{M}_{n,\rho}$ is reduced to the *binomial distribution* $\mathcal{B}_{n,p}$ on $\{0, \ldots, n\}$ with density

$$\mathcal{B}_{n,p}(\{k\}) = \binom{n}{k} p^k (1 - p)^{n-k}, \quad k \in \{0, \ldots, n\}.$$

The above reasoning is not restricted to the case of probability densities ρ with rational values, as it did in the urn example in Section 2.2.1. As a summary of this section we thus obtain the following.

(2.9) Theorem. Multinomial distribution of the sampling histogram. *Suppose E is a finite set with $|E| \geq 2$, ρ a discrete density on E and $P = \rho^{\otimes n}$ the associated n-fold product measure on $\Omega = E^n$. The random variable $S : \Omega \to \widehat{\Omega}$ defined by (2.7) then has the multinomial distribution $P \circ S^{-1} = \mathcal{M}_{n,\rho}$.*

If $E = \{0, 1\}$ and $\rho(1) = p \in [0, 1]$, this means the following: Relative to the Bernoulli distribution with parameter p, the random variable

$$S : \{0, 1\}^n \to \{0, \ldots, n\}, \quad \omega \to \sum_{i=1}^n \omega_i \quad (\text{'number of successes'})$$

has the binomial distribution $\mathcal{B}_{n,p}$.

(2.10) Example. *Children's birthday party.* At a birthday party 12 children come together, where 3 are from village A, 4 from village B and 5 from village C. They play a game of luck five times. The probability that one child from village A wins and two children from village B and village C each is given by

$$\mathcal{M}_{5;\frac{3}{12},\frac{4}{12},\frac{5}{12}}(\{(1,2,2)\}) = \frac{5!}{1!2!2!}\frac{3}{12}\left(\frac{4}{12}\right)^2\left(\frac{5}{12}\right)^2 = \frac{125}{864} \approx 0.14\,.$$

(2.11) Example. *The Maxwell–Boltzmann distribution.* Let us return to the framework of Example (2.2), where n indistinguishable particles must be allocated to N cells. We assume that the cells belong to finitely many energy levels from a set E, in such a way that there are N_a cells of level $a \in E$. Hence, $N = \sum_{a \in E} N_a$. If we suppose that the particles are indistinguishable only due to the lack of experimental technology, but 'in reality' can be labelled $1, \ldots, n$, we can assign to each particle its position by taking an ordered sample of size n – with replacement – from an urn containing N tickets for the different cells; N_a of these point to a cell with energy level a. The similarity of particles resp. cells justifies the assumption that the 'microstates' in the auxiliary sample space $\overline{\Omega} = \{1, \ldots, N\}^n$ are equally likely. As a matter of fact, however, we can only observe the respective macrostate in $\widehat{\Omega}$, which specifies the number of particles on each energy level $a \in E$. Now, Theorem (2.9) states that this macrostate has the multinomial distribution $\mathcal{M}_{n,\rho}$ with $\rho(a) = N_a/N$. This is a classical assumption in statistical physics, which dates back to J. C. Maxwell (1831–1879) and L. Boltzmann (1844–1906), but is not applicable if quantum effects have to be taken into account, see Examples (2.2) and (2.15).

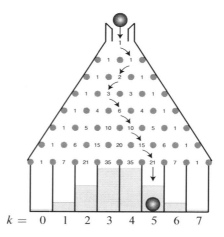

Figure 2.2. The Galton board for $n = 7$. The arrows mark a possible path for the ball. The numbers indicate the total number of paths leading to the respective position. Note that they form Pascal's triangle for the binomial coefficients. In the compartments at the bottom, the bar chart for $\mathcal{B}_{7,1/2}$ is indicated in grey.

As the last example shows, sampling with replacement from an urn containing coloured but otherwise identical balls is equivalent to distributing objects onto different positions which can, in part, be distinguished by certain characteristics and are open for multiple allocations.

The significance of the binomial distribution as the distribution of the number of successes in a Bernoulli experiment can be illustrated physically, in the case $p = 1/2$, by the *Galton board*, see Figure 2.2. A ball is dropped at the top of the board and passes n rows of pins, at which it is deflected either to the right or to the left, each with probability $1/2$. The probability that it takes the right option k times is given by $\mathcal{B}_{n,1/2}(\{k\})$, $0 \le k \le n$. In this case it ends up in compartment k. If the compartments are big enough to accommodate a large number of balls, the relative frequency of the balls in compartment k is close to $\mathcal{B}_{n,1/2}(\{k\})$, by the law of large numbers (Theorem (5.6)).

2.3 Urn Models without Replacement

2.3.1 Labelled Balls

Suppose an urn contains N labelled, but otherwise indistinguishable balls. We remove n balls successively from the urn, *without* replacing them after each draw. If we observe the order in which their labels appear, the appropriate sample space is

$$\overline{\Omega}_{\neq} = \left\{ \bar{\omega} \in \{1, \dots, N\}^n : \bar{\omega}_i \neq \bar{\omega}_j \text{ for } i \neq j \right\}.$$

By the similarity of the balls, it is natural to equip $\overline{\Omega}_{\neq}$ with the uniform distribution $\bar{P}_{\neq} = \mathcal{U}_{\overline{\Omega}_{\neq}}$.

In most applications, however, the order in the sample is not relevant, and we only observe the *set* of all labels drawn; think for instance of the lottery. In this case,

$$\widetilde{\Omega} = \left\{ \tilde{\omega} \subset \{1, \dots, N\} : |\tilde{\omega}| = n \right\}$$

is the space of outcomes. Does it make sense to equip $\widetilde{\Omega}$ with the uniform distribution as well? The transition from $\overline{\Omega}_{\neq}$ to $\widetilde{\Omega}$ is captured by the random variable

$$Y : \overline{\Omega}_{\neq} \to \widetilde{\Omega}, \quad Y(\bar{\omega}_1, \dots, \bar{\omega}_n) = \{\bar{\omega}_1, \dots, \bar{\omega}_n\},$$

which transforms an n-tuple into the corresponding (unordered) set. In accordance with intuition, its distribution $\bar{P}_{\neq} \circ Y^{-1}$ is indeed the uniform distribution on $\widetilde{\Omega}$, since

$$\bar{P}_{\neq}(Y = \omega) = \frac{|\{Y = \omega\}|}{|\overline{\Omega}_{\neq}|} = \frac{n(n-1)\dots 1}{N(N-1)\dots(N-n+1)} = \frac{1}{\binom{N}{n}} = \frac{1}{|\widetilde{\Omega}|}$$

for all $\omega \in \widetilde{\Omega}$. In other words, when a sample is taken without replacement and its order is disregarded, it does not matter whether the balls are picked successively, or all simultaneously with a single grip.

2.3.2   Coloured Balls

Now we return to the assumption that the balls have different colours taken from a set E, and we only observe the colour of the balls but not their labels. We also ignore the order in which the colours appear. As in Section 2.2.2, this leads us to the sample space

$$\widehat{\Omega} = \left\{ \vec{k} = (k_a)_{a \in E} \in \mathbb{Z}_+^E : \sum_{a \in E} k_a = n \right\}.$$

(Here we omit the condition $k_a \le N_a$, so that the sample space $\widehat{\Omega}$ is larger than necessary. But this does not matter because we can simply assign the probability 0 to the impossible outcomes.) Which probability measure on $\widehat{\Omega}$ describes the situation appropriately?

As we have seen above, we can imagine that the balls are drawn all together with a single grip. The transition from $\widetilde{\Omega}$ to $\widehat{\Omega}$ is described by the random variable

$$T : \widetilde{\Omega} \to \widehat{\Omega}, \quad T(\tilde{\omega}) := (|\tilde{\omega} \cap C_a|)_{a \in E},$$

where $C_a \subset \{1, \dots, N\}$ again denotes the family of (labels of) balls of colour a. For every $\vec{k} \in \widehat{\Omega}$, the set $\{T = \vec{k}\}$ contains as many elements as the product set

$$\prod_{a \in E} \{\tilde{\omega}_a \subset F_a : |\tilde{\omega}_a| = k_a\},$$

since the mapping $\tilde{\omega} \to (\tilde{\omega} \cap F_a)_{a \in E}$ is a bijection between these sets. Consequently, we deduce that

$$P(T = \vec{k}) = \frac{\prod_{a \in E} \binom{N_a}{k_a}}{\binom{N}{n}}.$$

Definition. Let E be a finite set (with at least two elements), $\vec{N} = (N_a)_{a \in E} \in \mathbb{Z}_+^E$, $N = \sum_{a \in E} N_a$, and $n \ge 1$. Then the probability measure $\mathcal{H}_{n,\vec{N}}$ on $(\widehat{\Omega}, \mathscr{P}(\widehat{\Omega}))$ with density

$$\mathcal{H}_{n,\vec{N}}(\{\vec{k}\}) = \frac{\prod_{a \in E} \binom{N_a}{k_a}}{\binom{N}{n}}, \quad \vec{k} \in \widehat{\Omega},$$

is called the (general) *hypergeometric distribution* with parameters n and $\vec{N}$.

In the special case $E = \{0, 1\}$, $\widehat{\Omega}$ can be identified with $\{0, \dots, n\}$, as shown in Section 2.2.2. The (classical) hypergeometric distribution then takes the form

$$\mathcal{H}_{n;N_1,N_0}(\{k\}) = \frac{\binom{N_1}{k}\binom{N_0}{n-k}}{\binom{N_1+N_0}{n}}, \quad k \in \{0, \dots, n\}.$$

So, if an urn contains N_a balls of colour $a \in E$, and if n balls are randomly selected (either successively without replacement, or all simultaneously), then the histogram of colours has the hypergeometric distribution $\mathcal{H}_{n,\vec{N}}$.

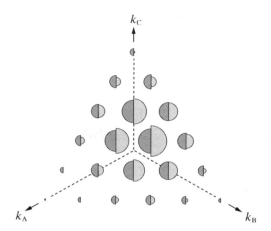

Figure 2.3. Comparison of the multinomial and the hypergeometric distribution for the urn in Examples (2.10) and (2.13). For every $\vec{k} \in \widehat{\Omega}$, the area of the left (darker) semicircle with centre $\vec{k}$ is proportional to $\mathcal{M}_{5;3/12,4/12,5/12}(\{\vec{k}\})$, whereas the right semicircle corresponds to $\mathcal{H}_{5,(3,4,5)}(\{\vec{k}\})$. Sampling without replacement gives a disadvantage to the 'vertex regions' of $\widehat{\Omega}$.

(2.12) Example. *Lottery.* In the '6 out of 49' lottery, the probability for exactly four correct numbers is $\mathcal{H}_{6;6,43}(\{4\}) = \binom{6}{4}\binom{43}{2}/\binom{49}{6} \approx 9.686 \cdot 10^{-4}$.

(2.13) Example. *Representation of interest groups.* A committee of 12 persons consists of 3 representatives of group A, 4 of group B and 5 of group C. By drawing lots, a subcommittee of 5 people is chosen. The probability that this subcommittee consists of exactly one member of group A and two members of groups B and C each is

$$\mathcal{H}_{5,(3,4,5)}\big(\{(1,2,2)\}\big) = \frac{\binom{3}{1}\binom{4}{2}\binom{5}{2}}{\binom{12}{5}} = \frac{5}{22} \approx 0.23 \, .$$

This probability is (obviously) not the same as in Example (2.10), where the same sample was taken from the same urn, but *with* replacement. Figure 2.3 gives a graphical comparison of both distributions. However, if the number N of balls is large enough, it should hardly make any difference whether the balls are being replaced, or not. This intuition is confirmed by the following result.

(2.14) Theorem. Multinomial approximation of hypergeometric distributions. *Let E be finite with $|E| \geq 2$, ρ a discrete density on E, and $n \geq 1$. Then, in the limit as $N \to \infty$, $N_a \to \infty$, and $N_a/N \to \rho(a)$ for all $a \in E$, the hypergeometric distribution $\mathcal{H}_{n,\vec{N}}$ converges (pointwise) to the multinomial distribution $\mathcal{M}_{n,\rho}$.*

Proof. Fix any $\vec{k} \in \widehat{\Omega}$. Obviously,

$$\binom{N_a}{k_a} = \frac{N_a^{k_a}}{k_a!} \frac{N_a(N_a-1)\ldots(N_a-k_a+1)}{N_a^{k_a}}$$

$$= \frac{N_a^{k_a}}{k_a!} 1\left(1-\frac{1}{N_a}\right)\left(1-\frac{2}{N_a}\right)\ldots\left(1-\frac{k_a-1}{N_a}\right) \underset{N_a\to\infty}{\sim} \frac{N_a^{k_a}}{k_a!}.$$

Here we use the notation '$a(\ell) \sim b(\ell)$ as $\ell \to \infty$' for asymptotic equivalence, meaning that $a(\ell)/b(\ell) \to 1$ as $\ell \to \infty$. It follows that

$$\mathcal{H}_{n,\vec{N}}(\{\vec{k}\}) \sim \left[\prod_{a\in E} \frac{N_a^{k_a}}{k_a!}\right] \bigg/ \frac{N^n}{n!} = \binom{n}{\vec{k}} \prod_{a\in E} \left(\frac{N_a}{N}\right)^{k_a} \to \mathcal{M}_{n,\rho}(\{\vec{k}\})$$

in the limit of the theorem. $\diamond$

(2.15) Example. *The Fermi–Dirac distribution (1926).* For a third time, we consider the particle situation of Examples (2.2) and (2.11), where n indistinguishable particles are to be distributed across N cells, which belong to certain levels of energy. Assume that there are N_a cells of level $a \in E$. We now impose 'Pauli's exclusion principle', which says that each cell can contain at most one particle; so, in particular, $N \geq n$ is required. This is a sensible assumption for the so-called fermions (particles with half-integer spin), which include the electrons, protons and neutrons. Then the cell arrangement of all particles corresponds to a sample taken without replacement from an urn containing N distinct seat reservations, of which N_a refer to seats of level a. In this case the above reasoning shows that the macrostate of the system, which for each a gives the number of particles of energy level a, is $\mathcal{H}_{n,\vec{N}}$-distributed with $\vec{N} = (N_a)_{a\in E}$. Moreover, Theorem (2.14) shows that Pauli's exclusion principle becomes irrelevant when each energy level offers much more cells than necessary to accommodate all particles.

   In particular, the last example shows that sampling without replacement from balls that differ in colour but are otherwise indistinguishable corresponds to distributing objects across boxes that carry different labels but are otherwise treated equally, provided multiple allocations are not allowed. This correspondence is illustrated in Figure 2.4.

Figure 2.4. Equivalence of drawing without replacement and distributing objects without multiple allocations. The situation corresponds to the urn in Example (2.13). The places that have been taken are coloured in black (and marked at the beginning of the row, since the places in each row are equivalent).

2.4 The Poisson Distribution

Let us begin with an example from everyday life.

(2.16) Example. *Insurance claims.* How many claims does an insurance company receive in a fixed time interval $]0, t]$, $t > 0$? Obviously, the number of claims is random, and the sample space is $\Omega = \mathbb{Z}_+$. But which P describes the situation? In order to find out, we develop a heuristic argument. We partition the interval $]0, t]$ into n subintervals of length t/n. If n is large (and so the subintervals small), we can assume that in each subinterval the company receives at most one claim. The probability for such a claim should be proportional to the length of the interval; so we assume it is equal to $\alpha t/n$ for some proportionality constant $\alpha > 0$. Moreover, it is plausible that the appearance of a claim in one subinterval does not depend on whether a claim appears in another subinterval. So we can pretend that the claims in the n subintervals are sampled by drawing n times with replacement from an urn that contains a proportion $\alpha t/n$ of 'liability balls'. Hence, if n is large, we can conclude from Theorem (2.9) that the probability of k claims during the time interval $]0, t]$ is approximately equal to $\mathcal{B}_{n,\alpha t/n}(\{k\})$. This leads us to the tentative assumption that

$$P(\{k\}) := \lim_{n\to\infty} \mathcal{B}_{n,\alpha t/n}(\{k\}), \quad k \in \mathbb{Z}_+,$$

for the probability measure P we are interested in. And indeed, this limit does exist.

(2.17) Theorem. *Poisson approximation of binomial distributions. Let $\lambda > 0$ and $(p_n)_{n\geq 1}$ a sequence in $[0, 1]$ with $np_n \to \lambda$. Then, for every $k \in \mathbb{Z}_+$,*

$$\lim_{n\to\infty} \mathcal{B}_{n,p_n}(\{k\}) = e^{-\lambda}\frac{\lambda^k}{k!}.$$

Proof. Exactly as in the proof of Theorem (2.14), we obtain for each $k \in \mathbb{Z}_+$ in the limit $n \to \infty$

$$\binom{n}{k} p_n^k (1-p_n)^{n-k} \sim \frac{n^k}{k!} p_n^k (1-p_n)^{n-k} \sim \frac{\lambda^k}{k!} (1-p_n)^n$$

$$= \frac{\lambda^k}{k!}\left(1 - \frac{np_n}{n}\right)^n \to \frac{\lambda^k}{k!} e^{-\lambda}.$$

In the second step we used that $(1-p_n)^k \to 1$; the last convergence follows from the well-known approximation formula for the exponential function. $\diamond$

Figure 2.5 illustrates the Poisson convergence; in Theorem (5.32) we will see that the deviation from the limit is of order $1/n$. The series expansion of the exponential function shows that the limit in Theorem (2.17) sums up to 1, and thus defines a discrete density on $\mathbb{Z}_+$. The corresponding probability measure is one of the fundamental distributions in stochastics.

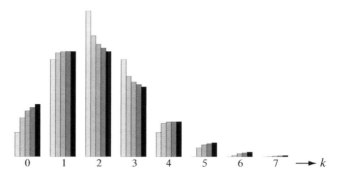

Figure 2.5. The Poisson approximation for $\lambda = 2$. For $k \in \mathbb{Z}_+$, the bars show $\mathcal{B}_{n,2/n}(\{k\})$ for $n = 4, 8, 16, 32$ (from light grey to dark grey), and the limit $\mathcal{P}_2(\{k\})$ (black).

Definition. For $\lambda > 0$, the probability measure $\mathcal{P}_\lambda$ on $(\mathbb{Z}_+, \mathscr{P}(\mathbb{Z}_+))$ defined by $\mathcal{P}_\lambda(\{k\}) = e^{-\lambda}\lambda^k/k!$ is called the *Poisson distribution* with parameter λ (in honour of Siméon-Denis Poisson, 1781–1840).

As a result of this subsection, we conclude that the family of Poisson distributions $\mathcal{P}_\lambda$ on $\mathbb{Z}_+$ provides natural models for the number of random time points in a fixed time interval. Typical applications, apart from the insurance claims, include the number of phone calls arriving at a telephone switchboard, the number of emails sent via a mail server, the number of particles emitted by a radioactive substance, the number of cars passing a check-point, and so on. In Section 3.5 we will discuss this situation in more detail.

2.5 Waiting Time Distributions

2.5.1 The Negative Binomial Distributions

Consider a Bernoulli experiment as discussed in Section 2.2.1. An urn contains red and white balls, where the proportion of red balls is $0 < p < 1$. The balls are drawn with replacement. What is the distribution of the waiting time until the rth success, i.e., until we draw the rth red ball? Here, $r \in \mathbb{N}$ is a fixed number. Since we must draw at least r times, we consider the remaining waiting time after r draws, or equivalently, the number of failures before the rth success; then the sample space is $\Omega = \mathbb{Z}_+$. Which P describes the situation?

On the infinite product space $\{0, 1\}^{\mathbb{N}}$ we could define the random variable

$$T_r(\omega) = \min \left\{ k : \sum_{i=1}^{k} \omega_i = r \right\} - r$$

and obtain P as the distribution of T_r. However, the existence of an infinite Bernoulli measure will only be shown later in Example (3.29). We therefore proceed in a more heuristic way here. For every k, $P(\{k\})$ is to be defined as the probability of the rth success on the $(r+k)$th draw, in other words the probability of a success at time $r+k$ and exactly k failures at some earlier times. Since there are exactly $\binom{r+k-1}{k}$ possible choices for the k failure times, the Bernoulli measure on $\{0, 1\}^{r+k}$ assigns to this event the probability

$$P(\{k\}) := \binom{r+k-1}{k} p^r (1-p)^k = \binom{-r}{k} p^r (p-1)^k .$$

Here,

$$(2.18) \qquad \binom{-r}{k} = \frac{(-r)(-r-1)\dots(-r-k+1)}{k!}$$

is the general binomial coefficient, and we have used the identity

$$k! \binom{r+k-1}{k} = (r+k-1)\dots(r+1)r$$

$$= (-1)^k (-r)(-r-1)\dots(-r-k+1) .$$

The expression for P thus obtained does define a probability measure on $\mathbb{Z}_+$, even when r is an arbitrary positive real number. Indeed, for $r > 0$ it is clear that $\binom{-r}{k}(-1)^k \geq 0$ for all $k \in \mathbb{Z}_+$, and the general binomial theorem implies that

$$\sum_{k \geq 0} \binom{-r}{k} p^r (p-1)^k = p^r (1+p-1)^{-r} = 1.$$

Definition. For $r > 0$ and $0 < p < 1$, the probability measure $\overline{\mathcal{B}}_{r,p}$ on $(\mathbb{Z}_+, \mathscr{P}(\mathbb{Z}_+))$ with density

$$\overline{\mathcal{B}}_{r,p}(\{k\}) = \binom{-r}{k} p^r (p-1)^k , \quad k \in \mathbb{Z}_+,$$

is called the *negative binomial distribution* or the *Pascal distribution* with parameters r and p (after Blaise Pascal, 1623–1662). In particular, $\mathcal{G}_p(\{k\}) = \overline{\mathcal{B}}_{1,p}(\{k\}) = p(1-p)^k$ is called the *geometric distribution* with parameter p.

To summarise, we have seen that $\overline{\mathcal{B}}_{r,p}$ is the distribution of the waiting time (after r) for the rth success in a Bernoulli experiment with parameter p. In particular, the waiting time for the first success has the geometric distribution $\mathcal{G}_p$. (In the terminology of some authors, the geometric distribution is not $\mathcal{G}_p$, but the distribution on $\mathbb{N} = \{1, 2, \dots\}$ that is obtained by shifting $\mathcal{G}_p$ one unit to the right.) Figure 2.6 shows the shape of $\overline{\mathcal{B}}_{r,p}$ for some values of r.

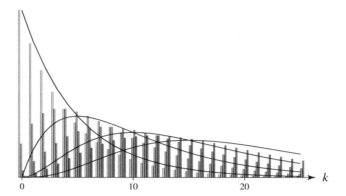

Figure 2.6. Bar charts of the negative binomial distributions $\overline{\mathcal{B}}_{r,p}$ for $p = 0.2$ and $r = 1, 2,$ 3, 4 (from light to darker hues of grey), as well as the densities $\gamma_{1,r}$ of the gamma distributions (on an adjusted scale). The connection between negative binomial and gamma distributions is discussed in Problem 2.13.

2.5.2 The Gamma Distributions

We now turn to continuous time. As in Section 2.4, we consider random time points in the interval $]0, \infty[$; we may again think of incoming claims in an insurance company, or of phone calls arriving at a telephone switchboard. The heuristic reasoning in Example (2.16) led to the conclusion that, for every $t > 0$, the number of points in $]0, t]$ is Poisson distributed with parameter αt, where $\alpha > 0$ represents the average number of points per time. In analogy with the previous subsection, we look for a model of the rth random time point. (In Example (2.16) this would correspond to the instant at which the rth claim comes in.) Obviously, $(\Omega, \mathscr{F}) = (]0, \infty[, \mathscr{B}_{]0,\infty[})$ is an appropriate event space. Which probability measure P describes the distribution of the rth random point? For this P, the number $P(]0, t])$ is the probability that the rth claim arrives no later than t, in other words the probability of at least r claims in $]0, t]$. In view of our Poisson assumption on the number of claims, we are thus led to the formula

$$P(]0, t]) = 1 - \mathcal{P}_{\alpha t}(\{0, \dots, r-1\})$$

(2.19)

$$= 1 - e^{-\alpha t} \sum_{k=0}^{r-1} \frac{(\alpha t)^k}{k!} = \int_0^t \frac{\alpha^r}{(r-1)!} x^{r-1} e^{-\alpha x} \, dx \, ;$$

the last equality is obtained by differentiating with respect to t. Hence, Remark (1.31) tells us that the P we are looking for is exactly the probability measure on $]0, \infty[$ with the Lebesgue density

$$\gamma_{\alpha,r}(x) = \frac{\alpha^r}{(r-1)!} x^{r-1} e^{-\alpha x}.$$

To check that $\gamma_{\alpha,r}$ is indeed a probability density, one can use *Euler's gamma function*

$$\Gamma(r) = \int_0^\infty y^{r-1} e^{-y} \, dy, \quad r > 0.$$

Clearly, $\Gamma(1) = 1$, and by partial integration one obtains the well-known recursive formula $\Gamma(r+1) = r\,\Gamma(r)$. In particular, for $r \in \mathbb{N}$ one finds $\Gamma(r) = (r-1)!$. By the substitution $\alpha x = y$, it follows that $\int_0^\infty \gamma_{\alpha,r}(x)\,dx = 1$. An analogous density function can be obtained for arbitrary real $r > 0$.

Definition. For every $\alpha, r > 0$, the probability measure $\Gamma_{\alpha,r}$ on $(]0,\infty[, \mathcal{B}_{]0,\infty[})$ with Lebesgue density

(2.20) $$\gamma_{\alpha,r}(x) = \frac{\alpha^r}{\Gamma(r)}\, x^{r-1} e^{-\alpha x}, \quad x > 0,$$

is called the *gamma distribution* with scale parameter α and shape parameter r. In particular, the probability measure $\mathcal{E}_\alpha = \Gamma_{\alpha,1}$ with the density $\gamma_{\alpha,1}(x) = \alpha e^{-\alpha x}$ is called the *exponential distribution* with parameter α.

Summarising, we see that for $r \in \mathbb{N}$ the distribution $\Gamma_{\alpha,r}$ describes the distribution of the rth point in a Poisson model of random points on the positive axis. In particular, the first point is exponentially distributed with parameter α. We will come back to this in Section 3.5. The gamma densities for some selected parameter values are plotted in Figures 2.6 and 9.2 (on p. 252).

2.5.3   The Beta Distributions

We will now take a different approach to the problem of choosing random time points and determining the waiting time for the rth point. Let us suppose that the number of points in a given time interval is not random, but fixed. Think for example of a supermarket which receives deliveries from n suppliers. At what time does the rth delivery arrive?

We choose the unit of time in such a way that the n deliveries arrive in the open unit interval $]0,1[$. If we label the suppliers with the numbers $1,\ldots,n$, we obtain the sample space $\Omega =]0,1[^n$, which is equipped with the Borel σ-algebra $\mathcal{B}^n_\Omega$ as usual. For every $1 \leq i \leq n$ and $\omega = (\omega_1,\ldots,\omega_n) \in \Omega$, we write $T_i(\omega) := \omega_i$ for the instant at which the ith supplier arrives. Assuming that there is no prior information about the delivery times, it is natural to choose the uniform distribution $P = \mathcal{U}_\Omega$ as the underlying probability measure.

How long does it typically take until the supermarket has received the rth delivery? To answer this, we must first order the n arrival times. This is possible because no two deliveries arrive at the same time, at least with probability 1. More precisely,

$$P\Big(\bigcup_{i \neq j} \{T_i = T_j\}\Big) = 0,$$

since this event is a finite union of hyperplanes in Ω and thus has n-dimensional volume 0. Hence, the following quantities are well-defined with probability 1.

Definition. The *order statistics* $T_{1:n}, \ldots, T_{n:n}$ of the random variables $T_1, \ldots, T_n$ are defined by the properties

$$T_{1:n} < T_{2:n} < \cdots < T_{n:n}, \quad \{T_{1:n}, \ldots, T_{n:n}\} = \{T_1, \ldots, T_n\}.$$

In other words, $T_{r:n}$ is the rth smallest of the times $T_1, \ldots, T_n$.

Let us now determine the distribution of $T_{r:n}$ for fixed $r, n \in \mathbb{N}$. Distinguishing the $n!$ possible orderings of the values of $T_1, \ldots, T_n$, we find that all give the same contribution because the order of integration can be interchanged arbitrarily by Fubini's theorem (see e.g. [56]). This means that for every $0 < c \le 1$

$$P(T_{r:n} \le c) = n! \int_0^1 dt_1 \ldots \int_0^1 dt_n \, 1_{\{t_1 < t_2 < \cdots < t_n\}} 1_{]0,c]}(t_r)$$

$$= n! \int_0^c dt_r \, a(r-1, t_r) \, a(n-r, 1-t_r),$$

where

$$a(r-1, s) = \int_0^s dt_1 \ldots \int_0^s dt_{r-1} \, 1_{\{t_1 < t_2 < \cdots < t_{r-1} < s\}} = \frac{s^{r-1}}{(r-1)!}$$

and

$$a(n-r, 1-s) = \int_s^1 dt_{r+1} \ldots \int_s^1 dt_n \, 1_{\{s < t_{r+1} < \cdots < t_n\}} = \frac{(1-s)^{n-r}}{(n-r)!}.$$

Altogether, we find

$$P(T_{r:n} \le c) = \frac{n!}{(r-1)! \, (n-r)!} \int_0^c ds \, s^{r-1} \, (1-s)^{n-r}.$$

Setting $c = 1$, it follows in particular that $(r-1)! \, (n-r)!/n! = \mathrm{B}(r, n-r+1)$, where

(2.21) $$\mathrm{B}(a, b) = \int_0^1 s^{a-1} (1-s)^{b-1} \, ds, \quad a, b > 0,$$

is *Euler's beta function*. By Remark (1.31) we conclude that $T_{r:n}$ has density

$$\beta_{r,n-r+1}(s) = \mathrm{B}(r, n-r+1)^{-1} s^{r-1} (1-s)^{n-r}$$

on $]0, 1[$. Density functions of this kind are also interesting for non-integer values of r and n.

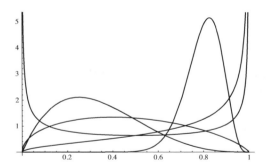

Figure 2.7. $\beta_{a,b}$ for $(a,b) = (1/2,1/2), (3/2,1/2), (3/2,7/4), (2,4), (20,5)$.

Definition. For $a,b > 0$, the probability measure $\beta_{a,b}$ on $]0,1[$ with density

$$(2.22) \qquad \beta_{a,b}(s) = B(a,b)^{-1} s^{a-1}(1-s)^{b-1}, \quad 0 < s < 1,$$

is called the *beta distribution* with parameters a,b.

We thus have seen that, for $r,n \in \mathbb{N}$, $\beta_{r,n-r+1}$ is the distribution of the rth smallest among n randomly chosen points in the unit interval. In particular, $\beta_{1,n}(s) = n(1-s)^{n-1}$ is the distribution density of the first point, and $\beta_{1,1} = \mathcal{U}_{]0,1[}$. Figure 2.7 shows the densities of beta distributions for various parameters; the reader should figure out which parameter values correspond to which graph. The distribution $\beta_{1/2,1/2}$ is called the *arcsine distribution*, cf. Problem 1.17.

For later purposes, we conclude this subsection with a characteristic property of the beta function. For $a,b > 0$ one can write

$$a\left(B(a,b) - B(a+1,b)\right) = \int_0^1 a s^{a-1}(1-s)^b \, ds$$

$$= \int_0^1 s^a b(1-s)^{b-1} \, ds = b\, B(a+1,b),$$

where the second equality follows by partial integration. This implies the recursion formula

$$(2.23) \qquad\qquad B(a+1,b) = \frac{a}{a+b} B(a,b).$$

So one might suspect that the beta function is closely related to the gamma function. This is indeed the case, see (9.8) on p. 251. The recursion can also be used to deduce the expression already derived in the line before (2.21) for the beta function with integer parameters.

2.6   The Normal Distributions

Is there any kind of uniform distribution on an infinite-dimensional ball with infinite radius? To find out, we consider the asymptotics of uniform distributions on large balls in high dimensions. Specifically, let $v > 0$ and

$$\Omega_N = \{x \in \mathbb{R}^N : |x|^2 \leq vN\}$$

be the N-dimensional ball with centre 0 and radius $\sqrt{vN}$. Furthermore, let $P_N = \mathcal{U}_{\Omega_N}$ be the (continuous) uniform distribution on $(\Omega_N, \mathcal{B}_{\Omega_N}^N)$ and $X_1 : \Omega_N \to \mathbb{R}$, $x \to x_1$, the projection onto the first coordinate. We are interested in the asymptotic distribution of X_1 under P_N in the limit as $N \to \infty$.

(2.24) Theorem. Normal distribution as a projection of an 'infinite-dimensional uniform distribution'; Henri Poincaré 1912, Émile Borel 1914. *For all $a < b$, we have*

$$\lim_{N \to \infty} P_N(a \leq X_1 \leq b) = \int_a^b \frac{1}{\sqrt{2\pi v}}\, e^{-x^2/2v}\, dx\,.$$

In statistical mechanics, an extension of this theorem justifies *Maxwell's velocity distribution* of the particles in a gas. If $x \in \mathbb{R}^N$ is the vector of the velocities of all particles, then $|x|^2$ is proportional to the kinetic energy, P_N corresponds to the Liouville measure in velocity space, and $X_1(x) = x_1$ is the first velocity coordinate of the first particle. (The theorem still holds if the kinetic energy per particle, instead of being bounded by v, is kept constant equal to v, which means that P_N is replaced by the uniform distribution on the surface of the ball.)

Proof. Let N be so large that $Nv > \max(a^2, b^2)$. Using Definition (1.17) of λ^N, and computing multiple integrals successively according to Fubini's theorem (see [56]), we can write

$$P_N(a \leq X_1 \leq b) = \lambda^N(\Omega_N)^{-1} \int \ldots \int 1_{\{a \leq x_1 \leq b,\ \sum_{i=1}^N x_i^2 \leq vN\}}\ dx_1 \ldots dx_N$$

$$= \lambda^N(\Omega_N)^{-1} \int_a^b \lambda^{N-1}\Big(B_{N-1}\big(\sqrt{vN - x_1^2}\big)\Big)\, dx_1$$

$$= \lambda^N(\Omega_N)^{-1} \int_a^b \big(vN - x_1^2\big)^{(N-1)/2}\, c_{N-1}\, dx_1\,.$$

Here, $B_{N-1}(r)$ stands for the centred ball in $\mathbb{R}^{N-1}$ with radius r, and $c_{N-1} = \lambda^{N-1}(B_{N-1}(1))$ is the volume of the $(N-1)$-dimensional unit ball; the second equation is obtained by integrating over the variables $x_2, \ldots, x_N$. Similarly, we get

$$\lambda^N(\Omega_N) = \int_{-\sqrt{vN}}^{\sqrt{vN}} \big(vN - x_1^2\big)^{(N-1)/2}\, c_{N-1}\, dx_1\,.$$

Thus by cancelling the constant $c_{N-1}(vN)^{(N-1)/2}$ we obtain

$$P_N(a \leq X_1 \leq b) = \int_a^b f_N(x)\,dx \Big/ \int_{-\sqrt{vN}}^{\sqrt{vN}} f_N(x)\,dx,$$

where $f_N(x) = (1 - x^2/vN)^{(N-1)/2}$.

We now perform the limit $N \to \infty$, treating the integrals in the numerator and the denominator separately. By the well-known approximation formula for the exponential function, $f_N(x)$ converges to $e^{-x^2/2v}$, and this convergence is uniform on every compact interval, in particular on $[a, b]$. The numerator therefore converges to $\int_a^b e^{-x^2/2v}\,dx$. To deal with the denominator, we take advantage of the estimate

$$f_N(x) \leq \left(e^{-x^2/vN}\right)^{(N-1)/2} \leq e^{-x^2/4v} \quad \text{for all } N \geq 2.$$

For any $c > 0$ and $N \geq \max(2, c^2/v)$, this gives us the sandwich inequality

$$\int_{-c}^c f_N(x)\,dx \leq \int_{-\sqrt{vN}}^{\sqrt{vN}} f_N(x)\,dx \leq \int_{-c}^c f_N(x)\,dx + \int_{\{|x|>c\}} e^{-x^2/4v}\,dx.$$

Now, if we let first N and then c tend to ∞, the terms on the left and the right, and therefore also the integral squeezed in between, converge towards $\int_{-\infty}^\infty e^{-x^2/2v}\,dx$. The theorem thus follows from the next result. $\diamond$

(2.25) Lemma. Gaussian integral. $\int_{-\infty}^\infty e^{-x^2/2v}\,dx = \sqrt{2\pi v}$.

Proof. Treating the square of the integral as a two-dimensional integral and introducing polar coordinates one finds

$$\left[\int_{-\infty}^\infty e^{-x^2/2v}\,dx\right]^2 = \int_{-\infty}^\infty dx \int_{-\infty}^\infty dy\, e^{-(x^2+y^2)/2v}$$

$$= \int_0^{2\pi} d\varphi \int_0^\infty dr\, r e^{-r^2/2v}$$

$$= -2\pi v\, e^{-r^2/2v}\Big|_{r=0}^\infty = 2\pi v. \qquad \diamond$$

Theorem (2.24) can be rephrased as follows. Projecting the uniform distribution on a large ball in high dimensions onto a single coordinate, one obtains in the limit a distribution density of the form $e^{-x^2/2v}/\sqrt{2\pi v}$. (The projection onto several coordinates is the subject of Problem 2.16.) These densities play a fundamental role in stochastics. The reason is that they also appear as asymptotic densities in another limit theorem, the so-called central limit theorem, which will be discussed in Sections 5.2 and 5.3. This will elucidate their significance as a universal approximation, and gives rise to their prominent role in statistics.

Definition. Let $m \in \mathbb{R}$ and $v > 0$. The probability measure $\mathcal{N}_{m,v}$ on $(\mathbb{R}, \mathscr{B})$ with density

$$\phi_{m,v}(x) = \frac{1}{\sqrt{2\pi v}} \, e^{-(x-m)^2/2v} \, , \quad x \in \mathbb{R},$$

is called the *normal distribution* or *Gauss distribution* with mean m and variance v. (The names of the parameters m and v will be justified in Chapter 4.) $\mathcal{N}_{0,1}$ is called the *standard normal distribution*, and $\phi_{m,v}$ is often called the *Gaussian bell curve*. It was pictured on the 10 Deutsche Mark banknote, which was valid until the euro was introduced in 2002; see Figure 2.8.

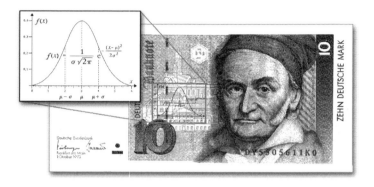

Figure 2.8. Carl Friedrich Gauss (1777–1855) and the bell curve on the 10 DEM note.

Problems

2.1 At a tombola, the winning lot shall be drawn by a 'good luck fairy' born on a Sunday. How many ladies must be present so that, with a probability of at least 99%, at least one of them is a Sunday child? Set up a suitable model.

2.2 [S] Consider a system of n indistinguishable particles, each of which can be located in one of N different cells. N_a of the cells belong to an energy level $a \in E$, E a finite set. Under the assumption of the Bose–Einstein distribution, determine the probability that, for each $a \in E$, k_a particles occupy a cell of energy a.

2.3 [S] Recall the situation of Bertrand's paradox, and let X be the distance of the random chord to the centre of the circle. Find the distribution density of X if

   (a) the midpoint of the chord is uniformly distributed on the disk Ω_1,

   (b) the angle under which the cord is seen from the centre of the circle is uniformly distributed on the interval Ω_2.

2.4 *Buffon's needle problem* (stated by G.-L. L. Comte de Buffon in 1733 and analysed in 1777). Think of (infinitely many) parallel straight lines at distance a, embedded in a plane. A needle of length $l < a$ is randomly placed onto the plane. What is the probability that the needle hits one of the straight lines? Set up an appropriate model. *Hint:* Describe the position of the needle by a suitable angle and the distance of its midpoint to the closest straight line.

2.5 *Light intensity.* A light source is placed at a distance $a > 0$ from a straight line. It radiates uniformly in all directions that eventually hit the line. Denote by X the point where a light ray hits the straight line. Show that X has the distribution density $c_a(x) = a/(\pi(a^2 + x^2))$ on $\mathbb{R}$. The corresponding distribution is called the *Cauchy distribution* with parameter a.

2.6 In the surroundings of each of 10 nuclear power stations, 100 people (chosen 'with replacement') are examined for a certain disease, which on average can be found in 1% of the nation's total population. The agreement is that a power station is considered suspicious if at least 3 out of the 100 people tested show the symptoms of the disease.

(a) What is the probability that at least one power station is identified as suspicious, even though the probability of having the disease in the neighbourhood of the power plants does not deviate from the national average?

(b) What is the probability that none of them will be detected as being suspicious, even though the probability of having the disease is 2% near the power plants?

2.7[S] *Simple symmetric random walk.* In the evening of an election day, the votes for two competing candidates A and B are being counted. Both candidates are equally popular, i.e., on each ballot A or B are chosen with equal probability $1/2$; in total $2N$ ballots have been cast. Set $X_i = 1$ or -1 depending on whether the ith ballot is for A or B. Then the sum $S_j = \sum_{i=1}^{j} X_i$ indicates by how much A is leading before B after j ballots have been counted (or by how much A is behind B); the sequence (S_j) is called the simple symmetric random walk. Let $1 \le n \le N$ and set $u_n := 2^{-2n} \binom{2n}{n}$. Specify the probability model and show the following:

(a) The event $G_n = \{S_{2n} = 0, S_{2k} \ne 0 \text{ for } 1 \le k < n\}$ ('the first tie appears when $2n$ ballots have been counted') satisfies

$$P(G_n) = 2^{-2n+1}\left[\binom{2n-2}{n-1} - \binom{2n-2}{n}\right] = u_{n-1} - u_n.$$

Hint: Each realisation of the sequence $(S_j)_{j \le 2n}$ can be visualised by its path, a broken line through the points (j, S_j). Find a bijection between the paths from $(1, 1)$ to $(2n-1, 1)$ that hit the horizontal axis, and all paths from $(1, -1)$ to $(2n-1, 1)$.

(b) The event $G_{>n} = \{S_{2k} \ne 0 \text{ for } 1 \le k \le n\}$ ('no tie during the count of the first $2n$ ballots') has the probability $P(G_{>n}) = u_n$.

2.8 The interval $[0, 2]$ is split in two parts by picking a point at random in $[0, 1]$ according to the uniform distribution. Let $X = l_1/l_2$ be the ratio of the length of the shorter part l_1 to the length l_2 of the longer one. Find the distribution density of X.

2.9 The genome of the fruit fly Drosophila melanogaster is divided into approximately $m = 7000$ sections (which can be identified by the colouring of the giant polytene chromosomes found in the salivary glands). As a simplification, suppose that each section contains the same number of $M = 23000$ base pairs. Hence, the genome contains $N = mM$ base pairs. Using high energy radiation, $n = 1000$ (randomly distributed) base pairs are destroyed. Find a stochastic model for the number of destroyed base pairs in all genome sections. Determine the distribution of the number Z_i of destroyed base pairs in section i ($1 \leq i \leq m$) and show that Z_i is approximately Poisson distributed.

2.10 ^{S} *Projecting the multinomial distribution.* Let E be a finite set, ρ a discrete density on E, $n \in \mathbb{N}$, and $X = (X_a)_{a \in E}$ a random variable with values in

$$\widehat{\Omega} = \left\{ \vec{k} = (k_a)_{a \in E} \in \mathbb{Z}_+^E : \sum_{a \in E} k_a = n \right\}$$

and multinomial distribution $\mathcal{M}_{n,\rho}$. Show that, for every $a \in E$, X_a has the binomial distribution $\mathcal{B}_{n,\rho(a)}$.

2.11 ^{S} *Fixed points of a random permutation.* Before a theatre performance, n people leave their coats in the cloak room. After the performance, due to a power cut, the coats are returned in the dark in random order. Let X be the random number of people who get their own coat back. Find the distribution of X, i.e., $P(X = k)$ for every $k \geq 0$. What happens in the limit as $n \to \infty$? (The case $k = 0$ corresponds to the rencontre problem of Problem 1.13.) *Hint:* Use Problem 1.7 again.

2.12 *Banach's matchbox problem.* The Polish mathematician Stefan Banach (1892–1945) always carried one matchbox in each of the two pockets of his coat. With equal probability, he used the matchbox in the left or in the right pocket. On finding an empty box, he replaced both boxes by new ones. Find the distribution of the remaining matches after one cycle (i.e., after finding an empty box), if each new box contains N matches.

2.13 *Gamma and negative binomial distributions.* Let $r \in \mathbb{N}$, $\alpha, t > 0$, $(p_n)_{n \geq 1}$ a sequence in $]0, 1[$ with $np_n \to \alpha$, and $(t_n)_{n \geq 1}$ a sequence in $\mathbb{Z}_+$ with $t_n/n \to t$. Show that

$$\Gamma_{\alpha,r}(]0, t]) = \lim_{n \to \infty} \overline{\mathcal{B}}_{r,p_n}(\{0, \dots, t_n\}),$$

and interpret this result in terms of waiting times. *Hint:* First show that $\overline{\mathcal{B}}_{r,p}(\{0, 1, \dots, m\}) = \mathcal{B}_{r+m,p}(\{r, r+1, \dots, r+m\})$.

2.14 ^{S} *Gamma and beta distributions.* In the situation of Section 2.5.3, let $(s_n)_{n \geq 1}$ be a sequence in $]0, \infty[$ with $n/s_n \to \alpha > 0$. Show that for all $r \in \mathbb{N}$ and $t > 0$,

$$\Gamma_{\alpha,r}(]0, t]) = \lim_{n \to \infty} P(s_n T_{r:n} \leq t).$$

Can you rephrase this result in terms of random points on the positive time axis?

2.15 *Affine transformations of normal distributions.* Let X be a real random variable with normal distribution $\mathcal{N}_{m,v}$, and $a, b \in \mathbb{R}$, $a \neq 0$. Show that the random variable $aX + b$ has the distribution $\mathcal{N}_{am+b,a^2v}$.

2.16 *Extending the Poincaré–Borel theorem.* Prove the following stronger version of Theorem (2.24). If $X_i : \Omega_N \to \mathbb{R}$ denotes the projection onto the ith coordinate, then

$$\lim_{N \to \infty} P_N \left(X_i \in [a_i, b_i] \text{ for } 1 \leq i \leq k \right) = \prod_{i=1}^{k} \mathcal{N}_{0,v}([a_i, b_i])$$

for all $k \in \mathbb{N}$ and all $a_i, b_i \in \mathbb{R}$ with $a_i < b_i$ for $1 \leq i \leq k$. That is, asymptotically, the projections are independent (as to be defined in Section 3.3) and normally distributed.

Chapter 3

Conditional Probabilities and Independence

The interdependence of events, or of subexperiments, is a central theme of stochastics. It can be expressed by means of the fundamental concept of conditional probability. After an introductory discussion of this notion, we will see how it can be used to construct multi-stage probability models that describe a sequence of subexperiments with a given dependence structure. The simplest case of dependence is independence, which will receive particular attention. In particular, we will consider a concrete model with 'quite a lot of independence', the Poisson process, as well as several algorithms for the simulation of independent random variables with given distributions. Finally, we will investigate the effect of independence on the long-term behaviour of a random experiment that is repeated infinitely often.

3.1 Conditional Probabilities

Let us start with an example.

(3.1) Example. *Sampling without replacement.* From an urn containing r red and w white balls, two balls are drawn successively without replacement. We imagine that the red balls are labelled with the numbers $1, \dots, r$, and the white balls with the numbers $r+1, \dots, r+w$. So, the model is $\Omega = \{(k,l) : 1 \leq k, l \leq r+w, k \neq l\}$ and $P = \mathcal{U}_\Omega$, the uniform distribution. We consider the events

$$A = \{\text{first ball is red}\} = \{(k,l) \in \Omega : k \leq r\},$$
$$B = \{\text{second ball is red}\} = \{(k,l) \in \Omega : l \leq r\}.$$

Before the start of the experiment, one expects B to occur with probability

$$P(B) = \frac{|B|}{|\Omega|} = \frac{r(r+w-1)}{(r+w)(r+w-1)} = \frac{r}{r+w}.$$

If A occurs in the first draw, do we still expect B to occur with probability $r/(r+w)$? Certainly not! Intuitively, we would argue that $r-1$ red balls and w white balls are left in the urn, so now the probability should be $\frac{r-1}{r+w-1}$. In other words, the occurrence of A has led us to revise the way we assign probabilities to the events. This means that the probability measure P must be replaced by a new probability measure P_A that takes account of the information that A has occurred. Sensibly, P_A should satisfy the following two conditions.

(a) $P_A(A) = 1$, i.e., the event A is now certain.

(b) For subevents of A, the new probability is proportional to the original one, i.e., there exists a constant $c_A > 0$ such that $P_A(B) = c_A P(B)$ for all $B \in \mathscr{F}$ with $B \subset A$.

The following proposition shows that P_A is uniquely determined by these properties.

(3.2) Proposition. Re-weighting of events. *Let $(\Omega, \mathscr{F}, P)$ be a probability space and $A \in \mathscr{F}$ with $P(A) > 0$. Then there is a unique probability measure P_A on $(\Omega, \mathscr{F})$ satisfying* (a) *and* (b), *which is given by*

$$P_A(B) := \frac{P(B \cap A)}{P(A)} \quad \text{for } B \in \mathscr{F}.$$

Proof. Suppose P_A satisfies (a) and (b). Then for every $B \in \mathscr{F}$ we have

$$P_A(B) = P_A(A \cap B) + P_A(B \setminus A) = c_A P(A \cap B),$$

since $P_A(B \setminus A) = 0$ by (a). For $B = A$, it follows that $1 = P_A(A) = c_A P(A)$, so $c_A = 1/P(A)$. Hence, P_A has the required form. Conversely, it is clear that P_A defined as above satisfies (a) and (b). $\diamond$

Definition. In the setting of Proposition (3.2), for every $B \in \mathscr{F}$, the expression

$$P(B|A) := \frac{P(B \cap A)}{P(A)}$$

is called the *conditional probability of B given A* with respect to P. (If $P(A) = 0$, it is sometimes convenient to set $P(B|A) := 0$.)

What does this imply for Example (3.1)? By definition, we obtain

$$P(B|A) = \frac{|B \cap A|}{|\Omega|} \Big/ \frac{|A|}{|\Omega|} = \frac{|A \cap B|}{|A|} = \frac{r(r-1)}{r(r+w-1)} = \frac{r-1}{r+w-1},$$

so the conditional probability takes exactly the value that corresponds to our intuition.

Let us now consider the following reversed situation. The first ball is drawn without looking at it, and the second one is red. How certain can one be that the first ball was red as well? Intuitively, one would argue that since B is known to occur, B is the set of all possible cases and $A \cap B$ the set of favourable cases, so with certainty

$$\frac{|A \cap B|}{|B|} = \frac{r(r-1)}{r(r+w-1)} = \frac{r-1}{r+w-1},$$

one can bet that A has occurred before. This is exactly the value of $P(A|B)$ according to the definition of the conditional probability. We thus find that, although the event B has definitely no influence on the occurrence of A, the information that B has occurred leads us to re-estimate the probability of A, and to use its conditional probability in place of its absolute probability. This observation entails the following conclusion about the

Interpretation of conditional probabilities.   The calculation of conditional probabilities does *not* allow to infer any causality relations among the events! Instead, conditional probabilities can be interpreted in either of the following ways.

(a) *The frequency interpretation.* If the random experiment is repeated many times, $P(B|A)$ gives the proportion of cases in which B occurs among those in which A occurs.

(b) *The subjective interpretation.* If P corresponds to my evaluation of the situation before the experiment, then $P(\cdot\,|A)$ is my evaluation after I have been informed that A has occurred (rather than: after A has occurred).

A naive interpretation of conditional probabilities would be dangerously close to an erroneous inference of causality relations, and is therefore omitted here.

　　Here are two elementary facts about conditional probabilities, which are visualised in Figure 3.1.

(3.3) Theorem. Case-distinction and Bayes' formula. *Let $(\Omega, \mathscr{F}, P)$ be a probability space and $\Omega = \bigcup_{i\in I} B_i$ an at most countable partition of Ω into pairwise disjoint events $B_i \in \mathscr{F}$. Then the following holds.*

(a) Case-distinction formula. *For every $A \in \mathscr{F}$,*

$$P(A) = \sum_{i\in I} P(B_i)P(A|B_i)\,.$$

(b) Bayes' formula. *For every $A \in \mathscr{F}$ with $P(A) > 0$ and every $k \in I$, the reverse conditional probability can be expressed as*

$$P(B_k|A) = \frac{P(B_k)P(A|B_k)}{\sum_{i\in I} P(B_i)P(A|B_i)}\,.$$

Proof. (a) By the definition of conditional probabilities and the σ-additivity of P, it follows that $\sum_{i\in I} P(B_i)\,P(A|B_i) = \sum_{i\in I} P(A \cap B_i) = P(A)$.
　　(b) This follows from (a) and the definition. ◇

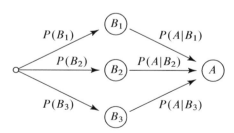

Figure 3.1. Illustration of Theorem (3.3). Part (a): $P(A)$ is split into the probabilities of the different paths leading to A. Part (b): $P(B_k|A)$ is the probability of the path via B_k in relation to the total probability of all paths to A. (The probabilities are multiplied along each path.)

Thomas Bayes (1702–1761) was a Presbyterian minister (and member of the Royal Society) in England. His mathematical work was only posthumously published in 1763. At that time Bayes' formula caused quite a stir, because people believed they could use it to deduce causes from their effects. But as we have seen above, this is impossible.

The following example presents a typical (and correct) application.

(3.4) Example. *Evaluation of medical tests.* A certain disease is present in 2% of the population (in medical jargon, 'the prevalence of the disease is 2%'). A test of the disease gives a positive result for 95% of the individuals that are affected by the disease ('the sensitivity of the test is 95%') and for 10% of those that are not ('specificity 90%'). What is the predictive value of a positive test result, in other words, what is the probability that a randomly chosen person has the disease when her test result is positive?

To answer this question we will set up the following stochastic model. Let Ω be the finite set of all individuals in the population and $P = \mathcal{U}_\Omega$ the uniform distribution on Ω. Let B_1 denote the set of individuals that are affected by the disease, and $B_2 = \Omega \setminus B_1$ the set of those that are not. Finally, let A be the set of individuals reacting positive to the test (as soon as they are tested). Then, by assumption, we have $P(B_1) = 0.02$, $P(B_2) = 0.98$, $P(A|B_1) = 0.95$ and $P(A|B_2) = 0.1$. By Bayes' formula, the predictive value is given by

$$P(B_1|A) = \frac{0.02 \cdot 0.95}{0.02 \cdot 0.95 + 0.98 \cdot 0.1} \lessapprox \frac{1}{6} .$$

So, the 'positive correctness' of the test is surprisingly low, and most of the test positives are actually false positive. On the other hand,

$$P(B_1|A^c) = \frac{P(B_1)P(A^c|B_1)}{P(B_1)P(A^c|B_1) + P(B_2)P(A^c|B_2)}$$
$$= \frac{0.02 \cdot 0.05}{0.02 \cdot 0.05 + 0.98 \cdot 0.9} \approx 0.001 .$$

That is, it is extremely unlikely that a person tested negative is, indeed, affected; so the 'negative correctness' is very high. Hence, the test is useful to rule out the possibility of having the disease, but a positive result does not mean very much without further investigation.

How can it be that a test with such a high sensitivity has such a low positive correctness? The reason is that the disease is so rare. Quite often, this effect is not well understood – not even by doctors –, and so patients with a positive test result are too readily told they have the disease, see the report in [22]. (Imagine what this means for the patients in the case of AIDS or breast cancer!) This effect is more readily understood if the percentages in the formulation of the problem (which each refer to a different reference population) are expressed so as to refer to the same population, see Table 3.1. It then becomes clear that, despite the low incidence of false negatives (one out of 20 affected), the majority of those

Table 3.1. Table for Bayes' formula, for a population of size 1000.

	affected	healthy	Σ
test positive	19	98	117
test negative	1	882	883
Σ	20	980	1000

tested positive (namely, 98 out of 117) are actually healthy (i.e., they are false positives). The picture changes if the individual tested belongs to a risk group, in which the disease shows up with prevalence 10%: Among 1000 members of this group, 185 are test positive, and 95 of these are affected. This means that the positive correctness within this group is slightly above 50%.

The next example is one of the best-known stochastic brainteasers, and its correct interpretation is still under discussion.

(3.5) Example. *The three doors paradox, or Monty Hall problem.* Inspired by Monty Hall's American TV show 'Let's Make a Deal', in 1990 a reader sent the following problem to the American journalist Marilyn vos Savant to be included in her column in the 'Parade Magazine':

Suppose you're on a game show, and you're given the choice of three doors. Behind one door is a car, behind the others, goats. You pick a door, say #1, and the host, who knows what's behind the doors, opens another door, say #3, which has a goat. He says to you, 'Do you want to pick door #2?' Is it to your advantage to switch your choice of doors?

Marilyn's answer was: 'Yes, you should switch. The first door has a 1/3 chance of winning, but the second door has a 2/3 chance'. She justified her answer by referring to the analogous problem of 1 million doors, where all but two doors are opened by the host. This started a lively public discussion of the problem. Often it was claimed that after opening door 3, the two remaining doors had the same probability 1/2 of winning.

So who is right? To find out, we must interpret the situation precisely. We label the three doors by the numbers 1, 2, 3. Since the doors look the same from the outside, we can assume without loss of generality that the winning door is number 1. Two doors are chosen randomly, one by the contestant and the other by the host. Their respective choices are recorded by two random variables C and H taking values in $\{1, 2, 3\}$. Since the contestant has no information about which one is the winning door, she will choose any door with the same probability, so C has the uniform distribution on $\{1, 2, 3\}$. In the problem, it is assumed that the event

$$A := \{C \neq H \neq 1\} = \{H \neq C\} \cap \{H \neq 1\}$$

Figure 3.2. The Monty Hall problem.

has occurred, and the contestant gets another chance to choose a door, either the same as before or the remaining third door; see Figure 3.2. In the first case, the conditional probability of winning is $P(C = 1|A)$. To calculate this, the contestant needs more information about A, in particular on how the host will behave. So what information is available?

First of all, it is obvious that the host will not open the winning door, because otherwise the game is immediately over, which would be rather pointless for the viewers. This justifies the assumption $P(H \neq 1) = 1$. Further, we can interpret the formulation 'opens another door' such that the host will not open the door chosen by the contestant. Then $P(H \neq C) = 1$ and by Theorem (1.11b) we also have $P(A) = 1$, and hence

$$P(C = 1|A) = P(C = 1) = \frac{1}{3}.$$

Correspondingly, if the contestant switches to the third door ($\neq C, H$), the conditional probability of winning is $P(C \neq 1|A) = 2/3$. This is precisely the answer that Marilyn vos Savant gave, and its justification is surprisingly simple.

The triviality of this answer comes from the assumption that the host behaves according to fixed rules, so that each time he carries out the game in exactly the same way as described in the problem, which implies that the event A will certainly occur. The deeper reason for this assumption is the implicit use of the frequency interpretation of conditional probabilities, which requires that the procedure can be repeated in the same way and thus needs fixed rules. (This becomes quite evident when people argue that Marilyn's answer can be justified by simulating the problem on a computer.) However, one might argue that the host will not behave in a predictable way to bring in an element of surprise for contestants and viewers. This point of view suggests the subjective interpretation, and everything depends on how the contestant predicts the behaviour of the host. As before, the contestant can guess that the host will not open the winning door, and assume that $P(H \neq 1) = 1$. Then, $P(A|C = 1) = P(H \neq 1|C = 1) = 1$ and by Bayes' formula (3.3b)

$$(3.6) \qquad P(C = 1|A) = \frac{P(C = 1)\, P(A|C = 1)}{P(C = 1)\, P(A|C = 1) + P(C \neq 1)\, P(A|C \neq 1)}$$

$$= \frac{1/3}{1/3 + (2/3)\, P(A|C \neq 1)} \,.$$

How should the contestant estimate the conditional probability $P(A|C \neq 1)$? As before, she might conclude that $P(H \neq C) = 1$ and hence $P(A) = 1$. Alternatively, she might assume that the host opens each of the doors concealing a goat with equal probability $1/2$, irrespective of which door C the contestant has chosen. (For example, if $H = C$, the host would say 'Look! You had bad luck. But I give you a second chance, you may pick another door!') Then $P(H = C|C = c) = 1/2$ for $c \in \{2, 3\}$ and thus by the case-distinction formula (3.3a) we also have

$$P(H = C|C \neq 1) = \sum_{c=2}^{3} \frac{1}{2}\, P(C = c|C \neq 1) = \frac{1}{2}\,.$$

By the assumption $P(H \neq 1) = 1$, it follows that

$$P(A|C \neq 1) = P(H \neq C|C \neq 1) = \frac{1}{2}$$

and thus, by (3.6), $P(C = 1|A) = 1/2$. This is exactly the answer of the critics!

Like in Bertrand's paradox, the different answers depend on different interpretations of a problem that is posed in an ambiguous way. The different viewpoints boil down to the question of whether or not the event A occurs according to a fixed rule, and are often intertwined with an uncertainty about the philosophical meaning of conditional probabilities. Various discussions of the problem can be found in [22, 26, 50, 54] and in the internet. It seems that some time has to pass until an unexcited general consensus will be reached.

In the above discussion, it has still been left open whether the random variables C and H with their respective properties actually do exist. But this is an immediate consequence of the next section.

3.2 Multi-Stage Models

Consider a random experiment that is performed in n successive stages, or subexperiments. We want to construct a probability space $(\Omega, \mathscr{F}, P)$ for the full experiment, as well as random variables $(X_i)_{1 \leq i \leq n}$ on Ω that describe the outcomes of the individual subexperiments. Suppose we are given

(a) the distribution of X_1,

(b) the conditional distribution of X_k when the values of $X_1, \ldots, X_{k-1}$ are known, for all $2 \leq k \leq n$.

In other words, we know how to describe the first subexperiment and, for each time point, how to describe the next subexperiment if all previous subexperiments have been performed so that their outcomes are known. The following proposition provides an idea of how to attack this problem.

(3.7) Proposition. Multiplication rule. *Let $(\Omega, \mathscr{F}, P)$ be a probability space and $A_1, \ldots, A_n \in \mathscr{F}$. Then,*

$$P(A_1 \cap \cdots \cap A_n) = P(A_1)\, P(A_2|A_1) \ldots P(A_n|A_1 \cap \cdots \cap A_{n-1}).$$

Proof. If the left-hand side vanishes, then the last factor on the right is zero as well. Otherwise, all the conditional probabilities on the right-hand side are defined and non-zero. They form a telescoping product, in which consecutive numerators and denominators cancel, and so only the left-hand side remains. $\diamond$

The following theorem describes the construction of random variables satisfying properties (a) and (b). For simplicity, we assume that every subexperiment has an at most countable sample space.

(3.8) Theorem. Construction of probability measures via conditional probabilities. *Suppose we are given n countable sample spaces $\Omega_1, \ldots, \Omega_n \neq \varnothing$, $n \geq 2$. Let ρ_1 be a probability density on Ω_1 and, for $k = 2, \ldots, n$ and arbitrary $\omega_i \in \Omega_i$ with $i < k$, let $\rho_{k|\omega_1,\ldots,\omega_{k-1}}$ be a probability density on Ω_k. Further, let $\Omega = \prod_{i=1}^{n} \Omega_i$ be the product space and $X_i : \Omega \to \Omega_i$ the ith projection. Then there exists a unique probability measure P on $(\Omega, \mathscr{P}(\Omega))$ with the properties*

(a) $P(X_1 = \omega_1) = \rho_1(\omega_1)$ *for all $\omega_1 \in \Omega_1$,*

(b) *for every $k = 2, \ldots, n$,*

$$P(X_k = \omega_k | X_1 = \omega_1, \ldots, X_{k-1} = \omega_{k-1}) = \rho_{k|\omega_1,\ldots,\omega_{k-1}}(\omega_k)$$

for all $\omega_i \in \Omega_i$ such that $P(X_1 = \omega_1, \ldots, X_{k-1} = \omega_{k-1}) > 0$.

This P is given by

(3.9) $$P(\{\omega\}) = \rho_1(\omega_1)\, \rho_{2|\omega_1}(\omega_2)\, \rho_{3|\omega_1,\omega_2}(\omega_3) \ldots \rho_{n|\omega_1,\ldots,\omega_{n-1}}(\omega_n)$$

for $\omega = (\omega_1, \ldots, \omega_n) \in \Omega$.

Equation (3.9) is illustrated by the tree diagram in Figure 3.3.

Proof. We claim that P can only be defined by (3.9). Indeed, writing $A_i = \{X_i = \omega_i\}$ and $\{\omega\} = \bigcap_{i=1}^{n} A_i$, and using assumptions (a) and (b), we see that equation (3.9) is identical to the multiplication rule. This proves the uniqueness of P.

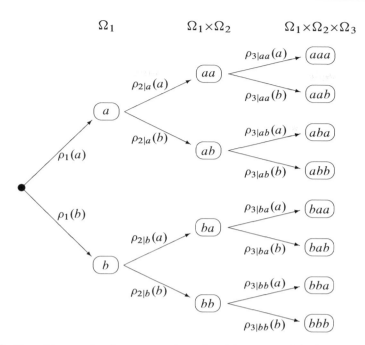

Figure 3.3. Tree diagram for the construction of multi-stage models, here for $n = 3$ and $\Omega_i = \{a, b\}$. The probability of a triple in Ω is the product of the transition probabilities along the path leading to this triple.

So define P by (3.9). Then, for all $1 \le k \le n$ and $\omega_1, \ldots, \omega_k$, we can sum over all possible values of $\omega_{k+1}, \ldots, \omega_n$ to obtain

$$P(X_1 = \omega_1, \ldots, X_k = \omega_k) = \sum_{\omega_{k+1} \in \Omega_{k+1}, \ldots, \omega_n \in \Omega_n} P(\{(\omega_1, \ldots, \omega_n)\})$$

$$= \rho_1(\omega_1) \cdots \rho_{k|\omega_1, \ldots, \omega_{k-1}}(\omega_k)$$

$$\cdot \sum_{\omega_{k+1} \in \Omega_{k+1}} \rho_{k+1|\omega_1, \ldots, \omega_k}(\omega_{k+1}) \cdots \sum_{\omega_n \in \Omega_n} \rho_{n|\omega_1, \ldots, \omega_{n-1}}(\omega_n).$$

Since $\rho_{n|\omega_1, \ldots, \omega_{n-1}}$ is a probability density, the last sum takes the value 1 and can thus be omitted. Evaluating the second last sum, we find that it also yields 1. Continuing in this way, we see that the total of all the sums in the last row is 1. For $k = 1$ we obtain (a), and another summation over ω_1 shows that the right-hand side of (3.9) is indeed a probability density. For $k > 1$ we get

$$P(X_1 = \omega_1, \ldots, X_k = \omega_k) = P(X_1 = \omega_1, \ldots, X_{k-1} = \omega_{k-1}) \rho_{k|\omega_1, \ldots, \omega_{k-1}}(\omega_k)$$

and thus (b). $\diamond$

(3.10) Example. *The game of skat.* Skat is a popular German card-game, which is played with a deck of 32 cards (including 4 aces). What is the probability that each of the three players gets exactly one ace? As we have seen in Section 2.3.1, we can assume that (after the cards have been well shuffled) the players, one by one, receive their batch of ten cards; the two remaining cards form the 'skat'. We are interested in the number of aces for each player, so we choose the sample spaces $\Omega_1 = \Omega_2 = \Omega_3 = \{0, \ldots, 4\}$ for the subexperiments, and the product space $\Omega = \{0, \ldots, 4\}^3$ for the full experiment. The probability measure P on Ω can be constructed by Theorem (3.8) via the hypergeometric transition probabilities

$$\rho_1(\omega_1) = \mathcal{H}_{10;4,28}(\{\omega_1\}) = \binom{4}{\omega_1}\binom{28}{10-\omega_1}\Big/\binom{32}{10}\,,$$

$$\rho_{2|\omega_1}(\omega_2) = \mathcal{H}_{10;4-\omega_1,18+\omega_1}(\{\omega_2\})\,,$$

$$\rho_{3|\omega_1,\omega_2}(\omega_3) = \mathcal{H}_{10;4-\omega_1-\omega_2,8+\omega_1+\omega_2}(\{\omega_3\})\,;$$

see Section 2.3.2. The event $\{(1, 1, 1)\}$ that each player receives one ace thus has the probability

$$P\big(\{(1, 1, 1)\}\big) = \rho_1(1)\,\rho_{2|1}(1)\,\rho_{3|1,1}(1)$$

$$= \frac{\binom{4}{1}\binom{28}{9}}{\binom{32}{10}}\,\frac{\binom{3}{1}\binom{19}{9}}{\binom{22}{10}}\,\frac{\binom{2}{1}\binom{10}{9}}{\binom{12}{10}} = 10^3\,\frac{2\cdot 4!}{32\ldots 29} \approx 0.0556\,.$$

(3.11) Example. *Population genetics.* Consider a gene with two alleles A and a. An individual with a diploid set of chromosomes thus has one of the three possible genotypes AA, Aa and aa. Suppose these genotypes are present in a population with relative frequencies u, $2v$, w respectively, where $u + 2v + w = 1$. We assume that for this gene there is no mutation or selection, and the gene is irrelevant for the choice of a partner. What is the distribution of the genotypes in the offspring generation?

As in Theorem (3.8), we construct a probability measure P on the product space $\{\mathrm{AA, Aa, aa}\}^3$, which contains all possible genotypes of mother, father and offspring. By assumption, the genotype ω_1 of the mother has the distribution $\rho_1 = (u, 2v, w)$, and the genotype ω_2 of the father has the same (conditional) distribution $\rho_{2|\omega_1} = \rho_1$, which by assumption does not depend on ω_1. The conditional distribution $\rho_{3|\omega_1\omega_2}(\omega_3)$ of the genotype ω_3 of the offspring can now be deduced from the fact that one gene from the mother is combined with one gene from the father, where both genes are chosen with equal probabilities; see Table 3.2.

What is the distribution $P \circ X_3^{-1}$ of the offspring genotype? Using (3.9), we obtain by summation over all possible genotypes for the parents

$$u_1 := P(X_3 = \mathrm{AA}) = u^2 + 2uv/2 + 2vu/2 + 4v^2/4 = (u+v)^2\,.$$

By symmetry, it follows that $w_1 := P(X_3 = \mathrm{aa}) = (w+v)^2$, and hence that

$$2v_1 := P(X_3 = \mathrm{Aa}) = 1 - u_1 - w_1$$

$$= \big((u+v) + (w+v)\big)^2 - (u+v)^2 - (w+v)^2 = 2(u+v)(w+v)\,.$$

Table 3.2. The transition probabilities $\rho_{3|\omega_1\omega_2}(AA)$ for the offspring genotype AA given the genotypes ω_1, ω_2 of mother and father.

ω_1	ω_2	$\rho_{3|\omega_1\omega_2}(AA)$
AA	AA	1
	Aa	1/2
Aa	AA	1/2
	Aa	1/4
otherwise		0

Similarly, the probability u_2 of the genotype AA in the second generation is obtained as

$$u_2 = (u_1 + v_1)^2 = \big((u+v)^2 + (u+v)(w+v)\big)^2$$
$$= (u+v)^2 \big((u+v) + (w+v)\big)^2 = (u+v)^2 = u_1$$

and, likewise, $w_2 = w_1$, $v_2 = v_1$. We thus arrive at the famous genetic law found independently in 1908 by the British mathematician G. H. Hardy and the German physician W. Weinberg, namely: Under the assumption of random mating, the genotype frequencies remain constant from the first offspring generation onwards.

Our next aim is to extend Theorem (3.8) to the case of a random experiment that consists of infinitely many subexperiments. Recall that the necessity to deal with an infinite sequence of subexperiments already turned up in Section 2.5.1 in the context of waiting times for Bernoulli trials, since one cannot know in advance how long it will take until the first success.

(3.12) Theorem. Construction of probability measures on infinite product spaces. *For every $i \in \mathbb{N}$, let $\Omega_i \neq \varnothing$ be a countable set. Let ρ_1 be a probability density on Ω_1 and, for every $k \geq 2$ and $\omega_i \in \Omega_i$ with $i < k$, let $\rho_{k|\omega_1,\ldots,\omega_{k-1}}$ be a probability density on Ω_k. Let $\Omega = \prod_{i \geq 1} \Omega_i$, $X_i : \Omega \to \Omega_i$ the projection onto the ith coordinate, and $\mathscr{F} = \bigotimes_{i \geq 1} \mathscr{P}(\Omega_i)$ the product σ-algebra on Ω. Then there exists a unique probability measure P on $(\Omega, \mathscr{F})$ such that*

$$(3.13) \qquad P(X_1 = \omega_1, \ldots, X_k = \omega_k) = \rho_1(\omega_1)\, \rho_{2|\omega_1}(\omega_2) \ldots \rho_{k|\omega_1,\ldots,\omega_{k-1}}(\omega_k)$$

for all $k \geq 1$ and $\omega_i \in \Omega_i$.

Equation (3.13) corresponds to equation (3.9) in Theorem (3.8) and is equivalent to the conditions (a) and (b) stated there.

Proof. The uniqueness follows from the uniqueness theorem (1.12), since

$$\mathscr{G} = \big\{\{X_1 = \omega_1, \ldots, X_k = \omega_k\} : k \geq 1,\ \omega_i \in \Omega_i\big\} \cup \{\varnothing\}$$

is an intersection-stable generator of $\mathscr{F}$; cf. Problem 1.5.

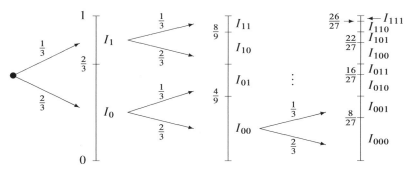

Figure 3.4. The intervals of the first up to the third level in the case $\Omega_i = \{0, 1\}$ and $\rho_{k|\omega_1,\dots,\omega_{k-1}}(1) = 1/3$, where the intervals are split into two parts with proportion $1:2$. The arrows indicate the analogy to the tree diagram. We have $Z(1/2) = (0, 1, 0, \dots)$.

The existence of P will be deduced from the existence of the Lebesgue measure $\lambda = \mathcal{U}_{[0,1[}$ on the half-open unit interval $[0, 1[$, cf. Remark (1.17). As illustrated in Figure 3.4, we partition the interval $[0, 1[$ into half-open intervals $(I_{\omega_1})_{\omega_1 \in \Omega_1}$ of length $\rho_1(\omega_1)$; they will be referred to as intervals of the first level. Each I_{ω_1} is then split into half-open intervals $(I_{\omega_1 \omega_2})_{\omega_2 \in \Omega_2}$ of length $\rho_1(\omega_1)\rho_{2|\omega_1}(\omega_2)$; these are the intervals of the second level. We continue in this way, that is, after defining the interval $I_{\omega_1\dots\omega_{k-1}}$ of the $(k-1)$st level, we split it further into disjoint subintervals $(I_{\omega_1\dots\omega_k})_{\omega_k \in \Omega_k}$ of the kth level with lengths $\lambda(I_{\omega_1\dots\omega_{k-1}})\,\rho_{k|\omega_1,\dots,\omega_{k-1}}(\omega_k)$. For each $x \in [0, 1[$ and arbitrary k there exists a unique interval of level k that contains x. In other words, there is a unique sequence $Z(x) = (Z_1(x), Z_2(x), \dots) \in \Omega$ such that $x \in I_{Z_1(x)\dots Z_k(x)}$ for all $k \geq 1$. (The sequence $Z(x)$ can be viewed as an infinitely long postcode that characterises the location x uniquely.) We claim that the mapping $Z : [0, 1[\to \Omega$ is a random variable; indeed, for $A = \{X_1 = \omega_1, \dots, X_k = \omega_k\} \in \mathcal{G}$ we have

$$Z^{-1}A = \{x : Z_1(x) = \omega_1, \dots, Z_k(x) = \omega_k\} = I_{\omega_1\dots\omega_k} \in \mathcal{B}_{[0,1[}\,,$$

so that the claim follows from Remark (1.25). Hence, by Theorem (1.28), the distribution $P := \lambda \circ Z^{-1}$ of Z is a well-defined probability measure on $(\Omega, \mathcal{F})$. By construction, P has the required properties. $\diamond$

(3.14) Example. *Pólya's urn model.* The following urn model goes back to the Hungarian mathematician G. Pólya (1887–1985). An urn contains r red and w white balls. An infinite sequence of balls is drawn from the urn, and after each draw the selected ball plus c additional balls of the same colour are placed into the urn. The case $c = 0$ corresponds to drawing with replacement. We are interested in the case $c \in \mathbb{N}$, when a self-reinforcing effect appears. Namely, the greater the proportion of red balls in the urn, the more likely it is that another red ball is drawn and so the proportion of red balls increases even further. This is a simple model for two competing populations (and maybe also for the career of politicians).

Which is the probability space for this urn model? We can proceed exactly as in Theorem (3.12). If we write 1 for 'red' and 0 for 'white', then $\Omega_i = \{0, 1\}$ and so $\Omega = \{0, 1\}^{\mathbb{N}}$. For the initial distribution ρ_1 we obviously have $\rho_1(0) = w/(r+w)$ and $\rho_1(1) = r/(r+w)$ corresponding to the initial proportion of white and red balls in the urn. For the transition densities at time $k > 1$, we obtain analogously

$$
\rho_{k|\omega_1,\dots,\omega_{k-1}}(\omega_k) =
\begin{cases}
\frac{r+c\ell}{r+w+c(k-1)} \\[2mm]
\frac{w+c(k-1-\ell)}{r+w+c(k-1)}
\end{cases}
\text{if } \sum_{i=1}^{k-1} \omega_i = \ell \text{ and } \omega_k =
\begin{cases}
1 \\[2mm]
0.
\end{cases}
$$

Indeed, if we have drawn ℓ red (and so $k-1-\ell$ white) balls in the first $k-1$ draws, then the urn contains $r+c\ell$ red and $w+c(k-1-\ell)$ white balls. We now build the product of the transition probabilities according to (3.9), and consider the numerator of the resulting product. At the ℓ different times k with $\omega_k = 1$, we obtain the successive factors $r, r+c, r+2c, \dots$, and at the remaining times with $\omega_k = 0$ the factors $w, w+c, w+2c, \dots$, respectively. The factors in the denominator do not depend on the ω_k. Combining all factors, we arrive at the following characterisation of the probability measure P that Theorem (3.12) provides for the Polya urn:

$$
P(X_1 = \omega_1, \dots, X_n = \omega_n) = \frac{\prod_{i=0}^{\ell-1}(r+ci) \prod_{j=0}^{n-\ell-1}(w+cj)}{\prod_{m=0}^{n-1}(r+w+cm)} \quad \text{if } \sum_{k=1}^{n} \omega_k = \ell.
$$

Remarkably, these probabilities do not depend on the order of the ω_i, but only on their sum. For this reason, the random variables $X_1, X_2, \dots$ are said to be *exchangeable* under P.

Now let $R_n = \sum_{k=1}^{n} X_k$ denote the number of red balls after n draws. Since all ω with $R_n(\omega) = \ell$ have the same probability, we obtain

$$
P(R_n = \ell) = \binom{n}{\ell} \frac{\prod_{i=0}^{\ell-1}(r+ci) \prod_{j=0}^{n-\ell-1}(w+cj)}{\prod_{m=0}^{n-1}(r+w+cm)}.
$$

For $c = 0$, this is just the binomial distribution, in agreement with Theorem (2.9). In the case $c \neq 0$, we can cancel the factor $(-c)^n$ in the fraction. Using the general binomial coefficient (2.18) and the abbreviations $a := r/c, b := w/c$ we then find

$$
P(R_n = \ell) = \frac{\binom{-a}{\ell}\binom{-b}{n-\ell}}{\binom{-a-b}{n}}.
$$

The probability measure $P \circ R_n^{-1}$ on $\{0, \dots, n\}$ thus obtained is known as the *Pólya distribution* with parameters $a, b > 0$ and $n \in \mathbb{N}$. In the case $c = -1$, i.e. $-a, -b \in \mathbb{N}$, which corresponds to sampling without replacement, these are exactly the hypergeometric distributions, as it should be according to Section 2.3.2. If $a = b = 1$, we obtain the uniform distribution. Figure 3.5 illustrates the Pólya distributions for various parameter values. Note their resemblance with the densities of the beta distributions in Figure 2.7 – this is no coincidence, see Problem 3.4. The long term behaviour of R_n/n is discussed in Problem 5.11.

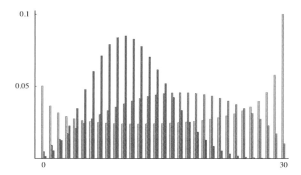

Figure 3.5. Bar charts of the Pólya distributions for $n = 30$ and $(a, b) = (5/7, 4/7)$ (light grey), $(8/4, 7/4)$ (grey), and $(5, 9)$ (dark grey).

3.3   Independence

Intuitively, the independence of two events A and B can be expressed as follows: The probability one assigns to A is not influenced by the information that B has occurred, and conversely the occurrence of A does not lead to a re-weighting of the probability of B. Explicitly, this means that

$$P(A|B) = P(A) \text{ and } P(B|A) = P(B), \text{ provided } P(A), P(B) > 0.$$

Writing these equations in a symmetric form one obtains:

Definition. Let $(\Omega, \mathscr{F}, P)$ be a probability space. Two events $A, B \in \mathscr{F}$ are called (stochastically) *independent* with respect to P if $P(A \cap B) = P(A)P(B)$.

Let us consider two examples for illustration.

(3.15) Example. *Sampling with and without replacement.* We take two samples with replacement from an urn with r red and w white (labelled) balls. A suitable model is $\Omega = \{1, \dots, r+w\}^2$ and $P = \mathcal{U}_\Omega$, the uniform distribution. We consider the events $A = \{$the first ball is red$\}$ and $B = \{$the second ball is red$\}$. Then, we have

$$P(A \cap B) = \frac{r^2}{(r+w)^2} = P(A)\,P(B),$$

so A and B are independent with respect to P. However, if we replace P by another probability measure P', the events A and B need not be independent. For instance, this happens for $P' = P(\cdot | \Omega_{\neq}) = \mathcal{U}_{\Omega_{\neq}}$, the uniform distribution on $\Omega_{\neq} = \{(k,l) \in \Omega : k \neq l\}$, which describes sampling without replacement, cf. Example (3.1). Then,

$$P'(A \cap B) = \frac{r(r-1)}{(r+w)(r+w-1)} < P'(A)P'(B).$$

This emphasises that independence is not only a property of events, but also of the underlying probability measure – a fact that is self-evident, though still sometimes overlooked.

(3.16) Example. *Independence despite causality.* Consider the experiment of rolling two distinguishable dice, which can be described by $\Omega = \{1,\ldots,6\}^2$ with the uniform distribution $P = \mathcal{U}_\Omega$. Let

$$A = \{\text{sum of points is } 7\} = \{(k,l) \in \Omega : k+l = 7\},$$
$$B = \{\text{first dice shows } 6\} = \{(k,l) \in \Omega : k = 6\}.$$

Then $|A| = |B| = 6$, $|A \cap B| = 1$, so

$$P(A \cap B) = \frac{1}{6^2} = P(A)\,P(B),$$

even though the sum is causally determined by the number shown by the first dice. In this case the independence follows because we have chosen 7 (instead of 12, say) for the value of the sum. Nevertheless it shows the following:

Independence should not be misunderstood as causal independence, although Example (3.15) seems to suggest just this. Rather, *independence means a proportional overlap of probabilities and does not involve any causality.* It depends essentially on the underlying probability measure. One should also note that A is independent of itself when $P(A) \in \{0,1\}$.

Next we proceed to the independence of more than just two events.

Definition. Let $(\Omega, \mathscr{F}, P)$ be a probability measure and $I \neq \varnothing$ an arbitrary index set. A family $(A_i)_{i \in I}$ of events in $\mathscr{F}$ is called *independent* with respect to P if, for every finite subset $\varnothing \neq J \subset I$, we have

$$P\Big(\bigcap_{i \in J} A_i\Big) = \prod_{i \in J} P(A_i).$$

(The trivial case $|J| = 1$ is included here just for simplicity of the formulation.)

The independence of a family of events is a stronger property than pairwise independence of any two events in the family, but it corresponds exactly to what one would intuitively understand as mutual independence. This becomes clear in the following example.

(3.17) Example. *Dependence despite pairwise independence.* In the model for tossing a coin twice (with $\Omega = \{0, 1\}^2$ and $P = \mathcal{U}_\Omega$) consider the three events

$$A = \{1\} \times \{0, 1\} = \{\text{first toss is heads}\},$$
$$B = \{0, 1\} \times \{1\} = \{\text{second toss is heads}\},$$
$$C = \{(0, 0), (1, 1)\} = \{\text{both tosses give the same result}\}.$$

Then $P(A \cap B) = 1/4 = P(A)\,P(B)$, $P(A \cap C) = 1/4 = P(A)\,P(C)$, and $P(B \cap C) = 1/4 = P(B)\,P(C)$, so A, B, C are pairwise independent. However,

$$P(A \cap B \cap C) = \frac{1}{4} \neq \frac{1}{8} = P(A)\,P(B)\,P(C),$$

i.e., A, B, C are dependent according to the definition above. This is exactly what one would expect, since in fact $C = (A \cap B) \cup (A^c \cap B^c)$.

We can take the generalisation even one step further. We are not only interested in the independence of events, but also in the independence of subexperiments, in other words: in the independence of random variables that describe such subexperiments.

Definition. Let $(\Omega, \mathscr{F}, P)$ be a probability space, $I \neq \varnothing$ an arbitrary index set, and for every $i \in I$ let $Y_i : \Omega \to \Omega_i$ be a random variable on $(\Omega, \mathscr{F})$ taking values in any event space $(\Omega_i, \mathscr{F}_i)$. The family $(Y_i)_{i \in I}$ is called *independent* with respect to P if, for an arbitrary choice of events $B_i \in \mathscr{F}_i$, the family $(\{Y_i \in B_i\})_{i \in I}$ is independent, i.e., if the product formula

(3.18) $$P\Big(\bigcap_{i \in J} \{Y_i \in B_i\}\Big) = \prod_{i \in J} P(Y_i \in B_i)$$

holds for any finite subset $\varnothing \neq J \subset I$ and for all $B_i \in \mathscr{F}_i$ (with $i \in J$).

How can one check that a family of random variables is independent? Is it really necessary to try each $B_i \in \mathscr{F}_i$? This is hardly possible, since in most cases all that is known explicitly about a σ-algebra is its generator. Hence, the following criterion is essential.

(3.19) Theorem. Independence criterion. *In the setting of the previous definition, suppose that, for every $i \in I$, an intersection-stable generator $\mathscr{G}_i$ of $\mathscr{F}_i$ is given, so that $\sigma(\mathscr{G}_i) = \mathscr{F}_i$. To show the independence of $(Y_i)_{i \in I}$, it is then sufficient to verify equation* (3.18) *for the events B_i in $\mathscr{G}_i$ only (rather than all events in $\mathscr{F}_i$).*

Readers who have skipped the proof of the uniqueness theorem (1.12) can also skip this proof and just take the result and its consequences for granted.

Proof. It is sufficient to prove the following statement by induction on n: Equation (3.18) holds for all finite $J \subset I$ and all $B_i \in \mathscr{F}_i$ with $|\{i \in J : B_i \notin \mathscr{G}_i\}| \leq n$. For $n \geq |J|$, the last condition holds automatically, and the required independence follows. The case $n = 0$ corresponds exactly to our hypothesis that (3.18) holds for any $B_i \in \mathscr{G}_i$. The inductive step $n \rightsquigarrow n+1$ runs as follows.

Suppose $J \subset I$ and $B_i \in \mathscr{F}_i$ are such that $|\{i \in J : B_i \notin \mathscr{G}_i\}| = n+1$. Pick any $j \in J$ with $B_j \notin \mathscr{G}_j$ and set $J' = J \setminus \{j\}$ and $A = \bigcap_{i \in J'} \{Y_i \in B_i\}$. The inductive hypothesis implies that $P(A) = \prod_{i \in J'} P(Y_i \in B_i)$. We can assume that $P(A) > 0$, because otherwise both sides of (3.18) vanish. We consider the probability measures $P(Y_j \in \cdot \,|A) := P(\cdot|A) \circ Y_j^{-1}$ and $P(Y_j \in \cdot) := P \circ Y_j^{-1}$ on $\mathscr{F}_j$. By the inductive hypothesis, these coincide on $\mathscr{G}_j$, so, by the uniqueness theorem (1.12), they are identical on all of $\mathscr{F}_j$. Multiplying by $P(A)$ shows that (3.18) holds for the required sets, and the inductive step is completed. $\diamond$

As a first application, we obtain a relation between the independence of events and their corresponding indicator functions; cf. (1.16).

(3.20) Corollary. *Independence of indicator functions. A family $(A_i)_{i \in I}$ of events is independent if and only if the corresponding family $(1_{A_i})_{i \in I}$ of indicator functions is independent. In particular, if $(A_i)_{i \in I}$ is independent and for each $i \in I$ an arbitrary $C_i \in \{A_i, A_i^c, \Omega, \varnothing\}$ is selected, then the family $(C_i)_{i \in I}$ is also independent.*

Proof. Every indicator function 1_A is a random variable taking values in the event space $(\{0, 1\}, \mathscr{P}(\{0, 1\}))$, and $\mathscr{P}(\{0, 1\})$ has the intersection-stable generator $\mathscr{G} = \{\{1\}\}$, which contains the singleton set $\{1\}$ as its only element. Moreover, we have $\{1_A = 1\} = A$. Hence, if $(A_i)_{i \in I}$ is independent, then $(1_{A_i})_{i \in I}$ satisfies the assumption of Theorem (3.19). Conversely, if $(1_{A_i})_{i \in I}$ is independent, then by definition the family $(\{1_{A_i} \in B_i\})_{i \in I}$ is independent no matter which sets $B_i \subset \{0, 1\}$ we choose. This proves the last statement, and for $B_i = \{1\}$ we see that $(A_i)_{i \in I}$ is independent. $\diamond$

Next, we state a criterion for independence of finite families of random variables.

(3.21) Corollary. *Independence of finitely many random variables. Let $(Y_i)_{1 \leq i \leq n}$ be a finite sequence of random variables on a probability space $(\Omega, \mathscr{F}, P)$. Then the following holds.*

(a) *Discrete case: If each Y_i takes values in a countable set Ω_i, then $(Y_i)_{1 \leq i \leq n}$ is independent if and only if*

$$P(Y_1 = \omega_1, \ldots, Y_n = \omega_n) = \prod_{i=1}^{n} P(Y_i = \omega_i)$$

for arbitrary $\omega_i \in \Omega_i$.

(b) Real case: *If each Y_i is real-valued, then $(Y_i)_{1 \le i \le n}$ is independent if and only if*

$$P(Y_1 \le c_1, \ldots, Y_n \le c_n) = \prod_{i=1}^{n} P(Y_i \le c_i)$$

for all $c_i \in \mathbb{R}$.

Proof. The implication 'only if' is trivial in both cases; the direction 'if' is obtained as follows. In case (a), Example (1.7) shows that $\mathscr{G}_i = \{\{\omega_i\} : \omega_i \in \Omega_i\} \cup \{\varnothing\}$ is an intersection-stable generator of $\mathscr{F}_i = \mathscr{P}(\Omega_i)$; the trivial events $\{Y_i \in \varnothing\} = \varnothing$ need not be considered since, in this case, both sides in (3.18) vanish. Our assumption thus corresponds exactly to the condition of Theorem (3.19) for the special case $J = I$. In the case $J \subsetneqq I$, the corresponding product formula is simply obtained by summation over all ω_i with $i \in I \setminus J$. Hence, the claim follows from Theorem (3.19).

Statement (b) follows in the same way, by using the intersection-stable generator $\{]-\infty, c] : c \in \mathbb{R}\}$ of the Borel σ-algebra $\mathscr{B}$ (cf. Example (1.8d)), and letting $c_i \to \infty$ for $i \in I \setminus J$. $\diamond$

Note that the cases (a) and (b) in the preceding corollary can in fact both apply at the same time, namely when each Y_i takes values in a countable set $\Omega_i \subset \mathbb{R}$. By Problem 1.4, it then does not make a difference whether Y_i is treated as a random variable with values in $(\Omega_i, \mathscr{P}(\Omega_i))$, or in $(\mathbb{R}, \mathscr{B})$, and both criteria in (a) and (b) can be used.

(3.22) Example. *Product measures.* Let E be a finite set, ρ a discrete density on E, $n \ge 2$ and $P = \rho^{\otimes n}$ the n-fold product measure on $\Omega = E^n$; this includes the situation of ordered samples with replacement taken from an urn, where the colours of the balls are distributed according to ρ, see Section 2.2.1. Let $X_i : \Omega \to E$ be the ith projection. Then, by definition, the equation

$$P(X_1 = \omega_1, \ldots, X_n = \omega_n) = P(\{(\omega_1, \ldots, \omega_n)\}) = \prod_{i=1}^{n} \rho(\omega_i)$$

holds for arbitrary $\omega_i \in E$, and summing over all $\omega_j \in E$ for all $j \ne i$ yields $P(X_i = \omega_i) = \rho(\omega_i)$. So, Corollary (3.21a) shows that, relative to $P = \rho^{\otimes n}$, the random variables X_i are independent with distribution ρ, as one would expect for sampling with replacement.

(3.23) Example. *Polar coordinates of a random point of the unit disc.* Let $K = \{x = (x_1, x_2) \in \mathbb{R}^2 : |x| \le 1\}$ be the unit disc and $Z = (Z_1, Z_2)$ a K-valued random variable (on an arbitrary probability space $(\Omega, \mathscr{F}, P)$) with uniform distribution $\mathcal{U}_K$ on K. Let $R = |Z| = \sqrt{Z_1^2 + Z_2^2}$ and $\Psi = \arg(Z_1 + iZ_2) \in [0, 2\pi[$ be the polar coordinates of Z. (Ψ is the argument of the complex number $Z_1 + iZ_2$, in other words the angle between the line segment from 0 to Z and the positive half-line.) Then we have for any $0 \le r \le 1$ and $0 \le \psi < 2\pi$

$$P(R \le r, \Psi \le \psi) = \frac{\pi r^2}{\pi} \frac{\psi}{2\pi} = P(R \le r) \, P(\Psi \le \psi) \, .$$

Thus, by Corollary (3.21b), R and Ψ are independent. In particular, Ψ is uniformly distributed on $[0, 2\pi[$, and R^2 is uniformly distributed on $[0, 1]$.

The next theorem shows that independence is conserved if independent random variables are grouped into disjoint classes and combined to form new random variables. Figure 3.6 illustrates the situation.

(3.24) Theorem. *Combining independent random variables. Let $(Y_i)_{i \in I}$ be an independent family of random variables on a probability space $(\Omega, \mathscr{F}, P)$, so that each Y_i takes values in an arbitrary event space $(\Omega_i, \mathscr{F}_i)$. Let $(I_k)_{k \in K}$ be a family of pairwise disjoint subsets of I, and for $k \in K$ let $(\widetilde{\Omega}_k, \widetilde{\mathscr{F}}_k)$ be an arbitrary event space and $\widetilde{Y}_k := \varphi_k \circ (Y_i)_{i \in I_k}$ for some random variable $\varphi_k : (\prod_{i \in I_k} \Omega_i, \bigotimes_{i \in I_k} \mathscr{F}_i) \to (\widetilde{\Omega}_k, \widetilde{\mathscr{F}}_k)$. Then the family $(\widetilde{Y}_k)_{k \in K}$ is independent.*

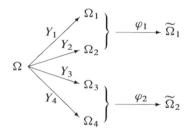

Figure 3.6. Disjoint classes of random variables are combined and 'processed further'.

Proof. For $k \in K$ consider the product space $(\widehat{\Omega}_k, \widehat{\mathscr{F}}_k) = (\prod_{i \in I_k} \Omega_i, \bigotimes_{i \in I_k} \mathscr{F}_i)$ and the corresponding vector-valued random variable $\widehat{Y}_k := (Y_i)_{i \in I_k} : \Omega \to \widehat{\Omega}_k$. The σ-algebra $\widehat{\mathscr{F}}_k$ has the intersection-stable generator

$$\widehat{\mathscr{G}}_k = \left\{ \bigcap_{i \in J} \{X_{k,i} \in B_i\} : \varnothing \ne J \text{ finite} \subset I_k, \ B_i \in \mathscr{F}_i \right\},$$

where $X_{k,i} : \widehat{\Omega}_k \to \Omega_i$ denotes the ith projection. We choose a finite $\varnothing \ne L \subset K$ and, for each $k \in L$, an arbitrary set $\widehat{B}_k = \bigcap_{i \in J_k} \{X_{k,i} \in B_i\} \in \widehat{\mathscr{G}}_k$ (with finite $J_k \subset I_k$ and $B_i \in \mathscr{F}_i$). Then we can write

$$P\left(\bigcap_{k \in L} \{\widehat{Y}_k \in \widehat{B}_k\} \right) = P\left(\bigcap_{k \in L} \bigcap_{i \in J_k} \{Y_i \in B_i\} \right)$$

$$= \prod_{k \in L} \prod_{i \in J_k} P(Y_i \in B_i) = \prod_{k \in L} P(\widehat{Y}_k \in \widehat{B}_k) \, ;$$

the last two equalities follow from the independence of $(Y_i)_{i \in I}$. So we conclude from Theorem (3.19) that $(\widehat{Y}_k)_{k \in K}$ is independent. Consequently, choosing arbitrary sets $\widetilde{B}_k \in \widetilde{\mathscr{F}}_k$ and considering

$$\widehat{B}_k := \varphi_k^{-1}\widetilde{B}_k \in \widehat{\mathscr{F}}_k\,,$$

we find that the product formula (3.18) holds for the events $\{\widetilde{Y}_k \in \widetilde{B}_k\} = \{\widehat{Y}_k \in \widehat{B}_k\}$. This means that $(\widetilde{Y}_k)_{k \in K}$ is independent. $\diamond$

(3.25) Example. *Partial sums when rolling a dice.* Let $M, N \geq 2$ and let $\Omega = \{1, \dots, 6\}^{MN}$, $P = \mathcal{U}_\Omega$ the uniform distribution, and let $X_i : \Omega \to \{1, \dots, 6\}$ denote the ith projection. Then, by Example (3.22) and Theorem (3.24), the random variables

$$\widetilde{X}_k = \sum_{i=(k-1)M+1}^{kM} X_i\,, \quad 1 \leq k \leq N,$$

(describing the sum of points during the kth run of length M) are independent.

3.4   Existence of Independent Random Variables, Product Measures

Do independent random variables actually exist? And if so, how can they be constructed? We have already seen some examples of finite families of independent random variables. But how about the existence of infinitely many independent random variables? This question naturally arises when one wants to set up a model for tossing a coin infinitely often, see Example (1.3). After the negative result in Theorem (1.5) (where we showed that we cannot simply use the power set as the σ-algebra), we will now obtain a positive result by using the product σ-algebra. We confine ourselves to the case of countably many random variables.

(3.26) Theorem. Construction of independent random variables with prescribed distributions. *Let I be a countable index set, and for each $i \in I$ let $(\Omega_i, \mathscr{F}_i, P_i)$ be an arbitrary probability space. Then there exists a probability space $(\Omega, \mathscr{F}, P)$ and independent random variables $Y_i : \Omega \to \Omega_i$ with $P \circ Y_i^{-1} = P_i$ for all $i \in I$.*

Proof. Since each subfamily of an independent family of random variables is, by definition, independent, we can assume that I is countably infinite, and so (via a suitable bijection) that $I = \mathbb{N}$. We proceed in stages by distinguishing several cases.

Case 1: All Ω_i are countable. Applying Theorem (3.12) to the transition densities $\rho_{k|\omega_1,\dots,\omega_{k-1}}(\omega_k) = P_k(\{\omega_k\})$ (which do not depend on the conditions), we obtain random variables $Y_i = X_i$, namely the projections on the product space $\Omega = \prod_{i \in \mathbb{N}} \Omega_i$ with a suitable probability measure, that satisfy (3.13). It is then evident that each Y_i has distribution P_i. This in turn implies that equation (3.13), for our choice of transition densities, is equivalent to the independence criterion in Corollary (3.21a), and the independence of $(Y_i)_{i \geq 1}$ follows.

Case 2: $\Omega_i = [0, 1]$ *and* $P_i = \mathcal{U}_{[0,1]}$ *for all* $i \in \mathbb{N}$. Since $\mathbb{N} \times \mathbb{N}$ is countable, the first case provides us with an auxiliary family $(Y_{i,j})_{i,j \geq 1}$ of independent $\{0, 1\}$-valued random variables on some suitable $(\Omega, \mathcal{F}, P)$ so that $P(Y_{i,j} = 1) = P(Y_{i,j} = 0) = 1/2$. Then, for $i \in \mathbb{N}$ let

$$Y_i = \sum_{j \geq 1} Y_{i,j} \, 2^{-j} = \varphi \circ (Y_{i,j})_{j \geq 1}$$

be the number with binary expansion $(Y_{i,j})_{j \geq 1}$. Here, $\varphi(y_1, y_2, \dots) = \sum_{j \geq 1} y_j \, 2^{-j}$ is the mapping from $\{0, 1\}^{\mathbb{N}}$ to $[0, 1]$ that assigns to each infinite binary sequence the corresponding real number. We claim that φ is a random variable with respect to the underlying σ-algebras $\mathcal{P}(\{0, 1\})^{\otimes \mathbb{N}}$ and $\mathcal{B}_{[0,1]}$. Indeed, if $X_i : \{0, 1\}^{\mathbb{N}} \to \{0, 1\}$ denotes the ith projection and if $0 \leq m < 2^n$ has the binary expansion $m = \sum_{k=1}^n y_k \, 2^{n-k}$ with $y_k \in \{0, 1\}$, then

$$\varphi^{-1}\left[\tfrac{m}{2^n}, \tfrac{m+1}{2^n}\right] = \{X_1 = y_1, \dots, X_n = y_n\} \in \mathcal{P}(\{0, 1\})^{\otimes \mathbb{N}} \, ,$$

and by Remark (1.25) this implies our claim. Furthermore, we have in this case that

$$P\left(Y_i \in \left[\tfrac{m}{2^n}, \tfrac{m+1}{2^n}\right]\right) = P(Y_{i,1} = y_1, \dots, Y_{i,n} = y_n) = 2^{-n} \, .$$

In particular, letting the intervals shrink to one point shows that $P(Y_i = m/2^n) = 0$ for all i, m, n and thus

$$P\left(Y_i \in \left]\tfrac{m}{2^n}, \tfrac{m+1}{2^n}\right]\right) = 2^{-n} = \mathcal{U}_{[0,1]}\left(\left]\tfrac{m}{2^n}, \tfrac{m+1}{2^n}\right]\right) \, .$$

The uniqueness theorem (1.12) thus yields the identity $P \circ Y_i^{-1} = \mathcal{U}_{[0,1]}$. Finally, Theorem (3.24) implies that the sequence $(Y_i)_{i \geq 1}$ is independent.

Case 3: $\Omega_i = \mathbb{R}$ *for all* $i \in \mathbb{N}$. According to the second case, there exist independent random variables $(Y_i)_{i \geq 1}$ on a probability space $(\Omega, \mathcal{F}, P)$ such that $P \circ Y_i^{-1} = \mathcal{U}_{[0,1]}$. In fact, we even know that $P(0 < Y_i < 1 \text{ for all } i \geq 1) = 1$ because $\mathcal{U}_{[0,1]}(]0, 1[) = 1$. According to Proposition (1.30), each probability measure P_i on $(\mathbb{R}, \mathcal{B})$ is the distribution of a random variable $\varphi_i :]0, 1[\to \mathbb{R}$, namely the quantile transformation of P_i; that is, $\mathcal{U}_{]0,1[} \circ \varphi_i^{-1} = P_i$. It follows that the family $(\varphi_i \circ Y_i)_{i \geq 1}$ has the required properties: It is independent by Theorem (3.24), and $P \circ (\varphi_i \circ Y_i)^{-1} = \mathcal{U}_{]0,1[} \circ \varphi_i^{-1} = P_i$.

General case. If $\Omega_i = \mathbb{R}^d$ or Ω_i is a complete separable metric space, then in some intricate way it is still possible to find a φ_i as in the third case. In the general case, however, the existence problem cannot be reduced to the existence of the

Lebesgue measure (as was done in the proof of Theorem (3.12) and hence in the particular cases above), and more measure theory is required. We do not pursue this further because the particular cases above are all we need later on. The full result can be found in Durrett [16, Section 1.4.c] or Dudley [15, Theorem 8.2.2]. $\diamond$

(3.27) Corollary. *Existence of infinite product measures. Let* $(\Omega_i, \mathscr{F}_i, P_i)$, $i \in I$, *be a countable family of probability spaces and* $\Omega = \prod_{i \in I} \Omega_i$, $\mathscr{F} = \bigotimes_{i \in I} \mathscr{F}_i$. *Then there is a unique probability measure P on* $(\Omega, \mathscr{F})$ *that satisfies the product formula*

$$P(X_i \in A_i \text{ for all } i \in J) = \prod_{i \in J} P_i(A_i)$$

for all finite $\varnothing \neq J \subset I$ *and* $A_i \in \mathscr{F}_i$, *which is to say that the projections* $X_i : \Omega \to \Omega_i$ *are independent with distribution* P_i.

Definition. The measure P in (3.27) is called the *product* of the measures P_i. It is denoted by $\bigotimes_{i \in I} P_i$ or, if $P_i = Q$ for all $i \in I$, by $Q^{\otimes I}$.

Proof. The uniqueness of P follows from the uniqueness theorem (1.12), since the sets $\{X_i \in A_i \text{ for } i \in J\}$ with finite $\varnothing \neq J \subset I$ and $A_i \in \mathscr{F}_i$ form an intersection-stable generator of $\mathscr{F}$.

As for the existence, we can apply Theorem (3.26) to obtain independent random variables $(Y_i)_{i \in I}$ defined on a probability space $(\Omega', \mathscr{F}', P')$ such that $P' \circ Y_i^{-1} = P_i$. Then, $Y = (Y_i)_{i \in I} : \Omega' \to \Omega$ is a random variable. Indeed, for each finite $\varnothing \neq J \subset I$ and all $A_i \in \mathscr{F}_i$ we have

$$Y^{-1}\{X_i \in A_i \text{ for } i \in J\} = \{Y_i \in A_i \text{ for } i \in J\} \in \mathscr{F}',$$

whence our claim follows from Remark (1.25). So the distribution $P = P' \circ Y^{-1}$ of Y is well-defined, and

$$P(X_i \in A_i \text{ for } i \in J) = P'(Y_i \in A_i \text{ for } i \in J) = \prod_{i \in J} P'(Y_i \in A_i) = \prod_{i \in J} P_i(A_i).$$

Thus P has the required property. $\diamond$

As the above proof shows, the concepts of independence and product measure are closely related:

(3.28) Remark. *Independence as a property of distributions.* A countable family $(Y_i)_{i \in I}$ of random variables taking values in arbitrary event spaces $(\Omega_i, \mathscr{F}_i)$ is independent if and only if

$$P \circ (Y_i)_{i \in I}^{-1} = \bigotimes_{i \in I} P \circ Y_i^{-1},$$

or, in words, if the *joint distribution* of the Y_i (namely the distribution of the random vector $Y = (Y_i)_{i \in I}$ with values in $\prod_{i \in I} \Omega_i$) coincides with the product of the single distributions $P \circ Y_i^{-1}$.

The typical case of interest is when all random variables have the same distribution.

(3.29) Example and Definition. *Canonical product models and the Bernoulli measures.* Let $I = \mathbb{N}$ and $(\Omega_i, \mathscr{F}_i, P_i) = (E, \mathscr{E}, Q)$ for each $i \in \mathbb{N}$. Then, by Corollary (3.27), one can define the infinite product probability space $(E^{\mathbb{N}}, \mathscr{E}^{\otimes \mathbb{N}}, Q^{\otimes \mathbb{N}})$, which serves as the so-called *canonical model* for the infinite independent repetition of an experiment described by $(E, \mathscr{E}, Q)$. The outcomes of the single experiments are then described by the projection variables $X_i : \Omega \to E$ onto the ith coordinate, and these are independent with identical distribution Q. As such sequences of random variables are a basic and frequent ingredient of stochastic modelling, it is common to use the abbreviation *i.i.d.* for *independent and identically distributed.*

In the case $E = \{0, 1\}$ and $Q(\{1\}) = p \in {]0, 1[}$, the product measure $Q^{\otimes \mathbb{N}}$ is called the (infinite) *Bernoulli measure* or *Bernoulli distribution* on $\{0, 1\}^{\mathbb{N}}$ with probability p of success; its existence on the product σ-algebra $\mathscr{P}(\{0, 1\})^{\otimes \mathbb{N}}$ is the positive counterpart of the 'no-go theorem' (1.5). Likewise, a sequence $(Y_i)_{i \geq 1}$ of $\{0, 1\}$-valued random variables is called a *Bernoulli sequence with parameter p* if it has the joint distribution $Q^{\otimes \mathbb{N}}$, in other words if

$$P(Y_i = x_i \text{ for all } i \leq n) = p^{\sum_{i=1}^{n} x_i} (1-p)^{\sum_{i=1}^{n}(1-x_i)}$$

for all $n \geq 1$ and $x_i \in \{0, 1\}$. The joint distribution of $(Y_1, \ldots, Y_n)$ is then the Bernoulli distribution on the finite product $\{0, 1\}^n$, as introduced in Section 2.2.1.

Finite products of probability measures on $\mathbb{R}$ have a density if and only if all factor measures have a density. More precisely, the following holds.

(3.30) Example. *Product densities.* For each $1 \leq i \leq n$, let P_i be a probability measure on $(\mathbb{R}, \mathscr{B})$ with existing Lebesgue density ρ_i. Then the (finite) product measure $P = \bigotimes_{i=1}^{n} P_i$ on $(\mathbb{R}^n, \mathscr{B}^n)$ is precisely the probability measure with Lebesgue density

$$\rho(x) := \prod_{i=1}^{n} \rho_i(x_i) \quad \text{for } x = (x_1, \ldots, x_n) \in \mathbb{R}^n.$$

Indeed, for arbitrary $c_i \in \mathbb{R}$ we have

$$P(X_1 \leq c_1, \ldots, X_n \leq c_n) = \prod_{i=1}^{n} P_i({]-\infty, c_i]}) = \prod_{i=1}^{n} \int_{-\infty}^{c_i} \rho_i(x_i)\, dx_i$$

$$= \int_{-\infty}^{c_1} \cdots \int_{-\infty}^{c_n} \rho_1(x_1) \ldots \rho_n(x_n)\, dx_1 \ldots dx_n = \int_{\{X_1 \leq c_1, \ldots, X_n \leq c_n\}} \rho(x)\, dx \,;$$

the third equality is justified by Fubini's theorem for multiple integrals, see e.g. [56]. Hence, by the uniqueness theorem (1.12), we find that $P(A) = \int_A \rho(x)\, dx$ for all $A \in \mathscr{B}^n$. (Note that an infinite product measure does not admit a density, even if all factors do. This follows already from the fact that there is no Lebesgue measure on $\mathbb{R}^{\mathbb{N}}$.)

We conclude this section by introducing the notion of *convolution*, which is closely related to that of a product measure. As a motivation, let Y_1 and Y_2 be two independent real-valued random variables with distributions Q_1 and Q_2, respectively. Then, by Remark (3.28), the pair (Y_1, Y_2) has distribution $Q_1 \otimes Q_2$ on $\mathbb{R}^2$. On the other hand, the sum $Y_1 + Y_2$ arises from the pair (Y_1, Y_2) by applying the addition map $A : (x_1, x_2) \to x_1 + x_2$ from $\mathbb{R}^2$ to $\mathbb{R}$. That is, $Y_1 + Y_2 = A \circ (Y_1, Y_2)$. Hence, $Y_1 + Y_2$ has distribution $(Q_1 \otimes Q_2) \circ A^{-1}$. (Note that A is continuous, and thus a random variable by (1.27).)

Definition. For any two probability measures Q_1, Q_2 on $(\mathbb{R}, \mathscr{B})$, the probability measure $Q_1 \star Q_2 := (Q_1 \otimes Q_2) \circ A^{-1}$ on $(\mathbb{R}, \mathscr{B})$ is called the *convolution* of Q_1 and Q_2.

In other words, $Q_1 \star Q_2$ is the distribution of the sum of any two independent random variables with distributions Q_1 and Q_2, respectively. In the case of probability measures with densities, the convolution also has a density, as is stated next.

(3.31) Remark. *Convolution of densities.* In the setting of the previous definition, the following holds.

(a) *Discrete case.* If Q_1 and Q_2 are probability measures on $(\mathbb{Z}, \mathscr{P}(\mathbb{Z}))$ with associated discrete densities ρ_1 and ρ_2 respectively, then $Q_1 \star Q_2$ is the probability measure on $(\mathbb{Z}, \mathscr{P}(\mathbb{Z}))$ with discrete density

$$\rho_1 \star \rho_2(k) := \sum_{l \in \mathbb{Z}} \rho_1(l)\, \rho_2(k-l), \quad k \in \mathbb{Z}.$$

(b) *Continuous case.* If Q_1 and Q_2 each have a Lebesgue density ρ_1 and ρ_2 respectively, then $Q_1 \star Q_2$ has the Lebesgue density

$$\rho_1 \star \rho_2(x) := \int \rho_1(y)\, \rho_2(x-y)\, dy, \quad x \in \mathbb{R}.$$

Proof. In case (a) we have for any $k \in \mathbb{Z}$

$$Q_1 \otimes Q_2(A = k) = \sum_{l_1, l_2 \in \mathbb{Z}: l_1 + l_2 = k} \rho_1(l_1)\, \rho_2(l_2) = \rho_1 \star \rho_2(k).$$

In case (b), we use Example (3.30), apply the substitutions $x_1 \rightsquigarrow y, x_2 \rightsquigarrow x = y + x_2$, and interchange the integrals to find

$$Q_1 \otimes Q_2(A \le c) = \int dy\, \rho_1(y) \int dx\, \rho_2(x-y)\, 1_{\{x \le c\}} = \int_{-\infty}^{c} \rho_1 \star \rho_2(x)\, dx$$

for arbitrary $c \in \mathbb{R}$. The result then follows from Remark (1.31). $\diamond$

In a few important cases, the type of distribution is preserved under convolution. One instance are the normal distributions.

(3.32) Example. *Convolution of normal distributions.* For each $m_1, m_2 \in \mathbb{R}$ and $v_1, v_2 > 0$, we have $\mathcal{N}_{m_1,v_2} \star \mathcal{N}_{m_2,v_2} = \mathcal{N}_{m_1+m_2,v_1+v_2}$. That is, the convolution of normal distributions is normal, and the parameters simply add up. One thus says that the normal distributions form a *two-parameter convolution semigroup*. For the proof, assume without loss of generality that $m_1 = m_2 = 0$. A short calculation then shows that, for every $x, y \in \mathbb{R}$,

$$\phi_{0,v_1}(y)\,\phi_{0,v_2}(x-y) = \phi_{0,v_1+v_2}(x)\,\phi_{xu,v_2u}(y)\,,$$

where $u = v_1/(v_1 + v_2)$. Applying Remark (3.31b) and integrating over y, the claim follows.

Further examples of convolution semigroups follow in Corollary (3.36), Problems 3.15 and 3.16, and Section 4.4.

3.5   The Poisson Process

Let us return to the model of random points on the time axis that was discussed in Sections 2.4 and 2.5.2. The existence of infinitely many independent random variables with given distribution now allows us to specify this model in more detail. From Section 2.5.2 we know that the waiting time for the first point has an exponential distribution, and the heuristic in Section 2.4 suggests that the gaps between two consecutive points are independent. Hence we use the following approach.

Let $\alpha > 0$ and $(L_i)_{i\geq 1}$ be a sequence of i.i.d. random variables that are exponentially distributed with parameter α; Theorem (3.26) guarantees the existence of such a sequence on a suitable probability space $(\Omega, \mathcal{F}, P)$. We interpret L_i as the gap between the $(i-1)$st and the ith point; then $T_k = \sum_{i=1}^{k} L_i$ is the kth random point in time; cf. Figure 3.7. Let

$$(3.33) \qquad\qquad N_t = \sum_{k\geq 1} 1_{]0,t]}(T_k)$$

be the number of points in the interval $]0, t]$. Thus, for $s < t$, the increment $N_t - N_s$ is the number of points in $]s, t]$.

Figure 3.7. Definition of the Poisson process N_t. The gaps L_i are independent and exponentially distributed. For $t \in [T_k, T_{k+1}[$ one has $N_t = k$.

(3.34) Theorem. Construction of the Poisson process. *Under the conditions above, the N_t are random variables, and, for any time points $0 = t_0 < t_1 < \cdots < t_n$, the increments $N_{t_i} - N_{t_{i-1}}$ with $1 \leq i \leq n$ are independent and Poisson distributed with parameters $\alpha(t_i - t_{i-1})$.*

Definition. A family $(N_t)_{t \geq 0}$ of random variables satisfying the properties in Theorem (3.34) is called a *Poisson process with intensity* $\alpha > 0$.

The independence of the number of points in disjoint intervals shows that the Poisson process does indeed provide a model of time points that are arranged as randomly as possible. One should note at this stage that the relation (3.33) can be inverted by writing $T_k = \inf\{t > 0 : N_t \geq k\}$ for $k \geq 1$. In other words, T_k is the kth time point at which the 'sample path' $t \to N_t$ of the Poisson process performs a jump of size 1. The times T_k are therefore also called the *jump times* of the Poisson process, and $(N_t)_{t \geq 0}$ and $(T_k)_{k \in \mathbb{N}}$ are two manifestations of the same mathematical object.

Proof. Since $\{N_t = k\} = \{T_k \leq t < T_{k+1}\}$ and the T_k are random variables by Problem 1.14, every N_t is a random variable. For the main part of the proof, we confine ourselves to the case $n = 2$ to keep the notation simple; the general case follows analogously. So let $0 < s < t$, $k, l \in \mathbb{Z}_+$. We show that

$$(3.35) \qquad P(N_s = k, N_t - N_s = l) = \left(e^{-\alpha s} \frac{(\alpha s)^k}{k!}\right)\left(e^{-\alpha(t-s)} \frac{(\alpha(t-s))^l}{l!}\right).$$

For, summing over l and k, respectively, we can then conclude that N_s and $N_t - N_s$ are Poisson distributed, and their independence follows from Corollary (3.21a). By Example (3.30), the (joint) distribution of $(L_j)_{1 \leq j \leq k+l+1}$ has the product density

$$x = (x_1, \ldots, x_{k+l+1}) \to \alpha^{k+l+1} e^{-\alpha \tau_{k+l+1}(x)},$$

where we set $\tau_j(x) = x_1 + \cdots + x_j$. Thus, in the case $k, l \geq 1$ we can write

$$P(N_s = k, N_t - N_s = l) = P(T_k \leq s < T_{k+1} \leq T_{k+l} \leq t < T_{k+l+1})$$

$$= \int_0^\infty \cdots \int_0^\infty dx_1 \ldots dx_{k+l+1}\, \alpha^{k+l+1} e^{-\alpha \tau_{k+l+1}(x)}$$

$$\cdot 1_{\{\tau_k(x) \leq s < \tau_{k+1}(x) \leq \tau_{k+l}(x) \leq t < \tau_{k+l+1}(x)\}};$$

for $k = 0$ or $l = 0$ we obtain an analogous formula. We integrate stepwise starting from the innermost integral and moving outwards. For fixed $x_1, \ldots, x_{k+l}$, the substitution $z = \tau_{k+l+1}(x)$ yields

$$\int_0^\infty dx_{k+l+1}\, \alpha e^{-\alpha \tau_{k+l+1}(x)} 1_{\{\tau_{k+l+1}(x) > t\}} = \int_t^\infty dz\, \alpha e^{-\alpha z} = e^{-\alpha t}.$$

Next, we fix $x_1, \ldots, x_k$ and make the substitutions $y_1 = \tau_{k+1}(x) - s$, $y_2 = x_{k+2}, \ldots,$ $y_l = x_{k+l}$ to obtain

$$\int_0^\infty \cdots \int_0^\infty dx_{k+1} \ldots dx_{k+l} \, 1_{\{s \, < \, \tau_{k+1}(x) \, \leq \, \tau_{k+l}(x) \, \leq \, t\}}$$

$$= \int_0^\infty \cdots \int_0^\infty dy_1 \ldots dy_l \, 1_{\{y_1 + \cdots + y_l \, \leq \, t - s\}} = \frac{(t-s)^l}{l!} \, .$$

The last equality can be proved by induction on l. (In the case $l = 0$, this integral does not appear and can formally be set equal to 1.) For the remaining integral, we find in the same way that

$$\int_0^\infty \cdots \int_0^\infty dx_1 \ldots dx_k \, 1_{\{\tau_k(x) \, \leq \, s\}} = \frac{s^k}{k!} .$$

Combining everything, we obtain

$$P(N_s = k, N_t - N_s = l) = e^{-\alpha t} \, \alpha^{k+l} \, \frac{s^k}{k!} \, \frac{(t-s)^l}{l!}$$

and hence (3.35). ◇

The Poisson process is the prototypical stochastic model for time points that are purely random. It is the rigorous version of the heuristic approach in Section 2.4. Along the way, it provides us with two further examples of convolution semigroups.

(3.36) Corollary. Convolution of Poisson and gamma distributions. *The convolution formulas* $\mathcal{P}_\lambda \star \mathcal{P}_\mu = \mathcal{P}_{\lambda+\mu}$ *and* $\Gamma_{\alpha,r} \star \Gamma_{\alpha,s} = \Gamma_{\alpha,r+s}$ *are valid for arbitrary parameters* $\lambda, \mu > 0$ *and* $\alpha > 0$, $r, s \in \mathbb{N}$, *respectively.*

Proof. Let $(N_t)_{t \geq 0}$ be the Poisson process with parameter $\alpha > 0$ as constructed above, and suppose first that $\alpha = 1$. By Theorem (3.34), the random variables N_λ and $N_{\lambda+\mu} - N_\lambda$ are independent with distribution $\mathcal{P}_\lambda$ and $\mathcal{P}_\mu$ respectively, and their sum $N_{\lambda+\mu}$ has distribution $\mathcal{P}_{\lambda+\mu}$. This proves the first claim.

Now let α be arbitrary and $T_r = \sum_{i=1}^r L_i$ the rth point of the Poisson process. As a sum of r independent exponentially distributed random variables, T_r has the distribution $\mathcal{E}_\alpha^{\star r}$. On the other hand, we have for any $t > 0$

$$P(T_r \leq t) = P(N_t \geq r) = 1 - \mathcal{P}_{\alpha t}(\{0, \ldots, r-1\}) = \Gamma_{\alpha,r}(\,]0, t]) \, .$$

Here, the first equality follows from the definition of N_t, the second from Theorem (3.34), and the third from (2.19). Hence, $\mathcal{E}_\alpha^{\star r} = \Gamma_{\alpha,r}$, which implies the second claim. ◇

In fact, it is easy to verify that the relation $\Gamma_{\alpha,r} \star \Gamma_{\alpha,s} = \Gamma_{\alpha,r+s}$ also holds for non-integer values of r and s, see Problem 3.15 and Corollary (9.9).

As one of the basic models of stochastics, the Poisson process is the starting point for a large variety of modifications and generalisations. Here, we will only mention two examples.

(3.37) Example. *The compound Poisson process.* Let $(N_t)_{t \geq 0}$ be a Poisson process with intensity $\alpha > 0$. As noticed above, the sample path $t \to N_t(\omega)$ associated with any $\omega \in \Omega$ is a piecewise constant function, which performs a jump of height 1 at the times $T_k(\omega)$, $k \geq 1$. We now modify the process by allowing the jump heights to vary randomly as well.

Let $(Z_i)_{i \geq 1}$ be a sequence of real random variables, which are independent of each other as well as of $(N_t)_{t \geq 0}$, and identically distributed according to Q on $(\mathbb{R}, \mathscr{B})$. (Such random variables exist by Theorem (3.26).) The process $(S_t)_{t \geq 0}$ defined by

$$ S_t = \sum_{i=0}^{N_t} Z_i , \quad t \geq 0, $$

is then called the *compound Poisson process with jump distribution Q and intensity α*. If each $Z_i \geq 0$, this process models, for example, the evolution of collective claims arriving at an insurance company. The ith claim arrives at the ith jump time T_i of the Poisson process (N_t), and Z_i is its amount. On the other hand, the regular premium income of the insurance company leads to a continuous increase in capital with rate $c > 0$. The net loss of the insurance company in the interval $[0, t]$ is thus described by the process $V_t = S_t - ct$, $t \geq 0$. This is the basic model of risk theory. Of special interest is the 'ruin probability' $r(a) = P(\sup_{t \geq 0} V_t > a)$, which is the probability that the total loss exceeds the capital reserve $a > 0$ at some time. To find the supremum over all V_t, it is clearly sufficient to consider only the jump times T_k. Using the L_i from Figure 3.7, we can therefore rewrite the ruin probability as

$$ r(a) = P\left(\sup_{k \geq 1} \sum_{i=1}^{k} (Z_i - cL_i) > a \right). $$

More on this topic can be found for instance in [19], Sections VI.5 and XII.5. Figure 3.8 shows a simulation of $(V_t)_{t \geq 0}$.

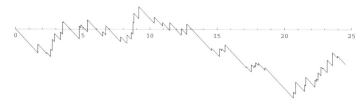

Figure 3.8. Simulation of the ruin process (V_t) for $Q = \mathcal{U}_{]0,1[}$, $\alpha = 2$, $c = 1.1$.

(3.38) Example. *The Poisson point process in* $\mathbb{R}^2$. Up to now, the Poisson process served as a model for random times in $[0, \infty[$. In many applications, however, there is also an interest in random points in higher-dimensional spaces. Think for instance of the particle positions in an ideal gas, or the positions of the pores in a liquid film. Here we confine ourselves to two dimensions and consider random points in a 'window' $\Lambda = [0, L]^2$ of the plane. Let $\alpha > 0$ and $(N_t)_{t \geq 0}$ be a Poisson process with intensity αL and jump times $(T_k)_{k \geq 1}$. Also, let $(Z_i)_{i \geq 1}$ be a sequence of i.i.d. random variables with uniform distribution $\mathcal{U}_{[0,L]}$ on $[0, L]$, which is independent of the Poisson process. Then, the random set of points

$$\xi = \big\{ (T_k, Z_k) : 1 \leq k \leq N_L \big\}$$

is called the *Poisson point process on* Λ *with intensity* α. (For an alternative construction of ξ see Problem 3.25.) As a variation of this model, one can draw a circle with random radius $R_k > 0$ around each point $(T_k, Z_k) \in \xi$. The union of these circles then forms a random set Ξ, which is called the *Boolean model* and is used in stochastic geometry as a basic model of random structures. More on this can be found for instance in [48, 61]. Two simulated realisations of Ξ for different intensities are shown in Figure 3.9.

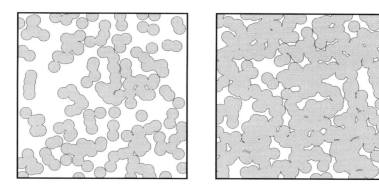

Figure 3.9. Simulations of the Boolean model for $L = 15$, $R_k = 0.5$, and $\alpha = 0.8$ and 2, respectively.

3.6 Simulation Methods

When developing a stochastic model for a specific application, one would often like to get a first impression of how the model behaves and if it exhibits the phenomena one expects. To this end, 'experimental stochastics' provides the useful tool of *Monte Carlo simulation*; two examples are Figures 3.8 and 3.9. In the following, we will discuss a few fundamental simulation methods.

Recall the proof of Theorem (3.26), which provided us with more than just the existence of independent random variables. In its second case, we saw how to construct independent $\mathcal{U}_{[0,1]}$-distributed random variables from a Bernoulli sequence (using the binary expansion), and in its third case, how to use these in combination with the quantile transformation (1.30) to generate random variables with arbitrary distribution in $\mathbb{R}$. These were two examples of the general problem: *How can one transform a given family of known random variables into new random variables with certain desired properties?* This is, in fact, the central question in computer simulation as well, where one starts with a sequence of i.i.d. random variables with uniform distribution on $[0, 1]$ and uses them to construct new random variables with prescribed distributions. We will demonstrate this by means of several examples. We start with two simple applications of the quantile transformation, which is also known as the *inversion method* in this context.

(3.39) Example. *Sampling from binomial distributions.* Let $0 < p < 1$ and $U_1, \ldots,$ U_n be independent $\mathcal{U}_{[0,1]}$-distributed random variables. By Theorem (3.24), the random variables $X_i = 1_{\{U_i \le p\}}$ then form a Bernoulli sequence with parameter p. Theorem (2.9) thus implies that the sum $S = \sum_{i=1}^{n} X_i$ has the binomial distribution $\mathcal{B}_{n,p}$. (Check that the construction of X_i is a special case of the quantile transformation.)

(3.40) Example. *Sampling from exponential distributions.* Let U_i, $i \ge 1$, be i.i.d. $\mathcal{U}_{[0,1]}$-distributed random variables and $\alpha > 0$. Then the random variables $X_i = (-\log U_i)/\alpha$ are exponentially distributed with parameter α (and by Theorem (3.24) also independent). To see this, observe that $P(X_i \ge c) = P(U_i \le e^{-\alpha c}) = e^{-\alpha c}$ for every $c > 0$. (Again, the reader should check that this is a simple modification of the quantile transformation.)

If we combine the last example with the construction of the Poisson process in Theorem (3.34), we obtain a convenient method of simulating Poisson random variables.

(3.41) Example. *Sampling from Poisson distributions.* Let U_i, $i \ge 1$, be independent $\mathcal{U}_{[0,1]}$-distributed random variables. By the previous example, the random variables $L_i = -\log U_i$ are then independent and exponentially distributed with parameter 1. Writing $T_k = \sum_{i=1}^{k} L_i$ for the kth partial sum and applying Theorem (3.34), we thus see that for every $\lambda > 0$ the random variable

$$N_\lambda = \min\{k \ge 0 : T_{k+1} > \lambda\} = \min\{k \ge 0 : U_1 \ldots U_{k+1} < e^{-\lambda}\}$$

(which coincides with the quantity in (3.33)) is $\mathcal{P}_\lambda$-distributed. This yields the following algorithm to simulate a Poisson distributed random variable N_λ, which is specified in pseudocode here:

```
v ← 1, k ← −1
repeat  v ← U v,  k ← k + 1
until  v < e^{−λ}
N_λ ← k
```

(Here, U stands for a uniform random number in $[0, 1]$ that is generated afresh at each call.)

Unfortunately, the inversion method is not always feasible, namely when the desired distribution function is not accessible numerically. In such cases, the following general principle, which goes back to John von Neumann (1903–1957), provides an alternative.

(3.42) Example. *Rejection sampling and conditional distributions.* Let $(Z_n)_{n \geq 1}$ be an i.i.d. sequence of random variables on a probability space $(\Omega, \mathscr{F}, P)$ taking values in an arbitrary event space $(E, \mathscr{E})$, and Q their identical distribution, i.e., $P \circ Z_n^{-1} = Q$ for all n. Let $B \in \mathscr{E}$ be an event such that $Q(B) > 0$. How can we construct a random variable Z^* with the conditional distribution $Q(\cdot \mid B)$, using only the Z_n? The basic idea is as follows: Observe the Z_n one after the other, ignore all n with $Z_n \notin B$, and then set $Z^* = Z_n$ for the first n with $Z_n \in B$. More precisely, let

$$\tau = \inf\{n \geq 1 : Z_n \in B\}$$

be the first hitting time of B or, put differently, the time of the first success of the Bernoulli sequence $1_B(Z_n)$, $n \geq 1$. From Section 2.5.1 we know that $\tau - 1$ has the geometric distribution with parameter $p = Q(B)$. In particular, $P(\tau < \infty) = 1$, which means that the hypothetical case that $Z_n \notin B$ for all n, and thus $\tau = \infty$, only occurs with probability zero. Define $Z^* = Z_\tau$, i.e., $Z^*(\omega) = Z_{\tau(\omega)}(\omega)$ for all ω with $\tau(\omega) < \infty$. (To be formal, one can set $Z^*(\omega) = Z_1(\omega)$ for the remaining ω, which do not play any role.) It then follows that for all $A \in \mathscr{E}$

$$P(Z^* \in A) = \sum_{n=1}^{\infty} P(Z_n \in A, \ \tau = n)$$

$$= \sum_{n=1}^{\infty} P(Z_1 \notin B, \ldots, Z_{n-1} \notin B, \ Z_n \in A \cap B)$$

$$= \sum_{n=1}^{\infty} (1 - Q(B))^{n-1} Q(A \cap B) = Q(A \mid B),$$

which shows that Z^* has distribution $Q(\cdot \mid B)$.

The rejection sampling method can be used as follows to simulate a random variable with a specified distribution density.

(3.43) Example. *Monte-Carlo simulation via rejection sampling.* Let $[a, b]$ be a compact interval and ρ a probability density function on $[a, b]$. Suppose ρ is bounded, so there exists a $c > 0$ with $0 \leq \rho(x) \leq c$ for all $x \in [a, b]$. Let $U_n, V_n, n \geq 1$, be i.i.d. with uniform distribution $\mathcal{U}_{[0,1]}$. Then the random variables

$$Z_n = (X_n, Y_n) := (a + (b-a)U_n, c V_n)$$

are independent with uniform distribution $\mathcal{U}_{[a,b] \times [0,c]}$. Let

$$\tau = \inf\{n \geq 1 : Y_n \leq \rho(X_n)\}$$

and $Z^* = (X^*, Y^*) = (X_\tau, Y_\tau)$. Example (3.42) then tells us that Z^* is uniformly distributed on $B = \{(x, y) : a \leq x \leq b, \ y \leq \rho(x)\}$. Consequently, for all $A \in \mathscr{B}_{[a,b]}$, we can write

$$P(X^* \in A) = \mathcal{U}_B\big((x, y) : \ x \in A, \ y \leq \rho(x)\big) = \int_A \rho(x)\, dx\,,$$

where the last equality follows from Fubini's theorem. This shows that X^* has distribution density ρ. The construction of X^* corresponds to the following simple algorithm written in pseudocode as follows.

```
repeat u ← U , v ← V
until  cv ≤ ρ(a + (b − a)u)
X* ← cv
```
(At each call, $U, V \in [0, 1]$ are two fresh random samples from $\mathcal{U}_{[0,1]}$, which are independent of each other and of everything else before.)

(If ρ is a discrete density on a finite set $\{0, \ldots, N-1\}$, we obtain an analogous algorithm by considering the Lebesgue density $x \to \rho(\lfloor x \rfloor)$ on the interval $[0, N[.$)

Combining rejection sampling with a suitable transformation, one can simulate random variables with normal distributions.

(3.44) Example. *Polar method for sampling from normal distributions.* Let $K = \{x \in \mathbb{R}^2 : |x| \leq 1\}$ be the unit disc and $Z = (Z_1, Z_2)$ a K-valued random variable with uniform distribution $\mathcal{U}_K$ on K; for instance, one can use rejection sampling as in Example (3.42) to obtain Z from a sequence of random variables with uniform distribution on $[-1, 1]^2$. Let

$$X = (X_1, X_2) := 2\sqrt{-\log|Z|}\ \frac{Z}{|Z|}\,.$$

Then the coordinate variables X_1 and X_2 are independent and $\mathcal{N}_{0,1}$-distributed.

To see this, consider the polar coordinates R, Ψ of Z. As was shown in Example (3.23), R^2 and Ψ are independent with distribution $\mathcal{U}_{[0,1]}$ resp. $\mathcal{U}_{[0,2\pi[}$. Let $S = \sqrt{-2 \log R^2}$. By Theorem (3.24), S and Ψ are independent. Moreover, S has the distribution density $s \to s\,e^{-s^2/2}$ on $[0, \infty[$ because

$$P(S \leq c) = P(R^2 \geq e^{-c^2/2}) = 1 - e^{-c^2/2} = \int_0^c s\,e^{-s^2/2}\,ds$$

for all $c > 0$. Now, obviously, $X = (S \cos \Psi, S \sin \Psi)$. Hence, for any $A \in \mathscr{B}^2$ transforming polar into Cartesian coordinates yields

$$\begin{aligned}
P(X \in A) &= \frac{1}{2\pi} \int_0^{2\pi} d\varphi \int_0^\infty ds\, s\,e^{-s^2/2}\, 1_A(s \cos \varphi, s \sin \varphi) \\
&= \frac{1}{2\pi} \int dx_1 \int dx_2\, e^{-(x_1^2 + x_2^2)/2}\, 1_A(x_1, x_2) = \mathcal{N}_{0,1} \otimes \mathcal{N}_{0,1}(A)\,;
\end{aligned}$$

the last equality comes from Example (3.30). Remark (3.28) thus gives the result.

The above construction yields the following algorithm (again written in pseudo-code) for generating two independent standard normal random variables X_1, X_2.

```
repeat
    u ← 2U − 1,  v ← 2V − 1,  w ← u² + v²        (At each call, U, V ∈ [0, 1] are
until  w ≤ 1                                       fresh and independent random
a ← √(−2 log w)/w,  X₁ ← au,  X₂ ← av             samples from U[0,1].)
```

Each of the previous examples has presented a method to construct new random variables with a desired distribution from independent uniform random numbers in $[0, 1]$. All these methods reduce the problem of simulating specific random variables to the simulation of independent $\mathcal{U}_{[0,1]}$-variables. But how can the latter problem be solved? This is the subject of our concluding remark.

(3.45) Remark. *Random numbers.* Random realisations of independent $\mathcal{U}_{[0,1]}$-distributed random variables are called *random numbers*. They can be found in tables or on the internet. Partly, these are generated by real chance, see e.g. `www.rand.org/pubs/monograph_reports/MR1418/index.html`. In practice, however, it is common to use so-called *pseudo-random numbers*, which are not random at all, but produced deterministically. A standard method to generate pseudo-random numbers is the following *linear congruence method*.

Choose a 'modulus' m (for example $m = 2^{32}$) as well as a factor $a \in \mathbb{N}$ and an increment $b \in \mathbb{N}$ (this requires a lot of skill). Next, pick a 'seed' $k_0 \in \{0, \ldots, m-1\}$ (e.g., tied to the internal clock of the processor), and define the recursive sequence $k_{i+1} = a\, k_i + b \bmod m$. The pseudo-random numbers then consist of the sequence $u_i = k_i/m$, $i \geq 1$. For an appropriate choice of a, b, the sequence (k_i) has period m (so it does not repeat itself after fewer iterations), and it survives several statistical tests of independence. (For example, this is the case for $a = 69069$, $b = 1$; G. Marsaglia 1972.) Pseudo-random variables are readily available as a standard feature of many compilers and software packages – but the user still has the responsibility to check whether they are suitable for the specific application at hand. It is a nice anecdote that the random generator `randu` by IBM, which was widely distributed in the 1960s, has the property that, for $a = 65539$, $b = 0$, $m = 2^{31}$ and period 2^{29}, the consecutive triples (u_i, u_{i+1}, u_{i+2}) lie in only 15 different parallel planes of $\mathbb{R}^3$, which is certainly not a characteristic of randomness! In fact, the existence of such grid structures is an inevitable consequence of the linear nature of the method; the art is to make them sufficiently fine. A standard reference on pseudo-random numbers is Knuth [37].

3.7 Tail Events

The existence of infinite models, as discussed above in Theorems (3.12) and (3.26), is not only of theoretical interest. It also opens up some new perspectives, since it allows us to define and describe events that capture the long-term behaviour of a random process. Let $(\Omega, \mathscr{F}, P)$ be a probability space and $(Y_k)_{k \geq 1}$ a sequence of random variables on $(\Omega, \mathscr{F})$ with values in arbitrary event spaces $(\Omega_k, \mathscr{F}_k)$.

Definition. An event $A \in \mathscr{F}$ is called an *asymptotic* or *tail event* for $(Y_k)_{k \geq 1}$ if, for every $n \geq 0$, A depends only on $(Y_k)_{k > n}$, in that there exists an event $B_n \in \bigotimes_{k > n} \mathscr{F}_k$ such that

(3.46) $A = \{(Y_k)_{k > n} \in B_n\}.$

Let $\mathscr{T}(Y_k : k \geq 1)$ be the collection of all such tail events.

We immediately note that $\mathscr{T}(Y_k : k \geq 1)$ is a sub-σ-algebra of $\mathscr{F}$; it is called the *asymptotic* or *tail σ-algebra* of the sequence $(Y_k)_{k \geq 1}$. Naively, one might think: Since an event $A \in \mathscr{T}(Y_k : k \geq 1)$ cannot depend on $Y_1, \ldots, Y_n$ and this is true for all n, A 'cannot depend on anything', and so one could conclude that $A = \Omega$ or $A = \varnothing$. This kind of reasoning, however, is completely mistaken! In fact, $\mathscr{T}(Y_k : k \geq 1)$ contains a particularly interesting class of events, namely those that describe the asymptotic behaviour of $(Y_k)_{k \geq 1}$. To see this, consider the following examples.

(3.47) Example. *Lim sup of events.* For arbitrary $A_k \in \mathscr{F}_k$ let

$$A = \{Y_k \in A_k \text{ for infinitely many k}\} = \bigcap_{m \geq 1} \bigcup_{k \geq m} \{Y_k \in A_k\}.$$

In view of the relation $1_A = \limsup_{k \to \infty} 1_{\{Y_k \in A_k\}}$ for the indicator functions, one writes $A = \limsup_{k \to \infty} \{Y_k \in A_k\}$ and calls A the lim sup of the events $\{Y_k \in A_k\}$. Alternatively, one writes $A = \{Y_k \in A_k \text{ i.o.}\}$, where 'i.o.' stands for 'infinitely often'. Now, whether or not something happens infinitely often does not depend on the first n instances. In other words, every such A is a tail event for $(Y_k)_{k \geq 1}$. To check this formally, take any $n \geq 0$ and let $X_i : \prod_{k > n} \Omega_k \to \Omega_i$ be the projection onto the ith coordinate. Then the event

$$B_n = \bigcap_{m > n} \bigcup_{k \geq m} \{X_k \in A_k\}$$

belongs to $\bigotimes_{k > n} \mathscr{F}_k$, and (3.46) holds.

(3.48) Example. *Existence of long-term averages.* Let $(\Omega_k, \mathscr{F}_k) = (\mathbb{R}, \mathscr{B})$ for all k, and $a < b$. Then the event

$$A = \left\{ \lim_{N \to \infty} \frac{1}{N} \sum_{k=1}^{N} Y_k \text{ exists and belongs to } [a, b] \right\}$$

is a tail event for $(Y_k)_{k \geq 1}$. This is because the existence and the value of a long-term average is not affected by a shift of all indices, so that (3.46) holds for the events

$$B_n = \left\{ \lim_{N \to \infty} \frac{1}{N} \sum_{k=1}^{N} X_{n+k} \text{ exists and belongs to } [a, b] \right\};$$

here, $X_i : \prod_{k>n} \mathbb{R} \to \mathbb{R}$ is again the ith projection. It only remains to be observed that B_n does indeed belong to $\bigotimes_{k>n} \mathscr{B}$; recall Problem 1.14.

Quite remarkably, if the random variables $(Y_k)_{k \geq 1}$ are *independent*, it turns out that the events in $\mathscr{T}(Y_k : k \geq 1)$ (although far from being trivial in general) are nevertheless 'trivial in probability'.

(3.49) Theorem. Kolmogorov's zero-one law. *Let $(Y_k)_{k \geq 1}$ be independent random variables on a probability space $(\Omega, \mathscr{F}, P)$ with values in arbitrary event spaces $(\Omega_k, \mathscr{F}_k)$. Then, for every $A \in \mathscr{T}(Y_k : k \geq 1)$, either $P(A) = 0$ or $P(A) = 1$.*

Proof. Fix $A \in \mathscr{T}(Y_k : k \geq 1)$ and let $\mathscr{G}$ be the collection of all sets $C \subset \prod_{k \geq 1} \Omega_k$ of the form

$$C = \{X_1 \in C_1, \dots, X_n \in C_n\}, \quad n \geq 1, \ C_k \in \mathscr{F}_k.$$

Then $\mathscr{G}$ is an intersection-stable generator of $\bigotimes_{k \geq 1} \mathscr{F}_k$; see the analogous result in Problem 1.5. For $C = \{X_1 \in C_1, \dots, X_n \in C_n\} \in \mathscr{G}$, Theorem (3.24) implies that the indicator function $1_{\{(Y_k)_{k \geq 1} \in C\}} = \prod_{k=1}^{n} 1_{C_k}(Y_k)$ is independent of $1_A = 1_{\{(Y_k)_{k > n} \in B_n\}}$. Hence, by Theorem (3.19) it follows that $(Y_k)_{k \geq 1}$ is independent of 1_A. Thus, $A = \{(Y_k)_{k \geq 1} \in B_0\}$ is independent of $A = \{1_A = 1\}$, which means that $P(A \cap A) = P(A)P(A)$ and thus $P(A) = P(A)^2$. But the equation $x = x^2$ only admits the solutions 0 and 1. $\diamond$

Knowing that a tail event for an independent sequence is either certain or impossible, one would like to decide which of these alternatives is true. In general, this can only be done ad hoc as the case arises. For events as in Example (3.47), however, an easy criterion is available.

(3.50) Theorem. Borel–Cantelli lemma, 1909/1917. *Let $(A_k)_{k \geq 1}$ be a sequence of events in a probability space $(\Omega, \mathscr{F}, P)$, and consider*

$$A := \{\omega \in \Omega : \omega \in A_k \text{ for infinitely many } k\} = \limsup_{k \to \infty} A_k.$$

(a) *If $\sum_{k \geq 1} P(A_k) < \infty$, then $P(A) = 0$.*

(b) *If $\sum_{k \geq 1} P(A_k) = \infty$ and $(A_k)_{k \geq 1}$ is independent, then $P(A) = 1$.*

Note that no independence is required for statement (a).

Proof. (a) We have that $A \subset \bigcup_{k \geq m} A_k$ and thus $P(A) \leq \sum_{k \geq m} P(A_k)$ for all m. The last sum is the tail of a convergent series, and thus tends to 0 as $m \to \infty$.

(b) Since $A^c = \bigcup_{m \geq 1} \bigcap_{k \geq m} A_k^c$, we can write

$$P(A^c) \leq \sum_{m \geq 1} P\left(\bigcap_{k \geq m} A_k^c\right) = \sum_{m \geq 1} \lim_{n \to \infty} P\left(\bigcap_{k=m}^{n} A_k^c\right)$$

$$= \sum_{m \geq 1} \lim_{n \to \infty} \prod_{k=m}^{n} [1 - P(A_k)]$$

$$\leq \sum_{m \geq 1} \lim_{n \to \infty} \exp\left[-\sum_{k=m}^{n} P(A_k)\right] = \sum_{m \geq 1} 0 = 0,$$

provided $\sum_{k \geq 1} P(A_k) = \infty$. Here we have used that the complements $(A_k^c)_{k \geq 1}$ are independent by Corollary (3.20), and that $1 - x \leq e^{-x}$. $\diamond$

The Borel–Cantelli lemma will become important later on in Sections 5.1.3 and 6.4. At this point, we will only present a simple application to number theory.

(3.51) Example. *Divisibility by prime numbers.* For each prime p let A_p be the set containing all multiples of p. Then there is *no* probability measure P on $(\mathbb{N}, \mathscr{P}(\mathbb{N}))$ for which the events A_p are independent with $P(A_p) = 1/p$. Indeed, suppose there were such a P. Since $\sum_{p \text{ prime}} 1/p = \infty$, part (b) of the Borel–Cantelli lemma would then imply that the impossible event

$$A = \{n \in \mathbb{N} : n \text{ is a multiple of infinitely many primes}\}$$

had probability 1.

Problems

3.1 A shop is equipped with an alarm system which in case of a burglary alerts the police with probability 0.99. During a night without a burglary, a false alarm is set off with probability 0.002 (e.g. by a mouse). The probability of a burglary on a given night is 0.0005. An alarm has just gone off. What is the probability that there is a burglary going on?

3.2 *Prisoners' paradox.* Three prisoners Andy, Bob and Charlie are sentenced to death. By drawing lots, where each had the same chance, one of the prisoners was granted pardon. The prisoner Andy, who has a survival probability of $1/3$, asks the guard, who knows the result, to tell him one of his fellow sufferers who has to die. The guard answers 'Bob'. Now Andy calculates: 'Since either me or Charlie are going to survive, I have a chance of 50%.' Would you agree with him? (For the construction of the probability space, suppose that the guard answers 'Bob' or 'Charlie' with equal probability, if he knows that Andy has been granted pardon.)

3.3 You are flying from Munich to Los Angeles and stop over in London and New York. At each airport, including Munich, your suitcase must be loaded onto the plane. During this process, it will get lost with probability p. In Los Angeles you notice that your suitcase hasn't arrived. Find the conditional probability that it got lost in Munich, resp. London, resp. New York. (As always, a complete solution includes a description of the probability model.)

3.4 ^{S} *Beta-binomial representation of the Pólya distribution.* Consider Pólya's urn model with parameters $a = r/c > 0$ and $b = w/c > 0$. Let R_n be the number of red balls obtained in n draws. Use the recursive formula (2.23) to show that

$$P(R_n = \ell) = \int_0^1 dp \; \beta_{a,b}(p) \; \mathcal{B}_{n,p}(\{\ell\})$$

for all $0 \leq \ell \leq n$. (Hence, the Pólya model is equivalent to an urn model with replacement where the initial fraction of red balls was determined by 'chance' according to a beta distribution.)

3.5 Generalise Pólya's urn model to the case when the balls can take colours from a finite set E (instead of only red and white), and find the distribution of the histogram R_n after n draws; cf. equation (2.7). Can you also generalise the previous Problem 3.4 to this case? (The corresponding generalisation of the beta distribution is called the *Dirichlet distribution*.)

3.6 Let $(\Omega, \mathscr{F}, P)$ be a probability space and $A, B, C \in \mathscr{F}$. Show directly (without using Corollary (3.20) and Theorem (3.24)):

(a) If A and B are independent, then so are A and B^c.

(b) If A, B, C are independent, then so are $A \cup B$ and C.

3.7 In number theory, *Euler's φ-function* is defined as the mapping $\varphi : \mathbb{N} \to \mathbb{N}$ such that $\varphi(1) = 1$ and $\varphi(n) =$ the number of integers in $\Omega_n = \{1, \ldots, n\}$ that are relatively prime to n, if $n \geq 2$. Show that if $n = p_1^{k_1} \ldots p_m^{k_m}$ is the prime factorisation of n into pairwise distinct primes $p_1, \ldots, p_m$ with powers $k_i \in \mathbb{N}$, then

$$\varphi(n) = n \left(1 - \frac{1}{p_1} \right) \ldots \left(1 - \frac{1}{p_m} \right).$$

Hint: Consider the events $A_i = \{p_i, 2p_i, 3p_i, \ldots, n\}$, $1 \leq i \leq m$.

3.8 ^{S} Let X be a real-valued random variable on a probability space $(\Omega, \mathscr{F}, P)$. Show that X is independent of itself if and only if X is constant with probability 1, i.e., if there exists a constant $c \in \mathbb{R}$ such that $P(X = c) = 1$. *Hint:* Consider the distribution function of X.

3.9 Let X, Y be independent random variables that are exponentially distributed with parameter $\alpha > 0$. Find the distribution density of $X/(X + Y)$.

3.10 A system consists of four components that are similar but work independently. To operate properly, it is necessary that (A and B) or (C and D) are working.

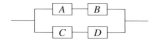

Let T be the failure time of the complete system, and T_k the failure time of component $k \in \{A, B, C, D\}$. Suppose the T_k have the exponential distribution with parameter α. Show that

$$P(T < t) = \left(1 - e^{-2\alpha t}\right)^2.$$

3.11 Consider a fair tetrahedral dice, whose faces are numbered by 1, 2, 3, 4, and which is thrown twice. Let X be the sum and Y the maximum of the two respective numbers on the faces falling downside.

(a) Find the joint distribution $P \circ (X, Y)^{-1}$ of X and Y.

(b) Construct two random variables X' and Y' on a suitable probability space $(\Omega', \mathscr{F}', P')$, which have the same distributions as X and Y (i.e., $P \circ X^{-1} = P' \circ X'^{-1}$ and $P \circ Y^{-1} = P' \circ Y'^{-1}$), but so that the distribution of $X' + Y'$ differs from that of $X + Y$.

3.12 ^{S} Let X, Y be i.i.d. random variables taking values in $\mathbb{Z}_+$. Suppose that either

(a) $P(X = k \mid X + Y = n) = 1/(n+1)$ for all $0 \le k \le n$, or

(b) $P(X = k \mid X + Y = n) = \binom{n}{k} 2^{-n}$ for all $0 \le k \le n$,

provided n is such that the condition has positive probability. Find the common distribution of X and Y.

3.13 *Coin tossing paradox.* Alice suggests the following game to Bob: 'You randomly choose two integers $X, Y \in \mathbb{Z}$ with $X < Y$. Then you toss a fair coin. If it shows tails, you tell me Y, otherwise X. Then I will guess whether the coin showed tails or heads. If my guess is right, you pay me €100, otherwise you get €100 from me.' Should Bob agree to play the game? (After all, he can freely dispose of the distribution β according to which he picks (X, Y), and isn't it clear that the chance of guessing the result of a fair coin toss is at best $50:50$?) To answer this, consider the following guessing strategy for Alice: Alice picks a random number $Z \in \mathbb{Z}$ with distribution α, where α is any discrete density on $\mathbb{Z}$ with $\alpha(k) > 0$ for every $k \in \mathbb{Z}$. She guesses that the coin was showing tails if the number Bob is announcing is at least Z, otherwise she guesses heads. Set up a stochastic model and find the winning probability for Alice when α and β are given.

3.14 ^{S} *Dice paradox.* Two dice D_1 and D_2 are labelled as follows.

$$D_1 : 6\,3\,3\,3\,3\,3, \qquad\qquad D_2 : 5\,5\,5\,2\,2\,2.$$

Andy and Beth roll D_1 and D_2 respectively. Whoever gets the higher number wins.

(a) Show that Andy has a higher probability of winning; we write this as $D_1 \succ D_2$.

(b) Beth notices this and suggests to Andy: 'I am now going to label a third dice. You can then choose an arbitrary dice and I will take one of the remaining two.' Can Beth label the third dice in such a way that she can always choose a dice with a better chance of winning, i.e., so that $D_1 \succ D_2 \succ D_3 \succ D_1$, meaning that the relation $\succ$ is not transitive?

3.15 *Convolutions of gamma and negative binomial distributions.* Show that

(a) for $\alpha, r, s > 0$, we have $\Gamma_{\alpha,r} \star \Gamma_{\alpha,s} = \Gamma_{\alpha,r+s}$;

(b) for $p \in {]0, 1[}$ and $r, s > 0$ we have $\overline{\mathcal{B}}_{r,p} \star \overline{\mathcal{B}}_{s,p} = \overline{\mathcal{B}}_{r+s,p}$. *Hint:* The Pólya distribution provides you with a useful identity for negative binomial coefficients.

3.16 ^{S} *Convolution of Cauchy distributions (Huygens' principle).* Consider the situation in Problem 2.5 and show that $c_a \star c_b = c_{a+b}$ for $a, b > 0$. In other words, the distribution of light on a straight line at distance $a+b$ from the light source is the same as if every light point X on the straight line at distance a is treated as a new and uniformly radiating light source that adds a contribution Y to the total deflection on the second line.

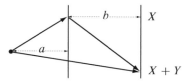

Hint: Verify the partial fraction decomposition

$$c_a(y)c_b(x-y)/c_{a+b}(x) = \frac{b}{a+b}\frac{x^2+b^2-a^2+2xy}{x^2+(a-b)^2}c_a(y) + \frac{a}{a+b}\frac{x^2+a^2-b^2+2x(x-y)}{x^2+(a-b)^2}c_b(x-y)$$

and use that $\lim_{n\to\infty}\int_{x-n}^{x+n} z\, c_b(z)\, dz = 0$ for all x.

3.17 *Thinning of a Poisson distribution.* Suppose that the number of eggs an insect lays is Poisson distributed with parameter λ. Out of each egg, a larva hatches with probability p, independently of all other eggs. Find the distribution of the number of larvae.

3.18 *Thinning of a Poisson process.* Let $\alpha > 0$, $(L_i)_{i\geq 1}$ be a sequence of i.i.d. random variables that are exponentially distributed with parameter α, and let $T_k = \sum_{i=1}^{k} L_i$, $k \geq 1$. Furthermore, let $(X_k)_{k\geq 1}$ be a Bernoulli sequence with parameter $p \in {]}0, 1{[}$, which is independent of the L_i. Show that the random variables

$$N_t^X := \sum_{k\geq 1} X_k\, 1_{]0,t]}(T_k), \quad t \geq 0,$$

form a Poisson process with parameter $p\alpha$. In particular, $T_1^X := \inf\{t > 0 : N_t^X \geq 1\}$ is exponentially distributed with parameter $p\alpha$.

3.19 *Telegraph process.* Let $(N_t)_{t\geq 0}$ be a Poisson process with intensity $\alpha > 0$ and $Z_t = (-1)^{N_t}$. Show that $P(Z_s = Z_t) = (1+e^{-2\alpha(t-s)})/2$ for $0 \leq s < t$.

3.20 [S] *Bernoulli sequence as a discrete analogue of the Poisson process.*
 (a) Let $(X_n)_{n\geq 1}$ be a Bernoulli sequence for $p \in {]}0, 1{[}$ and let

$$T_0 = 0, \quad T_k = \inf\{n > T_{k-1} : X_n = 1\}, \quad L_k = T_k - T_{k-1} - 1$$

for $k \geq 1$. (T_k is the time of the kth success and L_k the waiting time between the $(k-1)$st and the kth success.) Show that the random variables $(L_k)_{k\geq 1}$ are independent and have the geometric distribution with parameter p.
 (b) Let $(L_i)_{i\geq 1}$ be an i.i.d. sequence of random variables that are geometrically distributed with parameter $p \in {]}0, 1{[}$. For $k \geq 1$ let $T_k = \sum_{i=1}^{k} L_i + k$, and for $n \geq 1$ define

$$X_n = \begin{cases} 1 & \text{if } n = T_k \text{ for some } k \geq 1, \\ 0 & \text{otherwise.} \end{cases}$$

Show that the random variables $(X_n)_{n\geq 1}$ form a Bernoulli sequence with parameter p.

3.21 Let $(N_t)_{t\geq 0}$ be a Poisson process and $0 < s < t$. Find the conditional probability $P(N_s = k | N_t = n)$ for $0 \leq k \leq n$.

3.22 In a service centre with s different counters, customers arrive at the times of independent Poisson processes $(N_t^{(i)})_{t \geq 0}$ with intensities $\alpha(i) > 0$, $1 \leq i \leq s$. At time t, you observe that a total of n customers is waiting. What is the conditional distribution of the s-tuple of the customers waiting in front of each of the counters?

3.23 ^{S} *Comparison of independent Poisson processes.* Let $(N_t)_{t \geq 0}$ and $(\tilde{N}_t)_{t \geq 0}$ be two independent Poisson processes with intensities α resp. $\tilde{\alpha}$ and jump times (T_k) resp. $(\tilde{T}_k)$. Show the following.

  (a) $N_{\tilde{T}_1}$ has a geometric distribution (with which parameter?).

  (b) The random variables $N_{\tilde{T}_k} - N_{\tilde{T}_{k-1}}$, $k \geq 1$, form an i.i.d. sequence; here we set $\tilde{T}_0 := 0$.

  (c) Deduce from (b) that $X_n := 1_{\{\tilde{N}_{T_n} > \tilde{N}_{T_{n-1}}\}}$, $n \geq 1$, is a Bernoulli sequence. How does this provide an explanation for (a)?

3.24 Let $(S_t)_{t \geq 0}$ be the compound Poisson process with jump distribution Q and intensity $\alpha > 0$. Show that, for fixed $t > 0$, S_t has the distribution

$$Q_t := e^{-\alpha t} \sum_{n \geq 0} \frac{(\alpha t)^n}{n!} \, Q^{\star n} \, .$$

Here, $Q^{\star 0} = \delta_0$ is the Dirac distribution at 0.

3.25 ^{S} *Construction of the Poisson point process in $\mathbb{R}^d$.* Let $\Lambda \subset \mathbb{R}^d$ be a Borel set with $0 < \lambda^d(\Lambda) < \infty$, and $\alpha > 0$. Also, let $(X_i)_{i \geq 1}$ be a sequence of i.i.d. random variables with uniform distribution on Λ, and N_Λ a Poisson random variable with parameter $\alpha \lambda^d(\Lambda)$, which is independent of $(X_i)_{i \geq 1}$. Consider the random points $X_i \in \Lambda$ for $i \leq N_\Lambda$. For every Borel set $B \subset \Lambda$ let N_B be the number of points in B. That is,

$$N_B = \sum_{i=1}^{N_\Lambda} 1_B(X_i) \, .$$

Check that N_B is a random variable. Show further that, for every partition $\Lambda = \bigcup_{j=1}^n B_j$ of Λ into disjoint Borel sets $B_j \in \mathscr{B}_\Lambda^d$, the random variables $(N_{B_j})_{1 \leq j \leq n}$ are independent and Poisson distributed with parameter $\alpha \lambda(B_j)$. Finally, use a computer to produce a random sample of the Poisson point process on a square $\Lambda = [0, L]^2$, either by the above construction or that of Example (3.38).

  (To see that these constructions are equivalent, consider the process $N_t^{(1)} := N_{]0,t] \times [0,L]}$, which describes the first coordinates of all points. By the above, this is exactly the Poisson process on $[0, L]$ with intensity αL.)

3.26 *Box–Muller method for sampling from normal distributions, 1958.* Let U, V be independent random variables with uniform distribution on $]0, 1[$, and define $R = \sqrt{-2 \log U}$, $X = R \cos(2\pi V)$, and $Y = R \sin(2\pi V)$. Show that X, Y are independent and $\mathcal{N}_{0,1}$-distributed. *Hint:* First calculate the distribution density of R and then use the polar coordinate transformation of double integrals.

3.27 ^{S} *Failure times.* Determine the random life span of a wire rope (or any kind of technical appliance) as follows. For $t > 0$ let $F(t) := P(]0, t])$ be the probability that the rope fails in the time interval $]0, t]$, and suppose P has a Lebesgue density ρ. Suppose further that the conditional probability for failure of the rope in an infinitesimal time interval $[t, t+dt[$, provided it has not failed yet, is equal to $r(t)\,dt$ for some continuous function $r : [0, \infty[\to [0, \infty[$, the so-called *failure rate function*. Find a differential equation for F and solve it. Compute ρ. Which distribution do you obtain in the case of a constant failure rate r? If $r(t) = \alpha\beta t^{\beta-1}$ for some constants $\alpha, \beta > 0$, one obtains the so-called *Weibull distribution* with density

$$\rho(t) = \alpha\beta\, t^{\beta-1} \exp[-\alpha\, t^\beta], \quad t > 0.$$

3.28 Find all the probability measures P on $[0, \infty[$ satisfying the following property: If $n \in \mathbb{N}$ is arbitrary and $X_1, \ldots, X_n$ are independent random variables with identical distribution P, then the random variable $n\,\min(X_1, \ldots, X_n)$ also has distribution P. *Hint:* Start by finding an equation for $\overline{F}(t) := P(]t, \infty[)$.

3.29 ^{S} Let Y_k, $k \geq 1$, be $[0, \infty[$-valued random variables on a probability space $(\Omega, \mathscr{F}, P)$, and consider the events

$$A_1 = \left\{\sum_{k\geq 1} Y_k < \infty\right\}, \qquad A_2 = \left\{\sum_{k\geq 1} Y_k < 1\right\},$$
$$A_3 = \left\{\inf_{k\geq 1} Y_k < 1\right\}, \qquad A_4 = \left\{\liminf_{k\to\infty} Y_k < 1\right\}.$$

Which of these belong to the tail σ-algebra $\mathscr{T}(Y_k : k \geq 1)$? Decide and give proofs.

3.30 Let $(X_k)_{k\geq 1}$ be a Bernoulli sequence for $p \in]0, 1[$. For $n, l \in \mathbb{N}$, let A_n^l be the event $\{X_n = X_{n+1} = \cdots = X_{n+l-1} = 1\}$ that a run of luck of length at least l starts at time n, and let $A^l = \limsup_{n\to\infty} A_n^l$. Show that $P(\bigcap_{l\in\mathbb{N}} A^l) = 1$. Hence, with probability 1 there exist infinitely many runs of luck of arbitrary length.

3.31 ^{S} *Oscillations of the simple symmetric random walk*, cf. Problem 2.7. Let $(X_i)_{i\geq 1}$ be a sequence of independent random variables which are uniformly distributed on $\{-1, 1\}$, and set $S_n = \sum_{i=1}^n X_i$ for $n \geq 1$. Show that, for all $k \in \mathbb{N}$,

$$P\big(|S_{n+k} - S_n| \geq k \text{ for infinitely many } n\big) = 1.$$

Deduce that $P(|S_n| \leq m \text{ for all } n) = 0$ for all m, and further (using the symmetry of the X_i) that

$$P\Big(\sup_{n\geq 1} S_n = \infty,\ \inf_{n\geq 1} S_n = -\infty\Big) = 1.$$

That is, the random walk (S_n) fluctuates back and forth through $\mathbb{Z}$ up to its 'extreme ends'.

Chapter 4

Expectation and Variance

Real-valued random variables have two fundamental characteristics: the expectation, which gives the 'average' or 'typical' value of the random variable, and the variance, which measures how far the values of the random variable typically deviate from the expectation. These quantities and their properties are the subject of this chapter. As first applications of the notion of expectation, we treat the waiting time paradox and – as a taster from financial mathematics – the theory of option pricing. Moreover, we will consider generating functions of integer-valued random variables, a versatile analytical encoding of their distributions which can also be used to calculate their expectations and variances in a convenient way.

4.1   The Expectation

The expectation of real-valued random variables is defined in two stages. The first deals with the elementary case of random variables that take at most countably many different values.

4.1.1   The Discrete Case

Let $(\Omega, \mathscr{F}, P)$ be a probability space and $X : \Omega \to \mathbb{R}$ a real random variable. X is called *discrete* if its range $X(\Omega) := \{X(\omega) : \omega \in \Omega\}$ is at most countable.

Definition. A discrete random variable X is said to have an expectation if

$$\sum_{x \in X(\Omega)} |x| \; P(X = x) < \infty .$$

In this case, the sum

$$\mathbb{E}(X) = \mathbb{E}_P(X) := \sum_{x \in X(\Omega)} x \; P(X = x)$$

is well-defined and is called the *expectation* of X, and one writes $X \in \mathscr{L}^1(P)$ or, if P is understood, $X \in \mathscr{L}^1$.

Here are two basic observations.

(4.1) Remark. *Expectation as a distributional quantity.* The expectation depends solely on the distribution $P \circ X^{-1}$ of X. Indeed, by definition we have $X \in \mathscr{L}^1(P)$ if and only if the identity mapping $\mathrm{Id}_{X(\Omega)} : x \to x$ on $X(\Omega)$ belongs to $\mathscr{L}^1(P \circ X^{-1})$, and in this case we have $\mathbb{E}_P(X) = \mathbb{E}_{P \circ X^{-1}}(\mathrm{Id}_{X(\Omega)})$. In particular, this gives rise to the following physical interpretation of the expectation: If $P \circ X^{-1}$ is thought of as a discrete mass distribution (with total mass 1) on the (weightless) real axis $\mathbb{R}$, then $\mathbb{E}(X)$ is exactly the centre of gravity of $P \circ X^{-1}$; cf. Figure 4.1.

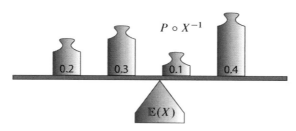

Figure 4.1. $\mathbb{E}(X)$ as the centre of gravity of the mass distribution $P \circ X^{-1}$.

(4.2) Remark. *Expectation of non-negative random variables.* If X is discrete and non-negative, then the sum $\mathbb{E}(X) = \sum_{x \in X(\Omega)} x\, P(X = x)$ is always well-defined, but it might be infinite. In other words, non-negative random variables always have an expectation, as long as we admit the value $+\infty$. This is still true if X is allowed to take the value $+\infty$. The condition for the existence of the expectation of an arbitrary discrete random variable can therefore be rewritten as follows: We have $X \in \mathscr{L}^1(P)$ if and only if $\mathbb{E}(|X|) < \infty$.

We now calculate the expectation in some special cases.

(4.3) Example. *Indicator functions.* For each $A \in \mathscr{F}$, 1_A belongs to $\mathscr{L}^1(P)$, and

$$\mathbb{E}(1_A) = 0 \cdot P(1_A = 0) + 1 \cdot P(1_A = 1) = P(A).$$

This relation connects the notions of expectation and probability.

(4.4) Example. *Countable domain of definition.* Let Ω be countable. Then, every random variable $X : \Omega \to \mathbb{R}$ is discrete, and a reordering of summation gives

$$\sum_{\omega \in \Omega} |X(\omega)|\, P(\{\omega\}) = \sum_{x \in X(\Omega)} |x| \sum_{\omega \in \{X = x\}} P(\{\omega\}) = \sum_{x \in X(\Omega)} |x|\, P(X = x).$$

Consequently, $X \in \mathscr{L}^1(P)$ if and only if $\sum_{\omega \in \Omega} |X(\omega)|\, P(\{\omega\}) < \infty$. In this case, the absolute convergence of all series allows to repeat the same calculation without the modulus signs, and one finds

$$\mathbb{E}(X) = \sum_{\omega \in \Omega} X(\omega)\, P(\{\omega\}).$$

(4.5) Example. *Expected number of successes in Bernoulli experiments.* Let $X_1, \ldots, X_n$ be a finite Bernoulli sequence with success probability p, and let $S = \sum_{i=1}^{n} X_i$ be the number of successes. Then, by Theorem (2.9), $P \circ S^{-1} = \mathcal{B}_{n,p}$, and thus

$$\mathbb{E}(S) = \sum_{k=0}^{n} k \binom{n}{k} p^k (1-p)^{n-k} = \sum_{k=1}^{n} np \binom{n-1}{k-1} p^{k-1}(1-p)^{n-k}$$

$$= np \sum_{k=1}^{n} \mathcal{B}_{n-1,p}(\{k-1\}) = np .$$

(4.6) Example. *Mean waiting time for the first success.* Let $(X_n)_{n \geq 1}$ be a Bernoulli sequence with success probability p, and let $T = \inf\{n \geq 1 : X_n = 1\}$ be the waiting time for the first success. We know from Section 2.5.1 that $T-1$ is geometrically distributed, so setting $q = 1-p$ we find

$$\mathbb{E}(T) = \sum_{k \geq 1} k p q^{k-1} = p \frac{d}{ds} \sum_{k \geq 0} s^k \Big|_{s=q} = p \frac{d}{ds} \frac{1}{1-s}\Big|_{s=q} = p \frac{1}{(1-q)^2} = \frac{1}{p} .$$

In the second step, we have used that power series can be differentiated term-by-term within their radius of convergence.

The following theorem summarises the most important properties of the expectation. For simplicity, they are stated here for random variables in $\mathscr{L}^1(P)$ only, but they hold analogously for non-negative random variables that possibly take the value $+\infty$, see Remark (4.2).

(4.7) Theorem. Expectation rules. *Let $X, Y, X_n, Y_n : \Omega \to \mathbb{R}$ be discrete random variables in $\mathscr{L}^1$. Then the following holds.*

(a) Monotonicity. *If $X \leq Y$, then $\mathbb{E}(X) \leq \mathbb{E}(Y)$.*

(b) Linearity. *For every $c \in \mathbb{R}$, we have $cX \in \mathscr{L}^1$ and $\mathbb{E}(cX) = c\,\mathbb{E}(X)$. Moreover, $X + Y \in \mathscr{L}^1$, and $\mathbb{E}(X+Y) = \mathbb{E}(X) + \mathbb{E}(Y)$.*

(c) σ-Additivity and monotone convergence. *If every $X_n \geq 0$ and $X = \sum_{n \geq 1} X_n$, then $\mathbb{E}(X) = \sum_{n \geq 1} \mathbb{E}(X_n)$. On the other hand, if $Y_n \uparrow Y$ for $n \uparrow \infty$, it follows that $\mathbb{E}(Y) = \lim_{n \to \infty} \mathbb{E}(Y_n)$.*

(d) Product rule in the case of independence. *If X and Y are independent, then $XY \in \mathscr{L}^1$, and $\mathbb{E}(XY) = \mathbb{E}(X)\,\mathbb{E}(Y)$.*

Proof. (a) Using the definition of expectation and the σ-additivity of P, one finds

$$\mathbb{E}(X) = \sum_{x \in X(\Omega),\, y \in Y(\Omega)} x\, P(X = x, Y = y),$$

where the order of summation is irrelevant due to the absolute convergence of the series. Since by assumption $P(X = x, Y = y) = 0$ unless $x \leq y$, the sum is not greater than

$$\sum_{x \in X(\Omega),\ y \in Y(\Omega)} y\, P(X = x, Y = y) = \mathbb{E}(Y).$$

(b) The first statement follows directly from the definition. Next, by Problem 1.14, the sum $X + Y$ is a random variable, and it is again discrete, since its range is contained in the image of the countable set $X(\Omega) \times Y(\Omega)$ under the addition of coordinates. To see that $X + Y \in \mathscr{L}^1$, we distinguish the possible values of X and apply the triangle inequality to get

$$\sum_z |z|\, P(X + Y = z) = \sum_{z,x} |z|\, P(X = x, Y = z - x)$$

$$\leq \sum_{z,x} (|x| + |z - x|)\, P(X = x, Y = z - x).$$

After substituting y for $z - x$, the sum splits into two parts containing $|x|$ resp. $|y|$. Summing over y in the first part and over x in the second one, we arrive at the sum

$$\sum_x |x|\, P(X = x) + \sum_y |y|\, P(Y = y) = \mathbb{E}(|X|) + \mathbb{E}(|Y|),$$

which is finite by assumption. In particular, all the above sums are absolutely convergent. Hence, one can repeat the same calculation without the modulus signs, and thus obtains the additivity of the expectation.

(c) We will prove the first statement; the second statement is then obtained by applying the first one to $X_n = Y_{n+1} - Y_n$ and $X = Y - Y_1$.

For every N, we know that $X \geq S_N := \sum_{n=1}^N X_n$, which by (a) and (b) gives that $\mathbb{E}(X) \geq \sum_{n=1}^N \mathbb{E}(X_n)$. In the limit $N \to \infty$ it follows that $\mathbb{E}(X) \geq \sum_{n \geq 1} \mathbb{E}(X_n)$. To prove the reverse inequality, we choose an arbitrary $0 < c < 1$ and consider the random variable $\tau = \inf\{N \geq 1 : S_N \geq cX\}$. Since $S_N \uparrow X < \infty$, we can conclude that $\tau < \infty$. The random sum $S_\tau = \sum_{n=1}^\tau X_n$ is a discrete random variable, since its range $S_\tau(\Omega)$ is clearly contained in the union $S(\Omega) := \bigcup_{N \geq 1} S_N(\Omega)$ of the ranges of the random variables S_N, which are by (b) discrete. Using (a), the definition of expectation and the σ-additivity of P, we obtain

$$c\,\mathbb{E}(X) \leq \mathbb{E}(S_\tau) = \sum_{x \in S(\Omega)} x \sum_{N \geq 1} P(\tau = N,\ S_N = x) = \sum_{N \geq 1} \mathbb{E}\big(1_{\{\tau = N\}} S_N\big).$$

The change of the order of summation in the last step is allowed since all the terms are non-negative. For the last expression we find, using (b) and interchanging summations again,

$$\sum_{N \geq 1} \sum_{n=1}^{N} \mathbb{E}\big(1_{\{\tau=N\}} X_n\big) = \sum_{n \geq 1} \sum_{N \geq n} \sum_{x \in X_n(\Omega)} x\, P(\tau = N,\, X_n = x)$$

$$= \sum_{n \geq 1} \sum_{x \in X_n(\Omega)} x\, P(\tau \geq n,\, X_n = x) \leq \sum_{n \geq 1} \mathbb{E}(X_n)\,.$$

In the limit $c \to 1$ we get the required inequality.

Occasionally we will use property (c) even when X is allowed to take the value $+\infty$. To see that this is possible, one can argue as follows. If $P(X = \infty) = 0$, it follows straight from the definition that X has the same expectation as the finite random variable $X\, 1_{\{X < \infty\}}$, and the above argument can be applied to the latter. In the alternative case $P(X = \infty) = a > 0$, it is clear from the definition that $\mathbb{E}(X) = \infty$. On the other hand, by the σ-continuity of P, for every $K > 0$ there exists an $N = N(K)$ such that $P(S_N \geq K) \geq a/2$. Together with property (a), this implies

$$\sum_{n \geq 1} \mathbb{E}(X_n) \geq \mathbb{E}(S_N) \geq K\, P(S_N \geq K) \geq Ka/2\,.$$

Since K is arbitrary, it follows that $\sum_{n \geq 1} \mathbb{E}(X_n) = \infty$, and thus the required equality holds.

(d) By the same argument as in (b), XY is a discrete random variable. It belongs to $\mathscr{L}^1$ because

$$\sum_{z} |z|\, P(XY = z) = \sum_{z \neq 0} |z| \sum_{x \neq 0} P(X = x,\, Y = z/x)$$

$$= \sum_{x \neq 0,\, y \neq 0} |x|\,|y|\, P(X = x) P(Y = y) = \mathbb{E}(|X|)\, \mathbb{E}(|Y|) < \infty\,.$$

In the second step, we made use of the independence of X and Y. By the absolute convergence of all series involved, we can again repeat the same calculation without modulus signs to get the required identity. $\diamond$

Using the linearity of the expectation, one can calculate the expectation of a binomially distributed random variable in a simple alternative way.

(4.8) Example. *Expected number of successes in Bernoulli experiments.* Consider the situation of Example (4.5). Applying Theorem (4.7b) and Example (4.3), one obtains

$$\mathbb{E}(S) = \sum_{i=1}^{n} \mathbb{E}(X_i) = \sum_{i=1}^{n} P(X_i = 1) = np\,,$$

just as before.

4.1.2   The General Case

How can one define the expectation of random variables that are not discrete? In the general case, the expectation cannot be written down directly as a sum. Instead, the idea is to approximate a given random variable X by discrete random variables $X_{(n)}$ and to define the expectation of X as the limit of the expectations of $X_{(n)}$.

For a real random variable $X : \Omega \to \mathbb{R}$ and arbitrary $n \in \mathbb{N}$, we consider the $1/n$-discretisation

$$X_{(n)} := \lfloor nX \rfloor / n$$

of X; here $\lfloor x \rfloor$ stands for the largest integer not exceeding x. That is, $X_{(n)}(\omega) = \frac{k}{n}$ if $\frac{k}{n} \le X(\omega) < \frac{k+1}{n}$ with $k \in \mathbb{Z}$; cf. Figure 4.2. Obviously, $X_{(n)}$ is a discrete random variable.

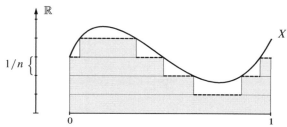

Figure 4.2. A real random variable X defined on $\Omega = [0, 1]$ and its $1/n$-discretisation $X_{(n)}$ (dashed). If $P = \mathcal{U}_{[0,1]}$, then $\mathbb{E}(X_{(n)})$ is exactly the area shaded in grey.

(4.9) Lemma. Discrete approximation of the expectation.

(a) *For every $n \ge 1$, the inequalities $X_{(n)} \le X < X_{(n)} + \frac{1}{n}$ are valid.*

(b) *If $X_{(n)} \in \mathcal{L}^1$ for some n, then $X_{(n)} \in \mathcal{L}^1$ for every n, and the expectations $\mathbb{E}(X_{(n)})$ form a Cauchy sequence.*

Proof. Statement (a) is obvious. In particular, for every $m, n \ge 1$, both the inequality $X_{(m)} < X_{(n)} + \frac{1}{n}$ and the reverse relation $X_{(n)} < X_{(m)} + \frac{1}{m}$ hold true. A first consequence is that $|X_{(n)}| < |X_{(m)}| + \max(\frac{1}{m}, \frac{1}{n})$. So if $X_{(m)} \in \mathcal{L}^1$ for some m, then Theorem (4.7a) allows to conclude that $X_{(n)} \in \mathcal{L}^1$ for every n, and further that $\mathbb{E}(X_{(m)}) \le \mathbb{E}(X_{(n)}) + \frac{1}{n}$. The last inequality also holds when the roles of m and n are interchanged. Hence $|\mathbb{E}(X_{(n)}) - \mathbb{E}(X_{(m)})| \le \max(\frac{1}{m}, \frac{1}{n})$, which proves (b). $\diamond$

The lemma suggests to proceed as follows in the general case.

Definition. Let $X : \Omega \to \mathbb{R}$ be an arbitrary real random variable. Then X is said to have an expectation if $X_{(n)} \in \mathcal{L}^1(P)$ for some (and thus every) $n \ge 1$. In this case,

$$\mathbb{E}(X) = \lim_{n \to \infty} \mathbb{E}(X_{(n)})$$

is called the *expectation* of X, and one says that X belongs to $\mathcal{L}^1 = \mathcal{L}^1(P)$.

In the theory of integration, the expectation defined in this way is called the integral of X with respect to P and it is denoted by $\int X\, dP$. As in Remark (4.1), one observes:

(4.10) Remark. *Expectation as a distributional quantity.* The relation $X \in \mathscr{L}^1(P)$ holds if and only if $\mathrm{Id}_{\mathbb{R}} \in \mathscr{L}^1(P \circ X^{-1})$, and then $\mathbb{E}_P(X) = \mathbb{E}_{P \circ X^{-1}}(\mathrm{Id}_{\mathbb{R}})$. In other words, the expectation depends only on the distribution $P \circ X^{-1}$ of X and can be interpreted as its centre of gravity. For the proof it suffices to write

$$\mathbb{E}(X_{(n)}) = \sum_{k \in \mathbb{Z}} \tfrac{k}{n}\, P\!\left(\tfrac{k}{n} \le X < \tfrac{k+1}{n}\right) = \mathbb{E}_{P \circ X^{-1}}\big((\mathrm{Id}_{\mathbb{R}})_{(n)}\big)$$

and let $n \to \infty$.

The point is now that the properties of the expectation carry over directly from the discrete to the general case.

(4.11) Theorem. *Expectation rules. The calculation rules* (a) *to* (d) *in Theorem* (4.7) *carry over from discrete to general real random variables.*

Proof. (a) The monotonicity of the expectation is evident, for if $X \le Y$, then $X_{(n)} \le Y_{(n)}$ for all n.

(b) For $c \in \mathbb{R}$, we have $|(cX)_{(n)} - cX| \le \tfrac{1}{n}$ as well as $|cX_{(n)} - cX| \le \tfrac{|c|}{n}$, so by the triangle inequality $|(cX)_{(n)} - cX_{(n)}| \le \tfrac{1+|c|}{n}$. Theorem (4.7a) thus shows that

$$\big|\mathbb{E}((cX)_{(n)}) - c\,\mathbb{E}(X_{(n)})\big| \le \tfrac{1+|c|}{n}\,,$$

and by sending $n \to \infty$ the first claim follows. The additivity of the expectation follows analogously by discrete approximation.

(c) Again, it is sufficient to prove the first statement. The inequality $\mathbb{E}(X) \ge \sum_{n \ge 1} \mathbb{E}(X_n)$ follows from (b), just as in the discrete case. To prove the reverse inequality, we choose an arbitrary $k \ge 1$ and consider the discrete random variables $Y_{n,k} = (X_n)_{(2^n + k)}$. Then we have $Y_{n,k} \le X_n < Y_{n,k} + 2^{-n-k}$ and thus

$$\mathbb{E}(X) \le \mathbb{E}\left(\sum_{n \ge 1}\big[Y_{n,k} + 2^{-n-k}\big]\right) = \sum_{n \ge 1} \mathbb{E}\big(Y_{n,k} + 2^{-n-k}\big) \le \sum_{n \ge 1} \mathbb{E}(X_n) + 2^{-k}\,.$$

Here, we first used the monotonicity property (a) in the general case, then the σ-additivity (c) in the discrete case (where it does not matter whether or not $\sum_{n \ge 1} Y_{n,k}$ is discrete, since the proof of Theorem (4.7c) can be repeated without this assumption by using the properties we have proved by now), and finally we used (a) and (b). In the limit $k \to \infty$ we obtain the required inequality.

(d) We note first that $|(XY)_{(n)} - X_{(n)} Y_{(n)}| \le \tfrac{1}{n} + \tfrac{1}{n}(|X| + |Y| + \tfrac{1}{n})$; this follows from the triangle inequality by comparing with XY and $XY_{(n)}$. Taking expectations and applying Theorem (4.7d) to the random variables $X_{(n)}$ and $Y_{(n)}$ (which are independent by Theorem (3.24)), we find that $\mathbb{E}((XY)_{(n)}) - \mathbb{E}(X_{(n)})\mathbb{E}(Y_{(n)}) \to 0$ as $n \to \infty$. This proves the claim. $\diamond$

If P has a Lebesgue density, the expectation $\mathbb{E}_P(X)$ can be expressed directly in terms of the Lebesgue integral in (1.14), and this expression is a direct analogue of the sum in Example (4.4).

(4.12) Theorem. *Expectation in the presence of a density. Let $\Omega \subset \mathbb{R}^d$ be a Borel set, P a probability measure on $(\Omega, \mathcal{B}_\Omega^d)$ with a Lebesgue density ρ, and X a real random variable on Ω. Then $X \in \mathcal{L}^1(P)$ if and only if $\int_\Omega |X(\omega)|\rho(\omega)\, d\omega < \infty$, and*

$$\mathbb{E}(X) = \int_\Omega X(\omega)\rho(\omega)\, d\omega\,.$$

Proof. For every n, we see that $X_{(n)} \in \mathcal{L}^1(P)$ if and only if the expression

$$\sum_{k \in \mathbb{Z}} \left|\tfrac{k}{n}\right| P\left(X_{(n)} = \tfrac{k}{n}\right) = \sum_{k \in \mathbb{Z}} \left|\tfrac{k}{n}\right| \int_{\{\frac{k}{n} \le X < \frac{k+1}{n}\}} \rho(\omega)\, d\omega$$

$$= \int_\Omega |X_{(n)}(\omega)|\, \rho(\omega)\, d\omega$$

is finite, and since $|X - X_{(n)}| \le \tfrac{1}{n}$ this is exactly the case if $\int_\Omega |X(\omega)|\rho(\omega)\, d\omega < \infty$. Furthermore, we have

$$\mathbb{E}(X_{(n)}) = \int_\Omega X_{(n)}(\omega)\rho(\omega)\, d\omega \to \int_\Omega X(\omega)\rho(\omega)\, d\omega\,,$$

because $\int_\Omega X_{(n)}(\omega)\rho(\omega)\, d\omega \le \int_\Omega X(\omega)\rho(\omega)\, d\omega < \int_\Omega X_{(n)}(\omega)\rho(\omega)\, d\omega + \tfrac{1}{n}$. $\diamond$

(4.13) Corollary. *Random variables with distribution density. Let X be a random variable with values in $\mathbb{R}^d$ and distribution density ρ, in other words, suppose $P \circ X^{-1}$ has the Lebesgue density ρ on $\mathbb{R}^d$. For any other random variable $f : \mathbb{R}^d \to \mathbb{R}$, it then follows that $f \circ X \in \mathcal{L}^1$ if and only if $\int_{\mathbb{R}^d} |f(x)|\, \rho(x)\, dx < \infty$, and then*

$$\mathbb{E}(f \circ X) = \int_{\mathbb{R}^d} f(x)\rho(x)\, dx\,.$$

Proof. If $f \ge 0$ or $f \circ X \in \mathcal{L}^1$, Remark (4.10) can be used twice and combined with Theorem (4.12) to obtain

$$\mathbb{E}_P(f \circ X) = \mathbb{E}_{P \circ (f \circ X)^{-1}}(\mathrm{Id}_\mathbb{R}) = \mathbb{E}_{P \circ X^{-1}}(f) = \int_{\mathbb{R}^d} f(x)\rho(x)\, dx\,.$$

In the general case, apply this result to $|f|$. $\diamond$

(4.14) Example. *Gamma distribution.* Let $X : \Omega \to [0, \infty[$ be gamma distributed with parameters $\alpha, r > 0$, which means that X has the distribution density

$$\gamma_{\alpha,r}(x) = \alpha^r x^{r-1} e^{-\alpha x} / \Gamma(r)$$

on $]0, \infty[$. Corollary (4.13) (with $f(x) = x$) then implies that

$$\mathbb{E}(X) = \int_0^\infty x\, \gamma_{\alpha,r}(x)\, dx = \frac{\Gamma(r+1)}{\alpha \Gamma(r)} \int_0^\infty \gamma_{\alpha,r+1}(x)\, dx = \frac{r}{\alpha} \, ;$$

here we have used that $\Gamma(r+1) = r\,\Gamma(r)$, and integrating over $\gamma_{\alpha,r+1}$ gives 1. In particular, since $X \geq 0$, this calculation shows that $X \in \mathscr{L}^1$.

Before completing this section, let us mention another notion that can be interpreted, just like the expectation, as the 'average value' of a real random variable X. In contrast to the expectation, this notion has the advantage of being always well-defined and being less sensitive to values of X that have a large modulus.

Definition. Let X be a real random variable with distribution Q on $(\mathbb{R}, \mathscr{B})$. A number $\mu \in \mathbb{R}$ is called a *median* of X, or of Q, if $P(X \geq \mu) \geq 1/2$ and $P(X \leq \mu) \geq 1/2$.

Hence, the median splits the distribution Q of X into two halves. Equivalently, a median μ is a point where the distribution function F_X of X crosses the level $1/2$ (or jumps over it). Thus, the existence of a median follows immediately from (1.29). Although F_X is increasing, it is not necessarily strictly increasing, so, in general, μ is not uniquely determined. Its meaning will become particularly clear in the following context.

(4.15) Example. *Radioactive decay.* The decay time of a radioactive particle is well approximated by an exponential random variable X. Thus we have $P(X \geq c) = e^{-\alpha c}$ for some constant $\alpha > 0$. Hence, the median of X is the number μ satisfying $e^{-\alpha\mu} = 1/2$, thus $\mu = \alpha^{-1} \log 2$. Now, if we anticipate the law of large numbers (5.6), it follows that μ is the time when a radioactive substance consisting of a very large number of (independently decaying) particles has decayed to half of its initial mass. So, μ is the so-called *half-life*.

4.2   Waiting Time Paradox and Fair Price of an Option

Intuitively, the expectation of a random variable X corresponds to its 'average value', or the price you would be prepared to pay in advance for a future pay-off X. The next two examples confirm this intuition, but they also show that surprises are waiting if one takes things too naively. (This section can be skipped at first reading.)

(4.16) Example. *The waiting time, or inspection, paradox.* At a bus stop, buses arrive at random times with average temporal distance $1/\alpha$, and there is a bus at time 0. We model this by making the following assumptions:

(a) $T_0 = 0$, and the interarrival times $L_k := T_k - T_{k-1}, k \geq 1$, are independent.

(b) For each $k \geq 1$, $\mathbb{E}(L_k) = 1/\alpha$, and L_k exhibits the 'lack of memory property' that $P(L_k > s + t \mid L_k > s) = P(L_k > t)$ for all $s, t \geq 0$.

We are interested in the expected waiting time for a passenger who arrives at the bus stop at time $t > 0$. Intuitively, there are two candidates for an answer:

(A) By the lack of memory property of the L_k, it does not matter when the last bus came before time t, hence the average waiting time is exactly the mean interarrival time between two buses, namely $1/\alpha$.

(B) The passenger's arrival time t is uniformly distributed in the random period between the arrivals of the last bus before t and the first one after t, hence the expected waiting time is only half as long as the mean interarrival time between two buses, namely $1/2\alpha$.

Which answer is right? Assumption (b) implies that L_k is exponentially distributed with parameter α. Indeed, the function $a(t) = P(L_k > t)$ is decreasing, and the assumption means that $a(t + s) = a(t)a(s)$. Hence, $a(t) = a(1)^t$ for all $t \geq 0$, which shows that L_k is exponentially distributed. Example (4.14) implies that α is the associated parameter. By Theorem (3.34), it follows that

$$N_s = \sum_{k \geq 1} 1_{]0,s]}(T_k)$$

is the Poisson process with parameter α. Let

$$W_t = \min\{T_k - t : k \geq 1, \, T_k \geq t\}$$

be the waiting time after t. Then, for all s, we obtain

$$P(W_t > s) = P(N_{t+s} - N_t = 0) = e^{-\alpha s},$$

i.e., W_t is exponentially distributed with parameter α, and thus $\mathbb{E}(W_t) = 1/\alpha$. Hence answer (A) is correct.

What is wrong with answer (B)? The time interval between the last bus before t and the first one after t has length $L_{(t)} := V_t + W_t$, where

$$V_t := \min\{t - T_k : k \geq 0, \, T_k < t\} \leq t - T_0 = t$$

denotes the time spent since the last bus before t has arrived, see Figure 4.3. For $s < t$ we find that $P(V_t > s) = P(N_t - N_{t-s} = 0) = e^{-\alpha s}$, and similarly $P(V_t = t) = e^{-\alpha t}$. By the uniqueness theorem (1.12), the distribution of V_t thus coincides with the distribution of $U \wedge t := \min(U, t)$, when U has the exponential distribution $\mathcal{E}_\alpha$. By

Figure 4.3. The spent interarrival time V_t, and the waiting time W_t.

Corollary (4.13), as applied to the function $f(x) = x \wedge t$, we get

$$
\mathbb{E}(V_t) = \mathbb{E}(U \wedge t) = \int_0^\infty x \wedge t \; \alpha e^{-\alpha x} \, dx
$$

$$
= \int_0^\infty x\alpha \, e^{-\alpha x} \, dx - \int_t^\infty (x - t)\alpha \, e^{-\alpha x} \, dx
$$

$$
= \frac{1}{\alpha} - \int_0^\infty y\alpha \, e^{-\alpha(y+t)} \, dy = \frac{1}{\alpha} - \frac{1}{\alpha} e^{-\alpha t} .
$$

Consequently, $\mathbb{E}(L_{(t)}) = \mathbb{E}(V_t) + \mathbb{E}(W_t) = \frac{2}{\alpha} - \frac{1}{\alpha} e^{-\alpha t}$. For large t, when the effect of the initial condition $T_0 = 0$ has more or less worn off, we thus find that $\mathbb{E}(L_{(t)}) \approx 2/\alpha$. So, what is wrong with answer (B) is not the intuition that the ratio $V_t/L_{(t)}$ is uniformly distributed on $]0, 1[$ (this is approximately correct for large t), but the tacit assumption that $L_{(t)}$ has expectation $1/\alpha$. Indeed, the time span $L_{(t)}$ is singled out from all other interarrival times by the fact that it is long enough to contain t, and this condition favours longer intervals and so roughly doubles the expected length!

This phenomenon is also important when, rather than bus arrival times, the T_k describe the times when a machine breaks down and is replaced by a new one. Then $\mathbb{E}(L_{(t)})$ is the expected life time of the machine that is in operation at time t, or that is inspected at that time. When observing $L_{(t)}$, the inspector is led to believe that the life time is almost twice as long as it really is. Therefore, instead of observing the life time of the machine operating at time t, it would be better to observe the first machine that is newly installed after time t.

(4.17) Example. *Theory of option pricing.* As a glimpse into financial mathematics, we will calculate here the fair price of an option in a simple market model. Let X_n be the (random) price of a certain share at time n. A European call option with maturity N and 'strike price' K is the right, sold by an investor (the so-called writer, e.g. a bank), to buy this share at time $N \geq 1$ at a price K. What is a fair price that the writer can or should demand for this right?

Apart from the random fluctuations in the share price, this price also depends on the market situation. We suppose that, apart from the share, there is a risk-free asset on the market, not depending on chance, a so called 'bond', which is freely available. We also suppose that the value of the bond is constant in time. (This does not mean that the interest rate is zero, but only that we are working in discounted units, which are adapted to the corresponding value of the bond.) For simplicity, we ignore all 'friction losses' (such as taxes and transaction costs) and additional profits (such as dividends), and assume that share and bond can be traded in arbitrary quantities at discrete time units. The profit for the buyer at time N is then

$$(X_N - K)_+ := \begin{cases} X_N - K & \text{if } X_N > K, \\ 0 & \text{otherwise.} \end{cases}$$

For, if $X_N > K$ the buyer makes use of his (or her) right and gains the difference between share price and strike price. Otherwise, he abandons his right and neither gains nor loses anything. Hence, the expected profit for the buyer is

$$\Pi := \mathbb{E}\big((X_N - K)_+\big).$$

Consequently, the writer can ask the price Π from the buyer.

This would be your first thought, but it is wrong! This was first noticed by F. Black and M. Scholes (1973), and they discovered the by now famous Black–Scholes formula for the fair price, see (5.27) on p. 138. In 1997, Scholes received the Nobel Prize in economics, together with R. Merton, who also made essential contributions to this topic (Black had just passed away).

> In order to understand what is wrong with Π, we consider a simple example. Let $N = 1$, $X_0 = K = 1$ and $X_1 = 2$ or $1/2$ each with probability $1/2$. Then $\Pi = (2-1)/2 = 1/2$. If the buyer is naive enough to buy the option at a price Π, then the writer can do the following: At time 0 he buys a 2/3-share at the unit share price 1 and sells 1/6 of a bond from his portfolio. Together with the income $1/2$ from selling the option, he has an even balance. At time 1, the writer rebuys his 1/6-bond and proceeds as follows with the share. If $X_1 = 2$, he buys an additional 1/3 share on the free market at the price of $X_1 = 2$ and sells the now complete share to the buyer at the agreed price of $K = 1$; then his balance is $-\frac{1}{6} - \frac{1}{3}2 + 1 = \frac{1}{6}$. If $X_1 = 1/2$, he sells his 2/3-share on the free market (since the buyer has no interest in exercising his option), and again his balance is $-\frac{1}{6} + \frac{2}{3}\frac{1}{2} = \frac{1}{6}$. This means the writer has exactly the same portfolio as before selling the option, and still he has made a risk-free profit of $1/6$! This 'arbitrage' opportunity shows that the price $\Pi = 1/2$ is not fair.

So what is the correct price for the option? And how can the writer find an appropriate strategy to eliminate his risk? To find out, we consider the following *binomial model by Cox–Ross–Rubinstein* (1979) for the time evolution of the share price. (This CRR model is far from realistic, but it is well-suited to demonstrate some useful concepts.) Let $\Omega = \{0, 1\}^N$, P_p the Bernoulli distribution on Ω with parameter $0 < p < 1$, and $Z_k : \Omega \to \{0, 1\}$ the kth projection. We suppose that the share price X_n at time n is given by the recursion

$$X_0 = 1, \quad X_n = X_{n-1} \exp[2\sigma Z_n - \mu] \quad \text{for } 1 \le n \le N;$$

the parameter $\sigma > 0$, the so-called *volatility*, determines the amplitude of the price changes per time unit, and μ depends on the choice of units (i.e., the increase in value

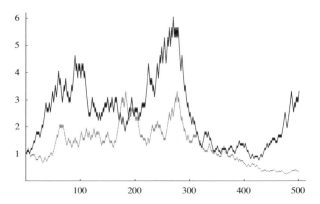

Figure 4.4. Two simulations of a geometric random walk.

of the bond). We assume that $0 < \mu < 2\sigma$, in other words, the share price X_n can increase as well as decrease. Of special interest is the case $\mu = \sigma$, where each price increase can be cancelled out by a subsequent price decrease. Then (X_n) is called a *geometric random walk*. Figure 4.4 shows two realisations that were obtained by using the `Mathematica` commands

```
GeomIrr[n_,s_]:=NestList[(#Exp[s(-1)^RandomInteger[]])&,1,n]
ListPlot[GeomIrr[500,0.05],Joined->True,AxesOrigin->{0,0}]
```

Now, let us consider the possible strategies for a market participant. For every $1 \leq n \leq N$, she can decide how many units of the share (say α_n) and how many of the bond (say β_n) she would like to keep in her portfolio during the time interval $]n-1, n]$. Here, α_n and β_n may be random, but can only depend on the previous price changes $\omega_1 = Z_1(\omega), \ldots, \omega_{n-1} = Z_{n-1}(\omega)$; in particular, α_1 and β_1 are constant. Any sequence of such mappings $\alpha_n, \beta_n : \Omega \to \mathbb{R}$ defines a strategy, which is denoted by the shorthand $\alpha\beta$. The initial capital in the portfolio is $V_0^{\alpha\beta} := \alpha_1 + \beta_1$, and the value of the portfolio at time $1 \leq n \leq N$ is

$$V_n^{\alpha\beta} = \alpha_n X_n + \beta_n \, .$$

A strategy $\alpha\beta$ is called *self-financing* if a rearrangement of the portfolio does not change the value for any time n, that is, if

$$(4.18) \qquad (\alpha_{n+1} - \alpha_n) X_n + (\beta_{n+1} - \beta_n) = 0$$

for every $1 \leq n < N$. In particular, β_{n+1} can be deduced from the other quantities, and $V_n^{\alpha\beta} = \alpha_{n+1} X_n + \beta_{n+1}$. A self-financing strategy $\alpha\beta$ is called a *hedging strategy* (for a writer) with initial value v, if

$$(4.19) \qquad V_0^{\alpha\beta} = v, \quad V_n^{\alpha\beta} \geq 0 \text{ for } 1 \leq n < N, \quad V_N^{\alpha\beta} \geq (X_N - K)_+ \, ;$$

that is, the value of the portfolio should never be negative, and at time N it should compensate for a possible loss. A natural fair price for the option is then given by

$$\Pi^* = \inf\{v > 0 : \text{there is a self-financing hedging strategy for } v\}\,.$$

This so-called Black–Scholes price Π^* can be calculated explicitly.

(4.20) Proposition. Black–Scholes price of a European option in the CRR model. *The Black–Scholes price Π^* of an option in the above CRR model is given by*

$$\Pi^* = \mathbb{E}^*\big((X_N - K)_+\big)\,.$$

Here, $\mathbb{E}^*$ is the expectation with respect to the Bernoulli distribution $P^* := P_{p^*}$ with parameter

$$p^* = \frac{e^\mu - 1}{e^{2\sigma} - 1}\,.$$

Moreover, there exists a self-financing hedging strategy with initial value Π^*.

Hence, the fair price is still the expected profit, but with respect to an appropriately modified parameter! Remarkably, the fair price depends only on the size of the up and down steps of the share price, rather than on the trend p. The parameter p^* is characterised by the fact that $\mathbb{E}^*(X_n) = 1$ for all n; see Figure 4.5. In other words, the expected value of the share remains constant in time, that is, it evolves in parallel to the value of the bond.

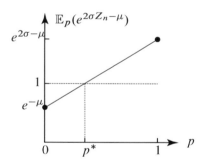

Figure 4.5. p^* is determined so that the growth factor $\exp[2\sigma Z_n - \mu]$ has expectation 1.

Proof. Let $\alpha\beta$ be an arbitrary self-financing hedging strategy with initial value $v > 0$. Then by (4.18) one finds

$$(4.21) \qquad V_n^{\alpha\beta} - V_{n-1}^{\alpha\beta} = \alpha_n(X_n - X_{n-1}) = \alpha_n X_{n-1}\big[e^{2\sigma Z_n - \mu} - 1\big]\,.$$

There are two points to observe here. First, $\alpha_n X_{n-1}$ depends only on $Z_1, \ldots, Z_{n-1}$ and, by Theorem (3.24), is thus independent of $e^{2\sigma Z_n - \mu} - 1$. Secondly, p^* is chosen so that

$$\mathbb{E}^*(e^{2\sigma Z_n - \mu} - 1) = p^* e^{2\sigma - \mu} + (1 - p^*)e^{-\mu} - 1 = 0\,.$$

Hence, by Theorem (4.7d),

$$\mathbb{E}^*(V_n^{\alpha\beta} - V_{n-1}^{\alpha\beta}) = \mathbb{E}^*(\alpha_n X_{n-1})\, \mathbb{E}^*(e^{2\sigma Z_n - \mu} - 1) = 0$$

and therefore, by (4.19) and Theorem (4.7a),

$$v = \mathbb{E}^*(V_0^{\alpha\beta}) = \mathbb{E}^*(V_N^{\alpha\beta}) \geq \mathbb{E}^*((X_N - K)_+) =: v^*.$$

So we obtain the inequality $\Pi^* \geq v^*$.

To prove the reverse inequality and the additional statement, we construct a self-financing hedging strategy $\alpha\beta$ with initial value v^*. We start by defining suitable candidates V_n^* for $V_n^{\alpha\beta}$. For $1 \leq n \leq N$ and $\omega \in \Omega$, we write $\omega_{\leq n} = (\omega_1, \dots, \omega_n)$ for the initial segment of ω describing the first n changes in the share price, and we define $\omega_{>n}$ and $Z_{>n}$ analogously; formally we also allow the empty sequence $\omega_{\leq 0}$. We set

$$V_n^*(\omega_{\leq n}) = \mathbb{E}^*\big((X_N(\omega_{\leq n}, Z_{>n}) - K)_+ \big)$$
$$= \sum_{\omega_{>n} \in \{0,1\}^{N-n}} (X_N(\omega_{\leq n}, \omega_{>n}) - K)_+ \, P^*(Z_{>n} = \omega_{>n})$$

for $0 \leq n \leq N$. So $V_n^*(\omega_{\leq n})$ is the expected profit of the buyer if $\omega_{\leq n}$ is already known and the trend is assumed to be given by the parameter p^*. In particular, $V_0^* = v^*$ and $V_N^*(\omega) = (X_N(\omega) - K)_+$. (We will sometimes treat V_n^* as a function defined on all of Ω that in fact only depends on the first n coordinates.) Due to the product structure of the Bernoulli distribution, we see that, for every $1 \leq n \leq N$,

$$V_{n-1}^*(\omega_{<n}) = p^* V_n^*(\omega_{<n}, 1) + (1 - p^*) V_n^*(\omega_{<n}, 0).$$

(The right-hand side is the conditional expectation of V_n^* given $\omega_{<n}$. The equation thus means that the sequence $(V_n^*)_{0 \leq n \leq N}$ is a 'martingale'. We refrain from discussing this important concept here.) Distinguishing the cases $\omega_n = 0$ and $\omega_n = 1$, we thus obtain

$$\gamma_n(\omega) := \big[V_n^*(\omega_{\leq n}) - V_{n-1}^*(\omega_{<n})\big] / \big(e^{2\sigma Z_n(\omega) - \mu} - 1\big)$$
$$= \big[V_n^*(\omega_{<n}, 1) - V_n^*(\omega_{<n}, 0)\big] / \big(e^{2\sigma - \mu} - e^{-\mu}\big).$$

This shows that $\gamma_n(\omega)$ depends only on $\omega_{<n}$, not on ω_n. In particular, γ_1 is constant. Now, we set $\alpha_n = \gamma_n / X_{n-1}$ and determine the corresponding β_n recursively from (4.18); β_1 is chosen so that $V_0^{\alpha\beta} = v^*$. Then $\alpha\beta$ is a self-financing strategy, and (4.21) yields that

$$V_n^{\alpha\beta} - V_{n-1}^{\alpha\beta} = \gamma_n\big[e^{2\sigma Z_n - \mu} - 1\big] = V_n^* - V_{n-1}^*.$$

Since $V_0^{\alpha\beta} = v^* = V_0^*$, we see that $V_n^{\alpha\beta} = V_n^*$ for every $1 \leq n \leq N$. In particular, $V_n^{\alpha\beta} \geq 0$ and $V_N^{\alpha\beta} = (X_N - K)_+$. This shows that $\alpha\beta$ is a hedging strategy with initial value v^*. $\diamond$

Note that the above proof is constructive. It does not only yield the formula for Π^*, but at the same time, also an optimal strategy. For instance, the strategy in the example on p. 103 is obtained this way, see Problem 4.11.

Finally, we will derive an alternative expression for Π^*. Expressing X_N in terms of $S_N := \sum_{k=1}^{N} Z_k$ and using the abbreviation $a_N = (\log K + \mu N)/2\sigma$, we can write

$$\begin{aligned}
\Pi^* &= \mathbb{E}^*\big((X_N - K)\,1_{\{X_N > K\}}\big) \\
&= \mathbb{E}^*\big(\exp[2\sigma S_N - \mu N]\,1_{\{S_N > a_N\}}\big) - K\,P^*(S_N > a_N).
\end{aligned}$$

We note further that, under P^*, S_N has distribution $\mathcal{B}_{n,p^*}$ by Theorem (2.9), and that

$$\mathcal{B}_{N,p^*}(\{k\})\,\exp[2\sigma k - \mu N] = \mathcal{B}_{N,p^\circ}(\{k\})$$

for $p^\circ = e^{2\sigma - \mu} p^*$ and all k. Setting $P^\circ = P_{p^\circ}$, we thus end up with the identity

(4.22) $$\Pi^* = P^\circ(S_N > a_N) - K\,P^*(S_N > a_N),$$

which is already close to the famous Black–Scholes formula that will be derived in Example (5.26).

4.3   Variance and Covariance

Let $(\Omega, \mathscr{F}, P)$ be a probability space and $X : \Omega \to \mathbb{R}$ a real random variable. If $X^r \in \mathscr{L}^1(P)$ for some $r \in \mathbb{N}$, the expectation $\mathbb{E}(X^r)$ is called the *rth moment* of X, and one writes $X \in \mathscr{L}^r = \mathscr{L}^r(P)$. Note that $\mathscr{L}^s \subset \mathscr{L}^r$ for $r < s$, since $|X|^r \le 1 + |X|^s$. We are mainly interested in the case $r = 2$.

Definition. Let $X, Y \in \mathscr{L}^2$.

(a) $\mathbb{V}(X) = \mathbb{V}_P(X) := \mathbb{E}\big([X - \mathbb{E}(X)]^2\big) = \mathbb{E}(X^2) - \mathbb{E}(X)^2$ is called the *variance*, and $\sqrt{\mathbb{V}(X)}$ the *standard deviation* of X with respect to P.

(b) $\mathrm{Cov}(X, Y) := \mathbb{E}\big([X - \mathbb{E}(X)][Y - \mathbb{E}(Y)]\big) = \mathbb{E}(XY) - \mathbb{E}(X)\,\mathbb{E}(Y)$ is called the *covariance* of X and Y (relative to P). It exists since $|XY| \le X^2 + Y^2$.

(c) If $\mathrm{Cov}(X, Y) = 0$, then X and Y are called *uncorrelated*.

The variance measures the average spread of the values of X. For instance, if X is uniformly distributed on $\{x_1, \ldots, x_n\} \subset \mathbb{R}$, then

$$\mathbb{E}(X) = \bar{x} := \frac{1}{n}\sum_{i=1}^{n} x_i \quad \text{and} \quad \mathbb{V}(X) = \frac{1}{n}\sum_{i=1}^{n}(x_i - \bar{x})^2,$$

i.e., the variance is exactly the mean squared deviation from the mean. In a physical interpretation, the variance corresponds to the moment of inertia of a bar with mass distribution $P \circ X^{-1}$ rotating around (an axis perpendicular to $\mathbb{R}$ through) the centre of gravity. Furthermore, let Y be uniformly distributed on $\{y_1, \ldots, y_n\} \subset \mathbb{R}$. Up to a factor of n, the covariance $\mathrm{Cov}(X, Y)$ is then the Euclidean scalar product of the centred vectors $(x_i - \bar{x})_{1 \le i \le n}$ and $(y_i - \bar{y})_{1 \le i \le n}$, and the so-called *correlation coefficient* $\rho(X, Y) := \mathrm{Cov}(X, Y)/\sqrt{\mathbb{V}(X)\mathbb{V}(Y)}$ corresponds to the cosine of the angle between these vectors. Hence, saying that X and Y are uncorrelated expresses a condition of orthogonality. The following theorem summarises the basic properties of variance and covariance.

(4.23) Theorem. Rules for variances and covariances. *Let $X, Y, X_i \in \mathcal{L}^2$ and $a, b, c, d \in \mathbb{R}$. Then the following holds.*

(a) $aX + b, \ cY + d \in \mathcal{L}^2$ and $\mathrm{Cov}(aX + b, cY + d) = ac \, \mathrm{Cov}(X, Y)$.

In particular, $\mathbb{V}(aX + b) = a^2 \, \mathbb{V}(X)$.

(b) $\mathrm{Cov}(X, Y)^2 \le \mathbb{V}(X) \, \mathbb{V}(Y)$.

(c) $\sum_{i=1}^{n} X_i \in \mathcal{L}^2$ and $\mathbb{V}\left(\sum_{i=1}^{n} X_i\right) = \sum_{i=1}^{n} \mathbb{V}(X_i) + \sum_{i \ne j} \mathrm{Cov}(X_i, X_j)$.

In particular, if $X_1, \ldots, X_n$ are pairwise uncorrelated, then

$$\mathbb{V}\left(\sum_{i=1}^{n} X_i\right) = \sum_{i=1}^{n} \mathbb{V}(X_i) \quad \text{(identity of I.-J. Bienaymé, 1853).}$$

(d) *If X, Y are independent, then X, Y are also uncorrelated.*

Proof. (a) Expand the product defining the covariance and use Theorem (4.11b).

(b) By (a), statement (b) is not affected by adding constants to X and Y. So we can assume that $\mathbb{E}(X) = \mathbb{E}(Y) = 0$. For discrete X, Y, the Cauchy–Schwarz inequality yields

$$\mathrm{Cov}(X, Y)^2 = \mathbb{E}(XY)^2 = \left(\sum_{x,y} xy \, P(X = x, Y = y)\right)^2$$

$$\le \left(\sum_{x,y} x^2 P(X = x, Y = y)\right)\left(\sum_{x,y} y^2 P(X = x, Y = y)\right) = \mathbb{V}(X) \, \mathbb{V}(Y).$$

The general case is obtained by a discrete approximation as in (4.9).

(c) Without loss of generality, we can again assume that $\mathbb{E}(X_i) = 0$. Then we can write, using Theorem (4.11b),

$$\mathbb{V}\left(\sum_{i=1}^{n} X_i\right) = \mathbb{E}\left(\left(\sum_{i=1}^{n} X_i\right)^2\right) = \mathbb{E}\left(\sum_{i,j=1}^{n} X_i X_j\right)$$

$$= \sum_{i,j=1}^{n} \mathbb{E}(X_i X_j) = \sum_{i=1}^{n} \mathbb{V}(X_i) + \sum_{i \ne j} \mathrm{Cov}(X_i, X_j).$$

(d) This follows directly from Theorem (4.11d). $\diamond$

As an immediate consequence of statement (a) we obtain:

(4.24) Corollary. Standardisation. *If $X \in \mathscr{L}^2$ with $\mathbb{V}(X) > 0$, then the random variable*

$$X^* := \frac{X - \mathbb{E}(X)}{\sqrt{\mathbb{V}(X)}}$$

is 'standardised', in that $\mathbb{E}(X^*) = 0$ and $\mathbb{V}(X^*) = 1$.

It is important to distinguish clearly between the notions of independence and un-correlatedness. Both are characterised by a product rule. But in the case of uncor-relatedness, the factorisation only involves the expectations, whereas in the case of independence it applies to the complete joint distribution of the random variables. The following counterexample confirms this distinction.

(4.25) Example. *Uncorrelated, but dependent random variables.* Let $\Omega = \{1, 2, 3\}$ and $P = \mathcal{U}_\Omega$ be the uniform distribution. Suppose the random variable X is defined by its three values $(1, 0, -1)$ on Ω, and Y takes the values $(0, 1, 0)$. Then, $XY = 0$ and $\mathbb{E}(X) = 0$, thus $\mathrm{Cov}(X, Y) = \mathbb{E}(XY) - \mathbb{E}(X)\,\mathbb{E}(Y) = 0$, but

$$P(X = 1, Y = 1) = 0 \neq \tfrac{1}{9} = P(X = 1)\,P(Y = 1),$$

so X and Y are not independent. That is, the converse of Theorem (4.23d) fails.

Since the expectation of a random variable only depends on its distribution (re-call Remark (4.10)), it makes sense to talk about the expectation and variance of a probability measure on $\mathbb{R}$.

Definition. If P is a probability measure on $(\mathbb{R}, \mathscr{B})$, then $\mathbb{E}(P) := \mathbb{E}_P(\mathrm{Id}_\mathbb{R})$ and $\mathbb{V}(P) = V_P(\mathrm{Id}_\mathbb{R})$ are called the *expectation* and the *variance* of P (provided they exist). Here $\mathrm{Id}_\mathbb{R} : x \to x$ is the identity mapping on $\mathbb{R}$.

Remark (4.10) can then be restated as follows.

(4.26) Remark. *Dependence of expectation and variance on the distribution.* By definition, $\mathbb{E}(X) = \mathbb{E}(P \circ X^{-1})$ when $X \in \mathscr{L}^1(P)$, and $\mathbb{V}(X) = \mathbb{V}(P \circ X^{-1})$ for $X \in \mathscr{L}^2(P)$. Expectation and variance of a real random variable X thus coincide with the expectation and variance of the distribution $P \circ X^{-1}$ of X.

We now calculate the variance of some standard distributions.

(4.27) Example. *Binomial distributions.* For every n and p, the binomial distribution $\mathcal{B}_{n,p}$ has the variance $\mathbb{V}(\mathcal{B}_{n,p}) = np(1-p)$. Indeed, we know from Theorem (2.9) that $\mathcal{B}_{n,p} = P \circ S_n^{-1}$, where $S_n = \sum_{i=1}^n X_i$ and $(X_i)_{1 \leq i \leq n}$ is a Bernoulli sequence with parameter p. Clearly, $\mathbb{E}(X_i) = \mathbb{E}(X_i^2) = p$ and therefore $\mathbb{V}(X_i) = p(1-p)$.

Hence, Theorem (4.23cd) yields

$$\mathbb{V}(\mathcal{B}_{n,p}) = \mathbb{V}(S_n) = \sum_{i=1}^{n} \mathbb{V}(X_i) = np(1-p).$$

(4.28) Example. *Normal distributions.* Let $\mathcal{N}_{m,v}$ be the normal distribution with parameters $m \in \mathbb{R}$, $v > 0$. By Corollary (4.13), the second moment of $\mathcal{N}_{m,v}$ exists if and only if $\int x^2 e^{-(x-m)^2/2v} \, dx < \infty$. Clearly, this is true. So, using the substitution $y = (x-m)/\sqrt{v}$ and the $y \leftrightarrow -y$ symmetry we find

$$\mathbb{E}(\mathcal{N}_{m,v}) = \frac{1}{\sqrt{2\pi v}} \int x \, e^{-(x-m)^2/2v} \, dx$$

$$= \frac{1}{\sqrt{2\pi}} \int (m + \sqrt{v} \, y) \, e^{-y^2/2} \, dy = m,$$

and also

$$\mathbb{V}(\mathcal{N}_{m,v}) = \frac{1}{\sqrt{2\pi v}} \int (x - m)^2 e^{-(x-m)^2/2v} \, dx = \frac{v}{\sqrt{2\pi}} \int y^2 e^{-y^2/2} \, dy$$

$$= \frac{v}{\sqrt{2\pi}} \left(\left[-y e^{-y^2/2} \right]_{-\infty}^{\infty} + \int e^{-y^2/2} \, dy \right) = v,$$

where we have used integration by parts and Lemma (2.25). Hence, the parameters of a normal distribution are simply its expectation and variance.

(4.29) Example. *Beta distributions.* For the beta distribution $\beta_{a,b}$ with parameters $a, b > 0$, Corollary (4.13) and the recursive formula (2.23) imply that

$$\mathbb{E}(\beta_{a,b}) = \int_0^1 s \, \beta_{a,b}(s) \, ds = \frac{\mathrm{B}(a+1, b)}{\mathrm{B}(a, b)} = \frac{a}{a+b}$$

and

$$\mathbb{V}(\beta_{a,b}) = \int_0^1 s^2 \, \beta_{a,b}(s) \, ds - \mathbb{E}(\beta_{a,b})^2$$

$$= \frac{\mathrm{B}(a+2, b)}{\mathrm{B}(a, b)} - \left(\frac{a}{a+b} \right)^2 = \frac{ab}{(a+b)^2(a+b+1)}.$$

4.4   Generating Functions

We now consider probability measures P on the set $\mathbb{Z}_+$ of non-negative integers. Any such P is uniquely determined by its discrete density $\rho(k) = P(\{k\})$, which in turn is determined by the power series with coefficients $\rho(k)$.

Definition. If P is a probability measure on $(\mathbb{Z}_+, \mathscr{P}(\mathbb{Z}_+))$ with density ρ, then the function

$$\varphi_P(s) = \sum_{k \geq 0} \rho(k)\, s^k\,, \quad 0 \leq s \leq 1,$$

is called the *generating function* of P or ρ.

Since $\sum_{k \geq 0} \rho(k) = 1$, the generating function is well-defined (at least) on its domain $[0, 1]$, and it is infinitely differentiable (at least) on $[0, 1[$. As all coefficients $\rho(k)$ are non-negative, φ_P is convex on $[0, 1]$. For some familiar distributions, φ_P can be calculated easily; see also Figure 4.6.

(4.30) Example. *Binomial distributions.* For $n \in \mathbb{N}$ and $0 < p < 1$, the generating function of the binomial distribution $\mathcal{B}_{n,p}$ is given by

$$\varphi_{\mathcal{B}_{n,p}}(s) = \sum_{k=0}^{n} \binom{n}{k} p^k q^{n-k} s^k = (q + ps)^n\,,$$

where $q := 1 - p$.

(4.31) Example. *Negative binomial distributions.* For the generating function of the negative binomial distribution $\overline{\mathcal{B}}_{r,p}$ with parameters $r > 0$ and $0 < p < 1$ we find, using the general binomial theorem,

$$\varphi_{\overline{\mathcal{B}}_{r,p}}(s) = p^r \sum_{k \geq 0} \binom{-r}{k}(-qs)^k = \left(\frac{p}{1-qs}\right)^r.$$

(4.32) Example. *Poisson distributions.* If $\mathcal{P}_\lambda$ is the Poisson distribution with parameter $\lambda > 0$, then

$$\varphi_{\mathcal{P}_\lambda}(s) = \sum_{k \geq 0} e^{-\lambda} \frac{\lambda^k}{k!}\, s^k = e^{-\lambda(1-s)}\,.$$

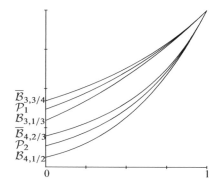

Figure 4.6. Generating functions of binomial distributions, negative binomial distributions, and Poisson distributions with expectation 1 and 2, respectively.

As the next theorem shows, generating functions can be used to compute moments. (Other applications will follow in Chapter 6.) We write $\varphi_P^{(k)}(0)$ for the kth derivative of φ_P at the point 0.

(4.33) Theorem. *Calculation of moments via generating functions. Let P be a probability measure on $\mathbb{Z}_+$ with density ρ. Then the following holds.*

(a) *For every k we have $\rho(k) = \varphi_P^{(k)}(0)/k!$, so P is uniquely determined by φ_P.*

(b) $\mathbb{E}(P)$ *exists if and only if $\varphi_P'(1)$ $(= \lim_{s\uparrow 1} \varphi_P'(s))$ exists, and then*

$$\mathbb{E}(P) = \varphi_P'(1) = \lim_{s\uparrow 1} \varphi_P'(s) \,.$$

(c) $\mathbb{V}(P)$ *exists if and only if $\varphi_P''(1)$ $(= \lim_{s\uparrow 1} \varphi_P''(s))$ exists, and then*

$$\mathbb{V}(P) = \varphi_P''(1) - \mathbb{E}(P)^2 + \mathbb{E}(P) \,.$$

Proof. (a) This follows directly from Taylor's theorem.

(b) All the following expressions exist in $[0, \infty]$:

$$\lim_{s\uparrow 1} \frac{\varphi(1)-\varphi(s)}{1-s} = \lim_{s\uparrow 1} \sum_{k\geq 0} \rho(k)\frac{1-s^k}{1-s} = \lim_{s\uparrow 1} \sum_{k\geq 0} \rho(k) \sum_{j=0}^{k-1} s^j$$

$$= \sup_{s<1} \sup_{n\geq 1} \sum_{k=0}^{n} \rho(k) \sum_{j=0}^{k-1} s^j = \sum_{k\geq 0} \rho(k)\,k = \lim_{s\uparrow 1} \sum_{k\geq 0} \rho(k)\,ks^{k-1} = \lim_{s\uparrow 1} \varphi'(s) \,.$$

The fourth equality comes from the fact that, by monotonicity, the suprema over s and n can be interchanged. Now, $\mathbb{E}(P)$ exists if and only if all these expressions are finite.

(c) As in (b) we obtain

$$\lim_{s\uparrow 1} \frac{\varphi'(1)-\varphi'(s)}{1-s} = \sum_{k\geq 0} \rho(k)\,k(k-1) \in [0, \infty] \,,$$

and the claim follows immediately. $\diamond$

Let us apply the theorem to the distributions considered above.

(4.34) Example. *Binomial distributions.* By Example (4.30),

$$\mathbb{E}(\mathcal{B}_{n,p}) = \frac{d}{ds}(q + ps)^n \Big|_{s=1} = np \,,$$

$$\mathbb{V}(\mathcal{B}_{n,p}) = \frac{d^2}{ds^2}(q + ps)^n \Big|_{s=1} - (np)^2 + np = npq \,,$$

in agreement with Examples (4.5) and (4.27).

(4.35) Example. *Negative binomial distributions.* In view of Example (4.31), Theorem (4.33bc) yields

$$
\mathbb{E}(\overline{B}_{r,p}) = \frac{d}{ds}\left(\frac{p}{1-qs}\right)^r\Big|_{s=1} = p^r rq(1-q)^{-r-1} = \frac{rq}{p}\,,
$$

$$
\mathbb{V}(\overline{B}_{r,p}) = \frac{d^2}{ds^2}\left(\frac{p}{1-qs}\right)^r\Big|_{s=1} - \frac{r^2q^2}{p^2} + \frac{rq}{p}
$$

$$
= p^r r(r+1)q^2 p^{-r-2} - \frac{r^2q^2}{p^2} + \frac{rq}{p} = \frac{rq}{p^2}\,.
$$

(4.36) Example. *Poisson distributions.* Example (4.32) shows that

$$
\mathbb{E}(P_\lambda) = \frac{d}{ds}e^{-\lambda+\lambda s}\Big|_{s=1} = \lambda\,,
$$

$$
\mathbb{V}(P_\lambda) = \frac{d^2}{ds^2}e^{-\lambda+\lambda s}\Big|_{s=1} - \lambda^2 + \lambda = \lambda\,.
$$

Hence, the parameter of a Poisson distribution determines both the expectation and the variance.

The generating function of a random variable is defined as the generating function of its distribution.

Definition. Let X be a $\mathbb{Z}_+$-valued random variable on a probability space $(\Omega, \mathscr{F}, P)$. The function

$$
\varphi_X(s) := \varphi_{P\circ X^{-1}}(s) = \sum_{k\geq 0} s^k\, P(X=k) = \mathbb{E}(s^X)\,, \quad 0 \leq s \leq 1,
$$

is then called the *generating function* of X.

(4.37) Theorem. *Generating function of a sum of independent random variables. If X, Y are independent $\mathbb{Z}_+$-valued random variables, then*

$$
\varphi_{X+Y}(s) = \varphi_X(s)\,\varphi_Y(s) \quad \text{for } 0 \leq s \leq 1.
$$

Proof. By Theorem (3.24) we know that s^X and s^Y are independent, so the claim follows from Theorem (4.7d). ◇

Using the notion of convolution (see p. 74), Theorem (4.37) can be rephrased as follows.

(4.38) Theorem. *Product rule for the generating function of a convolution. For any two probability measures P_1, P_2 on $(\mathbb{Z}_+, \mathscr{P}(\mathbb{Z}_+))$,*

$$
\varphi_{P_1 \star P_2}(s) = \varphi_{P_1}(s)\,\varphi_{P_2}(s) \quad \text{for } 0 \leq s \leq 1.
$$

As a consequence, the distributions considered above can be identified as convolutions.

(4.39) Example. *Binomial distributions.* Let $0 < p < 1$ and $\mathcal{B}_{1,p}$ be the binomial distribution on $\{0, 1\}$, i.e., $\mathcal{B}_{1,p}(\{1\}) = p$, $\mathcal{B}_{1,p}(\{0\}) = q := 1 - p$. Then, for every $n \geq 2$,

$$\mathcal{B}_{n,p} = \mathcal{B}_{1,p}^{\star n} := \underbrace{\mathcal{B}_{1,p} \star \cdots \star \mathcal{B}_{1,p}}_{n \text{ times}} .$$

We already know this from Theorem (2.9), since $\mathcal{B}_{n,p} = P \circ (X_1 + \cdots + X_n)^{-1}$ for a Bernoulli sequence $(X_i)_{1 \leq i \leq n}$. But we can also use Theorem (4.38). Namely, by Example (4.30) we obtain

$$\varphi_{\mathcal{B}_{n,p}}(s) = (q + ps)^n = \varphi_{\mathcal{B}_{1,p}}(s)^n = \varphi_{\mathcal{B}_{1,p}^{\star n}}(s), \quad 0 \leq s \leq 1.$$

Thus, by Theorem (4.33a) the probability measures $\mathcal{B}_{n,p}$ and $\mathcal{B}_{1,p}^{\star n}$ do coincide.

(4.40) Example. *Negative binomial distributions.* For $0 < p < 1$, $r \in \mathbb{N}$ we have $\overline{\mathcal{B}}_{r,p} = \mathcal{G}_p^{\star r}$, in other words, $\overline{\mathcal{B}}_{r,p}$ is the r-fold convolution of the geometric distribution. As in the last example, this follows from the relation

$$\varphi_{\overline{\mathcal{B}}_{r,p}}(s) = \left(\frac{p}{1-qs}\right)^r = \varphi_{\mathcal{G}_p}(s)^r = \varphi_{\mathcal{G}_p^{\star r}}(s).$$

Similarly, for arbitrary real $a, b > 0$, we obtain $\overline{\mathcal{B}}_{a+b,p} = \overline{\mathcal{B}}_{a,p} \star \overline{\mathcal{B}}_{b,p}$, which means that the negative binomial distributions form a convolution semigroup.

(4.41) Example. *Poisson distributions.* For arbitrary parameters $\lambda, \mu > 0$, the identity

$$\varphi_{\mathcal{P}_{\lambda+\mu}}(s) = e^{(\lambda+\mu)(s-1)} = \varphi_{\mathcal{P}_\lambda}(s)\, \varphi_{\mathcal{P}_\mu}(s) = \varphi_{\mathcal{P}_\lambda \star \mathcal{P}_\mu}(s)$$

holds for every s, whence $\mathcal{P}_{\lambda+\mu} = \mathcal{P}_\lambda \star \mathcal{P}_\mu$. Therefore, the Poisson distributions also form a convolution semigroup, as we have already seen in Corollary (3.36).

Problems

4.1   Let X be a random variable taking values in $[0, \infty]$. Show the following (considering the discrete case first).

(a) If $\mathbb{E}(X) < \infty$, then $P(X < \infty) = 1$.

(b) If $\mathbb{E}(X) = 0$, then $P(X = 0) = 1$.

4.2   Which of the following statements hold for arbitrary $X, Y \in \mathcal{L}^1$? Prove or disprove.

(a) $\mathbb{E}(X) = \mathbb{E}(Y) \Rightarrow P(X = Y) = 1$.

(b) $\mathbb{E}(|X - Y|) = 0 \Rightarrow P(X = Y) = 1$.

4.3 *Inclusion–exclusion principle.* Give an alternative proof of the inclusion–exclusion principle in Problem 1.7b by calculating the expectation of the product

$$\prod_{i \in J} 1_{A_i} \prod_{i \in I \setminus J} (1 - 1_{A_i}).$$

4.4 *Jensen's inequality (J. Jensen 1906).* Let $\varphi : \mathbb{R} \to \mathbb{R}$ be a convex function, $X \in \mathscr{L}^1$ and $\varphi \circ X \in \mathscr{L}^1$. Show that

$$\varphi\big(\mathbb{E}(X)\big) \leq \mathbb{E}\big(\varphi(X)\big).$$

Hint: Consider a tangent to φ at the point $\mathbb{E}(X)$.

4.5 ^{S} (a) Let X be a random variable taking values in $\mathbb{Z}_+$. Show that

$$\mathbb{E}(X) = \sum_{k \geq 1} P(X \geq k).$$

(Both sides can be equal to $+\infty$.)

(b) Let X be an arbitrary random variable taking values in $[0, \infty[$. Show that

$$\mathbb{E}(X) = \int_0^\infty P(X \geq s)\, ds.$$

(Again, both sides can be equal to $+\infty$.) *Hint:* Use discrete approximations.

4.6 Let $(\Omega, \mathscr{F}, P)$ be a probability space and $A_n \in \mathscr{F}$, $n \geq 1$. Define and interpret a random variable X with $\mathbb{E}(X) = \sum_{n \geq 1} P(A_n)$. In particular, discuss the special case that the A_n are pairwise disjoint.

4.7 Suppose $X, Y, X_1, X_2, \ldots \in \mathscr{L}^1$. Prove the following statements.

(a) *Fatou's lemma.* If $X_n \geq 0$ for every n and $X = \liminf_{n \to \infty} X_n$, then $\mathbb{E}(X) \leq \liminf_{n \to \infty} \mathbb{E}(X_n)$. *Hint:* By Problem 1.14, $Y_n := \inf_{k \geq n} X_k$ is a random variable, and $Y_n \uparrow X$.

(b) *Dominated convergence theorem.* If $|X_n| \leq Y$ for every n and $X = \lim_{n \to \infty} X_n$, then $\mathbb{E}(X) = \lim_{n \to \infty} \mathbb{E}(X_n)$. *Hint:* Apply (a) to the random variables $Y \pm X_n$.

4.8 ^{S} *Fubini's theorem.* Let X_1, X_2 be independent random variables taking values in some arbitrary event spaces $(E_1, \mathscr{E}_1)$ and $(E_2, \mathscr{E}_2)$ respectively, and suppose that $f : E_1 \times E_2 \to \mathbb{R}$ is a bounded random variable. For $x_1 \in E_1$ let $f_1(x_1) = \mathbb{E}(f(x_1, X_2))$, which is well-defined by Problem 1.6. Show: f_1 is a random variable, and $\mathbb{E}(f(X_1, X_2)) = \mathbb{E}(f_1(X_1))$. That is, the expectation of $f(X_1, X_2)$ can be calculated in two steps. *Hint:* First show the claim for $f = 1_A$, where $A = A_1 \times A_2$ with $A_i \in \mathscr{E}_i$. With the help of Theorem (1.12), proceed to $f = 1_A$ for arbitrary $A \in \mathscr{E}_1 \otimes \mathscr{E}_2$, and finally to discrete and arbitrary f.

4.9 The magazine of a consumer safety organisation supplements its test articles with the 'average price' of a product; this is often given as a sample median, i.e., a median of the empirical distribution $\frac{1}{n} \sum_{i=1}^n \delta_{x_i}$ of the prices $x_1, \ldots, x_n$ found in n shops. Why can the median be more useful for the reader than the arithmetic mean? Give an example to show that median and expectation can differ considerably.

4.10 ^{S} *Wald's identity.* Let $(X_i)_{i \geq 1}$ be i.i.d. real random variables in $\mathscr{L}^1$ and let τ be a $\mathbb{Z}_+$-valued random variable with $\mathbb{E}(\tau) < \infty$. Suppose that, for every $n \in \mathbb{N}$, the event $\{\tau \geq n\}$ is independent of X_n. Show that the random variable $S_\tau = \sum_{i=1}^\tau X_i$ has an expectation, and

$$\mathbb{E}(S_\tau) = \mathbb{E}(\tau)\, \mathbb{E}(X_1).$$

4.11 Consider the Cox–Ross–Rubinstein model with parameters $X_0 = K = 1$, $\sigma = \mu = \log 2$. For the maturity times $N = 1, 2, 3$, find the Black–Scholes price Π^* and the optimal self-financing hedging strategy $\alpha\beta$.

4.12 ^{S} *Maturity-dependence of the Black–Scholes price.* Let $\Pi^*(N) = \mathbb{E}^*((X_N - K)_+)$ be the Black–Scholes price for maturity N in the Cox–Ross–Rubinstein model with parameters $\mu, \sigma > 0$, see Example (4.17). Prove that $\Pi^*(N) \leq \Pi^*(N+1)$. *Hint:* Use Problem 4.4.

4.13 Let X be a real random variable, and consider the cases

(a) X is $\mathcal{U}_{[0,1]}$-distributed,

(b) X is Cauchy distributed with density $\rho(x) = \frac{1}{\pi}\frac{1}{1+x^2}$,

(c) $X = e^Y$ for an $\mathcal{N}_{0,1}$-distributed random variable Y.

Check whether the expectation $\mathbb{E}(X)$ and the variance $\mathbb{V}(X)$ exist. If so, calculate their values.

4.14 Let $X_1, \dots, X_n \in \mathcal{L}^2$ be i.i.d. random variables and $M = \frac{1}{n}\sum_{i=1}^n X_i$ their average. Find

$$\mathbb{E}\left(\sum_{i=1}^n (X_i - M)^2\right).$$

4.15 ^{S} Show the following. (a) *The expectation minimises the mean squared deviation.* If $X \in \mathcal{L}^2$ with expectation m and $a \in \mathbb{R}$, then

$$\mathbb{E}\left((X - a)^2\right) \geq \mathbb{V}(X)$$

with equality if and only if $a = m$.

(b) *Every median minimises the mean absolute deviation.* Let $X \in \mathcal{L}^1$, μ be a median of X and $a \in \mathbb{R}$. Then

$$\mathbb{E}(|X - a|) \geq \mathbb{E}(|X - \mu|)$$

with equality if and only if a is also a median. *Hint:* Suppose without loss of generality that $a < \mu$, and verify the equation

$$|X - a| - |X - \mu| = (\mu - a)\left(2\,1_{\{X \geq \mu\}} - 1\right) + 2(X - a)\,1_{\{a < X < \mu\}}.$$

4.16 *Best linear prediction.* Suppose $X, Y \in \mathcal{L}^2$ and (without loss of generality) $\mathbb{V}(X) = 1$. What is the best approximation of Y by an affine function $a + bX$ of X? Show that the quadratic deviation

$$\mathbb{E}\left((Y - a - bX)^2\right)$$

is minimised by $b = \mathrm{Cov}(X, Y)$ and $a = \mathbb{E}(Y - bX)$. What does this mean when X, Y are uncorrelated?

4.17 *Normal moments.* Let X be an $\mathcal{N}_{0,1}$-distributed random variable. Show that $\mathbb{E}(X^{2k}) = 2^k\,\Gamma(k + \frac{1}{2})/\Gamma(\frac{1}{2})$ for every $k \in \mathbb{N}$, and calculate the explicit value for $k = 1, 2, 3$.

4.18 ^{S} *Fixed points of a random permutation.* Let $\Omega = \mathcal{S}_n$ be the set of all permutations of $\{1, \dots, n\}$ and $P = \mathcal{U}_\Omega$ the uniform distribution on Ω. For every permutation $\omega \in \Omega$, let $X(\omega)$ be the number of fixed points of ω. Find $\mathbb{E}(X)$ and $\mathbb{V}(X)$ (without using Problem 2.11).

4.19 *Positive correlation of monotone random variables.* Let $(\Omega, \mathscr{F}, P)$ be a probability space, $f, g \in \mathscr{L}^2(P)$, and X, Y two independent Ω-valued random variables with distribution P. Show that

$$\mathrm{Cov}_P(f, g) = \tfrac{1}{2} \mathbb{E}\big([f(X) - f(Y)][g(X) - g(Y)]\big).$$

In the case $(\Omega, \mathscr{F}) = (\mathbb{R}, \mathscr{B})$, deduce that any two increasing functions f, g are positively correlated, in that $\mathrm{Cov}(f, g) \geq 0$.

4.20 ^{S} *Collectibles (coupon collector's problem).* Suppose a company attaches to its product a sticker showing one of the players of the national football team. How many items do you have to buy on average to collect all $N = 20$ stickers? To formalise the problem, let $(X_i)_{i \geq 1}$ be a sequence of independent random variables that are uniformly distributed on $E = \{1, \ldots, N\}$. Here, X_i represents the sticker you obtain on the ith purchase. Let $\Xi_n := \{X_1, \ldots, X_n\}$ be the random set containing the distinct stickers obtained after n purchases, and for $1 \leq r \leq N$ let

$$T_r = \inf\{n \geq 1 : |\Xi_n| = r\}$$

the purchase providing you with the rth distinct sticker. Finally, let $D_r = T_r - T_{r-1}$, where $T_0 := 0$.

(a) Let $r, d_1, \ldots, d_r, d \in \mathbb{N}, i \in E$, and $I \subset E$ with $|I| = r$, and define $n := d_1 + \cdots + d_r$. Show that

$$P\big(T_{r+1} = n + d, \ X_{n+d} = i \big| D_j = d_j \text{ for } 1 \leq j \leq r, \ \Xi_n = I\big) = \big(\tfrac{r}{N}\big)^{d-1} 1_{I^c}(i) \tfrac{1}{N}.$$

(b) Show that the random variables $D_1, \ldots, D_N$ are independent, and that $D_r - 1$ is geometrically distributed with parameter $1 - \frac{r-1}{N}$.

(c) Find $\mathbb{E}(T_N)$ and $\mathbb{V}(T_N)$.

4.21 *Collectibles (alternative approach).* In the situation of Problem 4.20, prove the recursive formula

$$P(T_r = n + 1) = \big(1 - \tfrac{r-1}{N}\big) \sum_{k=1}^{n} \big(P(T_{r-1} = k) - P(T_r = k)\big).$$

Use this to find the generating functions of the T_r and deduce that the distribution of T_N is the convolution $\delta_N \star \overset{N}{\underset{r=1}{\star}} \mathcal{G}_{r/N}$. Find the expectation and variance of T_N.

4.22 *Factorial moments.* Let X be a random variable with values in $\mathbb{Z}_+$ and generating function φ_X, and for each $k \in \mathbb{N}$ let $X_{(k)} = X(X-1) \ldots (X-k+1)$. Show that $\varphi_X^{(k)}(1) = \mathbb{E}(X_{(k)})$ when $X \in \mathscr{L}^k$.

4.23 Let $\tau, X_1, X_2, \ldots \in \mathscr{L}^1$ be independent random variables taking values in $\mathbb{Z}_+$, and let $X_1, X_2, \ldots$ be identically distributed. Suppose S_τ is defined as in Problem 4.10. Show that S_τ has generating function $\varphi_{S_\tau} = \varphi_\tau \circ \varphi_{X_1}$, and deduce Wald's identity again. Moreover, find the variance $\mathbb{V}(S_\tau)$ and express it in terms of the expectations and variances of X_1 and τ.

4.24 Consider the situation of Problem 4.23, and let $0 < p < 1$. Determine a discrete density ρ_p on $\mathbb{Z}_+$ such that for every $r > 0$ the following holds. If τ is Poisson distributed with parameter $-r \log p$ and the X_i have distribution density ρ_p, then S_τ has the negative binomial distribution $\overline{\mathcal{B}}_{r,p}$. This ρ_p is called the *logarithmic distribution* with parameter p.

4.25 In the situation of Problem 3.17, use Problem 4.23 to determine the generating function of the number of larvae and hence deduce their distribution.

4.26 [S] *Simple symmetric random walk.* In the situation of Problem 2.7, let

$$\tau = \inf\{2n \geq 2 : S_{2n} = 0\}$$

be the first time during the count at which there is a tie between the two candidates. Find the generating function and the expectation of τ.

Chapter 5

The Law of Large Numbers
and the Central Limit Theorem

In this chapter we discuss two fundamental limit theorems for long-term averages of independent, identically distributed real-valued random variables. The first theorem is the law of large numbers, which is about the convergence of the averages to the common expectation. Depending on the notion of convergence, one distinguishes between the weak and the strong law of large numbers. The second theorem, the central limit theorem, describes how far the averages typically deviate from their expectation in the long run. It is here that the universal importance of the normal distribution is revealed.

5.1 The Law of Large Numbers

5.1.1 The Weak Law of Large Numbers

Our experience shows: If we perform n independent, but otherwise identical experiments, for example measurements in physics, and if we denote by X_i the result of the ith experiment, then for large n the average $\frac{1}{n} \sum_{i=1}^{n} X_i$ is very close to a fixed number. Intuitively, we would like to call this number the expectation of the X_i. This experience serves also as a justification for the interpretation of probabilities as relative frequencies, which states that

$$P(A) \approx \frac{1}{n} \sum_{i=1}^{n} 1_{\{X_i \in A\}} .$$

In words: The probability of an event A coincides with the relative frequency of occurrences of A, provided we perform a large number of independent, identically distributed observations $X_1, \ldots, X_n$.

Does our mathematical model reflect this experience? To answer this, we first have to deal with the question: In which sense can we expect convergence to the mean? The next example shows that, even after a long time, *exact* agreement of the average and the expectation will only happen with a negligible probability. Let us first recall a fact from analysis (see for example [18, 57]), known as *Stirling's formula*:

$$(5.1) \qquad n! = \sqrt{2\pi n} \ n^n e^{-n + \eta(n)} \quad \text{with } 0 < \eta(n) < \frac{1}{12n}.$$

In particular we see that $n! \sim \sqrt{2\pi n} \ n^n e^{-n}$ for $n \to \infty$. Here the symbol '$\sim$' stands for asymptotic equivalence in the sense that the ratio of both sides tends to one, just as in the proof of Theorem (2.14).

(5.2) Example. *Bernoulli trials.* Let $(X_i)_{i \geq 1}$ be a Bernoulli sequence with success probability $p = 1/2$; for a concrete example, we can think of tossing a coin repeatedly. What is the probability it falls heads exactly with the relative frequency $1/2$? Obviously, this is only possible when the number of tosses is even. In this case Stirling's formula yields

$$P\left(\frac{1}{2n} \sum_{i=1}^{2n} X_i = \frac{1}{2}\right) = \mathcal{B}_{2n, \frac{1}{2}}(\{n\}) = \binom{2n}{n} 2^{-2n}$$

$$\underset{n \to \infty}{\sim} \frac{(2n)^{2n} \sqrt{2\pi 2n}}{(n^n \sqrt{2\pi n})^2} 2^{-2n} = \frac{1}{\sqrt{\pi n}} \underset{n \to \infty}{\longrightarrow} 0.$$

That is, for large n, the relative frequency is equal to $1/2$ with minor probability only. This is in accordance with our experience, which shows that the relative frequency settles down at best 'around $1/2$'.

The preceding example suggests that we weaken our conjecture as follows. For all $\varepsilon > 0$,

(5.3)
$$P\left(\left|\frac{1}{n} \sum_{i=1}^{n} X_i - \mathbb{E}(X_1)\right| \leq \varepsilon\right) \underset{n \to \infty}{\longrightarrow} 1;$$

in words, the average is *close* to the expectation with large probability when n is large. The corresponding notion of convergence of random variables has a name:

Definition. Let $Y, Y_1, Y_2, \ldots$ be arbitrary real random variables on a probability space $(\Omega, \mathscr{F}, P)$. The sequence (Y_n) is said to *converge to Y in probability*, abbreviated as $Y_n \overset{P}{\longrightarrow} Y$, if

$$P(|Y_n - Y| \leq \varepsilon) \underset{n \to \infty}{\longrightarrow} 1 \quad \text{for all } \varepsilon > 0.$$

The conjecture (5.3) will indeed prove correct. As a preparation we need the following result.

(5.4) Proposition. *Markov's inequality. Let Y be a real random variable and $f : [0, \infty[\to [0, \infty[$ an increasing function. Then, for all $\varepsilon > 0$ with $f(\varepsilon) > 0$,*

$$P(|Y| \geq \varepsilon) \leq \frac{\mathbb{E}(f \circ |Y|)}{f(\varepsilon)}.$$

Proof. Since $\{f \leq c\}$ is an interval for every c, f and hence $f \circ |Y|$ are random variables. As the latter is non-negative, its expectation is well-defined by Remark (4.2). Thus it follows from Theorem (4.11ab) that

$$f(\varepsilon) \, P(|Y| \geq \varepsilon) = \mathbb{E}(f(\varepsilon) \, 1_{\{|Y| \geq \varepsilon\}}) \leq \mathbb{E}(f \circ |Y|),$$

since $f(\varepsilon) \, 1_{\{|Y| \geq \varepsilon\}} \leq f \circ |Y|$. $\diamond$

(5.5) Corollary. *Chebyshev's inequality, 1867. For all* $Y \in \mathscr{L}^2$ *and* $\varepsilon > 0$,

$$P(|Y - \mathbb{E}(Y)| \geq \varepsilon) \leq \frac{\mathbb{V}(Y)}{\varepsilon^2}.$$

Proof. Apply Proposition (5.4) to $Y' = Y - \mathbb{E}(Y)$ and $f(x) = x^2$. ◇

The above inequalities are named after the Russian mathematicians Андрей А. Марков (1856–1922) and his teacher Пафнутий Л. Чебышёв (1821–1894). One also finds the transliteration Chebyshov, which indicates the correct pronunciation. In fact, Chebyshev used (5.4) already before Markov, and an earlier version of (5.5) appeared in 1853 in some work of Irénée-Jules Bienaymé (1796–1878).

Chebyshev's inequality is remarkably simple and general but, for this very reason, provides only a rough estimate. Here we take advantage of its generality, as it gives us the following general answer to the question posed at the beginning of this section.

(5.6) Theorem. *Weak law of large numbers,* $\mathscr{L}^2$*-version. Let* $(X_i)_{i \geq 1}$ *be pairwise uncorrelated (e.g. independent) random variables in* $\mathscr{L}^2$ *with bounded variance, in that* $v := \sup_{i \geq 1} \mathbb{V}(X_i) < \infty$. *Then for all* $\varepsilon > 0$

$$P\left(\left|\frac{1}{n}\sum_{i=1}^{n}(X_i - \mathbb{E}(X_i))\right| \geq \varepsilon\right) \leq \frac{v}{n\varepsilon^2} \xrightarrow[n \to \infty]{} 0,$$

and thus $\frac{1}{n}\sum_{i=1}^{n}(X_i - \mathbb{E}(X_i)) \xrightarrow{P} 0$. *In particular, if* $\mathbb{E}(X_i) = \mathbb{E}(X_1)$ *for all* i, *then*

$$\frac{1}{n}\sum_{i=1}^{n} X_i \xrightarrow{P} \mathbb{E}(X_1).$$

Proof. We set $Y_n = \frac{1}{n}\sum_{i=1}^{n}(X_i - \mathbb{E}(X_i))$. Then $Y_n \in \mathscr{L}^2$. By Theorem (4.11b), we have

$$\mathbb{E}(Y_n) = \frac{1}{n}\sum_{i=1}^{n} \mathbb{E}(X_i - \mathbb{E}(X_i)) = 0,$$

and by Theorem (4.23ac),

$$\mathbb{V}(Y_n) = \frac{1}{n^2}\sum_{i=1}^{n} \mathbb{V}(X_i) \leq \frac{v}{n}.$$

Hence Theorem (5.6) follows from Corollary (5.5). ◇

We will now present a second version of the weak law of large numbers, which does not require the existence of the variance. To compensate we must assume that the random variables, instead of being pairwise uncorrelated, are even pairwise independent and identically distributed. This paragraph can be skipped at first reading.

(5.7) Theorem. *Weak law of large numbers, $\mathscr{L}^1$-version. Let $(X_i)_{i\geq 1}$ be pairwise independent, identically distributed random variables in $\mathscr{L}^1$. Then,*

$$\frac{1}{n}\sum_{i=1}^{n} X_i \xrightarrow{P} \mathbb{E}(X_1).$$

Proof. We consider the truncated random variables

$$X_i^{\flat} = X_i \, 1_{\{|X_i|\leq i^{1/4}\}}$$

as well as the remainders

$$X_i^{\sharp} = X_i - X_i^{\flat} = X_i \, 1_{\{|X_i|>i^{1/4}\}}$$

and let $Y_n^* = \frac{1}{n}\sum_{i=1}^{n}(X_i^* - \mathbb{E}(X_i^*))$ for $* \in \{\flat, \sharp\}$. We will show that both $Y_n^{\flat} \xrightarrow{P} 0$ and $Y_n^{\sharp} \xrightarrow{P} 0$. Our claim then follows from Lemma (5.8a) below, since $\mathbb{E}(X_i^{\flat}) + \mathbb{E}(X_i^{\sharp}) = \mathbb{E}(X_i) = \mathbb{E}(X_1)$ for all i.

First, Theorem (3.24) shows that the truncated variables $X_i^{\flat}$ are still pairwise independent. Hence, using Bienaymé's identity (4.23c) and the inequality $\mathbb{V}(X_i^{\flat}) \leq \mathbb{E}((X_i^{\flat})^2) \leq i^{1/2}$, we obtain the estimate

$$\mathbb{V}(Y_n^{\flat}) = \frac{1}{n^2}\sum_{i=1}^{n}\mathbb{V}(X_i^{\flat}) \leq n^{-1/2}.$$

Together with Chebyshev's inequality (5.5), this implies the convergence in probability $Y_n^{\flat} \xrightarrow{P} 0$.

For the remainder variables we get

$$\mathbb{E}(|X_i^{\sharp}|) = \mathbb{E}(|X_1|\, 1_{\{|X_1|>i^{1/4}\}})$$
$$= \mathbb{E}(|X_1|) - \mathbb{E}(|X_1|\, 1_{\{|X_1|\leq i^{1/4}\}}) \xrightarrow[i\to\infty]{} 0.$$

Here we used first that X_i and X_1 are identically distributed, and then applied the monotone convergence theorem (4.11c). Consequently,

$$\mathbb{E}(|Y_n^{\sharp}|) \leq \frac{2}{n}\sum_{i=1}^{n}\mathbb{E}(|X_i^{\sharp}|) \xrightarrow[n\to\infty]{} 0.$$

Together with Markov's inequality (5.4), this gives $Y_n^{\sharp} \xrightarrow{P} 0$. ◇

We still have to supply a lemma about convergence in probability. Since $Y_n \xrightarrow{P} Y$ if and only if $Y_n - Y \xrightarrow{P} 0$, we can and will set the limit variable equal to zero.

(5.8) Lemma. *Stability properties of convergence in probability. Let Y_n, Z_n be real random variables and $a_n \in \mathbb{R}$. Then*

(a) $Y_n \xrightarrow{P} 0$ *and* $Z_n \xrightarrow{P} 0$ *jointly imply* $Y_n + Z_n \xrightarrow{P} 0$,

(b) *if* $Y_n \xrightarrow{P} 0$ *and* $(a_n)_{n \geq 1}$ *is bounded, then also* $a_n Y_n \xrightarrow{P} 0$.

Proof. For arbitrary $\varepsilon > 0$ we have, on the one hand,

$$P(|Y_n + Z_n| > \varepsilon) \leq P(|Y_n| > \varepsilon/2) + P(|Z_n| > \varepsilon/2),$$

and on the other hand, if say $|a_n| < c$,

$$P(|a_n Y_n| > \varepsilon) \leq P(|Y_n| > \varepsilon/c).$$

This immediately implies the two statements. ◇

Finally we note that the law of large numbers fails in general if the random variables $(X_i)_{i \geq 1}$ do not have an expectation. For example, let the X_i be independent and Cauchy distributed with parameter $a > 0$, see Problem 2.5. Then by Problem 3.16, for all n, the average $\frac{1}{n} \sum_{i=1}^{n} X_i$ also has the Cauchy distribution with parameter a, so it certainly does not converge to a constant.

5.1.2   Some Examples

We will now discuss a number of applications of the weak law.

(5.9) Example. *Ehrenfests' model in equilibrium; Paul and Tatiana Ehrenfest 1907.* Suppose a box contains $n = 0.25 \cdot 10^{23}$ gas molecules. (This is the order of magnitude of Avogadro's number.) Figure 5.1 shows the situation for small n. Due to the irregularity of the motion, at a fixed time each molecule may be located in the left or right half of the box, each with probability $1/2$ and independently of all other molecules. What is the probability that the proportion of molecules in the left half is slightly larger than that in the right half, say $\geq (1 + 10^{-8})/2$?

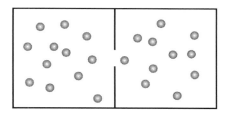

Figure 5.1. The model situation of P. and T. Ehrenfest.

Let $X_1, \ldots, X_n$ be Bernoulli variables with parameter $p = 1/2$. X_i indicates the position of the ith particle. Then, by symmetry and Theorem (5.6),

$$P\left(\frac{1}{n}\sum_{i=1}^{n}X_i \geq \frac{1+10^{-8}}{2}\right) = \frac{1}{2}P\left(\left|\frac{1}{n}\sum_{i=1}^{n}X_i - \frac{1}{2}\right| \geq 5\cdot 10^{-9}\right)$$

$$\leq \frac{1}{2}\frac{1/4}{n(5\cdot 10^{-9})^2} = \frac{1}{2}\frac{1}{n\cdot 10^{-16}} = 2\cdot 10^{-7}.$$

In view of this negligible probability, it is no surprise that such deviations are not observed in real life. (In fact, the preceding use of Chebyshev's inequality gives only a rather poor bound, and the probability in question is actually much smaller. Using the estimate of Problem 5.4, one obtains the incredibly small bound $10^{-500\,000}$.) We will return to the Ehrenfests' model in Examples (6.22) and (6.35), where its evolution in time will be considered.

(5.10) Example. *Bernstein polynomials.* Let $f : [0, 1] \to \mathbb{R}$ be continuous and

$$f_n(p) = \sum_{k=0}^{n} f\left(\frac{k}{n}\right)\binom{n}{k}p^k(1-p)^{n-k} \quad \text{for } p \in [0, 1]$$

the corresponding *Bernstein polynomial of degree n* (invented by the Ukrainian mathematician Sergei N. Bernstein, 1880–1968). Recalling Example (3.39) we can write

$$(5.11) \qquad f_n(p) = \mathbb{E}\left(f\left(\frac{1}{n}\sum_{i=1}^{n}1_{[0,p]}\circ U_i\right)\right),$$

where $U_1, \ldots, U_n$ are independent random variables with uniform distribution on $[0, 1]$. Choose any $\varepsilon > 0$. Since f is uniformly continuous on the compact interval $[0, 1]$, there exists some $\delta > 0$ such that $|f(x) - f(y)| \leq \varepsilon$ when $|x - y| \leq \delta$. Writing $\|f\|$ for the supremum norm of f we thus find

$$\left|f\left(\frac{1}{n}\sum_{i=1}^{n}1_{[0,p]}\circ U_i\right) - f(p)\right| \leq \varepsilon + 2\|f\|\,1_{\left\{\left|\frac{1}{n}\sum_{i=1}^{n}1_{[0,p]}\circ U_i - p\right| \geq \delta\right\}}$$

and hence by Theorems (4.7) and (5.6)

$$|f_n(p) - f(p)| \leq \mathbb{E}\left(\left|f\left(\frac{1}{n}\sum_{i=1}^{n}1_{[0,p]}\circ U_i\right) - f(p)\right|\right)$$

$$\leq \varepsilon + 2\|f\|\,P\left(\left|\frac{1}{n}\sum_{i=1}^{n}1_{[0,p]}\circ U_i - p\right| \geq \delta\right)$$

$$\leq \varepsilon + \frac{2\|f\|\,p(1-p)}{n\delta^2} \leq \varepsilon + \frac{\|f\|}{2n\delta^2}$$

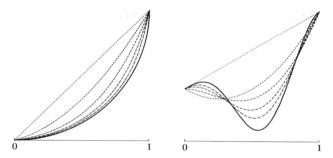

Figure 5.2. The Bernstein polynomials f_{2^k} for $k = 0, \ldots, 4$ (dashed) for two different functions f (solid lines).

for all $p \in [0, 1]$. This shows that f_n converges uniformly to f and therefore provides a constructive version of the Weierstrass approximation theorem. Moreover, the representation (5.11) implies that, for monotone f, the associated Bernstein polynomials are also monotone. Likewise, one can verify that the Bernstein polynomials for a convex f are again convex; see Problem 5.5 and Figure 5.2.

(5.12) Example. *Monte Carlo integration.* Let $f : [0, 1] \to [0, c]$ be a nonnegative bounded measurable function. We would like to find a numerical value for the Lebesgue integral $\int_0^1 f(x)\,dx$ (for instance because there is no method of evaluating this integral exactly). With the help of Theorem (5.6), we can take advantage of chance.

Let $U_1, \ldots, U_n$ be independent random variables with uniform distribution on $[0, 1]$. It then follows from Corollary (4.13) and Theorem (5.6) that

$$P\left(\left|\frac{1}{n}\sum_{i=1}^{n} f(U_i) - \int_0^1 f(x)\,dx\right| \geq \varepsilon\right) \leq \frac{\mathbb{V}(f \circ U_1)}{n\varepsilon^2} \leq \frac{c^2}{n\varepsilon^2}.$$

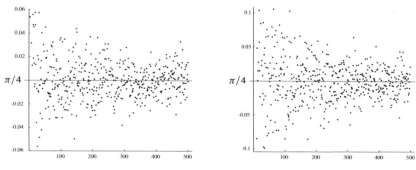

Figure 5.3. Monte-Carlo simulations of the two integrals $\int_0^1 \sqrt{1-x^2}\,dx$ (left) and $\int_0^1 \int_0^1 1_{\{x^2+y^2 \leq 1\}}\,dx\,dy$ (right) up to $n = 500$. The exact value is $\pi/4$.

So, choosing a large number of arguments at random and taking the average of the associated function values we obtain, with large probability, a good approximation of the true value of the integral. The simulation of $U_1, \ldots, U_n$ can be carried out as described in Remark (3.45). We can proceed analogously when f is defined on a bounded subset of $\mathbb{R}^d$; in the case $f = 1_A$, we can thus calculate the volume of A. Figure 5.3 shows two simulation results.

(5.13) Example. *Asymptotic equipartition property and entropy.* Consider a data source sending random signals $X_1, X_2, \ldots$ from an alphabet A. (The randomness is a simplifying assumption, which is reasonable as long as we do not know more about the type of information.) In the mathematical model, A is a finite set and $X_1, X_2, \ldots$ are i.i.d. random variables such that $P(X_i = a) = \rho(a)$ for all $a \in A$, where ρ is a probability density on A.

How can we measure the information content of the source? To answer this, we consider a 'data block' $X_n = (X_1, \ldots, X_n)$. How many Yes/No-questions do we need to determine X_n with error probability no larger than ε? In order to find out, we choose a smallest possible set C contained in the set A^n of all possible words of length n, such that $P(X_n \in C) \geq 1 - \varepsilon$. Next we determine the smallest l such that $|C| \leq 2^l$. Then there exists a bijection φ from C onto a subset C' of $\{0, 1\}^l$. The function φ converts every word $w \in C$ into a binary code word $\varphi(w) \in C'$, from which w can be uniquely recovered by applying the inverse $\psi = \varphi^{-1}$. The words in $A^n \setminus C$ are mapped to an arbitrary code word in C', i.e., we extend φ to an arbitrary mapping $A^n \to C'$, which is no longer injective. Figure 5.4 illustrates the situation. By construction,

$$P(X_n \neq \psi \circ \varphi(X_n)) = P(X_n \notin C) \leq \varepsilon .$$

That is, the word X_n sent out by the source can be recovered from the code word $\varphi(X_n)$ with an error probability no larger than ε. The code word consists of at most l bits, hence it can be identified by l Yes/No-questions. The next corollary – a little taster from information theory – tells us how many bits are needed. We write $\log_2$ for the logarithm to the base 2.

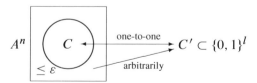

Figure 5.4. Encoding of the words in A^n with probability of error ε.

(5.14) Corollary. *Source coding theorem, Shannon 1948. In the above situation, let*

$$L(n, \varepsilon) = \min\{l \geq 0 : \text{there is a } C \subset A^n \text{ s.t. } |C| \leq 2^l, P(X_n \in C) \geq 1 - \varepsilon\}$$

be the minimum length of a binary code that encodes a data block of length n and can be decoded with an error probability not exceeding ε. Then,

$$\lim_{n \to \infty} \frac{L(n, \varepsilon)}{n} = H(\rho) := -\sum_{a \in A} \rho(a) \log_2 \rho(a).$$

Definition. $H(\rho)$ is called the *entropy* of the discrete density ρ. It measures the information content of the source per transmitted signal, freed from all redundancy.

Proof. Removing the superfluous letters of zero probability, we can achieve that $\rho(a) > 0$ for all $a \in A$.

Step 1. Consider the random variables $Y_i = -\log_2 \rho(X_i)$, $i \geq 1$. The Y_i are i.i.d. and only take finitely many values. We have $\mathbb{E}(Y_i) = H(\rho)$, and $v := \mathbb{V}(Y_i)$ exists. For arbitrary $\delta > 0$ it follows from (5.6) that

$$P\left(\left|\frac{1}{n}\sum_{i=1}^{n} Y_i - H(\rho)\right| > \delta\right) \leq \delta$$

when n is large enough. Since $\frac{1}{n}\sum_{i=1}^{n} Y_i = -\frac{1}{n}\log_2 \rho^{\otimes n}(X_n)$ and $P \circ X_n^{-1} = \rho^{\otimes n}$, this means that the set

$$B_n = \{w \in A^n : 2^{-n(H(\rho)+\delta)} \leq \rho^{\otimes n}(w) \leq 2^{-n(H(\rho)-\delta)}\}$$

is so large that $P(X_n \in B_n) \geq 1 - \delta$. This is called the *asymptotic equipartition property*, because it tells us that typical blocks of length n have roughly the same probability (on an exponential scale).

Step 2. If $\delta \leq \varepsilon$ and B_n is as above, then $C = B_n$ appears among the sets in the definition of $L(n, \varepsilon)$. Hence $L(n, \varepsilon) \leq \min\{l : |B_n| \leq 2^l\}$. As a consequence of the inequality

$$1 = \sum_{w \in A^n} \rho^{\otimes n}(w) \geq \sum_{w \in B_n} 2^{-n(H(\rho)+\delta)} = |B_n| \, 2^{-n(H(\rho)+\delta)},$$

we find that $L(n, \varepsilon) \leq n(H(\rho) + \delta) + 1$. Since $\delta > 0$ was arbitrary, we arrive at the inequality $\limsup_{n \to \infty} L(n, \varepsilon)/n \leq H(\rho)$.

Conversely, if $l = L(n, \varepsilon)$, there exists a set $C \subset A^n$ such that $|C| \leq 2^l$ and $P(X_n \in C) \geq 1 - \varepsilon$. Moreover $\rho^{\otimes n}(w) \, 2^{n(H(\rho)-\delta)} \leq 1$ for $w \in B_n$, which implies

$$2^{L(n,\varepsilon)} \geq |C| \geq |C \cap B_n| \geq \sum_{w \in C \cap B_n} \rho^{\otimes n}(w) \, 2^{n(H(\rho)-\delta)}$$

$$= P(X_n \in C \cap B_n) \, 2^{n(H(\rho)-\delta)} \geq (1 - \varepsilon - \delta) \, 2^{n(H(\rho)-\delta)},$$

so $L(n, \varepsilon) \geq \log_2(1 - \varepsilon - \delta) + n(H(\rho) - \delta)$ and thus $\liminf_{n \to \infty} L(n, \varepsilon)/n \geq H(\rho)$.

$\diamond$

Statistical applications of the weak law of large numbers follow in Sections 7.6, 8.2, 10.2, and others.

5.1.3   The Strong Law of Large Numbers

The weak law of large numbers alone does not quite fulfil our expectations. For example, if we flip a fair coin 100 times, then with some small probability it may happen that the relative frequency differs strongly from $1/2$, but this deviation should vanish gradually, provided we continue flipping the coin for long enough. This intuition is based on a further notion of convergence.

Definition. Let $Y, Y_1, Y_2, \ldots$ be real random variables on $(\Omega, \mathscr{F}, P)$. The sequence $(Y_n)_{n \geq 1}$ is said to *converge to Y P-almost surely* if

$$P\big(\omega \in \Omega : Y_n(\omega) \to Y(\omega)\big) = 1 .$$

(Quite generally, one says that a statement holds almost surely if it holds with probability 1.)

The event $\{Y_n \to Y\}$ is an element of $\mathscr{F}$, cf. Problem 1.14; so its probability is well-defined. Moreover, we note:

(5.15) Remark. *'almost sure'* $\Rightarrow$ *'in probability'*. Almost sure convergence implies convergence in probability, but the converse does not hold.

Proof. For each $\varepsilon > 0$ we can write, using Theorem (1.11e) in the second step,

$$P(|Y_n - Y| \geq \varepsilon) \leq P(\sup_{k \geq n} |Y_k - Y| \geq \varepsilon)$$

$$\xrightarrow[n \to \infty]{} P(|Y_k - Y| \geq \varepsilon \text{ for infinitely many } k) \leq P(Y_n \not\to Y) .$$

This implies the first assertion. Conversely, consider for instance the sequence Y_k (defined on $[0, 1]$ with the uniform distribution $\mathcal{U}_{[0,1]}$), where $Y_k = 1_{[m2^{-n},(m+1)2^{-n}]}$ if $k = 2^n + m$ with $0 \leq m < 2^n$; this sequence converges to zero in probability, but does not converge almost surely. $\diamond$

Our above intuitive picture says: For i.i.d. random variables $(X_i)_{i \geq 1}$ in $\mathscr{L}^2$, we have

$$\frac{1}{n} \sum_{i=1}^{n} X_i \to \mathbb{E}(X_1) \quad \text{almost surely.}$$

We know by Kolmogorov's zero-one law, Theorem (3.49), that

$$P\Big(\frac{1}{n} \sum_{i=1}^{n} X_i \to \mathbb{E}(X_1)\Big) = 0 \text{ or } 1 .$$

Is this probability really equal to 1, as we have intuitively guessed? Indeed, the following theorem gives a positive answer.

(5.16) Theorem. Strong law of large numbers. *If $(X_i)_{i\geq 1}$ are pairwise uncorrelated random variables in $\mathcal{L}^2$ with $v := \sup_{i\geq 1} \mathbb{V}(X_i) < \infty$, then*

$$\frac{1}{n}\sum_{i=1}^{n}(X_i - \mathbb{E}(X_i)) \to 0 \quad \textit{almost surely.}$$

Without proof, we mention that the strong law of large numbers also holds under the conditions of Theorem (5.7), see e.g. Durrett [16].

Proof. Without loss of generality, we can assume that $\mathbb{E}(X_i) = 0$ for all i; otherwise we consider $X_i' := X_i - \mathbb{E}(X_i)$. We set $Y_n = \frac{1}{n}\sum_{i=1}^{n} X_i$.

Step 1. We first show that $Y_{n^2} \to 0$ almost surely. For any $\varepsilon > 0$, Theorem (5.6) implies that $P(|Y_{n^2}| > \varepsilon) \leq v/n^2\varepsilon^2$. Therefore, these probabilities form a convergent series, and the Borel–Cantelli lemma (3.50a) shows that

$$P(\limsup_{n\to\infty}|Y_{n^2}| > \varepsilon) \leq P(|Y_{n^2}| > \varepsilon \text{ for infinitely many } n) = 0.$$

Letting $\varepsilon \to 0$ and using Theorem (1.11e) we therefore obtain

$$P(Y_{n^2} \not\to 0) = P(\limsup_{n\to\infty}|Y_{n^2}| > 0) = \lim_{\varepsilon\to 0} P(\limsup_{n\to\infty}|Y_{n^2}| > \varepsilon) = 0,$$

as desired.

Step 2. For $m \in \mathbb{N}$ let $n = n(m)$ be such that $n^2 \leq m < (n+1)^2$. We compare Y_m with Y_{n^2} and set $S_k = k\,Y_k = \sum_{i=1}^{k} X_i$. According to Chebyshev's inequality (5.5),

$$P\left(|S_m - S_{n^2}| > \varepsilon n^2\right) \leq \varepsilon^{-2} n^{-4}\, \mathbb{V}\left(\sum_{n^2 < i \leq m} X_i\right) \leq \frac{v(m - n^2)}{\varepsilon^2 n^4},$$

and further

$$\sum_{m\geq 1} P\left(|S_m - S_{n(m)^2}| > \varepsilon n(m)^2\right) \leq \frac{v}{\varepsilon^2}\sum_{n\geq 1}\sum_{m=n^2}^{(n+1)^2 - 1}\frac{m - n^2}{n^4}$$

$$= \frac{v}{\varepsilon^2}\sum_{n\geq 1}\sum_{k=1}^{2n}\frac{k}{n^4} = \frac{v}{\varepsilon^2}\sum_{n\geq 1}\frac{2n(2n+1)}{2n^4} < \infty,$$

so once again using Borel–Cantelli as in the first step, we obtain

$$P\left(\left|\frac{S_m}{n(m)^2} - Y_{n(m)^2}\right| \xrightarrow[m\to\infty]{} 0\right) = 1.$$

Since the intersection of two sets of probability 1 has again probability 1, we can combine this with the result of Step 1 and conclude that

$$P\left(\frac{S_m}{n(m)^2} \xrightarrow[m\to\infty]{} 0\right) = 1.$$

Since $|Y_m| = |S_m|/m \leq |S_m|/n(m)^2$, it follows that $P(Y_m \to 0) = 1$. $\diamond$

A classical application of the strong law of large numbers is the statement that 'most' real numbers are normal, in the sense that every digit in their decimal expansion appears with the relative frequency $1/10$. The following theorem presents an even stronger result.

(5.17) Corollary. *Borel's normal number theorem, 1909. If we randomly choose a number $x \in [0, 1]$ according to the uniform distribution $\mathcal{U}_{[0,1]}$ then, with probability 1, x is normal in the following sense:*

For all $q \geq 2$ and $k \geq 1$, every pattern $a = (a_1, \ldots, a_k) \in \{0, \ldots, q-1\}^k$ appears in the q-adic expansion $x = \sum_{i \geq 1} x_i q^{-i}$ with relative frequency q^{-k}. That is,

$$(5.18) \qquad \lim_{n \to \infty} \frac{1}{n} \sum_{i=1}^{n} 1_{\{(x_i, \ldots, x_{i+k-1}) = a\}} = q^{-k}.$$

Proof. Fix $q \geq 2$, $k \geq 1$ and $a \in \{0, \ldots, q-1\}^k$. It suffices to show that

$$\mathcal{U}_{[0,1]}\big(x \in [0, 1] : (5.18) \text{ holds for } q, k, a\big) = 1,$$

since the intersection of countably many events of probability 1 has itself probability 1 (form complements and use Theorem (1.11d)).

To this end, let $(X_n)_{n \geq 1}$ be a sequence of i.i.d. random variables that are uniformly distributed on $\{0, \ldots, q-1\}$. Then $X = \sum_{n \geq 1} X_n q^{-n}$ has the uniform distribution $\mathcal{U}_{[0,1]}$. This was shown in the proof of Theorem (3.26) for $q = 2$, and the proof for arbitrary q is analogous. Hence, we obtain

$$\mathcal{U}_{[0,1]}\big(x \in [0, 1] : (5.18) \text{ holds for } q, k, a\big) = P\big(\lim_{n \to \infty} R_n = q^{-k}\big),$$

where $R_n = \frac{1}{n} \sum_{i=1}^{n} 1_{\{\boldsymbol{X}_i = a\}}$ with $\boldsymbol{X}_i := (X_i, \ldots, X_{i+k-1})$.

For every j, the sequence $(\boldsymbol{X}_{ik+j})_{i \geq 0}$ is independent and $P(\boldsymbol{X}_{ik+j} = a) = q^{-k}$. Theorem (5.16) thus shows that the event

$$C = \bigcap_{j=1}^{k} \Big\{ \lim_{m \to \infty} \frac{1}{m} \sum_{i=0}^{m-1} 1_{\{\boldsymbol{X}_{ik+j} = a\}} = q^{-k} \Big\}$$

occurs with probability 1. Since

$$R_{mk} = \frac{1}{k} \sum_{j=1}^{k} \frac{1}{m} \sum_{i=0}^{m-1} 1_{\{\boldsymbol{X}_{ik+j} = a\}},$$

it follows that $C \subset \{\lim_{m \to \infty} R_{mk} = q^{-k}\}$, so that also this last event occurs with probability 1. But the sandwich inequality

$$\frac{m}{m+1} R_{mk} \leq R_n \leq \frac{m+1}{m} R_{(m+1)k} \quad \text{for } mk \leq n \leq (m+1)k$$

implies that this event coincides with the event $\{\lim_{n \to \infty} R_n = q^{-k}\}$ we are interested in. This gives the desired result. $\diamond$

5.2 Normal Approximation of Binomial Distributions

Let $(X_i)_{i \geq 1}$ be a Bernoulli sequence with parameter $0 < p < 1$; in the case $p = 1/2$ we may think of repeated coin flips, and for general (rational) p we can imagine sampling with replacement from an urn containing red and white balls. We consider the partial sums $S_n = \sum_{i=1}^{n} X_i$, which represent the 'number of successes' in n experiments. The question then is: How much do the S_n fluctuate around their expectation np, i.e., what is the order of magnitude of the deviations $S_n - np$ in the limit as $n \to \infty$? So far, we merely know that $S_n - np$ is of smaller order than n; this is simply another way of stating the law of large numbers.

A more precise version of our question is as follows: For which sequences (a_n) in $\mathbb{R}_+$ do the probabilities $P(|S_n - np| \leq a_n)$ remain non-trivial, meaning they do not tend to 0 or 1? The following remark tells us that (a_n) should grow neither faster nor slower than $\sqrt{n}$.

(5.19) Remark. *Order of magnitude of fluctuations.* In the above situation we have

$$P\big(|S_n - np| \leq a_n\big) \xrightarrow[n \to \infty]{} \begin{cases} 1 & \text{if } a_n/\sqrt{n} \to \infty, \\ 0 & \text{if } a_n/\sqrt{n} \to 0. \end{cases}$$

Proof. By Chebyshev's inequality (5.5), the probability in question is at least $1 - np(1-p)a_n^{-2}$, which converges to 1 in the first case. Conversely, by Theorem (2.9), the probability coincides with

$$\sum_{k:\, |k-np| \leq a_n} \mathcal{B}_{n,p}(\{k\}),$$

and this sum is at most $(2a_n + 1) \max_k \mathcal{B}_{n,p}(\{k\})$. To determine the maximum we note that $\mathcal{B}_{n,p}(\{k\}) > \mathcal{B}_{n,p}(\{k-1\})$ if and only if $(n-k+1)p > k(1-p)$ or, equivalently, $k < (n+1)p$; this follows from an easy computation. The maximum of $\mathcal{B}_{n,p}(\{k\})$ is thus attained at $k = k_{n,p} := \lfloor (n+1)p \rfloor$. Using Stirling's formula (5.1), we further find that

$$\mathcal{B}_{n,p}(\{k_{n,p}\}) \underset{n \to \infty}{\sim} \frac{1}{\sqrt{2\pi p(1-p)n}} \left(\frac{np}{k_{n,p}}\right)^{k_{n,p}} \left(\frac{n(1-p)}{n-k_{n,p}}\right)^{n-k_{n,p}},$$

and both the second and the third factor on the right are bounded by e. Indeed, since $|k_{n,p} - np| < 1$, the second factor is less than $(1 + 1/k_{n,p})^{k_{n,p}} \leq e$, and a similar estimate applies to the third factor. This means that $(2a_n + 1)\mathcal{B}_{n,p}(\{k_{n,p}\})$ is bounded by a multiple of $a_n/\sqrt{n}$, and thus tends to 0 in the second case. $\diamond$

Now that we know that the deviation between S_n and its expectation np typically grows with $\sqrt{n}$, we can focus on the case when $a_n = c\sqrt{n}$ for some $c > 0$. Our question thus takes the form: Do the probabilities $P(|S_n - np| \leq c\sqrt{n})$ really converge to a non-trivial limit, and if so, to which one?

Since S_n is $\mathcal{B}_{n,p}$-distributed by Theorem (2.9), we must analyse the probabilities $\mathcal{B}_{n,p}(\{k\})$ for $|k - np| \le c\sqrt{n}$. We can again use Stirling's formula (5.1), which gives, uniformly for all such k,

$$\mathcal{B}_{n,p}(\{k\}) = \binom{n}{k} p^k q^{n-k}$$

$$\underset{n \to \infty}{\sim} \frac{1}{\sqrt{2\pi}} \sqrt{\frac{n}{k(n-k)}} \left(\frac{np}{k}\right)^k \left(\frac{nq}{n-k}\right)^{n-k}$$

$$\underset{n \to \infty}{\sim} \frac{1}{\sqrt{2\pi npq}} e^{-n\,h(k/n)},$$

where we set $q = 1 - p$ and

$$h(s) = s \log \frac{s}{p} + (1-s) \log \frac{1-s}{q}, \quad 0 < s < 1.$$

(This h also appears in Problem 5.4 and is a special case of the relative entropy, which will be discussed in Remark (7.31).) As $|k/n - p| \le c/\sqrt{n}$, we are interested in the function h near p. Clearly $h(p) = 0$, and the derivatives $h'(s) = \log \frac{s}{p} - \log \frac{1-s}{q}$ and $h''(s) = \frac{1}{s(1-s)}$ exist. In particular, we have $h'(p) = 0$ and $h''(p) = 1/(pq)$. This leads to the Taylor approximation

$$h(k/n) = \frac{(k/n - p)^2}{2pq} + O(n^{-3/2})$$

uniformly for all specified k. (The Landau symbol $O(f(n))$ has its usual meaning, in that it represents a term satisfying $|O(f(n))| \le K f(n)$ for all n and some constant $K < \infty$.) With the help of the standardised variables

$$x_n(k) = \frac{k - np}{\sqrt{npq}}$$

we can rewrite this as

$$n\,h(k/n) = x_n(k)^2/2 + O(n^{-1/2}),$$

and altogether we get

$$\mathcal{B}_{n,p}(\{k\}) \underset{n \to \infty}{\sim} \frac{1}{\sqrt{npq}} \frac{1}{\sqrt{2\pi}} e^{-x_n(k)^2/2}$$

uniformly for all k with $|x_n(k)| \le c' = c/\sqrt{pq}$. So, after rescaling, the binomial probabilities approach the Gaussian bell curve

$$\phi(x) = \frac{1}{\sqrt{2\pi}} e^{-x^2/2},$$

the density function of the standard normal distribution $\mathcal{N}_{0,1}$. We have thus proved:

(5.20) Theorem. Normal approximation of binomial distributions, local form; Abraham de Moivre 1733, Pierre-Simon Laplace 1812. *Let* $0 < p < 1$, $q = 1-p$, *and* $c > 0$. *Then, setting* $x_n(k) = (k - np)/\sqrt{npq}$, *one has*

$$\lim_{n \to \infty} \max_{k:\, |x_n(k)| \le c} \left| \frac{\sqrt{npq}\; \mathcal{B}_{n,p}(\{k\})}{\phi(x_n(k))} - 1 \right| = 0.$$

As a matter of fact, the bell curve should be named after de Moivre and Laplace instead of Gauss; see for instance Dudley [15, p. 259] and the literature cited there. Incidentally, we could relax the condition $|x_n(k)| \le c$ to $|x_n(k)| \le c_n$, where (c_n) is a sequence with $c_n^3/\sqrt{n} \to 0$; this follows immediately from the above proof.

The *local limit theorem*, as the preceding result is often called for short, can be visualised by looking at the *histograms* of the binomial distributions. The actual histogram of $\mathcal{B}_{n,p}$ consists of columns of height $\mathcal{B}_{n,p}(\{k\})$ and width 1 over the intervals $[k - \frac{1}{2}, k + \frac{1}{2}]$, $k = 0, \ldots, n$. For large n it is of limited use since it consists of very many rectangles of hardly visible height, the rectangles of maximal (but still small) height being 'far to the right' near np. So, to see more structure, one should *standardise* the histogram by shifting it by np to the left, then squeezing its width by dividing by $\sigma_n := \sqrt{npq}$, and stretching its height by a factor of σ_n. In other words: The rectangle with base $[k - \frac{1}{2}, k + \frac{1}{2}]$ and height $\mathcal{B}_{n,p}(\{k\})$ in the histogram of the binomial distribution is replaced by the rectangle with base $[x_n(k - \frac{1}{2}), x_n(k + \frac{1}{2})]$ and height $\sigma_n \mathcal{B}_{n,p}(\{k\})$. During this transformation, the area of the individual rectangles

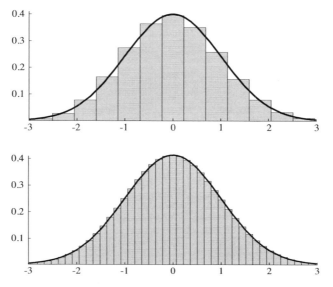

Figure 5.5. Standardised histograms of binomial distributions for $p = 0.4$ and $n = 20$ (top) resp. $n = 200$ (bottom), as well as the approximating normal density.

and the total area 1 remain unchanged. The local normal approximation now asserts that the 'upper boundary' of the standardised histogram approaches the bell-shaped Gaussian curve locally uniformly as $n \to \infty$. This is clearly seen in Figure 5.5.

It is also evident from the picture that the area of the standardised histogram over an interval $[a, b]$ approaches the corresponding area under the Gaussian curve. This is the content of the following corollary. For $c \in \mathbb{R}$ let

$$(5.21) \qquad \Phi(c) := \int_{-\infty}^{c} \phi(x)\, dx = \mathcal{N}_{0,1}(]-\infty, c])$$

be the (cumulative) distribution function of the standard normal distribution $\mathcal{N}_{0,1}$; for a plot see Figure 5.7 on p. 143.

(5.22) Corollary. Normal approximation of binomial distributions, integral form; de Moivre–Laplace. *Let $0 < p < 1$, $q = 1 - p$, and $0 \le k \le l \le n$. Defining the error terms $\delta_{n,p}(k, l)$ by*

$$(5.23) \qquad \mathcal{B}_{n,p}(\{k, \ldots, l\}) = \Phi\Big(\frac{l + \frac{1}{2} - np}{\sqrt{npq}}\Big) - \Phi\Big(\frac{k - \frac{1}{2} - np}{\sqrt{npq}}\Big) + \delta_{n,p}(k, l),$$

we have $\delta_{n,p} := \max_{0 \le k \le l \le n} |\delta_n(k, l)| \to 0$ for $n \to \infty$.

Depending on the viewpoint, the terms $\pm\frac{1}{2}$ on the right-hand side of (5.23) are known as *discreteness* or *continuity correction*. As we can infer from Figure 5.5 and the proof below, they account for the width of the columns in the standardised binomial histogram. They lead to a noticeable improvement in the approximation; see Problem 5.16. The actual maximal error $\delta_{n,p}$ is shown in Figure 5.6. In the limit $n \to \infty$, the correction terms $\pm\frac{1}{2}$ are obviously negligible due to the uniform continuity of Φ, and we will therefore often omit them later.

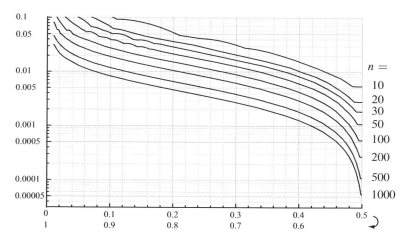

Figure 5.6. The maximal error $\delta_{n,p}$ as a function of p for various values of n. (Note the symmetry $\delta_{n,p} = \delta_{n,1-p}$.)

Proof. Let us introduce the shorthands $\sigma_n = \sqrt{npq}$ and $\delta_n = 1/(2\sigma_n)$. Let $\varepsilon > 0$ be given and $c > 0$ be so large that $\Phi(c - \delta_1) > 1 - \varepsilon$, and thus $\Phi(-c + \delta_1) < \varepsilon$. For $j \in \mathbb{Z}$ let

$$\Delta_j \Phi = \Phi\left(x_n(j + \tfrac{1}{2})\right) - \Phi\left(x_n(j - \tfrac{1}{2})\right)$$

be the Gaussian integral over the subinterval $[x_n(j - \tfrac{1}{2}), x_n(j + \tfrac{1}{2})]$ of length $2\delta_n$. Then

$$\max_{j : |x_n(j)| \le c} \left| \frac{\Delta_j \Phi}{\sigma_n^{-1} \phi(x_n(j))} - 1 \right| \xrightarrow[n \to \infty]{} 0 \,.$$

To see this, it suffices to write

$$\frac{\Delta_j \Phi}{\sigma_n^{-1} \phi(x_n(j))} = \frac{1}{2\delta_n} \int_{x_n(j) - \delta_n}^{x_n(j) + \delta_n} e^{(x_n(j) + x)(x_n(j) - x)/2} \, dx$$

and to note that, for all j with $|x_n(j)| \le c$ and all x in the domain of integration, the exponent in the integrand is bounded by $(2c + \delta_n)\delta_n/2 \to 0$. Combining this with Theorem (5.20), we find that

$$\max_{j : |x_n(j)| \le c} \left| \frac{\mathcal{B}_{n,p}(\{j\})}{\Delta_j \Phi} - 1 \right| < \varepsilon$$

for all sufficiently large n.

Now let $0 \le k \le l \le n$ and $\delta_n(k, l)$ be defined by (5.23). Consider first the case when $|x_n(k)| \le c$ and $|x_n(l)| \le c$. Then

$$\left| \delta_n(k, l) \right| \le \sum_{j=k}^{l} \left| \mathcal{B}_{n,p}(\{j\}) - \Delta_j \Phi \right| < \varepsilon \sum_{j=k}^{l} \Delta_j \Phi < \varepsilon \,,$$

since the last sum has the value $\Phi(x_n(l + \tfrac{1}{2})) - \Phi(x_n(k - \tfrac{1}{2})) < 1$. In particular, setting $k_c = \lceil np - c\sigma_n \rceil$ and $l_c = \lfloor np + c\sigma_n \rfloor$, we find

$$\mathcal{B}_{n,p}(\{k_c, \ldots, l_c\}) > \Phi\left(x_n(l_c + \tfrac{1}{2})\right) - \Phi\left(x_n(k_c - \tfrac{1}{2})\right) - \varepsilon > 1 - 3\varepsilon$$

due to our choice of c. That is, the tails of the binomial distribution to the left of k_c and to the right of l_c together have at most probability 3ε and are thus negligible up to an error term of at most 3ε. Similarly, the tails of the normal distribution to the left of $-c$ and to the right of c are negligible up to an error term of at most 2ε. Consequently, for all sufficiently large n and arbitrary $0 \le k \le l \le n$, one has the inequality $|\delta_n(k, l)| < 6\varepsilon$. $\diamond$

Corollary (5.22) has the practical advantage of saving tedious calculations: There is no need to explicitly calculate the binomial probabilities $\mathcal{B}_{n,p}(\{k, \ldots, l\})$ (which can be slightly laborious for large n due to the binomial coefficients), if instead we know the distribution function Φ. Although the latter cannot be represented in a closed form, there are good numerical approximations available. Also, since it does not depend on n and p anymore, we can easily summarise its values in a table, see Table A in the appendix. In Remark (5.30) we will provide a general though rough bound for the approximation error.

> One might argue that approximations and tables have lost their importance in times of efficient computer programs. For instance, in Mathematica you can type the command CDF[NormalDistribution[0,1],c] to obtain the value $\Phi(c)$ to any required precision. Similarly, the command CDF[BinomialDistribution[n,p],k] gives you the true probability $\mathcal{B}_{n,p}(\{0,\ldots,k\})$. Nevertheless, the normal approximation is of fundamental importance, simply because of its universality. As we will see in Theorem (5.29), it is not confined to the Bernoulli case, but works for a large class of distributions. And no computer program will be able to provide you with appropriate algorithms for any distribution you like.

To answer the question from the beginning of this section, we now only need to reformulate Corollary (5.22) in the language of random variables. In this way we arrive at a special case of the central limit theorem; the general case will follow in (5.29).

(5.24) Corollary. *Central limit theorem for Bernoulli sequences. Let* $(X_i)_{i\geq 1}$ *be a Bernoulli sequence with parameter* $p \in \,]0,1[$, *and let* $q = 1 - p$. *For every* $n \geq 1$ *set*

$$S_n = \sum_{i=1}^n X_i \quad and \quad S_n^* = \frac{S_n - np}{\sqrt{npq}}.$$

Then for all $a < b$ *the following limit exists and satisfies*

$$\lim_{n\to\infty} P(a \leq S_n^* \leq b) = \int_a^b \phi(x)\,dx = \Phi(b) - \Phi(a).$$

This convergence is uniform in a, b. *In particular, we have*

(5.25)
$$\lim_{n\to\infty} P(S_n^* \leq b) = \Phi(b)$$

uniformly in $b \in \mathbb{R}$.

Proof. In view of Theorem (2.9), we have $P(a \leq S_n^* \leq b) = \mathcal{B}_{n,p}(\{k_{n,a},\ldots,l_{n,b}\})$ with $k_{n,a} = \lceil np + \sigma_n a\rceil$, $l_{n,b} = \lfloor np + \sigma_n b\rfloor$, and $\sigma_n = \sqrt{npq}$. So, by (5.23), it only remains to be shown that $\Phi(x_n(k_{n,a} - \frac{1}{2}))$ converges to $\Phi(a)$ uniformly in a, and similarly for $l_{n,b} + \frac{1}{2}$. But this follows from the mean value theorem because $|x_n(k_{n,a} - \frac{1}{2}) - a| \leq 1/\sigma_n$ and $\Phi' = \phi \leq 1/\sqrt{2\pi}$. $\diamond$

We conclude this section with an application of the normal approximation to mathematical finance. Some statistical applications will follow in Part II, in particular in Section 8.2 and Chapter 11.

(5.26) Example. *The Black–Scholes formula.* We return to the problem of finding the fair price for a European call option, recall Example (4.17). There we worked under the assumption that the price changes at discrete time points. In a global economy, at any given time there is always a stock market where people are trading, so the assumption of continuous time might be more adequate. So let us consider an interval $[0, t]$ of continuous time. To make contact with Example (4.17) we set up a discrete approximation, by allowing to trade only at the discrete time points nt/N, $n = 0, \ldots, N$. We assume that the price evolution from point to point can be described by a geometric random walk with parameters $\sigma_N = \mu_N$. As a consequence of Bienaymé's identity, it seems plausible to assume that the variances of the price increments are proportional to the time increments. This means we can define the volatility per discrete time step to be $\sigma_N = \sigma \sqrt{t/N}$, where the new σ can be interpreted as volatility per unit time. In the N-step approximation, the Black–Scholes parameter $p^* = p^*(N)$ of Proposition (4.20) thus takes the form

$$p^*(N) = \frac{e^{\sigma \sqrt{t/N}} - 1}{e^{2\sigma \sqrt{t/N}} - 1},$$

and the parameter $p^\circ = p^\circ(N)$ in formula (4.22) reads

$$p^\circ(N) = \frac{e^{2\sigma \sqrt{t/N}} - e^{\sigma \sqrt{t/N}}}{e^{2\sigma \sqrt{t/N}} - 1} = 1 - p^*(N).$$

Moreover, $a_N = \frac{N}{2} + \sqrt{N} \, \frac{\log K}{2\sigma \sqrt{t}}$.

We now substitute these values into formula (4.22). To apply Corollary (5.24) we also replace S_N, which is binomially distributed, by the standardised variables $S_N^\bullet = (S_N - Np^\bullet)/\sqrt{Np^\bullet(1-p^\bullet)}$ with $\bullet \in \{*, \circ\}$. So we find that the fair price $\Pi^* = \Pi^*(N)$ in the N-step approximation is given by

$$\Pi^*(N) = P^\circ(S_N > a_N) - K\,P^*(S_N > a_N)$$
$$= P^\circ(S_N^\circ > a_N^\circ) - K\,P^*(S_N^* > a_N^*),$$

where

$$a_N^\bullet = \sqrt{N} \, \frac{\frac{1}{2} - p^\bullet(N)}{\sqrt{p^\bullet(N)(1 - p^\bullet(N))}} + \frac{\log K}{2\sigma \sqrt{t}\, p^\bullet(N)(1 - p^\bullet(N))}$$

for $\bullet \in \{*, \circ\}$. Further, by Taylor's approximation,

$$\sqrt{N}\left(\frac{1}{2} - p^*(N)\right) = \frac{\sqrt{N}}{2} \, \frac{\cosh \sigma \sqrt{t/N} - 1}{\sinh \sigma \sqrt{t/N}}$$
$$= \frac{\sqrt{N}}{2} \, \frac{\sigma^2 t/2N + O(N^{-2})}{\sigma \sqrt{t/N} + O(N^{-3/2})} \xrightarrow[N \to \infty]{} \frac{\sigma \sqrt{t}}{4},$$

and, in particular, $p^*(N) \to 1/2$. It follows that $a_N^* \to h + \sigma \sqrt{t}/2$ and $a_N^\circ \to h - \sigma \sqrt{t}/2$, where $h = (\log K)/\sigma \sqrt{t}$. By the uniform convergence in (5.25) (and the continuity of Φ), we finally obtain

$$\Pi^*(N) \xrightarrow[N \to \infty]{} 1 - \Phi(h - \sigma \sqrt{t}/2) - K \left(1 - \Phi(h + \sigma \sqrt{t}/2)\right)$$

(5.27)
$$= \Phi(-h + \sigma \sqrt{t}/2) - K \, \Phi(-h - \sigma \sqrt{t}/2) \, .$$

The last expression (5.27) is the famous *Black–Scholes formula* for the fair price of an option, which is implemented into every banking software. (Here we have chosen the currency unit such that the stock price X_0 at time 0 is just 1; in the general case, we have to replace K by K/X_0 and multiply the whole expression by X_0. If the value of the bond is not constant in time, but grows with interest rate $r > 0$, we must replace K by the discounted price Ke^{-rt}.) Unfortunately, our derivation of the Black–Scholes formula as the limit of the binomial model does not provide us with the corresponding hedging strategy; this would require working directly in the continuous time setting.

> The Black–Scholes formula should not prevent us from treating options with caution. One reason is that the underlying model for the stock price, namely the continuous limit of the geometric random walk (the so-called geometric Brownian motion), reflects reality only in a rough approximation, as can be seen by comparing the simulations on p. 104 with actual stock charts. On the other hand, the collapse of the LTCM (Long Term Capital Management) hedge fund in September 1998, which was consulted by the Nobel Laureates Merton and Scholes, has shown that even a recognised expertise in mathematical finance can fail when the market follows unexpected influences.

5.3   The Central Limit Theorem

The topic of this section is the universality of the normal approximation. We will show that the convergence result of Corollary (5.24) does not only hold for Bernoulli sequences, but in much greater generality. As a first step, we analyse the type of convergence in Corollary (5.24). Note that statement (5.25) just means that the distribution function of S_n^* converges uniformly to that of $\mathcal{N}_{0,1}$. The uniformity of this convergence is an immediate consequence of the continuity of the limit Φ and the monotonicity of the distribution functions $F_{S_n^*}$, see statement (5.28(c)) below. In the case of a discontinuous limit function, however, uniform convergence is much too strong, and the appropriate notion of convergence is the following.

Definition. Let $(Y_n)_{n \geq 1}$ be a sequence of real random variables defined on arbitrary probability spaces. Y_n is said to *converge in distribution* to a real random variable Y resp. to its distribution Q on $(\mathbb{R}, \mathscr{B})$, if

$$F_{Y_n}(c) \xrightarrow[n \to \infty]{} F_Y(c) = F_Q(c)$$

for all points $c \in \mathbb{R}$ at which F_Y is continuous. We then write $Y_n \xrightarrow{d} Y$ resp. Q. (One also speaks of *convergence in law* and then prefers to write $Y_n \xrightarrow{\mathscr{L}} Y$.)

Note that the notion of convergence in distribution does not involve the random variables themselves, but only their distributions. (The corresponding convergence concept for probability measures on $(\mathbb{R}, \mathscr{B})$ is called *weak convergence*.) The following characterisation of convergence in distribution shows why it is sensible to require the convergence of the distribution functions only at the continuity points of the limit. In particular, statement (5.28b) indicates how the notion of convergence in distribution can be extended to random variables taking values in arbitrary topological spaces. Further characterisations will follow in Remark (11.1) and Problem 11.1. We write $\|\cdot\|$ for the supremum norm.

(5.28) Remark. *Characterisation of convergence in distribution.* Under the same assumptions as in the above definition, the following statements are equivalent.

(a) $Y_n \xrightarrow{d} Y$.

(b) $\mathbb{E}(f \circ Y_n) \to \mathbb{E}(f \circ Y)$ for all continuous and bounded functions $f : \mathbb{R} \to \mathbb{R}$.

If F_Y is continuous, then the following is also equivalent.

(c) F_{Y_n} converges uniformly to F_Y, i.e., $\|F_{Y_n} - F_Y\| \to 0$.

Proof. (a) $\Rightarrow$ (b): Given any $\varepsilon > 0$, we can find points $c_1 < \cdots < c_k$ at which F_Y is continuous and such that $F_Y(c_1) < \varepsilon$, $F_Y(c_k) > 1 - \varepsilon$, and $|f(x) - f(c_i)| < \varepsilon$ for $c_{i-1} \leq x \leq c_i$, $1 < i \leq k$. This is possible because F_Y satisfies the limit relations (1.29) and has only countably many discontinuities by monotonicity, and f is uniformly continuous on every compact interval. Then we can write

$$\mathbb{E}(f \circ Y_n) = \sum_{i=2}^{k} \mathbb{E}\left(f \circ Y_n \, 1_{\{c_{i-1} < Y_n \leq c_i\}}\right) + \mathbb{E}\left(f \circ Y_n \, 1_{\{Y_n \leq c_1\} \cup \{Y_n > c_k\}}\right)$$

$$\leq \sum_{i=2}^{k} (f(c_i) + \varepsilon)\left[F_{Y_n}(c_i) - F_{Y_n}(c_{i-1})\right] + 2\varepsilon \, \|f\|,$$

and for $n \to \infty$ the last expression tends to a limit, which agrees with $\mathbb{E}(f \circ Y)$ up to an error term of $2\varepsilon(1 + 2\|f\|)$. For $\varepsilon \to 0$ it follows that $\limsup_{n \to \infty} \mathbb{E}(f \circ Y_n) \leq \mathbb{E}(f \circ Y)$, and using the same inequality for $-f$, we obtain statement (b).

(b) $\Rightarrow$ (a): For $c \in \mathbb{R}$ and $\delta > 0$, let f be continuous and bounded such that $1_{]-\infty,c]} \leq f \leq 1_{]-\infty,c+\delta]}$. Then,

$$\limsup_{n \to \infty} F_{Y_n}(c) \leq \lim_{n \to \infty} \mathbb{E}(f \circ Y_n) = \mathbb{E}(f \circ Y) \leq F_Y(c+\delta).$$

As F_Y is right-continuous, letting $\delta \to 0$ gives $\limsup_{n \to \infty} F_{Y_n}(c) \leq F_Y(c)$. If c is a point of continuity of F_Y, we know that also $F_Y(c-\delta) \to F_Y(c)$ as $\delta \to 0$. An analogous argument thus yields the reverse inequality $\liminf_{n \to \infty} F_{Y_n}(c) \geq F_Y(c)$.

(a) $\Rightarrow$ (c): Let $k \in \mathbb{N}$ be arbitrary and $\varepsilon = 1/k$. If F_Y is continuous, then, by the intermediate value theorem, we can find $c_i \in \mathbb{R}$ such that $F_Y(c_i) = i/k, 0 < i < k$. These c_i partition $\mathbb{R}$ in k intervals, on each of which F_Y increases by exactly ε. Since F_{Y_n} is also increasing, we obtain

$$\|F_{Y_n} - F_Y\| \le \varepsilon + \max_{0<i<k} |F_{Y_n}(c_i) - F_Y(c_i)|,$$

and the maximum tends to 0 as $n \to \infty$. (c) $\Rightarrow$ (a) holds trivially. $\diamond$

The following theorem is the generalisation of Corollary (5.24) promised at the beginning. It explains the central role played by the normal distribution in stochastics.

(5.29) Theorem. *Central limit theorem; A. M. Lyapunov 1901, J. W. Lindeberg 1922, P. Lévy 1922. Let $(X_i)_{i \ge 1}$ be a sequence of independent, identically distributed random variables in $\mathscr{L}^2$ with $\mathbb{E}(X_i) = m$, $\mathbb{V}(X_i) = v > 0$. Then,*

$$S_n^* := \frac{1}{\sqrt{n}} \sum_{i=1}^{n} \frac{X_i - m}{\sqrt{v}} \xrightarrow{d} \mathcal{N}_{0,1},$$

meaning that $\|F_{S_n^*} - \Phi\| \to 0$ for $n \to \infty$.

Before embarking on the proof, we add some comments.

(5.30) Remark. *Discussion of the central limit theorem.* (a) Why is it just the standard normal distribution $\mathcal{N}_{0,1}$ which appears as the limiting distribution? This can be made plausible by the following stability property of the normal distribution. If the (X_i) are independent and distributed according to $\mathcal{N}_{0,1}$ then, for arbitrary n, S_n^* has also the $\mathcal{N}_{0,1}$ distribution; this follows from Problem 2.15 and Example (3.32). In fact, $\mathcal{N}_{0,1}$ is the only probability measure Q on $\mathbb{R}$ with existing variance which exhibits this property. For, if the (X_i) are i.i.d. with distribution Q and each S_n^* has distribution Q then $S_n^* \xrightarrow{d} Q$, so $Q = \mathcal{N}_{0,1}$ by the central limit theorem (5.29).

(b) Without the assumption $X_i \in \mathscr{L}^2$, the central limit theorem fails in general. For an extreme counterexample, consider the case when the X_i are Cauchy distributed for any parameter $a > 0$. Then $\sum_{i=1}^{n} X_i / \sqrt{n}$ has the Cauchy distribution with parameter $a\sqrt{n}$, so it tails off to $\pm\infty$ for $n \to \infty$; see also the discussion at the end of Section 5.1.1. In contrast to the law of large numbers, even the condition $X_i \in \mathscr{L}^1$ does not suffice, and likewise we cannot replace the assumption of independence by pairwise independence or even pairwise uncorrelatedness. Counterexamples can be found in Stoyanov [62].

(c) If the third moments of the X_i exist (i.e., $X_i \in \mathscr{L}^3$), then we can bound the rate of convergence as follows:

$$\|F_{S_n^*} - \Phi\| \le 0.8 \, \frac{\mathbb{E}(|X_1 - \mathbb{E}(X_1)|^3)}{v^{3/2}} \, \frac{1}{\sqrt{n}} .$$

This is the Berry–Esséen theorem; a proof can be found in [16, 59], for example.

Proof of Theorem (5.29). Without loss of generality, suppose that $m = 0$ and $v = 1$; otherwise consider $X_i' = (X_i - m)/\sqrt{v}$. By Remark (5.28), it suffices to show that $\mathbb{E}(f \circ S_n^*) \to \mathbb{E}_{\mathcal{N}_{0,1}}(f)$ for every bounded continuous function $f : \mathbb{R} \to \mathbb{R}$. In fact, we can even assume f is twice continuously differentiable with bounded and uniformly continuous derivatives f' and f''; the reason being that in the proof of the implication (5.28b) $\Rightarrow$ (5.28a) we can approximate $1_{]-\infty,c]}$ by such f.

Now, let $(Y_i)_{i \geq 1}$ be an i.i.d. sequence of standard normal random variables, and suppose this sequence is also independent of $(X_i)_{i \geq 1}$. (By Theorem (3.26) we can find such Y_i; if necessary, we must switch to a new probability space.) According to Remark (5.30b), $T_n^* := \sum_{i=1}^n Y_i/\sqrt{n}$ also has the $\mathcal{N}_{0,1}$ distribution, so the required convergence statement takes the form $|\mathbb{E}(f \circ S_n^* - f \circ T_n^*)| \to 0$.

The advantage of this representation is that the difference $f \circ S_n^* - f \circ T_n^*$ can be expressed as a telescope sum. To simplify the notation, we set

$$X_{i,n} = \frac{X_i}{\sqrt{n}}, \quad Y_{i,n} = \frac{Y_i}{\sqrt{n}}, \quad W_{i,n} = \sum_{j=1}^{i-1} Y_{j,n} + \sum_{j=i+1}^{n} X_{j,n}.$$

Then we can write

$$(5.31) \qquad f \circ S_n^* - f \circ T_n^* = \sum_{i=1}^n \left[f(W_{i,n} + X_{i,n}) - f(W_{i,n} + Y_{i,n}) \right],$$

because $W_{i,n} + X_{i,n} = W_{i-1,n} + Y_{i-1,n}$ for $1 < i \leq n$. Since $X_{i,n}$ and $Y_{i,n}$ are small and f is smooth, the next step is a Taylor approximation for the terms on the right. We can write

$$f(W_{i,n} + X_{i,n}) = f(W_{i,n}) + f'(W_{i,n})\, X_{i,n} + \tfrac{1}{2} f''(W_{i,n})\, X_{i,n}^2 + R_{X,i,n}\,,$$

where

$$R_{X,i,n} = \tfrac{1}{2}\, X_{i,n}^2 \left[f''(W_{i,n} + \vartheta X_{i,n}) - f''(W_{i,n}) \right]$$

for some $0 \leq \vartheta \leq 1$. So, on the one hand, $|R_{X,i,n}| \leq X_{i,n}^2 \| f'' \|$. On the other hand, since f'' is uniformly continuous, for any given $\varepsilon > 0$ one can find some $\delta > 0$ such that $|R_{X,i,n}| \leq X_{i,n}^2\, \varepsilon$ for $|X_{i,n}| \leq \delta$. Altogether, this yields the estimate

$$|R_{X,i,n}| \leq X_{i,n}^2 \left[\varepsilon\, 1_{\{|X_{i,n}| \leq \delta\}} + \| f'' \|\, 1_{\{|X_{i,n}| > \delta\}} \right].$$

A similar Taylor approximation applies to $f(W_{i,n} + Y_{i,n})$.

Substituting these Taylor approximations into (5.31) and taking expectations, all terms vanish except the remainders. Indeed, on the one hand, $\mathbb{E}(X_{i,n}) = \mathbb{E}(Y_{i,n}) = 0$ and $\mathbb{E}(X_{i,n}^2) = \frac{1}{n} = \mathbb{E}(Y_{i,n}^2)$, and on the other hand, $X_{i,n}$ and $Y_{i,n}$ are independent of $W_{i,n}$ by Theorem (3.24), so that we can apply Theorem (4.11d); for instance we have

$$\mathbb{E}\big(f''(W_{i,n})\, [X_{i,n}^2 - Y_{i,n}^2] \big) = \mathbb{E}\big(f''(W_{i,n}) \big)\, \mathbb{E}\big(X_{i,n}^2 - Y_{i,n}^2 \big) = 0\,.$$

Therefore,

$$\left|\mathbb{E}(f \circ S_n^* - f \circ T_n^*)\right| \leq \sum_{i=1}^{n} \mathbb{E}\left(|R_{X,i,n}| + |R_{Y,i,n}|\right)$$

$$\leq \sum_{i=1}^{n} \left[\varepsilon\, \mathbb{E}(X_{i,n}^2 + Y_{i,n}^2) + \|f''\|\, \mathbb{E}\left(X_{i,n}^2\, 1_{\{|X_{i,n}|>\delta\}} + Y_{i,n}^2\, 1_{\{|Y_{i,n}|>\delta\}}\right)\right]$$

$$= 2\varepsilon + \|f''\|\, \mathbb{E}\left(X_1^2\, 1_{\{|X_1|>\delta\sqrt{n}\}} + Y_1^2\, 1_{\{|Y_1|>\delta\sqrt{n}\}}\right).$$

In the last step, we have pulled the factor $1/n$ in $X_{i,n}^2$ and $Y_{i,n}^2$ out of the expectation; we also used the fact that the X_i are identically distributed, and so are the Y_i. By Theorem (4.11c) we now get

$$\mathbb{E}\left(X_1^2\, 1_{\{|X_1|>\delta\sqrt{n}\}}\right) = 1 - \mathbb{E}\left(X_1^2\, 1_{\{|X_1|\leq\delta\sqrt{n}\}}\right) \to 0 \quad \text{for } n \to \infty$$

and similarly $\mathbb{E}(Y_1^2\, 1_{\{|Y_1|>\delta\sqrt{n}\}}) \to 0$, so that $\limsup_{n\to\infty}|\mathbb{E}(f \circ S_n^* - f \circ T_n^*)| \leq 2\varepsilon$. Since ε was chosen arbitrarily, we obtain the result. $\diamond$

There are several other ways to prove the theorem, e.g. using Fourier transforms; see for instance [3, 15, 16, 32, 59]. If the third moments of the X_i exist and f is three times continuously differentiable, then we get the improved estimate $|R_{X,i,n}| \leq \|f'''\|\, |X_{i,n}|^3/6$ and thus

$$\left|\mathbb{E}(f \circ S_n^*) - \mathbb{E}_{\mathcal{N}_{0,1}}(f)\right| \leq C\|f'''\|/\sqrt{n},$$

where $C = \mathbb{E}(|X_1|^3 + |Y_1|^3)/6$ (in the case $m = 0$, $v = 1$). So it seems we are close to a proof of the Berry–Esséen theorem mentioned in Remark (5.30c). This theorem, however, concerns the distribution functions, and if we try to approximate $1_{]-\infty,c]}$ by a smooth f as in the proof of the implication (5.28b) $\Rightarrow$ (5.28a), we run into trouble because invariably the supremum norm $\|f'''\|$ blows up; the proof of the Berry–Esséen theorem thus uses a different approach.

The central limit theorem can also be verified empirically. Fix some n, simulate i.i.d. real random variables $X_1, \ldots, X_n$ with your favourite distribution (as in Section 3.6), and determine the standardised partial sum S_n^*. Repeat this procedure independently k times; denote the result of the jth repetition by $S_n^*(j)$. Consider then the empirical distribution function

$$F_{n,k}(c) := \frac{1}{k} \sum_{j=1}^{k} 1_{]-\infty,c]} \circ S_n^*(j),$$

which converges to the distribution function of S_n^* for $k \to \infty$ by the law of large numbers. For sufficiently large n and k, you will find that $F_{n,k}$ is close to Φ. Figure 5.7 shows an example.

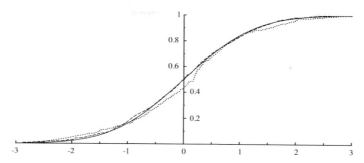

Figure 5.7. Simulation of the distribution of S_n^* for $\mathcal{U}_{[0,1]}$-distributed random variables X_i, in the cases $n = 10,\, k = 200$ (dotted line) as well as $n = 20,\, k = 500$ (dashed line). The distribution function Φ of $\mathcal{N}_{0,1}$ is shown as solid line.

5.4 Normal versus Poisson Approximation

In Sections 2.4 and 5.2 we discussed two different approximations of the binomial distribution, the Poisson and the normal approximation. The different limiting distributions rely on different approximation schemes, as can be seen as follows. Corollary (5.24) states the following: If $(X_i)_{1 \le i \le n}$ is a Bernoulli sequence with parameter p and

$$Y_i^{(n)} = \frac{X_i - p}{\sqrt{npq}}$$

the associated rescaled variables with mean zero and variance $1/n$, then the sum $\sum_{i=1}^n Y_i^{(n)}$ is approximately $\mathcal{N}_{0,1}$-distributed when n is large. Here all the summands have small modulus. In contrast, the Poisson approximation in Theorem (2.17) concerns random variables of a different nature, namely: If $(Y_i^{(n)})_{1 \le i \le n}$ is a Bernoulli sequence with parameter p_n such that $p_n \to 0$ for $n \to \infty$, then $\sum_{i=1}^n Y_i^{(n)}$ is approximately $\mathcal{P}_{np_n}$-distributed. In this case, it is not the values of the summands that are small, but rather the probabilities $p_n = P(Y_i^{(n)} = 1)$ of the events that the summands are *not* small. This is the distinguishing feature of the two approximations. Let us now present an alternative proof of the Poisson approximation, which gives us an explicit error bound.

(5.32) Theorem. Poisson approximation of binomial distributions. *For any $n \in \mathbb{N}$ and $0 < p < 1$,*

(5.33) $$\|\mathcal{B}_{n,p} - \mathcal{P}_{np}\| := \sum_{k \ge 0} \left| \mathcal{B}_{n,p}(\{k\}) - \mathcal{P}_{np}(\{k\}) \right| \le 2np^2 .$$

Proof. We represent the probability measures $\mathcal{B}_{n,p}$ and $\mathcal{P}_{np}$ as distributions of appropriately chosen random variables on the same probability space. (This is an example of a 'coupling argument', which is typical of modern stochastics.) Let $X_1, \ldots, X_n$ be a Bernoulli sequence with parameter p. By Theorem (2.9) or Example (4.39), the sum $S := \sum_{i=1}^n X_i$ has the binomial distribution $\mathcal{B}_{n,p}$. Furthermore, let $Y_1, \ldots, Y_n$ be independent random variables with Poisson distribution $\mathcal{P}_p$. The convolution property (4.41) of the Poisson distribution then implies that $T := \sum_{i=1}^n Y_i$ has distribution $\mathcal{P}_{np}$. The idea of the coupling argument derives from the observation that both X_i and Y_i take the value 0 with probability close to 1 when p is small (namely with probability $1-p$ resp. e^{-p}). To exploit this fact we will create a dependence between X_i and Y_i by prescribing that $Y_i = 0$ if $X_i = 0$ (rather than letting them be independent).

Specifically, let $Z_1, \ldots, Z_n$ be independent random variables taking values in $\{-1, 0, 1, \ldots\}$ such that $P(Z_i = k) = \mathcal{P}_p(\{k\})$ for $k \geq 1$, $P(Z_i = 0) = 1-p$, and $P(Z_i = -1) = e^{-p} - (1-p)$ for $i = 1, \ldots, n$. Note that $\mathcal{B}_{1,p}(\{0\}) = 1 - p \leq e^{-p} = \mathcal{P}_p(\{0\})$. We thus move the excess part of $\mathcal{P}_p(\{0\})$ to -1; see Figure 5.8. We further define
$$X_i = 1_{\{Z_i \neq 0\}}, \quad Y_i = \max(Z_i, 0).$$
By Theorem (3.24) it then follows that $X_1, \ldots, X_n$ are independent, and $Y_1, \ldots, Y_n$ in turn are also independent. Moreover, the X_i have distribution $\mathcal{B}_{1,p}$, and the Y_i have distribution $\mathcal{P}_p$, as desired. Finally, we consider the event
$$D = \{\text{there is an } i \text{ such that } X_i \neq Y_i\} = \bigcup_{1 \leq i \leq n} \{Z_i \notin \{0, 1\}\}.$$
Then we have $S = T$ on D^c, so that the left-hand side of (5.33) can be written as
$$\|\mathcal{B}_{n,p} - \mathcal{P}_{np}\| = \sum_{k \geq 0} |P(S = k) - P(T = k)|$$
$$= \sum_{k \geq 0} |P(\{S = k\} \cap D) - P(\{T = k\} \cap D)|.$$

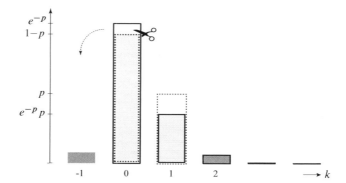

Figure 5.8. Coupling of $\mathcal{P}_p$ (columns with solid contours) and $\mathcal{B}_{1,p}$ (dotted contours). The light-grey 'intersection part' of the histograms and the darker 'difference part' define the distribution of the Z_i.

Now, the last expression is no larger than

$$2\,P(D) \leq 2 \sum_{1 \leq i \leq n} P\big(Z_i \notin \{0,1\}\big) = 2n\left(1 - (1-p) - e^{-p}p\right) \leq 2n\,p^2\,,$$

where the first inequality comes from subadditivity and the last from the estimate $1 - e^{-p} \leq p$. The result (5.33) follows. $\Diamond$

In the limit as $n \to \infty$ and $p = p_n \to 0$ such that $np_n \to \lambda > 0$, we get back our earlier Theorem (2.17). In fact we obtain the error bound $\|\mathcal{B}_{n,p_n} - \mathcal{P}_\lambda\| \leq 2(np_n^2 + |np_n - \lambda|)$; see the subsequent Remark (5.34). By the way, in the discrete case of probability measures on $\mathbb{Z}_+$, the convergence with respect to the so-called *total variation distance* $\|\cdot\|$ used here is equivalent to convergence in distribution, see Problem 5.24. We also note that the preceding proof can be directly generalised to the case of Bernoulli variables X_i with *distinct* success probabilities p_i. The distribution of the sum $S = \sum_{i=1}^n X_i$ is then no longer binomial, but can still be approximated by the Poisson distribution $\mathcal{P}_{\sum_i p_i}$ up to an error of at most $2 \sum_i p_i^2$.

(5.34) Remark. *Variation of the Poisson parameter.* If $\lambda, \delta > 0$ then

$$\|\mathcal{P}_{\lambda+\delta} - \mathcal{P}_\lambda\| \leq 2\delta\,.$$

For, if X and Y are independent with distribution $\mathcal{P}_\lambda$ and $\mathcal{P}_\delta$ respectively, $X + Y$ has the distribution $\mathcal{P}_{\lambda+\delta}$, and arguing as in the above proof one finds that $\|\mathcal{P}_{\lambda+\delta} - \mathcal{P}_\lambda\|$ is bounded by $2\,P(Y \geq 1) = 2(1 - e^{-\delta}) \leq 2\delta$.

When is the Poisson distribution appropriate to approximate the binomial distribution, and when the normal distribution? By (5.33) the Poisson approximation works well when np^2 is small. On the other hand, the Berry–Esséen theorem discussed in Remark (5.30c) shows that the normal approximation performs well when

$$\frac{p(1-p)^3 + (1-p)p^3}{(p(1-p))^{3/2}} \frac{1}{\sqrt{n}} = \frac{p^2 + (1-p)^2}{\sqrt{np(1-p)}}$$

is small, and since $1/2 \leq p^2 + (1-p)^2 \leq 1$, this is precisely the case when $np(1-p)$ is large. If p is very close to 1, one can use symmetry to obtain

$$\mathcal{B}_{n,p}(n-k) = \mathcal{B}_{n,1-p}(k) = \mathcal{P}_{n(1-p)}(\{k\}) + O\big(n(1-p)^2\big).$$

Problems

5.1 *The Ky Fan metric for convergence in probability.* For two real-valued random variables X, Y on an arbitrary probability space let

$$d(X, Y) = \min\{\varepsilon \geq 0 : P(|X - Y| > \varepsilon) \leq \varepsilon\}.$$

Show the following:

(a) The minimum is really attained, and d is a metric on the space of all real-valued random variables, with the convention that two random variables X, Y are regarded as equal if $P(X = Y) = 1$.

(b) For every sequence of real-valued random variables on Ω, it is the case that $Y_n \xrightarrow{P} Y$ if and only if $d(Y_n, Y) \to 0$.

5.2 *Collectibles.* Consider the problem of collecting a complete series of stickers, as described in Problem 4.20. How many items you have to buy so that with probability at least 0.95 you have collected all $N = 20$ stickers? Use Chebyshev's inequality to give a least possible bound.

5.3^S (a) A particle moves randomly on a plane according to the following rules. It moves one unit along a randomly chosen direction Ψ_1, then it chooses a new direction Ψ_2 and moves one unit along that new direction, and so on. We suppose that the angles Ψ_i are independent and uniformly distributed on $[0, 2\pi]$. Let D_n be the distance between the starting point and the location of the particle after the nth step. Calculate the mean square displacement $\mathbb{E}(D_n^2)$.

(b) Suppose at the centre of a large plane there are at time zero 30 particles, which move independently according to the rules set out in (a). For every step they take, the particles need one time unit. Determine, for every $n \geq 1$, the smallest number $r_n > 0$ with the following property: With probability at least 0.9 there are, at time n, more than 15 particles in a circle of radius r_n around the centre of the plane. *Hint:* Determine first some $p_0 > 1/2$ such that $P(\sum_{i=1}^{30} Z_i > 15) \geq 0.9$ for any Bernoulli sequence $Z_1, \ldots, Z_{30}$ with parameter $p \geq p_0$.

5.4 *Large deviations of empirical averages from the mean.* Let $(X_i)_{i \geq 1}$ be a Bernoulli sequence with $0 < p < 1$. Show that for all $p < a < 1$ we have

$$P\left(\frac{1}{n} \sum_{i=1}^{n} X_i \geq a\right) \leq e^{-n h(a;p)},$$

where $h(a; p) = a \log \frac{a}{p} + (1 - a) \log \frac{1-a}{1-p}$. *Hint:* Show first that for all $s \geq 0$

$$P\left(\frac{1}{n} \sum_{i=1}^{n} X_i \geq a\right) \leq e^{-nas} \, \mathbb{E}(e^{sX_1})^n.$$

5.5^S *Convexity of the Bernstein polynomials.* Let $f : [0, 1] \to \mathbb{R}$ be continuous and convex. Show that, for every $n \geq 1$, the corresponding Bernstein polynomial f_n is also convex. *Hint:* Let $p_1 < p_2 < p_3$, consider the frequencies $Z_k = \sum_{i=1}^{n} 1_{[0, p_k]} \circ U_i$, $k = 1, 2, 3$, and represent Z_2 as a convex combination of Z_1 and Z_3. Use that the vector $(Z_1, Z_2 - Z_1, Z_3 - Z_2, n - Z_3)$ has a multinomial distribution.

5.6 *Law of large numbers for random variables without expectation.* Let $(X_i)_{i\geq 1}$ be i.i.d. real-valued random variables having no expectation, i.e., $X_i \notin \mathscr{L}^1$. Let $a \in \mathbb{N}$ be arbitrary. Show the following:

(a) $P(|X_n| > an$ infinitely often$) = 1$. *Hint:* Use Problem 4.5.

(b) For the sums $S_n = \sum_{i=1}^{n} X_i$ we have $P(|S_n| > an$ infinitely often$) = 1$ and thus $\limsup_{n\to\infty} |S_n|/n = \infty$ almost surely.

(c) If all $X_i \geq 0$, we even obtain $S_n/n \to \infty$ almost surely.

5.7[S] *Renewals of, say, light bulbs.* Let $(L_i)_{i\geq 1}$ be i.i.d. non-negative random variables with finite or infinite expectation. One can interpret L_i as the life time of the ith light bulb (which is immediately replaced when it burns out); see also Figure 3.7. For $t > 0$ let

$$N_t = \sup\Big\{k \geq 1 : \sum_{i=1}^{k} L_i \leq t\Big\}$$

be the number of bulbs used up to time t. Show that $\lim_{t\to\infty} N_t/t = 1/\mathbb{E}(L_1)$ almost surely; here we set $1/\infty = 0$ and $1/0 = \infty$. (In the case $\mathbb{E}(L_1) = \infty$ use Problem 5.6c; the case $\mathbb{E}(L_1) = 0$ is trivial.) What does the result mean in the case of a Poisson process?

5.8[S] *Inspection, or waiting time, paradox.* As in the previous problem, let $(L_i)_{i\geq 1}$ be i.i.d. non-negative random variables representing the life times of machines, or light bulbs, which are immediately replaced when defect. In a waiting time interpretation, one can think of the L_i as the time spans between the arrivals of two consecutive buses at a bus stop. We assume $0 < \mathbb{E}(L_i) < \infty$. For $s > 0$, let $L_{(s)}$ be the life time of the machine working at instant s; so $L_{(s)} = L_k$ for $s \in [T_{k-1}, T_k[$, where the T_k are as in Figure 3.7 on p. 75. Use Problem 5.7 and the strong law of large numbers to show that, for every random variable $f : [0, \infty[\to [0, \infty[$,

$$\frac{1}{t} \int_0^t f(L_{(s)}) \, ds \xrightarrow[t\to\infty]{} \frac{\mathbb{E}(L_1 f(L_1))}{\mathbb{E}(L_1)} \quad \text{almost surely.}$$

For $f = \mathrm{Id}$ this means that, for large t, the life time of the machine inspected at a random instant in $[0, t]$ is approximately equal to $\mathbb{E}(L_1^2)/\mathbb{E}(L_1)$, which is larger than $\mathbb{E}(L_1)$ unless L_1 is almost surely constant. Compare this result with Example (4.16), where the L_i are exponentially distributed. The probability measure $Q(A) := \mathbb{E}(L_1 1_{\{L_1 \in A\}})/\mathbb{E}(L_1)$ on $([0, \infty[, \mathscr{B}_{[0,\infty[})$, which shows up in the above limit when $f = 1_A$, is called the *size-biased distribution* of L_i, and with slightly more effort one can even show that

$$\frac{1}{t} \int_0^t \delta_{L_{(s)}} \, ds \xrightarrow{d} Q \quad \text{almost surely for } t \to \infty.$$

5.9 Let $(X_n)_{n\geq 1}$ be a sequence of independent random variables that are exponentially distributed with parameter $\alpha > 0$. Show that

$$\limsup_{n\to\infty} X_n/\log n = 1/\alpha \quad \text{and} \quad \liminf_{n\to\infty} X_n/\log n = 0 \quad \text{almost surely.}$$

5.10 *Expectation versus probability.* Bob suggests the following game to Alice: 'Here is a biased coin, which shows heads with probability $p \in {]1/3, 1/2[}$. Your initial stake is €100; each time the coin shows heads, I double your capital, otherwise you pay me half of your capital. Let X_n denote your capital after the nth coin flip. As you can easily see, $\lim_{n\to\infty} \mathbb{E}(X_n) = \infty$, so your expected capital will grow beyond all limits.' Is it advisable for Alice to play this game? Verify Bob's claim and show that $\lim_{n\to\infty} X_n = 0$ almost surely.

5.11 [S] *Asymptotics of the Pólya model.* Consider Pólya's urn model with parameters $a = r/c$ and $b = w/c$, as introduced in Example (3.14). Let R_n be the number of red balls drawn after n iterations.

(a) Use Problem 3.4 and the law of large numbers to show that R_n/n converges in distribution to the beta distribution $\beta_{a,b}$. (For an illustration, compare Figures 3.5 and 2.7.)

(b) What does this mean for the long-term behaviour of the competing populations? Consider the cases (i) $a, b > 1$, (ii) $b < 1 < a$, (iii) $a, b < 1$, (iv) $a = b = 1$.

5.12 Give a sequence of random variables in $\mathscr{L}^2$ for which neither the (strong or weak) law of large numbers nor the central limit theorem holds.

5.13 *Decisive power of determined minorities.* In an election between two candidates A and B one million voters cast their votes. Among these, $2\,000$ know candidate A from his election campaign and vote unanimously for him. The remaining $998\,000$ voters are more or less undecided and make their decision independently of each other by tossing a fair coin. What is the probability p_A of a victory of candidate A?

5.14 [S] *Local normal approximation of Poisson distributions.* Let $x_\lambda(k) = (k - \lambda)/\sqrt{\lambda}$ for all $\lambda > 0$ and $k \in \mathbb{Z}_+$. Show that, for any $c > 0$,

$$\lim_{\lambda \to \infty} \ \max_{k \in \mathbb{Z}_+ : |x_\lambda(k)| \leq c} \left| \frac{\sqrt{\lambda} \, \mathcal{P}_\lambda(\{k\})}{\phi(x_\lambda(k))} - 1 \right| = 0 \,.$$

5.15 *Asymptotics of* Φ. Establish the sandwich estimate

$$\phi(x)\left(\frac{1}{x} - \frac{1}{x^3}\right) \leq 1 - \Phi(x) \leq \phi(x)\,\frac{1}{x} \quad \text{for all } x > 0,$$

and hence the asymptotics $1 - \Phi(x) \sim \phi(x)/x$ for $x \to \infty$. *Hint:* Compare the derivatives of the functions on the left- and right-hand sides with ϕ.

5.16 *Effect of the discreteness corrections.* Determine a lower bound for the error term in (5.23) when the discreteness corrections $\pm 1/2$ are omitted, by considering the case $k = l = np \in \mathbb{N}$. Compare the result with Figure 5.6.

5.17 [S] *No-Shows.* Frequently, the number of passengers turning up for their flight is smaller than the number of bookings made. This is the reason why airlines overbook their flights (i.e., they sell more tickets than seats are available), at the risk of owing compensation to an eventual surplus of people. Suppose the airline has an income of $a = 300\,\text{€}$ per person flying, and for every person that cannot fly it incurs a loss of $b = 500\,\text{€}$; furthermore, suppose that every person that has booked shows up for the flight independently with probability $p = 0.95$. How many places would you sell for an

(a) Airbus A319 with $S = 124$ seats,

(b) Airbus A380 with $S = 549$ seats

to maximise the expected profit? *Hint:* Let $(X_n)_{n \geq 1}$ be a Bernoulli sequence with parameter p, and $S_N = \sum_{k=1}^N X_k$. The profit G_N by selling N places then satisfies the recursion

$$G_{N+1} - G_N = a\,1_{\{S_N < S\}} X_{N+1} - b\,1_{\{S_N \geq S\}} X_{N+1} \,.$$

Deduce that $\mathbb{E}(G_{N+1}) \geq \mathbb{E}(G_N)$ if and only if $P(S_N < S) \geq b/(a+b)$, and then use the normal approximation.

5.18 *Rounding errors.* Estimate the error of a sum of rounded numbers as follows. The numbers $R_1, \ldots, R_n \in \mathbb{R}$ are rounded to the next integer, i.e., they can be represented as $R_i = Z_i + U_i$ with $Z_i \in \mathbb{Z}$ and $U_i \in [-1/2, 1/2[$. The deviation of the sum of rounded numbers $\sum_{i=1}^n Z_i$ from the true sum $\sum_{i=1}^n R_i$ is $S_n = \sum_{i=1}^n U_i$. Suppose the $(U_i)_{1 \le i \le n}$ are independent random variables having uniform distribution on $[-1/2, 1/2[$. Using the central limit theorem, determine a bound $k > 0$ with the property $P(|S_n| < k) \approx 0.95$ for $n = 100$.

5.19 In a sales campaign, a mail order company offers their first 1000 customers a complimentary ladies' respectively men's watch with their order. Suppose that both sexes are equally attracted by the offer. How many ladies' and how many men's watches should the company keep in stock to ensure that, with a probability of at least 98%, all 1000 customers receive a matching watch? Use (a) Chebyshev's inequality, (b) the normal approximation.

5.20 A company has issued a total of $n = 1000$ shares. At a fixed time, every shareholder decides for each share with probability $0 < p < 1$ to sell it. These decisions are independent for all shares. The market can take in $s = 50$ shares without the price falling. What is the largest value of p such that the price remains stable with 90% probability?

5.21 *Error propagation for transformed observations.* Let $(X_i)_{i \ge 1}$ be a sequence of i.i.d. random variables taking values in a (possibly unbounded) interval $I \subset \mathbb{R}$, and suppose the variance $v = \mathbb{V}(X_i) > 0$ exists. Let $m = \mathbb{E}(X_i)$ and $f : I \to \mathbb{R}$ be twice continuously differentiable with $f'(m) \ne 0$ and bounded f''. Show that

$$\frac{\sqrt{n/v}}{f'(m)} \left(f\left(\tfrac{1}{n} \sum_{i=1}^n X_i\right) - f(m) \right) \xrightarrow{d} \mathcal{N}_{0,1} \quad \text{for } n \to \infty.$$

Hint: Use a Taylor's expansion of f at the point m and control the remainder term by means of Chebyshev's inequality.

5.22 ^{S} *Brownian motion.* A heavy particle is randomly hit by light particles, so that its velocity is randomly reversed at equidistant time points. That is, its spatial coordinate (in a given direction) at time $t > 0$ satisfies $X_t = \sum_{i=1}^{\lfloor t \rfloor} V_i$ with independent velocities V_i, where $P(V_i = \pm v) = 1/2$ for some $v > 0$. If we pass to a macroscopic scale, the particle at time t can be described by the random variable $B_t^{(\varepsilon)} = \sqrt{\varepsilon}\, X_{t/\varepsilon}$, where $\varepsilon > 0$. Determine the distributional limit B_t of $B_t^{(\varepsilon)}$ for $\varepsilon \to 0$ and also the distribution density ρ_t of B_t. Verify that the family (ρ_t) of densities satisfies the heat equation

$$\frac{\partial \rho_t(x)}{\partial t} = \frac{D}{2} \frac{\partial^2 \rho_t(x)}{\partial x^2} \qquad (x \in \mathbb{R}, \ t > 0)$$

for some appropriately chosen diffusion coefficient $D > 0$.

5.23 *Convergence in probability versus convergence in distribution.* Let X and $(X_n)_{n \ge 1}$ be real-valued random variables on the same probability space. Show the following:

(a) $X_n \xrightarrow{P} X$ implies $X_n \xrightarrow{d} X$.

(b) The converse of (a) does not hold in general, but it does when X is almost surely constant.

5.24 *Convergence in distribution of discrete random variables.* Let X and X_n, $n \geq 1$, be random variables on the same probability space that take values in $\mathbb{Z}$. Prove that the following statements are equivalent:

(a) $X_n \xrightarrow{d} X$ for $n \to \infty$.

(b) $P(X_n = k) \to P(X = k)$ for $n \to \infty$ and every $k \in \mathbb{Z}$.

(c) $\sum_{k \in \mathbb{Z}} |P(X_n = k) - P(X = k)| \to 0$ for $n \to \infty$.

5.25 ^{S} *The arcsine law.* Consider the simple symmetric random walk $(S_j)_{j \leq 2N}$ introduced in Problem 2.7, for fixed $N \in \mathbb{N}$. Let $L_{2N} = \max\{2n \leq 2N : S_{2n} = 0\}$ be the last time both candidates have the same number of votes before the end of the count. (In a more general context, one would speak of the last time of return to 0 before time $2N$.) Show the following.

(a) For all $0 \leq n \leq N$, $P(L_{2N} = 2n) = u_n u_{N-n}$, where $u_k = 2^{-2k}\binom{2k}{k}$.

(b) For $0 < a < b < 1$,

$$\lim_{N \to \infty} P(a \leq L_{2N}/2N \leq b) = \int_a^b \frac{1}{\pi \sqrt{x(1-x)}} \, dx \,,$$

meaning that $L_{2N}/2N$ converges in distribution to the arcsine distribution $\beta_{1/2,1/2}$.

(The arcsine distribution assigns large probabilities to the values near 0 and 1, see Figure 2.7 on p. 45. Hence, it is quite likely that a candidate either gains the lead right at the beginning or just before the end of the count of votes.)

5.26 Let $(X_i)_{i \geq 1}$ be independent standard normal random variables and

$$M_n = \max(X_1, \ldots, X_n), \quad a_n = \sqrt{2 \log n - \log(\log n) - \log(4\pi)} \,.$$

Show that the sequence $a_n M_n - a_n^2$ converges in distribution to the probability measure Q on $\mathbb{R}$ with distribution function

$$F_Q(c) = \exp(-e^{-c}), \quad c \in \mathbb{R} \,.$$

Q is known as the *doubly exponential distribution* or, after Emil J. Gumbel (1891–1966), the *(standard) Gumbel distribution*. It is one of the so-called *extreme value distributions* that appear as the asymptotic distribution of rescaled maxima.

5.27 Let $(X_i)_{i \geq 1}$ be independent and Cauchy distributed with parameter $a > 0$ (cf. Problem 2.5), and define $M_n = \max(X_1, \ldots, X_n)$. Show that M_n/n converges in distribution to a random variable $Y > 0$, and Y^{-1} has a Weibull distribution (which one?), see Problem 3.27. (The inverse Weibull distributions form a second class of typical extreme value distributions.)

Chapter 6

Markov Chains

Independence is the simplest assumption about the joint behaviour of random variables, and so it occupies a lot of space in an introduction to stochastics. However, this should not give the impression that stochastics only deals with this simple case; the opposite is true. This chapter is devoted to a simple case of dependence: A Markov chain is a sequence of random variables with a short memory span; the behaviour at the next point in time depends only on the current value and does not depend on what happened before. Of special interest is the long-term behaviour of such a sequence – for instance the absorption in a 'trap', or the convergence to an equilibrium.

6.1 The Markov Property

Let $E \neq \varnothing$ be an at most countable set and $\Pi = (\Pi(x, y))_{x,y \in E}$ a *stochastic matrix*, i.e., a matrix for which each row $\Pi(x, \cdot)$ is a probability density on E. We consider a random process on E that at each time step moves from x to y with probability $\Pi(x, y)$.

Definition. A sequence $X_0, X_1, \ldots$ of random variables defined on a probability space $(\Omega, \mathscr{F}, P)$ and taking values in E is called a *Markov chain* with *state space* E and *transition matrix* Π, if

$$(6.1) \qquad P(X_{n+1} = x_{n+1} | X_0 = x_0, \ldots, X_n = x_n) = \Pi(x_n, x_{n+1})$$

for every $n \geq 0$ and every $x_0, \ldots, x_{n+1} \in E$ with $P(X_0 = x_0, \ldots, X_n = x_n) > 0$. The distribution $\alpha = P \circ X_0^{-1}$ of X_0 is called the *initial distribution* of the Markov chain. (The name refers to Andrei A. Markov, cf. p. 121.)

We add a couple of explanations.

(6.2) Remark. *Existence and illustration of Markov chains.* (a) Equation (6.1) consists of two statements: First, the conditional distribution of X_{n+1} given the past values $x_0, \ldots, x_n$ depends only on the present state x_n and not on the past; this so-called *Markov property* is the decisive assumption. Secondly, these conditional distributions do not depend on the time n. This time invariance of the transition law Π is known as the case of *stationary transition probabilities*; it is an additional assumption that is usually made for simplicity. In other words, a Markov chain $(X_n)_{n \geq 0}$ is a stochastic process with a short memory of exactly one time unit and without an internal clock.

(b) Substituting $\rho_{k|\omega_0,\dots,\omega_{k-1}}(\omega_k) = \Pi(\omega_{k-1}, \omega_k)$ in Theorem (3.8), one finds that (6.1) is a special case of condition (b) there. Theorem (3.12) thus implies that, for every initial distribution $\rho_1 = \alpha$ on E, there exists a unique probability measure P^α on the product space $(\Omega, \mathscr{F}) := (\prod_{k\geq 0} E, \bigotimes_{k\geq 0} \mathscr{P}(E))$ such that the projections $X_n : (\omega_k)_{k\geq 0} \to \omega_n$ from Ω to E form a Markov chain for Π and α. In the following we will, without loss of generality, only consider Markov chains of this type, which are also known as canonical Markov chains.

If $\alpha = \delta_x$ for some $x \in E$ ('deterministic start in x'), it is convenient to write P^x instead of P^α. In analogy to equation (3.9), P^x is characterised by the identities

$$(6.3) \qquad P^x(X_1 = x_1, \dots, X_n = x_n) = \Pi(x, x_1)\Pi(x_1, x_2)\dots\Pi(x_{n-1}, x_n)$$

for arbitrary $n \geq 1$ and $x_1, \dots, x_n \in E$. Summing over $x_1, \dots, x_{n-1} \in E$, one obtains in particular

$$(6.4) \qquad\qquad P^x(X_n = y) = \Pi^n(x, y) \quad \text{for all } x, y \in E,$$

where Π^n denotes the nth matrix power of Π. In other words, the powers of the transition matrix Π play a crucial role, since they specify the probabilities that the Markov chain started at a given point can be found in a certain state at a particular time.

(c) Since only the present state is relevant for the future, a Markov chain can be illustrated by its transition graph instead of the tree diagram in Figure 3.3. For example, the transition graph associated with the matrix

$$\Pi = \begin{pmatrix} 1/2 & 1/2 & 0 \\ 1/3 & 1/3 & 1/3 \\ 1 & 0 & 0 \end{pmatrix}$$

on the state space $E = \{1, 2, 3\}$ is shown in Figure 6.1.

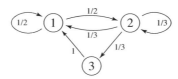

Figure 6.1. Example of a transition graph.

(d) The transition graph of a Markov chain $(X_n)_{n\geq 0}$ suggests the idea that X_{n+1} is obtained from X_n by applying a random mapping. This idea is correct and can be formalised as follows. Let E^E be the set of all mappings from E to E, and let $\varphi_1, \varphi_2, \dots \in E^E$ be a sequence of i.i.d. random mappings such that $\varphi_n(x)$ has distribution $\Pi(x, \cdot)$ for all $x \in E$ and $n \geq 1$. (For example, one can assume that in fact the family $(\varphi_n(x))_{x \in E, n \geq 1}$ of E-valued random variables is independent with the required distributions.) Define $(X_n)_{n \geq 0}$ recursively by $X_0 = x \in E$ and $X_n = \varphi_n(X_{n-1})$ for $n \geq 1$. Then $(X_n)_{n \geq 0}$ is a Markov chain with transition matrix Π and starting point x. To prove this, it suffices to verify equation (6.3). But there, the product structure of the right-hand side corresponds exactly to the independence of the random mappings $(\varphi_n)_{n \geq 1}$. This shows that there is still a lot of independence hidden in a Markov chain, and also provides a second existence proof for Markov chains.

The picture of realising a Markov chain by iterating a sequence of i.i.d. random mappings is also useful for the *simulation of Markov chains*. The random mappings are then constructed as follows. Numbering the elements of E in an arbitrary way, we can assume that E is a discrete subset of $\mathbb{R}$. For $x \in E$ and $u \in \,]0, 1[$, let $f(x, u)$ be the smallest $z \in E$ with $\sum_{y \leq z} \Pi(x, y) \geq u$. In other words, $f(x, \cdot)$ is the quantile transformation of $\Pi(x, \cdot)$, cf. (1.30). The function f is known as the *update function*. Next, let $U_1, U_2, \ldots$ be a sequence of independent random variables that are uniformly distributed on $]0, 1[$. (In practice, these are simulated by pseudo-random numbers; cf. (3.45).) The random mappings $\varphi_n : x \to f(x, U_n)$ from E to itself are then i.i.d. with the required distributional property, and so the recursion $X_n = f(X_{n-1}, U_n)$ yields a Markov chain with transition matrix Π.

Here are two classical examples of Markov chains.

(6.5) Example. *Random walk on* $E = \mathbb{Z}$. Let $(Z_i)_{i \geq 1}$ be independent $\mathbb{Z}$-valued random variables with common distribution density ρ, and set $X_0 = 0$, $X_n = \sum_{i=1}^{n} Z_i$ for $n \geq 1$. (Recall that such sums of independent random variables already showed up in the law of large numbers and the central limit theorem.) Then $(X_n)_{n \geq 0}$ is a Markov chain with transition matrix $\Pi(x, y) = \rho(y - x)$, $x, y \in \mathbb{Z}$. Indeed, the left-hand side of (6.1) is then given by

$$P(Z_{n+1} = x_{n+1} - x_n | X_0 = x_0, \ldots, X_n = x_n) = \rho(x_{n+1} - x_n),$$

where the last equality comes from the fact that Z_{n+1} is independent of $X_0, \ldots, X_n$. A Markov chain of this form is called a *random walk*.

(6.6) Example. *Coin tossing game, or random walk with absorbing barriers.* Two gamblers A and B each own a and b euros, respectively. They flip a fair coin repeatedly, and depending on the outcome one pays the other € 1. The game is over as soon as one of the gamblers has lost his capital. Let X_n be the cumulated gain of gambler A (and so the loss of gambler B) after the games $1, \ldots, n$. The random variables X_n take values in $E = \{-a, \ldots, b\}$ and form a Markov chain with transition matrix

$$\Pi(x, y) = \begin{cases} 1/2 & \text{if } -a < x < b, |y - x| = 1, \\ 1 & \text{if } x = y \in \{-a, b\}, \\ 0 & \text{otherwise.} \end{cases}$$

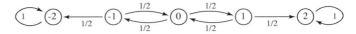

Figure 6.2. Transition graph for the coin tossing game when $a = b = 2$.

The transition graph in Figure 6.2 illustrates the situation. One seeks to determine the probability that the game ends with the ruin of gambler A. This is the so-called ruin problem, which will be solved in Example (6.10).

More examples follow later on. The Markov property (6.1) can be tightened to the following more general statement, which at time n is not restricted to the limited perspective of looking only to the next time point $n + 1$, but refers to the entire future after time n. If the current state is known, the Markov chain 'forgets' both the past and the time and starts completely afresh.

(6.7) Theorem. Markov property. *Let $(X_n)_{n \geq 0}$ be a Markov chain for Π and α. Then*

$$P^\alpha\big((X_n, X_{n+1}, \dots) \in A \,\big|\, (X_0, \dots, X_{n-1}) \in B, X_n = x\big) = P^x(A)$$

for every $n \geq 0$ and all $A \in \mathscr{F} = \mathscr{P}(E)^{\otimes \mathbb{Z}_+}$, $B \subset E^n$, and $x \in E$, provided that $P^\alpha((X_0, \dots, X_{n-1}) \in B, X_n = x) > 0$.

Proof. As agreed in Remark (6.2b), we will assume without loss of generality that $(\Omega, \mathscr{F}) = (E^{\mathbb{Z}_+}, \mathscr{P}(E)^{\otimes \mathbb{Z}_+})$ and that X_n is the nth projection. For arbitrary $k \geq 0$ and $x_0, \dots, x_k \in E$, the multiplication formula (6.3) yields

$$P^\alpha((X_0, \dots, X_{n-1}) \in B, X_n = x, X_{n+i} = x_i \text{ for } 0 \leq i \leq k)$$

$$= \sum_{(y_0, \dots, y_{n-1}) \in B} \alpha(y_0) \Pi(y_0, y_1) \dots \Pi(y_{n-1}, x) \delta_{x, x_0} \Pi(x_0, x_1) \dots \Pi(x_{k-1}, x_k)$$

$$= P^\alpha((X_0, \dots, X_{n-1}) \in B, X_n = x) \, P^x(X_i = x_i \text{ for } 0 \leq i \leq k).$$

Therefore, the claim follows for $A = \{X_i = x_i \text{ for } 0 \leq i \leq k\}$. The general case is obtained from the uniqueness theorem (1.12), since the sets A of this specific form (together with $\varnothing$) constitute an intersection-stable generator of $\mathscr{F}$; cf. Problem 1.5. $\diamond$

As the above proof shows, it is essential for the validity of Theorem (6.7) that the Markov chain is conditioned on a precise state x at time n.

6.2   Absorption Probabilities

In the coin tossing game (6.6), the chain cannot leave the two boundary states $-a$ and b. That is, they are absorbing in the following sense.

Definition. A state $z \in E$ with $\Pi(z, z) = 1$ is called *absorbing* with respect to the transition matrix Π. In this case,

$$h_z(x) := P^x(X_n = z \text{ eventually}) = P^x\Big(\bigcup_{N \geq 0} \bigcap_{n \geq N} \{X_n = z\} \Big)$$

is called the *absorption probability* in z for the starting point $x \in E$.

In other words, an absorbing state is a 'trap' from which the Markov chain cannot escape. It is intuitively clear that the absorption probability coincides with the hitting probability of the trap. So, let us introduce the following concepts.

(6.8) Remark and Definition. *Hitting and stopping times.* For arbitrary (not necessarily absorbing) $z \in E$, the random time

$$\tau_z = \inf\{n \geq 1 : X_n = z\}$$

(with the convention $\inf \varnothing := \infty$) is called the *hitting (or entrance) time* of the state z. When z is the starting point, it is called the *return time* to z. For every $n \geq 1$, the identity

$$\{\tau_z = n\} = \{(X_0, \ldots, X_{n-1}) \in B, X_n = z\} \quad \text{for } B = E \times (E \setminus \{z\})^{n-1}$$

shows that $\{\tau_z = n\}$ depends only on $(X_0, \ldots, X_n)$. A mapping $\tau : \Omega \to \mathbb{Z}_+ \cup \{\infty\}$ with this property of not anticipating the future is called a *stopping time* or *optional time* with respect to $(X_n)_{n \geq 0}$. Stopping times play an important role in stochastics. Here, however, the only stopping times we will encounter are hitting times.

The following theorem will allow us, for instance, to determine the ruin probability in the coin tossing game (6.6).

(6.9) Theorem. Characterisation of absorption probabilities. *For absorbing $z \in E$ and every starting point $x \in E$, the absorption probability can equivalently be expressed as*

$$h_z(x) = P^x(\tau_z < \infty) = \lim_{n \to \infty} P^x(X_n = z),$$

and h_z is the smallest non-negative function such that $h_z(z) = 1$ and

$$\sum_{y \in E} \Pi(x, y) h_z(y) = h_z(x) \quad \text{for all } x \in E.$$

Treating the function h_z as a column vector, we can rewrite the last equation as $\Pi h_z = h_z$. Thus, h_z is a right eigenvector of Π with corresponding eigenvalue 1.

Proof. By the σ-continuity of P^z and equation (6.3),

$$P^z(X_i = z \text{ for every } i \geq 1) = \lim_{k \to \infty} P^z(X_i = z \text{ for every } i \leq k)$$

$$= \lim_{k \to \infty} \Pi(z, z)^k = 1.$$

Combining this with Theorem (6.7), one finds for every $x \in E$

$$P^x(X_n = z) = P^x(X_n = z)\, P^z(X_i = z \text{ for every } i \geq 1)$$

$$= P^x(X_{n+i} = z \text{ for every } i \geq 0)$$

$$\xrightarrow[n \to \infty]{} P^x\Big(\bigcup_{n \geq 1} \bigcap_{i \geq n} \{X_i = z\}\Big) = h_z(x).$$

On the other hand, Theorem (6.7) and Remark (6.8) imply that

$$P^x(X_n = z | \tau_z = k) = P^z(X_{n-k} = z) = 1$$

for every $k \leq n$, so that, by the case-distinction formula (3.3a),

$$P^x(X_n = z) = \sum_{k=1}^{n} P^x(\tau_z = k)\, P^x(X_n = z | \tau_z = k)$$

$$= P^x(\tau_z \leq n) \xrightarrow[n \to \infty]{} P^x(\tau_z < \infty).$$

This proves the first two equations.

To prove the second statement, we first note that, trivially, $h_z \geq 0$ and $h_z(z) = 1$. For arbitrary $x \in E$, Theorem (6.7) and the case-distinction formula (3.3a) yield

$$\sum_{y \in E} \Pi(x, y)\, h_z(y) = \sum_{y \in E} P^x(X_1 = y)\, P^y(X_i = z \text{ eventually})$$

$$= \sum_{y \in E} P^x(X_1 = y)\, P^x(X_{i+1} = z \text{ eventually} \mid X_1 = y) = h_z(x),$$

i.e., $\Pi h_z = h_z$. For any other function $h \geq 0$ satisfying $\Pi h = h$ and $h(z) = 1$ we get

$$h(x) = \Pi^n h(x) \geq \Pi^n(x, z) = P^x(X_n = z),$$

and for $n \to \infty$ it follows that $h(x) \geq h_z(x)$. $\diamond$

As an aside let us note that a function h on E satisfying $\Pi h = h$ is called *harmonic* (with respect to Π). This is because, in the situation of the following example, such functions satisfy a discrete analogue of the mean value property that characterises harmonic functions in the sense of analysis.

(6.10) Example. *Gambler's ruin problem.* In the coin tossing game of Example (6.6), we are interested in the 'ruin probability' (maybe a rather drastic word) of gambler A, namely the probability $r_A := h_{-a}(0)$ that gambler A loses his capital. According to Theorem (6.9) we find for $-a < x < b$:

$$h_{-a}(x) = \Pi h_{-a}(x) = \tfrac{1}{2}\, h_{-a}(x-1) + \tfrac{1}{2}\, h_{-a}(x+1).$$

Therefore, the difference $c := h_{-a}(x+1) - h_{-a}(x)$ does not depend on x. Hence, we obtain $h_{-a}(x) - h_{-a}(-a) = (x+a)c$. Since $h_{-a}(-a) = 1$ and $h_{-a}(b) = 0$, it follows that $c = -1/(a+b)$ and thus

$$r_A = 1 - \frac{a}{a+b} = \frac{b}{a+b}.$$

Likewise, the ruin probability of gambler B is $r_B = \frac{a}{a+b}$. In particular, $r_A + r_B = 1$, which means that the game certainly ends with the ruin of one of the gamblers (and the win of the other) at a finite random time, and cannot go on forever.

(6.11) Example. *Branching processes.* We consider a population of organisms that propagate asexually and independently of each other at discrete time units. Every nth generation individual is replaced in the next generation by $k \geq 0$ offspring with probability $\rho(k)$, independently of all other individuals. What is the probability that the descendants of a certain progenitor die out? This question was first examined by I.-J. Bienaymé (1845), but fell into oblivion. Later on in 1873/74, the same question was studied by F. Galton and H. W. Watson, so that the underlying model became known as the *Galton–Watson process.* At that time, the question was motivated by the extinction of family names along the male family tree. Today, the model is of central biological interest as a particularly simple, prototypical model of the dynamics of a population. Moreover, it can be applied fairly realistically to cell assemblies and bacterial colonies.

Let X_n be the number of individuals in the nth generation. We model $(X_n)_{n \geq 0}$ as a Markov chain on $E = \mathbb{Z}_+$ with transition matrix

$$\Pi(n, k) = \rho^{\star n}(k) = \sum_{k_1 + \cdots + k_n = k} \rho(k_1) \dots \rho(k_n).$$

In words, if n individuals live in a given generation, then the distribution $\Pi(n, \cdot)$ of the number of offspring in the next generation is exactly the n-fold convolution of ρ. (This assumption incorporates the fact that the individuals propagate independently of each other. For $n = 0$ we set $\rho^{\star 0}(k) = \delta_{0,k}$, the Kronecker-delta.) Only the case $0 < \rho(0) < 1$ is interesting, since otherwise the population dies out immediately or it keeps growing forever. Clearly, 0 is the only absorbing state.

Our aim is to calculate the extinction probability $q := h_0(1)$ for the descendants of a single progenitor. For this purpose, we observe the following:

▷ For every $k, n \geq 0$, we find that $P^k(X_n = 0) = P^1(X_n = 0)^k$. (Intuitively, this comes from the fact that the progenies of different individuals evolve independently.) Indeed, this equation holds for $n = 0$, since then both sides agree with $\delta_{k,0}$. For the induction step one can write

$$P^k(X_{n+1} = 0) = \sum_{l \geq 0} \Pi(k, l) \, P^l(X_n = 0)$$

$$= \sum_{l_1, \dots, l_k \geq 0} \rho(l_1) \dots \rho(l_k) \, P^1(X_n = 0)^{l_1 + \dots + l_k}$$

$$= \left(\sum_{l \geq 0} \Pi(1, l) \, P^l(X_n = 0) \right)^k = P^1(X_{n+1} = 0)^k .$$

Here, the first and the last step follow from the case-distinction formula (3.3a) and Theorem (6.7), while the second step comes from the definition of Π and the induction hypothesis. Theorem (6.9) thus shows that $h_0(k) = q^k$ for all $k \geq 0$.

▷ $q = h_0(1)$ is the smallest fixed point of the generating function

$$\varphi_\rho(s) = \sum_{k \geq 0} \rho(k) \, s^k$$

of ρ. This is again a consequence of Theorem (6.9). Indeed, the previous observation shows that $q = \Pi h_0(1) = \sum_{k \geq 0} \rho(k) h_0(k) = \varphi_\rho(q)$. On the other hand, if s is an arbitrary fixed point of φ_ρ, then the function $h(k) := s^k$ satisfies $\Pi h = h$, so that $s = h(1) \geq h_0(1) = q$.

▷ φ_ρ is either affine (namely when $\rho(0) + \rho(1) = 1$) or strictly convex. We also know that $\varphi_\rho(0) = \rho(0) < 1$ and $\varphi_\rho(1) = 1$, and Theorem (4.33b) shows that $\varphi_\rho'(1) = \mathbb{E}(\rho)$. So, in the case $\mathbb{E}(\rho) \leq 1$, φ_ρ has 1 as its only fixed point, whereas otherwise there is an additional, non-trivial fixed point in $]0, 1[$; see Figure 6.3.

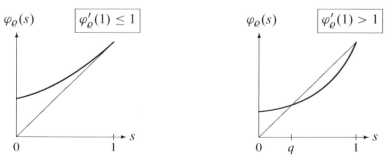

Figure 6.3. Extinction probability q of a Galton–Watson process as the smallest fixed point of the generating function φ_ρ of the offspring distribution.

To summarise our results: *If* $\mathbb{E}(\rho) \leq 1$, *the population dies out eventually with probability* 1. *For* $\mathbb{E}(\rho) > 1$, *the extinction probability q is strictly between* 0 *and* 1, *and equals the smallest fixed point of the generating function* φ_ρ *of the offspring distribution* ρ. Therefore, the population has a positive chance of survival if and only if each individual has more than one offspring on average. For example, in the case $\rho(0) = 1/4$, $\rho(2) = 3/4$ (where each individual has either 0 or 2 offspring), we get that $\varphi_\rho(s) = (1 + 3s^2)/4$ and thus $q = 1/3$.

To conclude this section, let us note that Theorem (6.9) can also be applied to the hitting times of non-absorbing states.

(6.12) Remark. *Hitting probabilities.* Let $z \in E$ be an arbitrary state and $h_z(x) = P^x(\tau_z < \infty)$ for $x \neq z$ and $h_z(z) = 1$. Then h_z is the smallest non-negative function for which $\Pi h(x) = h(x)$ for all $x \neq z$.

Indeed, let $\tilde{X}_n = X_{\min(n,\tau_z)}$, $n \geq 0$, be the 'Markov chain stopped at time τ_z'. It is easy to check that $(\tilde{X}_n)$ is a Markov chain with the modified transition matrix

$$\widetilde{\Pi}(x, y) = \begin{cases} \Pi(x, y) & \text{if } x \neq z \,, \\ 1 & \text{if } x = y = z \,, \\ 0 & \text{otherwise,} \end{cases}$$

for which the state z has been made absorbing. For $x \neq z$, $h_z(x)$ is exactly the probability that the Markov chain $(\tilde{X}_n)$, when started in x, is absorbed in z. Thus, the claim follows from Theorem (6.9).

6.3 Asymptotic Stationarity

As we have seen in Theorem (6.9), the limit $\lim_{n\to\infty} P^x(X_n = z)$ exists for every absorbing state $z \in E$. We are now interested in the existence of this limit in the opposite case of a 'very communicative' Markov chain that can get from any state to any other in a given number of steps. We restrict ourselves to the case of a finite state space E; the results do carry over to infinite E when the Markov chain is positive recurrent in the sense of Section 6.4.2 below, see e.g. [16].

6.3.1 The Ergodic Theorem

For a 'communicative' Markov chain the distribution settles down to a time-invariant equilibrium distribution after a long time. This is the assertion of the following theorem, which is known as the ergodic theorem.

The name ergodic theorem traces back to the famous ergodic hypothesis by Ludwig Boltzmann (1887) about the long-term behaviour of a mechanical system described by a point in the phase space; he coined the word 'ergode' (inspired by Greek) for the equilibrium distribution of a system with constant energy, see [21] for a historical analysis. The following ergodic theorem should not be confused with Birkhoff's ergodic theorem, which is a generalisation of the strong law of large numbers to the case of stationary sequences of random variables and applies directly to Boltzmann's situation (but will not be discussed here). It is now commonplace to use the same term also for the theorem below, because the Ehrenfests used it to analyse Boltzmann's ergodic hypothesis.

(6.13) Theorem. *Ergodic theorem for Markov chains. Let E be finite, and suppose there exists some $k \geq 1$ such that $\Pi^k(x, y) > 0$ for all $x, y \in E$. Then, for every $y \in E$, the limit*

$$\lim_{n \to \infty} \Pi^n(x, y) = \alpha(y) > 0$$

exists and does not depend on the choice of the starting point $x \in E$. Furthermore, the limit α is the unique probability density on E satisfying

(6.14) $$\sum_{x \in E} \alpha(x)\, \Pi(x, y) = \alpha(y) \quad \textit{for all } y \in E.$$

We postpone the proof for a moment and first comment on the the meaning of the theorem and deduce two consequences. To begin, recall equation (6.4), which can be written in short as $\Pi^n(x, \cdot) = P^x \circ X_n^{-1}$, i.e., the xth row of Π^n is the distribution of the value of the Markov chain at time n when started in x. Therefore, the ergodic theorem tells us that this distribution at time n tends to a limit α as $n \to \infty$, which does not depend on the starting point x and is characterised by the invariance property (6.14). What is the meaning of this property?

(6.15) Remark and Definition. *Stationary distributions.* If we interpret α as a row vector, equation (6.14) can be written as $\alpha \Pi = \alpha$; so the limiting distribution in Theorem (6.13) is a left eigenvector of Π with corresponding eigenvalue 1. (Compare this with Theorem (6.9), where h_z turned out to be a right eigenvector for the eigenvalue 1.) If we take such an α as the initial distribution, then the corresponding Markov chain is time invariant ('stationary') in the sense that

(6.16) $$P^\alpha((X_n, X_{n+1}, \dots) \in A) = P^\alpha(A)$$

for every $A \in \mathscr{F} = \mathscr{P}(E)^{\otimes \mathbb{Z}_+}$ and $n \geq 0$. Therefore, a probability density α with $\alpha \Pi = \alpha$ is called a *stationary (initial) distribution.*

Proof. The case-distinction formula (3.3a) and Theorem (6.7) imply that

$$P^\alpha((X_n, X_{n+1}, \dots) \in A) = \sum_{x \in E} P^\alpha(X_n = x)\, P^x(A)\,.$$

Using the case-distinction formula and (6.4), we find further that

$$P^\alpha(X_n = x) = \sum_{x_0 \in E} \alpha(x_0) \Pi^n(x_0, x) = \alpha \Pi^n(x)\,.$$

So, if α is stationary in the sense of (6.14), then $\alpha \Pi^n = \alpha$, and (6.16) follows. $\diamond$

Not only does the distribution of the Markov chain at a single time n converge to α as $n \to \infty$; in fact, in the limit, the entire process from time n onwards approaches the stationary process with initial distribution α. In physical terminology, this is 'convergence towards equilibrium', and can be stated as follows.

(6.17) Corollary. Asymptotic stationarity. *Under the conditions of Theorem (6.13),*

$$\sup_{A \in \mathscr{F}} \left| P^x((X_n, X_{n+1}, \dots) \in A) - P^\alpha(A) \right| \xrightarrow[n \to \infty]{} 0$$

for all $x \in E$. That is, after sufficiently long time the Markov chain is nearly stationary, no matter where it starts.

Proof. For every $A \in \mathscr{F}$, Theorem (6.7) shows that

$$\left| P^x((X_n, X_{n+1}, \dots) \in A) - P^\alpha(A) \right| = \left| \sum_{y \in E} [P^x(X_n = y) - \alpha(y)] \, P^y(A) \right|$$

$$\leq \sum_{y \in E} \left| P^x(X_n = y) - \alpha(y) \right|,$$

and by Theorem (6.13) the last expression converges to 0 as $n \to \infty$. $\diamond$

As a consequence of asymptotic stationarity, we obtain a zero-one law for the tail events of a Markov chain. This extends Kolmogorov's zero-one law (3.49), which concerns the case of independence (i.e., of complete forgetfulness), to the Markovian case of a short-term memory.

(6.18) Corollary. Orey's zero-one law. *Under the assumptions of Theorem (6.13), the equation*

$$P^x(A) = P^\alpha(A) = 0 \text{ or } 1$$

holds for every A in the tail σ-algebra $\mathscr{T}(X_n : n \geq 0)$ and every $x \in E$.

Proof. Consider any $A \in \mathscr{T} := \mathscr{T}(X_n : n \geq 0)$. By definition, for every $n \geq 0$ there exists a $B_n \in \mathscr{F}$ such that $A = \{(X_n, X_{n+1}, \dots) \in B_n\} = E^n \times B_n$; for the last equality, we used the agreement (6.2b) to always work with the canonical model. In particular, $B_n = E^k \times B_{n+k}$ for every $k \geq 0$, and thus $B_n \in \mathscr{T}$.

Now, equation (6.16) shows first that $P^\alpha(A) = P^\alpha(B_n)$. Corollary (6.17) thus yields that

$$\left| P^x(A) - P^\alpha(A) \right| = \left| P^x((X_n, X_{n+1}, \dots) \in B_n) - P^\alpha(B_n) \right| \to 0$$

as $n \to \infty$, whence $P^x(A) = P^\alpha(A)$ for all $A \in \mathscr{T}$ and $x \in E$. Since also $B_n \in \mathscr{T}$, it follows in particular that $P^x(B_n) = P^\alpha(B_n)$ for every n and x. By Theorem (6.7), this means that

$$P^\alpha(A) = P^\alpha(B_n) = P^{x_n}(B_n) = P^\alpha(A | X_0 = x_0, \dots, X_n = x_n)$$

for every $x_0, \ldots, x_n \in E$. Hence, with respect to P^α, A is independent of $X_0, \ldots, X_n$ for every n, so by Theorem (3.19) it is also independent of the entire sequence $(X_i)_{i \geq 0}$ and consequently of A itself. This means that $P(A) = P(A)^2$, and the claim follows. $\diamond$

We now turn to the proof of the ergodic theorem.

Proof of Theorem (6.13). *Step 1: Contractivity of* Π. We measure the distance between two probability densities ρ_1 and ρ_2 by the sum-norm

$$\|\rho_1 - \rho_2\| = \sum_{z \in E} |\rho_1(z) - \rho_2(z)|,$$

which in stochastics and measure theory is called the *total variation distance* between ρ_1 and ρ_2. It is clear that

$$(6.19) \qquad \|\rho_1 \Pi - \rho_2 \Pi\| \leq \sum_{x \in E} \sum_{y \in E} |\rho_1(x) - \rho_2(x)| \, \Pi(x, y) = \|\rho_1 - \rho_2\|.$$

We will show that this inequality can be tightened to a strict inequality if Π is replaced by Π^k.

As all entries of Π^k are positive and E is finite, there exists some $\delta > 0$ such that $\Pi^k(x, y) \geq \delta/|E|$ for all $x, y \in E$. Let U be the stochastic matrix whose rows contain the uniform distribution, i.e., $U(x, y) = |E|^{-1}$ for all $x, y \in E$. Our choice of δ means that $\Pi^k \geq \delta U$ elementwise. Since Π^k and U are stochastic matrices, it follows that either $\delta < 1$ or $\Pi^k = U$. Let us assume the former, since the latter case is trivial. Then we can define the matrix $V = (1-\delta)^{-1}(\Pi^k - \delta U)$, which is also stochastic, and obtain the identity $\Pi^k = \delta U + (1-\delta)V$. By the linearity of matrix multiplication and the norm properties of $\|\cdot\|$, we find

$$\|\rho_1 \Pi^k - \rho_2 \Pi^k\| \leq \delta \|\rho_1 U - \rho_2 U\| + (1-\delta) \|\rho_1 V - \rho_2 V\|.$$

But now $\rho_1 U = \rho_2 U$, since for every $y \in E$

$$\rho_1 U(y) = \sum_{x \in E} \rho_1(x) |E|^{-1} = |E|^{-1} = \rho_2 U(y).$$

Combining this with the inequality (6.19) for V in place of Π, we thus obtain

$$\|\rho_1 \Pi^k - \rho_2 \Pi^k\| \leq (1-\delta) \|\rho_1 - \rho_2\|$$

and, by iteration,

$$(6.20) \qquad \|\rho_1 \Pi^n - \rho_2 \Pi^n\| \leq \|\rho_1 \Pi^{km} - \rho_2 \Pi^{km}\| \leq 2(1-\delta)^m;$$

here $m = \lfloor n/k \rfloor$ is the integer part of n/k, and we have applied (6.19) to Π^{n-km} and also used that $\|\rho_1 - \rho_2\| \leq 2$.

Step 2: Convergence and characterisation of the limit. For an arbitrary density ρ we consider the sequence $(\rho\Pi^n)$. Since the set of all probability densities on E is a closed subset of the compact set $[0,1]^E$, there exists a subsequence (n_k) such that $\rho\Pi^{n_k}$ converges to a probability density α. Together with (6.20) as applied to $\rho_1 = \rho$ and $\rho_2 = \rho\Pi$, we therefore obtain

$$\alpha = \lim_{k\to\infty} \rho\Pi^{n_k} = \lim_{k\to\infty} \rho\Pi^{n_k+1} = \alpha\Pi\,.$$

Hence, α is a stationary distribution, and $\alpha(y) = \alpha\Pi^k(y) \geq \delta/|E|$ for every $y \in E$. Furthermore, another application of (6.20) (now with $\rho_2 = \alpha$) shows that $\rho\Pi^n \to \alpha$ as $n \to \infty$. In particular, if ρ is any stationary distribution, then $\rho \to \alpha$, so $\rho = \alpha$, which means that α is the only stationary distribution. On the other hand, if $\rho = \delta_x$ for some $x \in E$, then $\rho\Pi^n = \Pi^n(x,\cdot)$, and consequently $\Pi^n(x,\cdot) \to \alpha$ in the limit as $n \to \infty$. $\Diamond$

As the inequality (6.20) shows, the convergence $\Pi^n(x,\cdot) \to \alpha$ even takes place at exponential speed. Namely, we have $\|\Pi^n(x,\cdot) - \alpha\| \leq C\,e^{-cn}$ for $C = 2e^\delta$ and $c = \delta/k > 0$.

6.3.2 Applications

Let us now present some examples to illustrate the convergence to equilibrium as shown above. The first is a tutorial warm-up.

(6.21) Example. *The matrix from Remark* (6.2c). Let $E = \{1,2,3\}$ and

$$\Pi = \begin{pmatrix} 1/2 & 1/2 & 0 \\ 1/3 & 1/3 & 1/3 \\ 1 & 0 & 0 \end{pmatrix};$$

Figure 6.1 shows the corresponding transition graph. By inspection of the graph (or an immediate calculation), we see that Π^3 has positive entries only. Moreover, the linear equation $\alpha\Pi = \alpha$ has the solution $\alpha = (1/2, 3/8, 1/8)$. By Theorem (6.13) it follows that α is the only stationary distribution, and

$$\begin{pmatrix} 1/2 & 1/2 & 0 \\ 1/3 & 1/3 & 1/3 \\ 1 & 0 & 0 \end{pmatrix}^n \xrightarrow[n\to\infty]{} \begin{pmatrix} 1/2 & 3/8 & 1/8 \\ 1/2 & 3/8 & 1/8 \\ 1/2 & 3/8 & 1/8 \end{pmatrix}.$$

Now we move on to some more interesting examples.

(6.22) Example. *The urn model by P. and T. Ehrenfest.* Let us come back to the Ehrenfests' model for the exchange of gas molecules between two adjacent containers; see Example (5.9) and Figure 5.1 on p. 123. We will now investigate a time evolution of the model.

Suppose that N labelled balls are distributed across two urns. At every time step, a label is randomly chosen, and the corresponding ball is moved to the other urn with probability $p \in]0, 1[$; it is left in place with probability $1 - p$. (This is a variant of the original model, where $p = 1$.) Let X_n be the number of balls in urn 1 at time n. Then the sequence $(X_n)_{n \geq 0}$ is modelled by the Markov chain on $E = \{0, \ldots, N\}$ with transition matrix

$$\Pi(x, y) = \begin{cases} p\, x/N & \text{if } y = x - 1, \\ 1 - p & \text{if } y = x, \\ p\, (N - x)/N & \text{if } y = x + 1, \\ 0 & \text{otherwise.} \end{cases}$$

What is the distribution of X_n for large n? For every $x, y \in E$ with $x \leq y$, the Markov chain can move from x to y in N steps by staying in x for $N - |x - y|$ steps and then moving upwards to y in the remaining steps. This shows that

$$\Pi^N(x, y) \geq \Pi(x, x)^{N - |x - y|}\, \Pi(x, x+1)\, \Pi(x+1, x+2) \ldots \Pi(y-1, y) > 0,$$

and an analogous inequality holds when $x \geq y$. Therefore, Π satisfies the assumptions of Theorem (6.13) with $k = N$. It follows that the limit $\alpha(y) = \lim_{n \to \infty} \Pi^n(x, y)$ exists for all $x, y \in E$, and α is the unique solution of $\alpha \Pi = \alpha$. We can guess α. After a long time, every ball will be in urn 1 with probability $1/2$, independently of all others; so presumably $\alpha = \beta := \mathcal{B}_{N, 1/2}$. Indeed, in the case $x > 0$, $y = x - 1$ we find

$$\beta(x)\, \Pi(x, y) = 2^{-N} \binom{N}{x} p\, \frac{x}{N} = p\, 2^{-N}\, \frac{(N-1)!}{(x-1)!\, (N-x)!}$$

$$= 2^{-N} \binom{N}{x-1} p \left(1 - \frac{x-1}{N}\right) = \beta(y)\, \Pi(y, x)$$

and thus $\beta(x) \Pi(x, y) = \beta(y) \Pi(y, x)$ for every $x, y \in E$. This symmetry of β is called 'detailed balance equation', and β is also known as a *reversible distribution*. This is because, for every n, the distribution of $(X_i)_{0 \leq i \leq n}$ under P^β is invariant under time reversal. Namely, for arbitrary $x_0, \ldots, x_n \in E$ one can write

$$P^\beta(X_i = x_i \text{ for } 0 \leq i \leq n) = \beta(x_0)\, \Pi(x_0, x_1) \ldots \Pi(x_{n-1}, x_n)$$

$$= \beta(x_n)\, \Pi(x_n, x_{n-1}) \ldots \Pi(x_1, x_0)$$

$$= P^\beta(X_{n-i} = x_i \text{ for } 0 \leq i \leq n).$$

Summing over x in the detailed balance equation, one finds that $\beta \Pi = \beta$. By the uniqueness of the stationary distribution, this means that $\alpha = \beta$.

Combining this result with Example (5.9), we see the following. Suppose N has the realistic order of magnitude of 10^{23}. Then, after a sufficiently long time, it is almost certain that nearly half of the balls are contained in urn 1. This holds in any case, no matter how many balls were in urn 1 at time 0 – even if urn 1 was initially empty. This agrees with the physical observations.

(6.23) Example. *Shuffling a deck of cards.* Suppose we are given a deck of $N \geq 3$ playing cards. We think of the cards as being numbered. The order of the cards is then described by a permutation π, an element of the permutation group $E := \mathscr{S}_N$ of $\{1, \ldots, N\}$. The commonly used procedures for shuffling the cards take the form

$$\pi_0 \to X_1 = \xi_1 \circ \pi_0 \to X_2 = \xi_2 \circ X_1 \to X_3 = \xi_3 \circ X_2 \to \cdots,$$

where π_0 is the initial order and $(\xi_i)_{i \geq 1}$ are independent, identically distributed random permutations (i.e., $\mathscr{S}_N$-valued random variables). Comparing this situation to random walks on $\mathbb{Z}$, we see that the addition there is replaced here by the group operation '$\circ$'. In other words, $(X_n)_{n \geq 0}$ is a random walk on the finite group $E = \mathscr{S}_N$. The deck of cards is well-shuffled at time n if X_n is nearly uniformly distributed. Does this happen eventually, and how long does it take?

Let $\rho(\pi) = P(\xi_i = \pi)$ be the distribution density of the ξ_i. Then we can write

$$P(X_1 = \pi_1, \ldots, X_n = \pi_n) = \rho(\pi_1 \circ \pi_0^{-1}) \ldots \rho(\pi_n \circ \pi_{n-1}^{-1}),$$

where $n \geq 1$, $\pi_1, \ldots, \pi_n \in E$ are arbitrary. This means that $(X_n)_{n \geq 0}$ is a Markov chain with transition matrix $\Pi(\pi, \pi') = \rho(\pi' \circ \pi^{-1})$ for $\pi, \pi' \in E$. This matrix Π is *doubly stochastic*, in that both the row sums and the column sums are equal to 1. Explicitly, the identity

$$\sum_{\pi' \in E} \Pi(\pi, \pi') = \sum_{\pi' \in E} \rho(\pi' \circ \pi^{-1}) = 1$$

holds for any $\pi \in E$ (since $\pi' \to \pi' \circ \pi^{-1}$ is a bijection from E onto itself), and also

$$\sum_{\pi \in E} \Pi(\pi, \pi') = \sum_{\pi \in E} \rho(\pi' \circ \pi^{-1}) = 1$$

for all $\pi' \in E$ (because $\pi \to \pi' \circ \pi^{-1}$ is a bijection from E onto itself). It follows that the uniform distribution $\alpha = \mathcal{U}_{\mathscr{S}_N}$ is a stationary distribution for Π. Indeed,

$$\sum_{\pi \in E} \alpha(\pi) \Pi(\pi, \pi') = \frac{1}{N!} = \alpha(\pi') \quad \text{for every } \pi' \in E.$$

As a consequence, if Π satisfies the assumptions of Theorem (6.13), then

$$\lim_{n \to \infty} P^{\pi_0}(X_n = \pi) = \frac{1}{N!} \quad \text{for every } \pi, \pi_0 \in E,$$

i.e., in the long run the deck will be well-shuffled. For example, this applies to the popular shuffling technique in which a bunch of cards is taken from the top and inserted

Figure 6.4. Illustration of the equation $\pi_{1,i} \circ \pi_{i,i+1} = (i, i+1)$ for $N = 7, i = 3$.

somewhere below in the deck. This corresponds to a (randomly chosen) permutation of the form

$$\pi_{i,j} : (1, \ldots, N) \to (i+1, \ldots, j, 1, \ldots, i, j+1, \ldots, N), \quad 1 \le i < j \le N.$$

We claim that each permutation can be represented as a composition of a fixed number of these $\pi_{i,j}$. To see this, we note first that the transposition $(i, i+1)$ that swaps i and $i+1$ can be represented as

$$(i, i+1) = \begin{cases} \pi_{1,2} & \text{if } i = 1, \\ \pi_{1,i} \circ \pi_{i,i+1} & \text{if } i > 1; \end{cases}$$

see Figure 6.4. Since every permutation can be represented as the composition of finitely many transpositions, it follows that every permutation is composed of finitely many $\pi_{i,j}$. Since $\pi_{1,2}^2$ and $\pi_{1,3}^3$ are the identical permutations, we can add these if required to achieve that every permutation is represented by the same number k of $\pi_{i,j}$; see also Problem 6.13. So if $\rho(\pi_{i,j}) > 0$ for every $1 \le i < j \le N$, then $\Pi^k(\pi, \pi') > 0$ for all $\pi, \pi' \in E$, and Theorem (6.13) can be applied. Hence, if the deck of cards is shuffled repeatedly in this way, its distribution approaches the uniform distribution with an exponential rate. (Unfortunately, however, this rate is very small.)

(6.24) Example. *The Markov chain Monte Carlo method (MCMC).* Let α be a probability density on a finite, but very large set E. How can one simulate a random variable with distribution α? The rejection sampling method from Example (3.43) is not suitable in this case. For illustration, let us consider the problem of *image processing*. A (black and white) image is a configuration of white and black pixels arranged in an array Λ. Therefore, the state space is $E = \{-1, 1\}^{\Lambda}$, where -1 and 1 stand for the colours 'white' and 'black'. Even for relatively small dimensions of Λ, for instance 1000×1000 pixels, the cardinality of E is astronomical, namely $2^{1000 \times 1000} \approx 10^{301030}$. In order to reconstruct a randomly created image (for example captured by a satellite and transmitted by noisy radio communication), it turns out that a density α of the form

$$(6.25) \quad \alpha(x) = Z^{-1} \exp\left[\sum_{i,j \in \Lambda : i \ne j} J(i, j)\, x_i\, x_j + \sum_{i \in \Lambda} h(i)\, x_i \right] \quad \text{for } x = (x_i)_{i \in \Lambda} \in E$$

is useful. Here, J (the couplings between different, for instance adjacent pixels) and h (the local tendency towards 'black') are suitable functions that depend on the image received, and the normalisation constant Z is defined so that $\sum_{x \in E} \alpha(x) = 1$. (This approach is actually taken from statistical mechanics, and corresponds to the *Ising model* – first studied by E. Ising in 1924 – for ferromagnetic materials such as iron or nickel. There, the values ± 1 describe the 'spin' of an elementary magnet.)

If we were to simulate α using the rejection sampling method from Example (3.43), we would have to calculate $\alpha(x)$ for arbitrary $x \in E$, and so in particular Z, the sum over astronomically many summands. Numerically this is hopeless, and explains why direct methods of the type of rejection sampling are out of question. Fortunately, however, it is rather simple to determine the ratio $\alpha(x)/\alpha(y)$, if the configurations $x = (x_i)_{i \in \Lambda}$ and $y = (y_i)_{i \in \Lambda}$ only differ at one site $j \in \Lambda$.

This leads to the idea to simulate a Markov chain with the following properties: (i) the transition matrix depends only on the α-ratios that are easy to calculate, (ii) α is its unique stationary distribution, and (iii) the assumptions of Theorem (6.13) are satisfied. Since the convergence in Theorem (6.13) takes place at exponential rate by (6.20), one can hope that after sufficiently long time one ends up with a realisation that is typical of α.

In general, there are many possibilities to choose a suitable transition matrix Π. A classical choice is the following *algorithm by N. Metropolis* (1953). We describe it in a slightly more general setting. Let S and Λ be finite sets, $E = S^{\Lambda}$, and α a strictly positive probability density on E. Also let $d = |\Lambda|(|S| - 1)$ and define

$$(6.26) \qquad \Pi(x, y) = \begin{cases} d^{-1} \min(\alpha(y)/\alpha(x), 1) & \text{if } y \sim x, \\ 1 - \sum_{y: y \sim x} \Pi(x, y) & \text{if } y = x, \\ 0 & \text{otherwise.} \end{cases}$$

Here we write $y \sim x$ if y differs from x at exactly one coordinate, i.e., if there exists some $j \in \Lambda$ such that $y_j \neq x_j$, but $y_i = x_i$ for all $i \neq j$. Every x has exactly d such 'neighbours'. Clearly, Π is a stochastic matrix, and as in Example (6.22), we easily obtain the detailed balance equation $\alpha(x)\Pi(x, y) = \alpha(y)\Pi(y, x)$ for all $x, y \in E$. Summing over y we obtain the stationarity equation $\alpha \Pi = \alpha$. Unless α is the uniform distribution (which we will ignore here), every $y \in E$ can be reached from every $x \in E$ in $m := 2|\Lambda|$ admissible steps. Indeed, if α is not constant, there exist $\tilde{x}, \tilde{y} \in E$ such that $\alpha(\tilde{y}) < \alpha(\tilde{x})$ and $\tilde{y} \sim \tilde{x}$, and thus $\Pi(\tilde{x}, \tilde{x}) > 0$. Now, if x differs from $\tilde{x}$ in k coordinates and $\tilde{x}$ from y in l coordinates, then the Markov chain can first move in k steps from x to $\tilde{x}$, then stay there for $m - k - l$ steps, and finally reach y in l steps. Hence, we get $\Pi^m(x, y) > 0$ for all $x, y \in E$. Therefore, Π satisfies the assumption of Theorem (6.13), and the distribution of the corresponding Markov chain converges to α in the long run.

The Metropolis algorithm with n iterations can be written in pseudocode as follows:

$k \leftarrow 0, \ x \leftarrow x^0$
repeat
    $y \leftarrow Y_x, \ k \leftarrow k + 1$
    if $U < \min(\alpha(y)/\alpha(x), 1)$
       then $x \leftarrow y$
until $k > n$
$X_n \leftarrow x$

($x^0 \in E$ is an arbitrary initial configuration. $Y_x \in E$ is a random neighbour of x and $U \in [0, 1]$ is a (pseudo) random number; these are generated afresh at each step according to the corresponding uniform distributions.)

It is immediate that the Metropolis algorithm does not need the product structure of $E = S^\Lambda$, but also works when E has only some graph structure. If $x \in E$ has d_x neighbours, then the constant d in the right-hand side of (6.26) must be replaced by d_x, and the ratio $\alpha(y)/\alpha(x)$ must be multiplied by d_x/d_y.

An alternative to the Metropolis algorithm (6.26) is the *Gibbs sampler* (introduced by S. and D. Geman in 1984). It takes advantage of the fact that, for a distribution on a product space $E = S^\Lambda$ of the form (6.25), it is easy to calculate the conditional probability

$$\alpha_i(a|x_{\Lambda \setminus \{i\}}) := \alpha(ax_{\Lambda \setminus \{i\}}) \Big/ \sum_{b \in S} \alpha(bx_{\Lambda \setminus \{i\}})$$

for the value $a \in S$ at position $i \in \Lambda$ given the values $x_{\Lambda \setminus \{i\}}$ at the remaining positions, at least if the interdependencies expressed by J do not reach very far. Therefore, at every step, the rule for the Gibbs sampler reads:

1. Choose an arbitrary $i \in \Lambda$ according to the uniform distribution.

2. Replace x_i by a random $a \in S$ with distribution $\alpha_i(\cdot |x_{\Lambda \setminus \{i\}})$.

As a concrete example we consider the 'hard-core lattice gas' on the square grid $\Lambda = \{1, \dots, N\}^2$. Let $E = \{0, 1\}^\Lambda$. Each $x \in E$ is considered as a configuration of spheres located at the sites $i \in \Lambda$ with $x_i = 1$. The spheres are supposed to be so large that neighbouring sites on the grid cannot be occupied. Consequently, we call a configuration $x \in E$ admissible if $x_i x_j = 0$ for all $i, j \in \Lambda$ with $|i - j| = 1$. The *hard-core lattice gas* is then specified by the distribution

$$\alpha(x) = Z^{-1} \lambda^{n(x)} 1_{\{x \text{ is admissible}\}}, \quad x \in E.$$

Here $n(x) = \sum_{i \in \Lambda} x_i$ is the number of occupied grid sites, $\lambda > 0$ a parameter that controls the average number of spheres, and Z a normalising constant. In this case, the pseudocode of the Gibbs sampler is

$k \leftarrow 0, \ x \leftarrow x^0$
repeat
    $i \leftarrow I, \ k \leftarrow k + 1$
    if $U < \frac{\lambda}{\lambda+1}$ and $\sum_{j:|j-i|=1} x_j = 0$
       then $x_i \leftarrow 1$
      else $x_i \leftarrow 0$
until $k > n$
$X_n \leftarrow x$

($x^0 \in E$ is an admissible starting configuration, e.g., identically zero. The random quantities $I \in \Lambda$ and $U \in [0, 1]$ are generated afresh at each step according to the corresponding uniform distributions.)

Figure 6.5 shows a realisation.

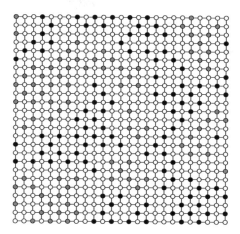

Figure 6.5. A realisation of the hard-core lattice gas model for $\lambda = 2$ obtained with the Gibbs
sampler. The occupied sites are marked grey or black, depending on whether the sum of their
coordinates is even or odd. In this way the grid splits into grey and black regions. The larger
λ, the more likely it is that one such region spans the whole square Λ. In the limit $\Lambda \uparrow \mathbb{Z}^2$,
one can then construct two different probability measures that describe a grey and a black
'phase'. This is an example of a phase transition.

6.3.3 Return Times and the Renewal Theorem

Let us now come back to the stationary distribution α we encountered in the ergodic
theorem (6.13). We will now show that there is a striking connection between α and
the return times τ_x introduced in Remark (6.8). This relationship also holds when the
state space E is infinite, and the assumption of Theorem (6.13) can be weakened as
follows.

Definition. A transition matrix Π is called *irreducible* if, for any $x, y \in E$, there
exists some $k = k(x, y) \geq 0$ such that $\Pi^k(x, y) > 0$.

Hence, irreducibility means that, with positive probability, each state can be reached
from any other state in a finite number of steps. (For a comparison with the assumption
of Theorem (6.13) see Problem 6.13.) We write $\mathbb{E}^x$ and $\mathbb{E}^\alpha$ for the expectation with
respect to P^x and P^α, respectively.

(6.27) Theorem. Stationarity and return times. *Let E be any countable set and Π an
irreducible transition matrix. If Π admits a stationary distribution α, then α is unique
and satisfies*

$$0 < \alpha(x) = 1/\mathbb{E}^x(\tau_x) \quad \text{for all } x \in E.$$

For a Markov chain in equilibrium, the probability of being in a given state thus coincides with the inverse mean return time. This fact will be discussed further in (and after) Theorem (6.34).

Proof. We proceed in two steps.

Step 1. Here we prove a more general result, namely *Mark Kac's recurrence theorem* (1947): Every stationary distribution α satisfies the identity

$$(6.28) \qquad \alpha(x)\, \mathbb{E}^x(\tau_x) = P^\alpha(\tau_x < \infty)$$

for all $x \in E$. This follows from the stationarity of the sequence $(X_n)_{n \geq 0}$ with respect to P^α, cf. Remark (6.15). Indeed, we can write

$$\alpha(x)\, \mathbb{E}^x(\tau_x) = \mathbb{E}^\alpha \big(1_{\{X_0 = x\}}\, \tau_x \big) = \mathbb{E}^\alpha \big(1_{\{X_0 = x\}} \sum_{k \geq 0} 1_{\{\tau_x > k\}} \big)$$

$$= \sum_{k \geq 0} P^\alpha \big(X_0 = x,\, \tau_x > k \big)$$

$$= \lim_{n \to \infty} \sum_{k=0}^{n-1} P^\alpha \big(X_0 = x,\, X_i \neq x \text{ for } 1 \leq i \leq k \big)$$

$$= \lim_{n \to \infty} \sum_{k=0}^{n-1} P^\alpha \big(X_{n-k} = x,\, X_i \neq x \text{ for } n-k+1 \leq i \leq n \big).$$

The third equation relies on the σ-additivity of the expectation (see Theorem (4.7c)), and the last equation follows from the stationarity of $(X_n)_{n \geq 0}$ by shifting time by $n - k$ steps. But the event in the last sum reads: '$n - k$ is the last time before n at which the Markov chain visits state x'. For different $k \in \{0, \ldots, n-1\}$, these events are disjoint, and their union is exactly the event $\{\tau_x \leq n\}$ that the state x is visited at least once in the time interval $\{1, \ldots, n\}$. Hence we obtain

$$\alpha(x)\, \mathbb{E}^x(\tau_x) = \lim_{n \to \infty} P^\alpha(\tau_x \leq n) = P^\alpha(\tau_x < \infty).$$

Step 2. It remains to be shown that $\alpha(x) > 0$ and $P^\alpha(\tau_x < \infty) = 1$ for all $x \in E$. Since α is a probability density, there is at least one $x_0 \in E$ with $\alpha(x_0) > 0$. For arbitrary $x \in E$ there exists, by the irreducibility of Π, some $k \geq 0$ such that $\Pi^k(x_0, x) > 0$, whence $\alpha(x) = \alpha \Pi^k(x) \geq \alpha(x_0) \Pi^k(x_0, x) > 0$. With the help of equation (6.28) one finds that $\mathbb{E}^x(\tau_x) \leq 1/\alpha(x) < \infty$, and so a fortiori $P^x(\tau_x < \infty) = 1$. As will be shown in Theorem (6.30) below, this implies that, in fact, $P^x(X_n = x \text{ infinitely often}) = 1$. Combining this with the case-distinction formula (3.3a) and the Markov property (6.7), we obtain for arbitrary $k \geq 1$

$$0 = P^x \big(X_n \neq x \text{ for all } n > k \big) = \sum_{y \in E} \Pi^k(x, y)\, P^y(\tau_x = \infty)$$

and therefore $P^y(\tau_x = \infty) = 0$ when $\Pi^k(x, y) > 0$. From the irreducibility of Π, we can conclude that $P^y(\tau_x < \infty) = 1$ for *all* $y \in E$. It follows that $P^\alpha(\tau_x < \infty) = \sum_{y \in E} \alpha(y) P^y(\tau_x < \infty) = 1$, and the proof of the theorem is complete. $\diamond$

The preceding theorem can be combined with the ergodic theorem to obtain the following application.

(6.29) Example. *Renewal of technical appliances.* We consider technical appliances (such as bulbs, machines, and the like) that, as soon as they fail, are immediately replaced by a new appliance of the same type. Suppose their life spans are given by i.i.d. random variables $(L_i)_{i \geq 1}$ taking values in $\{1, \ldots, N\}$, where N is the maximal life span. For simplicity we assume that $P(L_1 = l) > 0$ for all $1 \leq l \leq N$, in other words, the appliances can fail at any age.

Let $T_k = \sum_{i=1}^k L_i$ be the time at which the kth appliance is replaced, $T_0 = 0$, and

$$X_n = n - \max\{T_k : k \geq 1,\ T_k \leq n\}, \quad n \geq 0,$$

the age of the appliance used at time n. It is easy to check that $(X_n)_{n \geq 0}$ is a Markov chain on $E = \{0, \ldots, N-1\}$ with transition matrix

$$\Pi(x, y) = \begin{cases} P(L_1 > y \mid L_1 > x) & \text{if } y = x+1 < N, \\ P(L_1 = x+1 \mid L_1 > x) & \text{if } y = 0, \\ 0 & \text{otherwise.} \end{cases}$$

This Markov chain increases by one step at a time for some random period and then collapses to zero again, whence it is also known as the *house-of-cards process*. Since $\Pi(x, y) > 0$ if $y = x+1 < N$ or $y = 0$, one can easily verify that $\Pi^N(x, y) > 0$ for all $x, y \in E$. Therefore, Theorems (6.13) and (6.27) yield

$$\lim_{n \to \infty} P\left(\sum_{i=1}^k L_i = n \text{ for some } k \geq 1 \right) = \lim_{n \to \infty} P^0(X_n = 0) = 1/\mathbb{E}(L_1).$$

In words, the probability of a renewal at time n is asymptotically reciprocal to the mean life span. This is (a simplified version of) the so-called *renewal theorem*.

6.4 Recurrence

Here we consider the general situation of a Markov chain with arbitrary, at most countable state space E and a given transition matrix Π. Motivated by Theorem (6.27) we now investigate the recurrence to a fixed state $x \in E$. Does the Markov chain certainly return, and if so, how long does it take on average until the first return? Let us start with the first question.

6.4.1   Recurrence and Transience

For a given $x \in E$ let $F_1(x, x) = P^x(\tau_x < \infty)$ be the probability of at least one return to x at some finite (random) time,

$$F_\infty(x, x) = P^x(X_n = x \text{ for infinitely many } n)$$

the probability of returning infinitely often, and

$$G(x, x) = \sum_{n \geq 0} \Pi^n(x, x) = \sum_{n \geq 0} P^x(X_n = x) = \mathbb{E}^x\left(\sum_{n \geq 0} 1_{\{X_n = x\}}\right)$$

the expected number of visits in x. Here, the last identity follows from Theorem (4.7c), and again $\mathbb{E}^x$ denotes the expectation with respect to P^x.

(6.30) Theorem. Recurrence of Markov chains. *For every $x \in E$ the following alternative is valid.*

(a) *Either $F_1(x, x) = 1$. Then $F_\infty(x, x) = 1$ and $G(x, x) = \infty$.*

(b) *Or $F_1(x, x) < 1$. Then $F_\infty(x, x) = 0$ and $G(x, x) = (1 - F_1(x, x))^{-1} < \infty$.*

Definition. In case (a), x is called *recurrent*, whereas in case (b) it is called *transient* with respect to Π.

Proof. Let $\sigma = \sup\{n \geq 0 : X_n = x\}$ be the time of the last visit to x. (Since $X_0 = x$, we know that σ is well-defined, but possibly infinite.) We will use three identities. First, by definition,

$$1 - F_\infty(x, x) = P^x(\sigma < \infty).$$

Secondly, for every $n \geq 0$ we conclude from Theorem (6.7) that

$$P^x(\sigma = n) = P^x(X_n = x) \, P^x(X_i \neq x \text{ for all } i \geq 1) = \Pi^n(x, x) \, (1 - F_1(x, x)).$$

Summing over n and using the first equation, we finally get

$$1 - F_\infty(x, x) = G(x, x) \, (1 - F_1(x, x)).$$

In case (a), the second equation implies that $P^x(\sigma = n) = 0$ for all n, and thus $F_\infty(x, x) = 1$ by the first equation. The Borel–Cantelli lemma (3.50a) then yields $G(x, x) = \infty$.

In case (b), we deduce from the third equation that $G(x, x) < \infty$, and thus by the Borel–Cantelli lemma that $F_\infty(x, x) = 0$. Another application of the third equation yields the relation between G and F_1. $\Diamond$

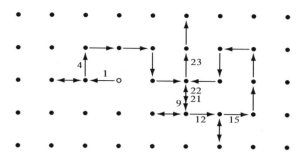

Figure 6.6. A realisation of the first few steps of a simple symmetric random walk on $\mathbb{Z}^2$ started at the origin. The numbers indicate the order of the steps when this is not evident.

(6.31) Example. *The simple symmetric random walk on $\mathbb{Z}^d$.* This is the Markov chain on $E = \mathbb{Z}^d$ with transition matrix

$$\Pi(x, y) = \begin{cases} 1/2d & \text{if } |x - y| = 1, \\ 0 & \text{otherwise.} \end{cases}$$

One associates with this random walk the idea of an aimless stroller (or a drunk person) who (for $d = 1$) moves from tree to tree on an infinitely long avenue, or (for $d = 2$) from junction to junction on a large, chessboard-like park or road network; cf. Figure 6.6. For $d = 3$ one may think of a child in a cubic jungle gym. What is the probability of an eventual return to the starting point? First note that, by the homogeneity of Π, the return probability $F_1(x, x)$ does not depend on x. Therefore, all $x \in \mathbb{Z}^d$ have the same type of recurrence. But which one? This depends on the dimension! Namely, the following holds:

▷ *For $d = 1$, every $x \in \mathbb{Z}^d$ is recurrent.*

Indeed, since a return is only possible at even times and the walker must take the same number of left and right steps, we get

$$G(x, x) = \sum_{n \geq 0} \Pi^{2n}(x, x) = \sum_{n \geq 0} 2^{-2n} \binom{2n}{n} = \infty.$$

The last equality comes from the asymptotic equivalence $2^{-2n} \binom{2n}{n} \sim 1/\sqrt{\pi n}$ proved in Example (5.2), and the divergence of the series $\sum_{n \geq 1} 1/\sqrt{n}$. The drunkard on the long avenue will therefore eventually come back to his starting point. (However, this might take quite long, see Example (6.36) below.) Similarly:

▷ *For $d = 2$, every $x \in \mathbb{Z}^d$ is recurrent.*

Indeed, for a return in $2n$ steps one has to take the same number (say k) of steps to the right and to the left and the same number (so $n - k$) up and down. Hence, we get

$$\Pi^{2n}(x,x) = 4^{-2n} \sum_{k=0}^{n} \frac{(2n)!}{k!^2\,(n-k)!^2} = 4^{-2n} \binom{2n}{n} \sum_{k=0}^{n} \binom{n}{k}\binom{n}{n-k}$$

$$= 4^{-2n} \binom{2n}{n}^2 \sum_{k=0}^{n} \mathcal{H}_{n;n,n}(\{k\}) = \left[2^{-2n}\binom{2n}{n}\right]^2 \underset{n\to\infty}{\sim} \frac{1}{\pi n}\,,$$

exactly as in the case $d = 1$ above. Since $\sum_{n\geq 1} 1/n = \infty$, we can conclude that $G(x,x) = \infty$. Therefore, in Manhattan, the drunkard will also return to his starting point after a finite time. (But again, this will take very long on average, see (6.36).)

Is the same true for higher dimensions? No! As noticed by G. Pólya in 1921, three dimensions suffice to get lost:

▷ *For $d \geq 3$, every $x \in \mathbb{Z}^d$ is transient.*

Indeed, for every $n \geq 1$ we obtain in the same way as above

$$\Pi^{2n}(x,x) = (2d)^{-2n} \sum_{\vec{k}} \frac{(2n)!}{k_1!^2\ldots k_d!^2} = 2^{-2n} \binom{2n}{n} \sum_{\vec{k}}\left[d^{-n}\binom{n}{\vec{k}}\right]^2;$$

the sums extend over all $\vec{k} = (k_1,\ldots,k_d) \in \mathbb{Z}_+^d$ with $\sum k_i = n$, and we have used the notation (2.8) for the multinomial coefficient. Furthermore, a comparison with the multinomial distribution shows that $\sum_{\vec{k}} d^{-n}\binom{n}{\vec{k}} = 1$. It follows that

$$\Pi^{2n}(x,x) \leq 2^{-2n} \binom{2n}{n} \; \max_{\vec{k}} \; d^{-n}\binom{n}{\vec{k}}.$$

The maximum is attained when $|k_i - n/d| \leq 1$ for all i; otherwise we have $k_i \geq k_j + 2$ for some i, j, and the substitution $k_i \rightsquigarrow k_i - 1$, $k_j \rightsquigarrow k_j + 1$ increases the multinomial coefficient. Using Stirling's formula (5.1) we conclude that

$$\max_{\vec{k}} \; d^{-n}\binom{n}{\vec{k}} \underset{n\to\infty}{\sim} d^{d/2}\big/(2\pi n)^{(d-1)/2}$$

and thus $\Pi^{2n}(x,x) \leq c\, n^{-d/2}$ for some $c < \infty$. But now $\sum_{n\geq 1} n^{-d/2} < \infty$ for $d \geq 3$, so it follows that $G(x,x) < \infty$. Hence, with positive probability a child on an infinitely large jungle gym will never return to its starting point (and should always be supervised)!

(6.32) Example. *A queueing model.* At a service counter, customers arrive randomly and expect to be served, for example at the supermarket check-out, at a chair lift, at a telephone hotline, etc. For simplicity, we consider a discrete time model and suppose that it takes exactly one time unit to serve one customer; this applies, for example, to the chair lift. Let X_n be the number of waiting customers at time n (so for the chair lift, immediately before chair number n can be mounted).

We model the sequence $(X_n)_{n\geq 0}$ by a Markov chain on $E = \mathbb{Z}_+$ with the matrix

$$\Pi(x, y) = \begin{cases} \rho(y) & \text{if } x = 0, \\ \rho(y - x + 1) & \text{if } x \geq 1, \ y \geq x - 1, \\ 0 & \text{otherwise.} \end{cases}$$

Here, ρ is a probability density on E, namely the distribution of the number of new customers that arrive per time unit. It is helpful to introduce the customer influx explicitly. Let Z_n be the number of customers that arrive during the nth time unit. The sequences $(X_n)_{n\geq 0}$ and $(Z_n)_{n\geq 0}$ are then related by

$$X_n - X_{n-1} = \begin{cases} Z_n - 1 & \text{if } X_{n-1} \geq 1, \\ Z_n & \text{if } X_{n-1} = 0, \end{cases}$$

so that they define each other. Now, it is easy to see that $(X_n)_{n\geq 0}$ is a Markov chain with transition matrix Π if and only if the random variables Z_n are independent with identical distribution ρ. In other words, apart from a boundary effect at the state 0 (an 'upward shift by 1', so that the process does not become negative), $(X_n)_{n\geq 0}$ is a random walk; cf. Example (6.5). The boundary effect can be described more precisely. Let $S_n = X_0 + \sum_{k=1}^{n}(Z_k - 1)$ be the random walk on $\mathbb{Z}$ without the boundary effect. Then one has the decomposition

$$(6.33) \hspace{3cm} X_n = S_n + V_n \, ,$$

where $V_n = \sum_{k=0}^{n-1} 1_{\{X_k = 0\}}$ is the idle time before time n. (V_n can also be expressed in terms of the S_k, viz. $V_n = \max(0, \max_{0 \leq k < n}(1 - S_k))$, as can be checked by induction on n.)

Now we ask: When is the state 0 recurrent, in other words, when is it sure that all customers will be served at some finite time (so that a chair in the lift remains empty)? The only interesting case is when $0 < \rho(0) < 1$; for $\rho(0) = 0$ there is always at least one customer arriving, so the queue never gets shorter, and if $\rho(0) = 1$ not a single customer arrives.

Suppose first that $\mathbb{E}(\rho) > 1$. (Since ρ lives on $\mathbb{Z}_+$, $\mathbb{E}(\rho)$ is always well-defined, but possibly infinite.) Then there exists some $c \in \mathbb{N}$ such that $\sum_{k=1}^{c} k \, \rho(k) > 1$, so $\mathbb{E}(\min(Z_n, c) - 1) > 0$ for all $n \geq 1$. By (6.33) and the strong law of large numbers (namely Theorem (5.16)), we get

$$P^0\Big(\lim_{n\to\infty} X_n = \infty \Big) \geq P^0\Big(\lim_{n\to\infty} S_n = \infty \Big)$$

$$\geq P^0\Big(\lim_{n\to\infty} \tfrac{1}{n} \sum_{k=1}^{n} (\min(Z_k, c) - 1) > 0 \Big) = 1 \, .$$

This eventual escape to infinity does not exclude that X_n hits the origin several times, but it makes certain that X_n does not return to 0 infinitely often. Hence $F_\infty(0, 0) = 0$, and we have proved that 0 is transient when $\mathbb{E}(\rho) > 1$.

Conversely, let us show that 0 is recurrent when $\mathbb{E}(\rho) \leq 1$. Note first that by the case-distinction formula (3.3a) and Theorem (6.7),

$$F_1(0,0) = \rho(0) + \sum_{y \geq 1} \rho(y) \, P^y(\tau_0 < \infty).$$

Therefore, it suffices to show that $h(y) := P^y(\tau_0 < \infty) = 1$ for all $y \geq 1$. For this purpose we derive two equations:

▷ $h(1) = \rho(0) + \sum_{y \geq 1} \rho(y) \, h(y)$. This is simply the equation $h(1) = \Pi h(1)$ (with the convention $h(0) = 1$), which holds by Remark (6.12).

▷ $h(y) = h(1)^y$ for all $y \geq 1$. Indeed, to get from y to 0, $(X_n)_{n \geq 0}$ has to pass $y - 1$ first, and therefore $h(y) = P^y(\tau_{y-1} < \infty, \ \tau_0 < \infty)$. Theorem (6.7) and Remark (6.8) therefore imply that

$$h(y) = \sum_{k \geq 1} P^y(\tau_{y-1} = k) \, h(y-1) = P^y(\tau_{y-1} < \infty) \, h(y-1).$$

Moreover, $P^y(\tau_{y-1} < \infty) = h(1)$. This is because X_n coincides with the random walk S_n for $n \leq \tau_0$ by (6.33), and S_n is homogeneous. Hence, $h(y) = h(1) \, h(y-1)$, whence the claim follows inductively.

Both equations together yield

$$h(1) = \sum_{y \geq 0} \rho(y) \, h(1)^y = \varphi_\rho(h(1)) \, ;$$

that is, $h(1)$ is a fixed point of the generating function φ_ρ of ρ. But if $\mathbb{E}(\rho) = \varphi_\rho'(1) \leq 1$, then φ_ρ has 1 as its only fixed point; cf. Example (6.11). (It is no coincidence that we find this analogy to branching processes, see Problem 6.27.) Hence, it follows that $h \equiv 1$ and therefore $F_1(0,0) = 1$. Altogether, we get the (intuitively obvious) result, which is illustrated in Figure 6.7:

The state 0 is recurrent if and only if $\mathbb{E}(\rho) \leq 1$, in other words, if on average the customer influx is not larger than the number of customers that can be served.

The queueing model considered above is essentially identical to the so-called M/G/1-model in queueing theory, which is formulated in continuous time. The '1' says that there is only one counter to serve the customers. The 'M' stands for 'Markov' and means that the customers' arrival times form a Poisson process. (The lack of memory property of the exponential distribution underlying the Poisson process implies that the Poisson process is a Markov process in continuous time.) The 'G' stands for 'general' and means that the serving times of the customers, which are assumed to be i.i.d., can have an arbitrary distribution β on $[0, \infty[$.

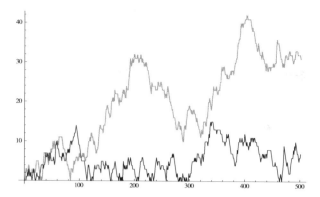

Figure 6.7. Two realisations of the queueing process (X_n) with Poisson influx of mean 0.95 (black) resp. 1.05 (grey).

Now, let X_n be the length of the queue at the random time point at which the nth customer has just been served and so has just left the queue, and let Z_n be the number of customers that arrived while she was being served. Due to the independence properties of the Poisson arrival process, the Z_n are i.i.d. with distribution

$$\rho(k) := \int_0^\infty \beta(dt)\, \mathcal{P}_{\alpha t}(\{k\}) = \int_0^\infty \beta(dt)\, e^{-\alpha t}\, (\alpha t)^k / k!\,;$$

here $\alpha > 0$ is the intensity of the arrival process (see also Problem 3.23). Therefore, as long as the queue is never empty, it follows that X_n coincides with the discrete-time Markov chain considered above, and its recurrence properties also hold for the continuous-time model. Hence, we have recurrence if and only if $\mathbb{E}(\rho) = \alpha\, \mathbb{E}(\beta) \le 1$.

6.4.2 Positive Recurrence and Null Recurrence

We now ask the question: How long does it take on average until the Markov chain returns to its starting point?

Definition. A recurrent state $x \in E$ is called *positive recurrent*, if the mean return time is finite, in other words if $\mathbb{E}^x(\tau_x) < \infty$. Otherwise x is called *null recurrent*.

As we have seen already in Theorem (6.27), the stationary distribution of an irreducible chain is given by the reciprocal mean return times. The following theorem sheds some further light on this relationship.

(6.34) Theorem. Positive recurrence and existence of stationary distributions. *If $x \in E$ is positive recurrent, then there exists a stationary distribution α satisfying $\alpha(x) = 1/\mathbb{E}^x(\tau_x) > 0$. Conversely, if $\alpha(x) > 0$ for a stationary distribution α, then x is positive recurrent.*

Proof. Suppose first that $\alpha(x) > 0$ for some α such that $\alpha \Pi = \alpha$. Then it follows immediately from Kac's recurrence theorem (6.28) that $\mathbb{E}^x(\tau_x) \le 1/\alpha(x) < \infty$, so x is positive recurrent. Conversely, suppose that x positive recurrent, so that $\mathbb{E}^x(\tau_x) < \infty$. Let

$$\beta(y) := \sum_{n \ge 1} P^x(X_n = y, n \le \tau_x) = \mathbb{E}^x\left(\sum_{n=1}^{\tau_x} 1_{\{X_n = y\}}\right)$$

be the expected number of visits to y on an 'excursion' from x. Since $1_{\{X_{\tau_x} = y\}} = \delta_{x,y} = 1_{\{X_0 = y\}}$ P^x-almost surely, we can also write

$$\beta(y) = \mathbb{E}^x\left(\sum_{n=0}^{\tau_x - 1} 1_{\{X_n = y\}}\right) = \sum_{n \ge 0} P^x(\tau_x > n, X_n = y).$$

For all $z \in E$ we therefore obtain

$$\begin{aligned}
\beta\Pi(z) &= \sum_{y \in E} \sum_{n \ge 0} P^x(\tau_x > n, X_n = y)\, \Pi(y, z) \\
&= \sum_{n \ge 0} \sum_{y \in E} P^x(\tau_x > n, X_n = y, X_{n+1} = z) \\
&= \sum_{n \ge 0} P^x(\tau_x > n, X_{n+1} = z) \\
&= \beta(z)\,;
\end{aligned}$$

in the second step we applied once again Remark (6.8) and Theorem (6.7). Since $\sum_y \beta(y) = \mathbb{E}^x(\tau_x)$, we see that $\alpha := \mathbb{E}^x(\tau_x)^{-1}\beta$ is a stationary distribution with $\alpha(x) = 1/\mathbb{E}^x(\tau_x) > 0$. $\diamond$

An alternative approach to the formula $\alpha(x) = 1/\mathbb{E}^x(\tau_x)$ runs as follows: Let x be positive recurrent, $T_0 = 0$ and $T_k = \inf\{n > T_{k-1} : X_n = x\}$ the time of the kth visit to x. Then $L_k = T_k - T_{k-1}$ is the length of the kth excursion from x. By Problem 6.26, the sequence $(L_k)_{k \ge 1}$ is i.i.d. The strong law of large numbers thus shows that

$$1/\mathbb{E}^x(\tau_x) = \lim_{k \to \infty} k \Big/ \sum_{j=1}^{k} L_j = \lim_{k \to \infty} \frac{k}{T_k} = \lim_{N \to \infty} \frac{1}{N} \sum_{n=1}^{N} 1_{\{X_n = x\}}$$

almost surely, since the last sum takes the value k when $T_k \le N < T_{k+1}$. Consequently, $1/\mathbb{E}^x(\tau_x)$ is exactly the relative frequency of the visits to x. Analogously, $\beta(y)/\mathbb{E}^x(\tau_x)$ is almost surely the relative frequency of the visits to y.

(6.35) Example. *The Ehrenfests' model, cf. Example* (6.22). Suppose that at time 0 an experimenter has created a vacuum in the left part of the container, has filled the right part with N molecules and only then has opened the connection between the two parts. Is it possible that at some later time all molecules are back in the right part and so create a vacuum in the left chamber? According to Example (6.22), the Ehrenfest Markov chain for N particles has stationary distribution $\alpha = \mathcal{B}_{N,1/2}$, and so $\alpha(0) = 2^{-N} > 0$. By Theorem (6.34) it follows that the state 0 is positive recurrent, which means the left chamber will be empty again at some *random* time τ_0. However, to observe this, you would have to be very patient, since for $N = 0.25 \cdot 10^{23}$ we get $\mathbb{E}^0(\tau_0) = 1/\alpha(0) = 2^N \approx 10^{7.5 \cdot 10^{21}}$. So the gas would have to be observed for this immensely long time span (exceeding by far the expected life time of the universe), and even continuously so, since the time τ_0 is random. At every *fixed* (and sufficiently large) observation time the result of Example (6.22) holds, which said that, with overwhelming certainty, the fraction of particles in both parts is nearly one half.

(6.36) Example. *The simple symmetric random walk on $\mathbb{Z}^d$, cf. Example* (6.31). For $d \le 2$, every $x \in \mathbb{Z}^d$ is null recurrent. Indeed, suppose some $x \in \mathbb{Z}^d$ were positive recurrent. By Theorem (6.34) there would then exist a stationary distribution α. Since $\sum_{y \in \mathbb{Z}^d} \alpha(y) = 1$, there would also exist some $z \in \mathbb{Z}^d$ such that $\alpha(z)$ is maximal, in that $\alpha(z) = m := \max_{y \in \mathbb{Z}^d} \alpha(y) > 0$. In view of the equation

$$ m = \alpha(z) = \alpha \Pi(z) = \frac{1}{2d} \sum_{y: \, |y-z|=1} \alpha(y), $$

we could conclude that $\alpha(y) = m$ for all y with $|y - z| = 1$. By iteration, we would find that $\alpha(y) = m$ for all $y \in \mathbb{Z}^d$, which is impossible because $\sum_{y \in \mathbb{Z}^d} \alpha(y) = 1$.

(6.37) Example. *The queueing model, cf. Example* (6.32). We show that state 0 is positive recurrent if and only if $\mathbb{E}(\rho) < 1$, i.e., if on average less customers arrive than can be served. For the proof we must examine $\mathbb{E}^0(\tau_0)$. We claim first that

(6.38) $$ \mathbb{E}^0(\tau_0) = 1 + \mathbb{E}(\rho) \, \mathbb{E}^1(\tau_0). $$

Indeed, distinguishing the possible values of X_1 and using Theorem (6.7), we find

$$ \begin{aligned} P^0(\tau_0 > k) &= P^0(X_1 \ge 1, \ldots, X_k \ge 1) \\ &= \sum_{y \ge 1} \rho(y) \, P^y(X_1 \ge 1, \ldots, X_{k-1} \ge 1) \\ &= \sum_{y \ge 1} \rho(y) \, P^y(\tau_0 > k-1) \end{aligned} $$

for all $k \ge 1$, and thus by Theorem (4.7c)

(6.39) $$ \begin{aligned} \mathbb{E}^0(\tau_0) = \sum_{k \ge 0} P^0(\tau_0 > k) &= 1 + \sum_{y \ge 1} \rho(y) \sum_{k \ge 1} P^y(\tau_0 > k-1) \\ &= 1 + \sum_{y \ge 1} \rho(y) \, \mathbb{E}^y(\tau_0). \end{aligned} $$

We note further that the queue can only get shorter by at most one customer at a time.
For $y \geq 2$ we thus obtain, applying again Theorem (6.7),

$$
\begin{aligned}
\mathbb{E}^y(\tau_0) &= \sum_{n \geq 1} \sum_{k \geq 1} (n+k)\, P^y(\tau_{y-1} = n, \tau_0 = n+k) \\
&= \sum_{n \geq 1} n\, P^y(\tau_{y-1} = n) + \sum_{k \geq 1} k \sum_{n \geq 1} P^y(\tau_{y-1} = n)\, P^{y-1}(\tau_0 = k) \\
&= \mathbb{E}^y(\tau_{y-1}) + \mathbb{E}^{y-1}(\tau_0)\,.
\end{aligned}
$$

Finally we use the spatial homogeneity, which is expressed by the fact that X_n coincides with the random walk S_n for $n \leq \tau_0$ (by (6.33)). This implies that $\mathbb{E}^y(\tau_{y-1}) = \mathbb{E}^1(\tau_0)$. Altogether, we find that $\mathbb{E}^y(\tau_0) = y\,\mathbb{E}^1(\tau_0)$ for $y \geq 1$. Substituting this into (6.39) we arrive at equation (6.38).

Suppose now that 0 is positive recurrent. Then, (6.33) yields

$$
\begin{aligned}
0 = \mathbb{E}^1(X_{\tau_0}) &= \mathbb{E}^1\Big(1 + \sum_{k=1}^{\tau_0} (Z_k - 1)\Big) \\
&= 1 + \sum_{k \geq 1} \mathbb{E}^1\big(Z_k\, 1_{\{\tau_0 \geq k\}}\big) - \mathbb{E}^1(\tau_0) \\
&= 1 + \sum_{k \geq 1} \mathbb{E}(\rho)\, P^1(\tau_0 \geq k) - \mathbb{E}^1(\tau_0) \\
&= 1 + (\mathbb{E}(\rho) - 1)\, \mathbb{E}^1(\tau_0)\,.
\end{aligned}
$$

The third step follows because $\mathbb{E}^1(\tau_0) < \infty$ by (6.38), and because the Z_k are non-negative so that Theorem (4.7c) can be applied. The fourth identity uses the fact that the event $\{\tau_0 \geq k\}$ can be expressed in terms of $Z_1, \ldots, Z_{k-1}$, and so, by Theorem (3.24), it is independent of Z_k. The equation just proved shows that $\mathbb{E}(\rho) < 1$. In particular we get $\mathbb{E}^1(\tau_0) = 1/(1 - \mathbb{E}(\rho))$ and, by (6.38), $\mathbb{E}^0(\tau_0) = 1/(1 - \mathbb{E}(\rho))$ as well. That is, the closer the average influx of customers per time approaches the serving rate 1, the longer the average 'busy period'.

Conversely, suppose that $\mathbb{E}(\rho) < 1$. The same calculation as above, with $\tau_0 \wedge n := \min(\tau_0, n)$ in place of τ_0, then shows that

$$
0 \leq \mathbb{E}^1(X_{\tau_0 \wedge n}) = 1 + (\mathbb{E}(\rho) - 1)\, \mathbb{E}^1(\tau_0 \wedge n)\,,
$$

and therefore $\mathbb{E}^1(\tau_0 \wedge n) \leq 1/(1 - \mathbb{E}(\rho))$. Letting $n \to \infty$, we can conclude that $\mathbb{E}^1(\tau_0) \leq 1/(1 - \mathbb{E}(\rho)) < \infty$, and (6.38) then gives that 0 is positive recurrent.

Problems

6.1 *Iterated random functions.* Let E be a countable set, $(F, \mathscr{F})$ an arbitrary event space, $f : E \times F \to E$ a measurable function, and $(U_i)_{i \geq 1}$ a sequence of i.i.d. random variables taking values in $(F, \mathscr{F})$. Let $(X_n)_{n \geq 0}$ be recursively defined by $X_0 = x \in E$, and $X_{n+1} = f(X_n, U_{n+1})$ for $n \geq 0$. Show that $(X_n)_{n \geq 0}$ is a Markov chain and determine the transition matrix.

6.2[S] *Functions of Markov chains.* Let $(X_n)_{n \geq 0}$ be a Markov chain with countable state space E and transition matrix Π, and $f : E \to F$ a mapping from E to another countable set F.

(a) Show by example that $(f \circ X_n)_{n \geq 0}$ is not necessarily a Markov chain.

(b) Find a (non-trivial) condition on f and Π under which $(f \circ X_n)_{n \geq 0}$ is a Markov chain.

6.3 *Embedded jump chain.* Let E be countable and $(X_n)_{n \geq 0}$ a Markov chain on E with transition matrix Π. Let $T_0 = 0$ and $T_k = \inf\{n > T_{k-1} : X_n \neq X_{n-1}\}$ be the time of the kth jump of $(X_n)_{n \geq 0}$. Show that the sequence $X_k^* := X_{T_k}$, $k \geq 0$, is a Markov chain with transition matrix

$$\Pi^*(x, y) = \begin{cases} \Pi(x, y)/(1 - \Pi(x, x)) & \text{if } y \neq x, \\ 0 & \text{otherwise.} \end{cases}$$

Show further that, conditional on $(X_k^*)_{k \geq 0}$, the differences $T_{k+1} - T_k - 1$ are independent and geometrically distributed with parameter $1 - \Pi(X_k^*, X_k^*)$.

6.4 *Self-fertilisation.* Suppose the gene of a plant can come in two 'versions', the alleles A and a. A classical procedure to grow pure-bred (i.e. homozygous) plants of genotype AA respectively aa is self-fertilisation. The transition graph

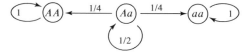

describes the transition from one generation to the next. Let $(X_n)_{n \geq 0}$ be the corresponding Markov chain. Calculate the probability $p_n = P^{Aa}(X_n = Aa)$ for arbitrary n.

6.5[S] *Hoppe's urn model and Ewens' sampling formula.* Imagine a gene in a population that can reproduce and mutate at discrete time points, and assume that every mutation leads to a new allele (this is the so-called infinite alleles model). If we consider the genealogical tree of n randomly chosen individuals at the times of mutation or birth, we obtain a picture as in Figure 6.8. Here, every bullet marks a mutation, which is the starting point of a new 'clan' of individuals with the new allele. Let us now ignore the family structure of the clans and only record their sizes. The reduced evolution is then described by the following urn model introduced by F. Hoppe (1984).

Let $\vartheta > 0$ be a fixed parameter that describes the mutation rate. Suppose that at time 0 there is a single black ball with weight ϑ in the urn, whereas outside there is an infinite reservoir of balls of different colours and weight 1. At each time step, a ball is drawn from the urn with a probability proportional to its weight. If it is black (which is certainly the case in the first draw), then a ball of a colour that is not yet present in the urn is put in. If the chosen ball is coloured, then it is returned together with another ball of the same colour. The number of

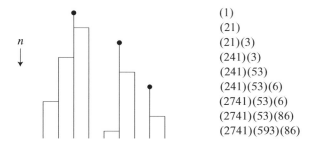

$$
\begin{aligned}
&(1)\\
&(21)\\
&(21)(3)\\
&(241)(3)\\
&(241)(53)\\
&(241)(53)(6)\\
&(2741)(53)(6)\\
&(2741)(53)(86)\\
&(2741)(593)(86)
\end{aligned}
$$

Figure 6.8. A genealogical tree in the infinite alleles model, with corresponding description in terms of cycles as in the Chinese restaurant process from Problem 6.6.

balls in the urn thus increases by 1 at each draw, and the coloured balls can be decomposed into clans of the same colour. The size distribution of these clans is described by a sequence of the form $x = (x_i)_{i \geq 1}$, where x_i specifies the number of clans of size i. The total number of coloured balls after the nth draw is $N(x) := \sum_{i \geq 1} i\, x_i = n$. Formally, the model is described by the Markov chain $(X_n)_{n \geq 0}$ with state space $E = \{x = (x_i)_{i \geq 1} : x_i \in \mathbb{Z}_+, N(x) < \infty\}$ and transition matrix

$$
\Pi(x, y) = \begin{cases}
\vartheta/(\vartheta + N(x)) & \text{if } y = x + (1, 0, 0, \ldots),\\
j\, x_j/(\vartheta + N(x)) & \text{if } y = x + (0, \ldots 0, -1, 1, 0, \ldots)\\
& \quad \text{with } -1 \text{ and } 1 \text{ at positions } j \text{ and } j + 1,\\
0 & \text{otherwise.}
\end{cases}
$$

(The first case corresponds to drawing a black ball, the second to drawing a coloured ball from one of the x_j clans of size j, so that the size of this clan increases to $j + 1$.) Let $\mathbf{0} = (0, 0, \ldots)$ be the initial state, in which the urn does not contain any coloured balls.

(a) Show by induction on $n \geq 1$ for arbitrary $x \in E$ with $N(x) = n$:

$$
P^{\mathbf{0}}(X_n = x) = \rho_{n,\vartheta}(x) := \frac{n!}{\vartheta^{(n)}} \prod_{i \geq 1} \frac{(\vartheta/i)^{x_i}}{x_i!},
$$

where $\vartheta^{(n)} := \vartheta(\vartheta + 1) \ldots (\vartheta + n - 1)$. Hence, $\rho_{n,\vartheta}$ is the size distribution of the clans of a random sample of n individuals from a population with mutation rate ϑ. This is the sampling formula by W. J. Ewens (1972).

(b) Verify that $\rho_{n,\vartheta}$ has a conditional Poisson structure as follows. If $Y = (Y_i)_{i \geq 1}$ is a sequence of independent random variables with Poisson distributions $P \circ Y_i^{-1} = \mathcal{P}_{\vartheta/i}$, then $\rho_{n,\vartheta}(x) = P(Y = x \mid N(Y) = n)$.

6.6 *Chinese restaurant process and random permutations.* The Hoppe model from Problem 6.5 can be slightly refined by taking the family structure of the clans into consideration. The balls in Hoppe's urn are labelled in the order in which they arrived in the urn. The state of the urn after the nth draw is written as a permutation in cycle notation as in Figure 6.8: For a new colour at time n we add (n) as a new cycle and otherwise the label of the ball is written to the left of the label of the 'mother ball' in its respective cycle. Let Z_n be the permutation created after n draws. The sequence $(Z_n)_{n \geq 0}$ was introduced by D. J. Aldous (1984) as the 'Chinese restaurant process'; the labels are interpreted as the guests of a Chinese restaurant (in the order of their appearance), and each cycle as the seating order at a (round) table. Show the following:

(a) $(Z_n)_{n \geq 0}$ is a Markov chain. For which E and Π?

(b) For each permutation π of $\{1, \ldots, n\}$ with k cycles, $P(Z_n = \pi) = \vartheta^k / \vartheta^{(n)}$, where $\vartheta^{(n)}$ is as in Problem 6.5a. So, Z_n is uniformly distributed when $\vartheta = 1$.

(c) Deduce that the number of all permutations of $\{1, \ldots, n\}$ that have, for each $1 \leq i \leq n$, x_i cycles of length i is equal to $n! / \prod_{i=1}^{n} (i^{x_i} x_i!)$.

6.7 ^{S} *The Wright–Fisher model in population genetics.* Consider a particular gene with the two alleles A and a in a population of constant size N. Suppose for simplicity that the individuals have a haploid set of chromosomes. So the gene occurs N times in each generation. Assume each generation is created from the previous generation by random mating: Each gene of the offspring generation 'selects', independently of all others, a gene of the parental generation and adopts its allele.

(a) Let Ξ_n be the random set of A-individuals in the nth generation. Describe the evolution of Ξ_n by a sequence of i.i.d. random mappings and conclude that (Ξ_n) is a Markov chain. Determine the transition matrix.

(b) Show that also $X_n := |\Xi_n|$ is a Markov chain, and find its transition matrix.

(c) Show that $\lim_{N \to \infty} P^x(X_n = N) = x/N$ for each $x \in \{0, \ldots, N\}$.

6.8 Let $(X_n)_{n \geq 0}$ be a Markov chain with transition matrix Π on a countable set E, and suppose that $P^x(\tau_y < \infty) = 1$ for all $x, y \in E$. Let $h : E \to [0, \infty[$ be harmonic, in that $\Pi h = h$. Show that h must be constant.

6.9 ^{S} *The asymmetric ruin problem.* A well-known dexterity game consists of a ball in a 'maze' of N concentric rings (numbered from the centre to the outside) that have an opening to the next ring on alternating sides. The aim of the game is to get the ball to the centre ('ring no. 0') by suitably tilting the board. Suppose that the ball is initially in the mth ring $(0 < m < N)$, and that with probability $0 < p < 1$ the player manages to get the ball from the kth to the $(k-1)$st ring, but that with probability $1 - p$ the ball rolls back to the $(k+1)$st ring. The player stops if the ball enters either into ring 0 (so that the player succeeds), or into the Nth ring (due to demoralisation). Describe this situation as a Markov chain and find the probability of success.

6.10 Find the extinction probability for a Galton–Watson process with offspring distribution ρ in the cases

(a) $\rho(k) = 0$ for all $k > 2$,

(b) $\rho(k) = ba^{k-1}$ for all $k \geq 1$ and $a, b \in]0, 1[$ with $b \leq 1 - a$. (According to empirical studies by Lotka in the 1930s, for $a = 0.5893$ and $b = 0.2126$ this ρ describes the distribution of the number of sons of American men quite well, whereas, according to Keyfitz [33], the parameters $a = 0.5533$ and $b = 0.3666$ work best for the number of daughters of Japanese women.)

6.11 *Total size of a non-surviving family tree.* Consider a Galton–Watson process $(X_n)_{n \geq 0}$ with offspring distribution ρ, and suppose that $\mathbb{E}(\rho) \leq 1$ and $X_0 = 1$. Let $T = \sum_{n \geq 0} X_n$ be the total number of descendants of the progenitor. (Note that $T < \infty$ almost surely.) Show that the generating function φ_T of T satisfies the functional equation $\varphi_T(s) = s \varphi_\rho \circ \varphi_T(s)$, and determine $\mathbb{E}^1(T)$, the expected total number of descendants.

6.12 [S] *Branching process with migration and annihilation.* Consider the following modification of the Galton–Watson process. Given $N \in \mathbb{N}$, assume that at each site $n \in \{1, \ldots, N\}$
there is a certain number of 'particles' that behave independently of each other as follows.
During a time unit, a particle at site n first moves to $n-1$ or $n+1$, each with probability
$1/2$. There it dies and produces k offspring with probability $\rho(k)$, $k \in \mathbb{Z}_+$. If $n-1 = 0$ or
$n+1 = N+1$, the particle is annihilated and does not produce any offspring. Let $\varphi(s) =
\sum_{k \geq 0} \rho(k) s^k$ be the generating function of $\rho = (\rho(k))_{k \geq 0}$, and for $1 \leq n \leq N$ let $q(n)$ be
the probability that the progeny of a single particle at site n becomes eventually extinct. By
convention, $q(0) = q(N+1) = 1$. Here are your tasks:

(a) Describe the evolution of all particles by a Markov chain on $\mathbb{Z}_+^N$ and find the transition
matrix.

(b) Justify the equation $q(n) = \frac{1}{2} \varphi(q(n-1)) + \frac{1}{2} \varphi(q(n+1))$, $1 \leq n \leq N$.

(c) In the subcritical case $\varphi'(1) \leq 1$, show that $q(n) = 1$ for all $1 \leq n \leq N$.

(d) On the other hand, suppose that $\varphi(s) = (1 + s^3)/2$. Show that $q(1) = q(2) = 1$ when
$N = 2$, whereas $q(n) < 1$ for all $1 \leq n \leq 3$ when $N = 3$.

6.13 Let E be finite and Π a stochastic matrix on E. Show that Π satisfies the assumptions
of the ergodic theorem (6.13) if and only if Π is irreducible and *aperiodic* in the sense that for
one (and thus all) $x \in E$ the greatest common divisor of the set $\{k \geq 1 : \Pi^k(x, x) > 0\}$ is 1.

6.14 *Random replacement I.* Consider an urn containing initially N balls. Let X_n be the
number of balls in the urn after performing the following procedure n times. If the urn is
non-empty, one of the balls is removed at random; by flipping a fair coin, it is then decided
whether or not the ball is returned to the urn. If the urn is empty, the fair coin is used to decide
whether or not the urn is filled afresh with N balls. Describe this situation as a Markov chain
and find the transition matrix. What is the distribution of X_n as $n \to \infty$?

6.15 *Random replacement II.* As in the previous problem, consider an urn holding at most N
balls, but now they come in two colours, either white or red. If the urn is non-empty, a ball is
picked at random and is or is not replaced according to the outcome of the flip of a fair coin.
If the urn is empty, the coin is flipped to decide whether the urn should be filled again; if so,
it is filled with N balls, each of which is white or red depending on the outcomes of further
independent coin flips. Let W_n and R_n be the numbers of white and red balls, respectively,
after performing this procedure n times. Show that $X_n = (W_n, R_n)$ is a Markov chain, and
determine its asymptotic distribution.

6.16 *A variant of Pólya's urn model.* Consider again an urn containing no more than $N > 2$
balls in the colours white and red, but at least one ball of each colour. If there are less than N
balls, one of them is chosen at random and returned together with a further ball of the same
colour (taken from an external reserve). If there are already N balls in the urn, then by tossing
a coin it is decided whether the urn should be modified. If so, all balls are removed and one
ball of each colour is put in. Let W_n and R_n be the respective numbers of white and red balls
after performing this procedure n times. Show the following:

(a) The total number $Y_n := W_n + R_n$ of balls is a Markov chain. Find the transition matrix.
Will the chain eventually come to an equilibrium? If so, which one?

(b) $X_n := (W_n, R_n)$ is also a Markov chain. Find the transition matrix and (if it exists) the
asymptotic distribution.

6.17 *A cycle condition for reversibility.* Under the assumptions of the ergodic theorem (6.13), show that Π has a reversible distribution if and only if the probability of running through a cycle does not depend on the direction, in that

$$\Pi(x_0, x_1)\,\Pi(x_1, x_2)\ldots\Pi(x_{n-1}, x_0) = \Pi(x_0, x_{n-1})\,\Pi(x_{n-1}, x_{n-2})\ldots\Pi(x_1, x_0)$$

for all $n \geq 1$ and $x_0, \ldots, x_{n-1} \in E$. Does this hold for the house-of-cards process in (6.29)?

6.18 ^{S} *Time reversal for renewals.* In addition to the age process (X_n) in Example (6.29), consider the process $Y_n = \min\{T_k - n : k \geq 1, T_k \geq n\}$ that indicates the remaining life span of the appliance used at time n.

(a) Show that $(Y_n)_{n\geq 0}$ is also a Markov chain, find its transition matrix $\tilde{\Pi}$ and re-derive the renewal theorem.

(b) Find the stationary distribution α of $(X_n)_{n\geq 0}$ and show that α is also a stationary distribution of $(Y_n)_{n\geq 0}$.

(c) Which connection does α create between the transition matrices Π and $\tilde{\Pi}$ of $(X_n)_{n\geq 0}$ and $(Y_n)_{n\geq 0}$?

6.19 (a) *Random walk on a finite graph.* Let E be a finite set and $\sim$ a symmetric relation on E. Here, E is interpreted as the vertex set of a graph, and the relation $x \sim y$ means that x and y are connected by an (undirected) edge. Suppose that each vertex is connected by an edge to at least one other vertex or to itself. Let $d(x) = |\{y \in E : x \sim y\}|$ be the degree of the vertex $x \in E$, and set $\Pi(x, y) = 1/d(x)$ if $y \sim x$, and $\Pi(x, y) = 0$ otherwise. The Markov chain with transition matrix Π is called the random walk on the graph $(E, \sim)$. Under which conditions on the graph $(E, \sim)$ is Π irreducible? Find a reversible distribution for Π.

(b) *Random walk of a knight.* Consider a knight on an (otherwise empty) chess board, which chooses each possible move with equal probability. It starts (i) in a corner, (ii) in one of the 16 squares in the middle of the board. How many moves does it need on average to get back to its starting point?

6.20 Let $0 < p < 1$ and consider the stochastic matrix Π on $E = \mathbb{Z}_+$ defined by

$$\Pi(x, y) = \mathcal{B}_{x,p}(\{y\}), \quad x, y \in \mathbb{Z}_+ .$$

Find Π^n for arbitrary $n \geq 1$. Can you imagine a possible application of this model?

6.21 ^{S} *Irreducible classes.* Let E be countable, Π a stochastic matrix on E, and E_{rec} the set of all recurrent states. Let us say a state y is accessible from x, written as $x \to y$, if there exists some $k \geq 0$ such that $\Pi^k(x, y) > 0$. Show the following:

(a) The relation '$\to$' is an equivalence relation on E_{rec}. The corresponding equivalence classes are called *irreducible classes*.

(b) If x is positive recurrent and $x \to y$, then y is also positive recurrent, and

$$\mathbb{E}^x\left(\sum_{n=1}^{\tau_x} 1_{\{X_n=y\}}\right) = \mathbb{E}^x(\tau_x)/\mathbb{E}^y(\tau_y) .$$

In particular, all states within an irreducible class are of the same recurrence type.

6.22 *Extinction or unlimited growth of a population.* Consider a Galton–Watson process $(X_n)_{n\geq 0}$ with supercritical offspring distribution ρ, i.e., suppose $\mathbb{E}(\rho) > 1$. Show that all states $k \neq 0$ are transient, and that

$$P^k\left(X_n \to 0 \text{ or } X_n \to \infty \text{ for } n \to \infty\right) = 1.$$

6.23^S *Birth-and-death processes.* Let Π be a stochastic matrix on $E = \mathbb{Z}_+$. Suppose that $\Pi(x, y) > 0$ if and only if either $x \geq 1$ and $|x - y| = 1$, or $x = 0$ and $y \leq 1$. Find a necessary and sufficient condition on Π under which Π has a stationary distribution α. If α exists, express it in terms of the entries of Π.

6.24 *A migration model.* Consider the following simple model of an animal population in an open habitat. Each animal living there leaves the habitat, independently of all others, with probability $1 - p$, and it stays with probability p. At the same time, a Poisson number (with parameter $a > 0$) of animals immigrates from the outside world.

(a) Describe the number X_n of animals living in the habitat by a Markov chain and find the transition matrix Π.

(b) Calculate the distribution of X_n when the initial distribution is $\mathcal{P}_\lambda$, the Poisson distribution with parameter $\lambda > 0$.

(c) Determine a reversible distribution α.

6.25 Generalise Theorem (6.30) as follows. For $x, y \in E$, let $F_1(x, y) = P^x(\tau_y < \infty)$ be the probability that y can eventually be reached from x, $N_y = \sum_{n\geq 1} 1_{\{X_n=y\}}$ the number of visits to y (from time 1 onwards), $F_\infty(x, y) = P^x(N_y = \infty)$ the probability for infinitely many visits, and $G(x, y) = \delta_{xy} + \mathbb{E}^x(N_y)$ the expected number of visits (including at time 0), the so-called *Green function*. Show that

$$P^x(N_y \geq k + 1) = F_1(x, y)\, P^y(N_y \geq k) = F_1(x, y)\, F_1(y, y)^k$$

for all $k \geq 0$, and therefore

$$F_\infty(x, y) = F_1(x, y)\, F_\infty(y, y), \quad G(x, y) = \delta_{xy} + F_1(x, y)\, G(y, y).$$

What does this mean when y is recurrent and transient, respectively?

6.26 *Excursions from a recurrent state.* Consider a Markov chain with a countable state space E and transition matrix Π that starts in a recurrent state $x \in E$. Let $T_0 = 0$ and, for $k \geq 1$, let $T_k = \inf\{n > T_{k-1} : X_n = x\}$ be the time of the kth return to x and $L_k = T_k - T_{k-1}$ the length of the kth 'excursion' from x. Show that, under P^x, the random variables L_k are (almost surely well-defined and) i.i.d.

6.27 *Busy period of a queue viewed as a branching process.* Recall Example (6.32) and the random variables X_n and Z_n defined there. Interpret the queue as a population model, by interpreting the customers newly arriving at time n as the children of the customer waiting at the front of the queue; a generation is complete if the last member of the previous generation has been served. Correspondingly, define $Y_0 = X_0$, $Y_1 = \sum_{n=1}^{Y_0} Z_n$, and for general $k \geq 1$ set

$$Y_{k+1} = \sum_{n\geq 1} 1_{\left\{\sum_{i=0}^{k-1} Y_i < n \leq \sum_{i=0}^{k} Y_i\right\}} Z_n.$$

Show the following:

(a) (Y_k) is a Galton–Watson process with offspring distribution ρ.

(b) P^x-almost surely for every $x \geq 1$, one has $Y_{k+1} = X_{T_k}$ for all $k \geq 0$, and therefore

$$\{X_n = 0 \text{ for some } n \geq 1\} = \{Y_k = 0 \text{ for all sufficiently large } k\}.$$

Here, the random times T_k are recursively defined by setting $T_0 = X_0$ and $T_{k+1} = T_k + X_{T_k}$. (Verify first that these times are not larger than the first time τ_0 at which the queue is empty.)

Deduce (without using the result from Example (6.32)) that the queue is recurrent if and only if $\mathbb{E}(\rho) \leq 1$. (In this case, Problem 6.11 yields the average number of customers that are served during a busy period, and so gives an alternative proof for the result from Example (6.37).)

6.28 ^{S} *Markov chains in continuous time.* Let E be countable and $G = (G(x, y))_{x,y\in E}$ a matrix satisfying the properties

(i) $G(x, y) \geq 0$ for $x \neq y$,

(ii) $-a(x) := G(x, x) < 0$, $\sum_{y\in E} G(x, y) = 0$ for all x, and

(iii) $a := \sup_{x\in E} a(x) < \infty$.

We construct a Markov process $(X_t)_{t\geq 0}$ that 'jumps with rate $G(x, y)$ from x to y' by using the stochastic matrix

$$\Pi(x, y) = \delta_{xy} + G(x, y)/a, \quad x, y \in E.$$

Suppose that for $x \in E$ we are given

▷ a Markov chain $(Z_k)_{k\geq 0}$ on E with starting point x and transition matrix Π, and

▷ a Poisson process $(N_t)_{t\geq 0}$ with intensity a independent of the Markov chain;

these can be defined on a suitable probability space $(\Omega, \mathcal{F}, P^x)$. Let $X_t = Z_{N_t}$ for $t \geq 0$ and prove the following:

(a) For all $t \geq 0$, X_t is a random variable, and

$$P^x(X_t = y) = e^{tG}(x, y) := \sum_{n\geq 0} t^n G^n(x, y)/n!$$

for all $x, y \in E$. Note that the right-hand side is well defined since $|G^n(x, y)| \leq (2a)^n$. Conclude in particular that $\frac{d}{dt} P^x(X_t = y)|_{t=0} = G(x, y)$.

(b) $(X_t)_{t\geq 0}$ is a Markov process with transition semigroup $\Pi_t := e^{tG}$, $t \geq 0$. By definition, this means that

$$P^x(X_{t_1} = x_1, \ldots, X_{t_n} = x_n) = \prod_{k=1}^{n} \Pi_{t_k - t_{k-1}}(x_{k-1}, x_k)$$

for all $n \geq 1$, $0 = t_0 < t_1 < \cdots < t_n$ and $x_0 = x, x_1, \ldots, x_n \in E$.

(c) Let $T_0^* = 0$, $Z_0^* = x$ and, recursively for $n \geq 1$, let

$$T_n^* = \inf\{t > T_{n-1}^* : X_t \neq Z_{n-1}^*\}$$

be the time and $Z_n^* = X_{T_n^*}$ the target of the nth jump of $(X_t)_{t\geq 0}$. Then,

▷ the sequence $(Z_n^*)_{n\geq 0}$ is a Markov chain on E with starting point x and transition matrix $\Pi^*(x, y) = \delta_{xy} + G(x, y)/a(x)$,

▷ conditional on $(Z_n^*)_{n\geq 0}$, the holding times $T_{n+1}^* - T_n^*$ in the states Z_n^* are independent and exponentially distributed with respective parameters $a(Z_n^*)$, $n \geq 0$.

Hint: Recall the construction (3.33) and combine the Problems 6.3, 3.20 and 3.18.

$(X_t)_{t\geq 0}$ is called the *Markov chain on E in continuous time with infinitesimal generator G,* and $(Z_n^*)_{n\geq 0}$ the *embedded discrete jump chain.*

6.29 *Explosion in finite time.* Without the assumption (iii) in Problem 6.28, a Markov chain with generator G does not exist in general. For instance, suppose $E = \mathbb{N}$ and

$$G(x, x+1) = -G(x, x) = x^2$$

for all $x \in \mathbb{N}$. Determine the discrete jump chain $(Z_n^*)_{n\geq 0}$ with starting point 0 and jump times $(T_n^*)_{n\geq 1}$ as in Problem 6.28c and show that $\mathbb{E}^0(\sup_{n\geq 1} T_n^*) < \infty$, i.e., the Markov chain $(X_t)_{t\geq 0}$ 'explodes' at the almost surely finite time $\sup_{n\geq 1} T_n^*$.

6.30[S] *Ergodic theorem for Markov chains in continuous time.* In the situation of Problem 6.28, suppose that E is finite and G irreducible, in that for all $x, y \in E$ there is a $k \in \mathbb{N}$ and $x_0, \ldots, x_k \in E$ such that $x_0 = x$, $x_k = y$ and $\prod_{i=1}^{k} G(x_{i-1}, x_i) \neq 0$. Show the following:

(a) For all $x, y \in E$ there is a $k \in \mathbb{Z}_+$ such that $\lim_{t\to 0} \Pi_t(x, y)/t^k > 0$. Consequently, for all $t > 0$, all entries of Π_t are positive.

(b) For all $x, y \in E$, $\lim_{t\to\infty} \Pi_t(x, y) = \alpha(y)$, where α is the unique probability density on E that satisfies one of the equivalent conditions $\alpha G = 0$ or $\alpha \Pi_s = \alpha$ for all $s > 0$.

Part II

Statistics

Chapter 7

Estimation

Statistics is the art of finding rational answers to the following key question: Given a specific phenomenon of chance, how can one uncover its underlying probability law from a number of random observations? The laws that appear to be possible are described by a family of suitable probability measures, and one intends to identify the true probability measure that governs the observations. In this chapter, we will first give an overview over the basic methods and then discuss the most elementary one, the method of estimation, where one simply tries to guess the true probability measure. The problem is to find sensible ways of guessing.

7.1 The Approach of Statistics

Suppose you are faced with a situation governed by chance, and you make a series of observations. What can you then say about the type and the features of the underlying random mechanism? Let us look at an example.

(7.1) Example. *Quality checking.* An importer of oranges receives a delivery of $N = 10\,000$ oranges. Quite naturally, he would like to know how many of these have gone bad. To find out, he takes a sample of $n = 50$ oranges. A random number x of these is rotten. What can the importer then conclude about the true number r of rotten oranges? The following three procedures suggest themselves, and each of them corresponds to a fundamental statistical method.

Approach 1: Naive estimation. As a rule of thumb, one would suspect that the proportion of bad oranges in the sample is close to the proportion of bad oranges in the total delivery, in other words that $x/n \approx r/N$. Therefore, the importer would guess that approximately $R(x) := N\,x/n$ oranges are rotten. That is, the number $R(x) = N\,x/n$ (or more precisely, the nearest integer) is a natural estimate of r resulting from the observed value x. We thus come up with a mapping R that assigns to the observed value x an estimate $R(x)$ of the unknown quantity. Such a mapping is called an *estimator*.

The estimate $R(x)$ obviously depends on chance. If the importer draws a second sample he will, in general, get a different result x', and hence a different estimate $R(x')$. Which estimate should he trust more? This question cannot be answered unless the underlying randomness is taken into consideration, as follows.

Approach 2: Estimation with an error bound. Given an observed value x, one does not guess a particular value $R(x)$. Instead, one determines an interval $C(x)$ depending on x, which includes the true value r with sufficient certainty. Since x is random, it is clear that $C(x)$ is also random. Therefore, one requires that the interval contains the true value r with large probability. This means that

$$P_r(x : C(x) \ni r) \approx 1$$

for the true r and the right probability measure P_r. Now, the sample taken by the orange importer corresponds to drawing n balls without replacement from an urn containing r red and $N - r$ white balls; thus the number of rotten oranges in the sample has the hypergeometric distribution $P_r = \mathcal{H}_{n;r,N-r}$. However, the true value r (the number of rotten oranges) is unknown. After all, it is supposed to be deduced from the sample x! So the properties of $C(x)$ cannot depend on r. This leads to the requirement that

$$\mathcal{H}_{n;r,N-r}(C(\cdot) \ni r) \geq 1 - \alpha$$

for *all* $r \in \{0, \dots, N\}$ and some (small) $\alpha > 0$. Such an interval $C(x)$, which depends on the observed value x, is called a *confidence interval with error level α.*

Approach 3: Decision making. The orange importer is less interested in the precise value of r, but has to worry about the financial implications. Suppose, for instance, that he has a contract with the supplier which says that the agreed price is due only if less than 5% of the oranges are bad. On the basis of the sample x he must decide whether the quality is sufficient. He has the choice between

the 'null hypothesis' $H_0 : r \in \{0, \dots, 500\}$

and the 'alternative' $H_1 : r \in \{501, \dots, 10\,000\}$

and thus needs a decision rule of the type

$x \leq c \Rightarrow$ decision for the null hypothesis 'all is fine',

$x > c \Rightarrow$ decision for the alternative 'the delivery is bad'.

Here, c must be determined so that $\mathcal{H}_{n;r,N-r}(x : x > c)$ is small for $r \leq 500$, and as large as possible for $r > 500$. The first requirement means that a mistake that is embarrassing for the importer should be very unlikely, and the second, that the importer is not deceived – after all he really wants to recognise insufficient quality. A decision rule of this type is called a *test*.

We will eventually deal with all three methods; this chapter is devoted to the first one. To support imagination, the reader might already want to take a first glance at the Figures 7.1, 8.1, and 10.1 (on pages 196, 228, and 262). We continue with a second example.

(7.2) Example. *Material testing.* In a nuclear power station the brittleness of a cooling pipe has to be monitored. So n independent measurements with (random) outcomes $x_1, \ldots, x_n$ are taken. Since during the measurement many small perturbations add up, the central limit theorem suggests to assume that the measured values $x_1, \ldots, x_n$ follow a normal distribution (in a first approximation, at least). Let us assume further that the variance $v > 0$ is known. (It determines the quality of the gauge and so is hopefully small!) The expectation $m \in \mathbb{R}$, however, which corresponds to the real brittleness of the pipe, is unknown. Now, the aim is to determine m from the data set $x = (x_1, \ldots, x_n) \in \mathbb{R}^n$. Again, a statistician can proceed in one of the following three ways:

1) *Estimation.* One simply provides a number that is the most plausible according to the observed values x. The first idea that comes to mind is to use the average

$$M(x) = \bar{x} := \frac{1}{n} \sum_{i=1}^{n} x_i \,.$$

But this estimate is subject to chance, so one cannot trust it too much.

2) *Confidence interval.* One determines an interval $C(x)$ depending on x, for instance of the form $C(x) =]M(x) - \varepsilon, M(x) + \varepsilon[$, which contains the true value m with sufficient certainty. Because of the assumption of the normal distribution, the latter means that

$$\mathcal{N}_{m,v}^{\otimes n}\big(C(\cdot) \ni m\big) \geq 1 - \alpha$$

for some (small) $\alpha > 0$ and all m that come into question.

3) *Test.* If it is to be decided whether the brittleness of the pipe remains below a threshold m_0, then a confidence interval is not the right approach. Rather one needs a decision rule. A sensible rule takes the form

$$M(x) \leq c \Rightarrow \text{decision for the null hypothesis } H_0 : m \leq m_0,$$
$$M(x) > c \Rightarrow \text{decision for the alternative } H_1 : m > m_0,$$

for an appropriate threshold level c. The latter should be chosen in such a way that

$$\mathcal{N}_{m,v}^{\otimes n}(M > c) \leq \alpha \text{ for } m \leq m_0$$

and

$$\mathcal{N}_{m,v}^{\otimes n}(M > c) \text{ as large as possible for } m > m_0 \,.$$

That is, on the one hand, with certainty $1 - \alpha$, one wants to avoid that the pipe is declared corrupt if it still works safely, and on the other hand, one wants to make sure that a brittle pipe will be recognised with a probability as large as possible. Observe that this procedure is not symmetric in H_0 and H_1. If security has higher priority than costs, the roles of null hypothesis and alternative should be interchanged.

What is the general structure behind the above examples?

▷ The possible outcomes x of the observation form a set $\mathcal{X}$, the sample space. An observation selects a random element from $\mathcal{X}$. (In Example (7.1) we had $\mathcal{X} = \{0, \dots, n\}$, and in (7.2) $\mathcal{X} = \mathbb{R}^n$.)

The notation $\mathcal{X}$ instead of Ω is based on the idea that the observation is given by a random variable $X : \Omega \to \mathcal{X}$, where Ω yields a detailed description of the randomness, whereas $\mathcal{X}$ contains only the actually observable outcomes. However, since only the distribution of X (rather than X itself) plays a role, Ω does not appear explicitly here.

▷ The probability measure on $\mathcal{X}$ that describes the distribution of the observation is unknown; it is to be identified from the observed values. Therefore, it is not sufficient to consider a single probability measure on $\mathcal{X}$. Instead, a whole class of possible probability measures has to be taken into account. (In Example (7.1) this was the class of hypergeometric distributions $\mathcal{H}_{n;w,N-w}$ with $w \leq N$, and in (7.2) the class of products of normal distributions $\mathcal{N}_{m,v}{}^{\otimes n}$ with m in the relevant range.)

We are thus led to the following general definition.

Definition. A *statistical model* is a triple $(\mathcal{X}, \mathscr{F}, P_\vartheta : \vartheta \in \Theta)$ consisting of a sample space $\mathcal{X}$, a σ-algebra $\mathscr{F}$ on $\mathcal{X}$, and a class $\{P_\vartheta : \vartheta \in \Theta\}$ of (at least two) probability measures on $(\mathcal{X}, \mathscr{F})$, which are indexed by an index set Θ.

Since we have to deal with many (or at least two) different probability measures, we must indicate the respective probability measure when taking expectations. Therefore, we write $\mathbb{E}_\vartheta$ for the expectation and $\mathbb{V}_\vartheta$ for the variance with respect to P_ϑ.

Although self-evident, it should be emphasised here that *the first basic task of a statistician is to choose the right model!* For, it is evident that a statistical procedure can only make sense if the underlying class of probability measures is appropriate for (or, at least, an acceptable approximation to) the application at hand.

The statistical models that will be considered later on will typically have one of the following additional properties.

Definition.   (a) A statistical model $\mathscr{M} = (\mathcal{X}, \mathscr{F}, P_\vartheta : \vartheta \in \Theta)$ is called a *parametric model* if $\Theta \subset \mathbb{R}^d$ for some $d \in \mathbb{N}$. For $d = 1$, $\mathscr{M}$ is called a *one-parameter model*.

(b) $\mathscr{M}$ is called a *discrete model* if $\mathcal{X}$ is discrete (i.e., at most countable), and $\mathscr{F} = \mathscr{P}(\mathcal{X})$; then every P_ϑ has a discrete density, namely $\rho_\vartheta : x \to P_\vartheta(\{x\})$. $\mathscr{M}$ is called a *continuous model* if $\mathcal{X}$ is a Borel subset of $\mathbb{R}^n$, $\mathscr{F} = \mathscr{B}_\mathcal{X}^n$ the Borel σ-algebra restricted to $\mathcal{X}$, and every P_ϑ has a Lebesgue density ρ_ϑ. If either of these two cases applies, we say that $\mathscr{M}$ is a *standard model*.

Here are some comments on this definition.

(a) The case of an at most countable index set Θ is included in the parametric case, since then Θ can be identified with a subset of $\mathbb{R}$. But the typical parametric case would be that Θ is an interval or, for $d > 1$, an open or closed convex subset of $\mathbb{R}^d$.

(b) The key point of the concept of a standard model is the existence of densities with respect to a so-called dominating measure. In the discrete case, the dominating measure is the counting measure that assigns to each point of $\mathcal{X}$ the mass 1, and in the continuous case it is the Lebesgue measure λ^n on $\mathbb{R}^n$. Clearly, these two cases are completely analogous, except that sums appear in the discrete case where integrals appear in the continuous case. For instance, for $A \in \mathcal{F}$ we either get $P_\vartheta(A) = \sum_{x \in A} \rho_\vartheta(x)$ or $P_\vartheta(A) = \int_A \rho_\vartheta(x)\,dx$. *In the following, we will not treat both cases separately, but we will always write integrals, which in the discrete case have to be replaced by sums.*

We will often consider statistical models that describe the independent repetition of identical experiments. This already appeared in Example (7.2). For this purpose we will introduce the following notion.

Definition. If $(E, \mathcal{E}, Q_\vartheta : \vartheta \in \Theta)$ is a statistical model and $n \geq 2$ an integer, then

$$(\mathcal{X}, \mathcal{F}, P_\vartheta : \vartheta \in \Theta) = (E^n, \mathcal{E}^{\otimes n}, Q_\vartheta^{\otimes n} : \vartheta \in \Theta)$$

is called the corresponding *n-fold product model*. In this case we write $X_i : \mathcal{X} \to E$ for the projection onto the ith coordinate. It describes the outcome of the ith sub-experiment. With respect to every P_ϑ, the random variables $X_1, \ldots, X_n$ are i.i.d. with distribution Q_ϑ.

Obviously, the product model of a parametric model is again parametric, since it has the same index set Θ. Moreover, the product model of a standard model is again a standard model; compare Example (3.30).

We conclude this section with a remark concerning concepts and terminology. The process of observing a random phenomenon that yields a random outcome is described by a suitable random variable, and will typically be denoted by X, or X_i (as in the above product model). To emphasise this interpretation, such random variables will be denoted as (random) *observations* or *measurements*. In contrast, a realisation of X or X_i, i.e., the specific value x that was obtained by an observation, will be called the *outcome*, *sample*, or the *measured* or *observed value*.

7.2   Facing the Choice

We now turn to the estimation problem. After the above informal introduction to estimators, we now provide their general definition.

Definition. Let $(\mathcal{X}, \mathscr{F}, P_\vartheta : \vartheta \in \Theta)$ be a statistical model and $(\Sigma, \mathscr{S})$ an arbitrary event space.

(a) Any function $S : \mathcal{X} \to \Sigma$ that is measurable with respect to $\mathscr{F}$ and $\mathscr{S}$ is called a *statistic*.

(b) Let $\tau : \Theta \to \Sigma$ be a mapping that assigns to each $\vartheta \in \Theta$ a certain characteristic $\tau(\vartheta) \in \Sigma$. (In the parametric case suppose for instance that $(\Sigma, \mathscr{S}) = (\mathbb{R}, \mathscr{B})$ and $\tau(\vartheta) = \vartheta_1$ is the first coordinate of ϑ.) Then, a statistic $T : \mathcal{X} \to \Sigma$ is called an *estimator of* τ.

These definitions will cause some surprise:

▷ Why is the notion of a statistic introduced if it is nothing but a random variable? The reason is that although these two notions are mathematically identical, they are interpreted differently. In our perception, a random variable describes the unpredictable outcomes that chance presents us with. By way of contrast, a statistic is a mapping a statistician has cleverly constructed to extract some essential information from the observed data.

▷ Why is the notion of an estimator introduced if it is the same as a statistic? And why does the definition of an estimator not tell us whether T is somehow related to τ? Again, this is because of the interpretation. An estimator is a statistic that is specifically tailored for the task of estimating τ. But this is not formalised any further to avoid an unnecessary narrowing of this notion.

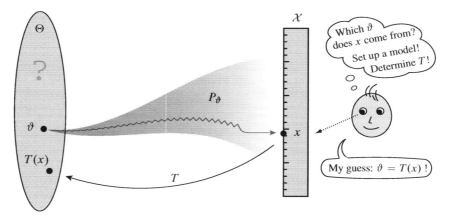

Figure 7.1. The principle of estimation: An unknown quantity ϑ is to be determined; the possible ϑ form a set Θ. The true ϑ determines a probability measure P_ϑ, which governs the random experiment and which leads to the observed value x in the concrete case. To infer ϑ, the statistician first has to analyse the probability measure P_ϑ for each ϑ – this is part of the set-up of the model. Then he must use the detected structure of all P_ϑ to construct a mapping $T : \mathcal{X} \to \Theta$ in a such way that the value $T(x)$ is typically close to the true ϑ – no matter which ϑ this happens to be in the end.

An estimator is often called a *point estimator*, in order to emphasise that for every $x \in \mathcal{X}$ it yields a single estimate instead of an entire confidence region. Figure 7.1 illustrates the basic principle of estimation.

In Examples (7.1) and (7.2) the construction of a good estimator seemed to be a matter of routine. However, this is not always the case. The following example shows some of the problems one might have to deal with.

(7.3) Example. *Guessing the range of random numbers.* On a TV show the host presents a machine that generates random numbers in the interval $[0, \vartheta]$, if the host has set it to the value $\vartheta > 0$. Two players are allowed to use the machine $n = 10$ times and then they are supposed to guess ϑ. Whoever gets closest wins.

First of all we have to ask which underlying statistical model should be used. Since the machine generates n non-negative random numbers, the sample space is clearly the product space $\mathcal{X} = [0, \infty[^n$. The parameter ϑ set by the host belongs to $\Theta =]0, \infty[$, and the possible probability measures are – by the implicit assumption of independence and uniform distribution of the generated random numbers – the product measures $P_\vartheta = \mathcal{U}_{[0,\vartheta]}^{\otimes n}$. Therefore, we use the n-fold product model

$$(\mathcal{X}, \mathcal{F}, P_\vartheta : \vartheta \in \Theta) = \left([0, \infty[^n, \mathcal{B}_{[0,\infty[}^{\otimes n}, \mathcal{U}_{[0,\vartheta]}^{\otimes n} : \vartheta > 0\right)$$

of scaled uniform distributions. We now consider two guessing strategies for the two players.

(A) Player A recalls the law of large numbers. Since $\mathbb{E}(\mathcal{U}_{[0,\vartheta]}) = \vartheta/2$, he considers the doubled sample mean $T_n := 2\,M := \frac{2}{n}\sum_{k=1}^{n} X_k$ and notices that

$$P_\vartheta\left(|T_n - \vartheta| > \varepsilon\right) \xrightarrow[n\to\infty]{} 0 \quad \text{for all } \varepsilon > 0\,.$$

That is, T_n converges in P_ϑ-probability to ϑ. (Since the underlying model also depends on the number n of observations, the notion of convergence in probability is used here in a slightly more general sense than before.) Therefore, player A chooses T_n as an estimator and hopes that this turns out to be reasonable already for $n = 10$.

(B) Player B considers the following: Although the observed maximum $\tilde{T}_n := \max(X_1, \ldots, X_n)$ is always less than ϑ, it will be close to ϑ when n is large. Indeed, for all $\varepsilon > 0$ one has

$$P_\vartheta\left(\tilde{T}_n \leq \vartheta - \varepsilon\right) = P_\vartheta\left(X_1 \leq \vartheta - \varepsilon, \ldots, X_n \leq \vartheta - \varepsilon\right) = \left(\tfrac{\vartheta-\varepsilon}{\vartheta}\right)^n \xrightarrow[n\to\infty]{} 0\,.$$

This means that $\tilde{T}_n$, too, converges in P_ϑ-probability to ϑ, and so $\tilde{T}_n$ is also a reasonable estimator of ϑ.

Which player has a better chance of winning, i.e., which of the two estimators is better? In order to find out we have to decide on how to measure their performance. We have already seen:

▷ Both (sequences of) estimators are *consistent* in the sense that

$$T_n \xrightarrow{P_\vartheta} \vartheta \quad \text{and} \quad \tilde{T}_n \xrightarrow{P_\vartheta} \vartheta \quad \text{as } n \to \infty.$$

But this only gives information about the asymptotic behaviour. Which criteria are relevant for small n already, such as $n = 10$?

▷ T_n is *unbiased* in the sense that

$$\mathbb{E}_\vartheta(T_n) = \frac{2}{n} \sum_{i=1}^n \mathbb{E}_\vartheta(X_i) = \vartheta \quad \text{for all } \vartheta \in \Theta.$$

However, $P_\vartheta(\tilde{T}_n < \vartheta) = 1$, so $\tilde{T}_n$ is certainly not unbiased. But for large n, $\tilde{T}_n$ turns out to be 'nearly unbiased'. Indeed, we find that $P_\vartheta(\tilde{T}_n \le c) = (c/\vartheta)^n$ for all $c \in [0, \vartheta]$. So, under P_ϑ, $\tilde{T}_n$ has the distribution density $\frac{d}{dx}(x/\vartheta)^n = n\,x^{n-1}\vartheta^{-n}$ on $[0, \vartheta]$. Together with Corollary (4.13) we thus obtain

$$\mathbb{E}_\vartheta(\tilde{T}_n) = \int_0^\vartheta x\,n\,x^{n-1}\vartheta^{-n}\,dx = n\vartheta^{-n}\int_0^\vartheta x^n\,dx = \frac{n}{n+1}\,\vartheta$$

for all $\vartheta \in \Theta$. Therefore, $\tilde{T}_n$ is asymptotically unbiased in the sense that $\mathbb{E}_\vartheta(\tilde{T}_n) \to \vartheta$ for all $\vartheta \in \Theta$. However, it would be a good idea for player B to choose the slightly modified estimator $T_n^* := \frac{n+1}{n}\tilde{T}_n$ instead of $\tilde{T}_n$. Indeed, T_n^* has the advantage of being unbiased, and so avoids a systematic underestimation of ϑ. It also inherits from $\tilde{T}_n$ the property of being consistent.

▷ The unbiasedness of an estimator ensures that its values are typically centred around the true ϑ. But this does not exclude the possibility that they fluctuate wildly around ϑ, which would render the estimator fairly useless. We are therefore led to the question: Which of the above estimators has the smallest variance? By Theorem (4.23) and Corollary (4.13) we get for the variance of player A's estimator

$$\mathbb{V}_\vartheta(T_n) = \left(\frac{2}{n}\right)^2 \mathbb{V}_\vartheta\left(\sum_{k=1}^n X_k\right) = \frac{4}{n}\,\mathbb{V}_\vartheta(X_1) = \frac{4}{n\vartheta}\int_0^\vartheta \left(x - \frac{\vartheta}{2}\right)^2 dx = \frac{\vartheta^2}{3n}.$$

In contrast, the variance of the estimator used by player B is

$$\mathbb{V}_\vartheta(\tilde{T}_n) = \mathbb{E}_\vartheta(\tilde{T}_n^2) - \mathbb{E}_\vartheta(\tilde{T}_n)^2$$

$$= \int_0^\vartheta x^2\,n\,x^{n-1}\vartheta^{-n}\,dx - \left(\frac{n\vartheta}{n+1}\right)^2$$

$$= \left(\frac{n}{n+2} - \frac{n^2}{(n+1)^2}\right)\vartheta^2 = \frac{n\,\vartheta^2}{(n+1)^2(n+2)},$$

and for its unbiased modification we get $\mathbb{V}_\vartheta(T_n^*) = \vartheta^2/n(n+2)$. These variances even tend to 0 as fast as $1/n^2$. The estimator $\tilde{T}_n$ varies even less than T_n^*, but unfortunately around the wrong value $\frac{n}{n+1}\vartheta$. The total mean squared error for $\tilde{T}_n$ is

$$\mathbb{E}_\vartheta\left((\tilde{T}_n - \vartheta)^2\right) = \mathbb{V}_\vartheta(\tilde{T}_n) + \left(\mathbb{E}_\vartheta(\tilde{T}_n) - \vartheta\right)^2 = \frac{2\vartheta^2}{(n+1)(n+2)},$$

so (for large n) it is almost twice as big as the mean squared error $\mathbb{V}_\vartheta(T_n^*)$ of T_n^*. For $n = 10$, for example, we obtain $\mathbb{V}_\vartheta(T_{10}) = \vartheta^2/30$, $\mathbb{E}_\vartheta((\tilde{T}_{10} - \vartheta)^2) = \vartheta^2/66$, and $\mathbb{V}_\vartheta(T_{10}^*) = \vartheta^2/120$. Therefore, neither of the players uses an optimal strategy, and if in their place, you would obviously use T_n^*.

Summarising, we see that the naive estimation using the doubled sample mean is certainly not optimal, and T_n^* is better. But whether T_n^* is best, or if a best estimator does exist at all, has not been answered yet and cannot be answered in general. Different requirements on the efficiency of estimators are often not compatible, and which requirement is most suitable may depend on the situation. Therefore, the choice of an appropriate estimator may need a good deal of instinct. Nevertheless, the specific qualifications discussed above will be investigated later in this chapter in more detail.

7.3 The Maximum Likelihood Principle

Despite the difficulties discussed above, there is a universal and intuitively plausible principle for the choice of an estimator. In many cases, the resulting estimators satisfy the above criteria of efficiency, although sometimes only approximately. Let us return to the introductory example of this chapter.

(7.4) Example. *Quality checking, cf.* (7.1). Recall the orange importer who wants to assess the quality of a delivery of N oranges on the basis of a sample of size n. The statistical model is the hypergeometric model with $\mathcal{X} = \{0, \ldots, n\}$, $\Theta = \{0, \ldots, N\}$, and $P_\vartheta = \mathcal{H}_{n;\vartheta,N-\vartheta}$. The aim is to find an estimator $T : \mathcal{X} \to \Theta$ of ϑ.

As opposed to the naive approach in Example (7.1), we will now get our inspiration from the properties of the statistical model. Suppose we have observed x. Then we can calculate the probability $\rho_\vartheta(x) = P_\vartheta(\{x\})$ of this outcome if ϑ was the correct parameter, and we can argue as follows: A value of ϑ leading to a very small $\rho_\vartheta(x)$ cannot be the true parameter, because otherwise the observed outcome x would be an extremely exceptional case. Rather than believing that chance has led to such an atypical outcome, we would rather guess that x comes from a ϑ giving x a more plausible probability. This reasoning leads to the following estimation rule: The estimate $T(x)$ based on x should be determined in such a way that

$$\rho_{T(x)}(x) = \max_{\vartheta \in \Theta} \rho_\vartheta(x),$$

that is, a reasonable candidate for the true ϑ is the (or any) parameter for which the observed outcome x has the largest probability.

What does this mean in our case? Suppose $x \in \mathcal{X} = \{0, \ldots, n\}$ has been observed. For which ϑ is the hypergeometric density

$$\rho_\vartheta(x) = \frac{\binom{\vartheta}{x}\binom{N-\vartheta}{n-x}}{\binom{N}{n}}$$

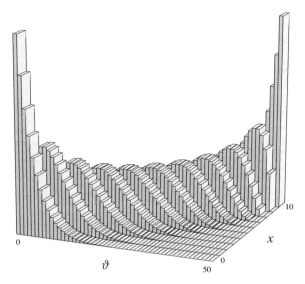

Figure 7.2. The likelihood function $\rho_x(\vartheta) = \mathcal{H}_{n;\vartheta,N-\vartheta}(\{x\})$ of Example (7.4) for $N = 50$ and $n = 10$. For each x, ρ_x attains its maximum at an integer $T(x)$ with $T(x)/N \approx x/n$.

maximal? For $\vartheta \in \mathbb{N}$ we find

$$\frac{\rho_\vartheta(x)}{\rho_{\vartheta-1}(x)} = \frac{\binom{\vartheta}{x}}{\binom{\vartheta-1}{x}} \frac{\binom{N-\vartheta}{n-x}}{\binom{N-\vartheta+1}{n-x}} = \frac{\vartheta(N-\vartheta+1-n+x)}{(\vartheta-x)(N-\vartheta+1)},$$

and this ratio is at least 1 if and only if $\vartheta n \leq (N+1)x$, so if $\vartheta \leq \frac{N+1}{n}x$. Therefore, the function $\rho_x : \vartheta \to \rho_\vartheta(x)$ is increasing on the set $\{0, \ldots, \lfloor \frac{N+1}{n}x \rfloor\}$ and decreasing for larger values of ϑ. (Here we write $\lfloor s \rfloor$ for the integer part of a real number s.) In the case $x < n$ we thus see that

$$T(x) := \left\lfloor \frac{N+1}{n}x \right\rfloor$$

is a maximiser of ρ_x (and it is unique, unless $\frac{N+1}{n}x \in \mathbb{N}$); in the case $x = n$ we obtain the maximiser $T(x) = N$. The estimator T determined in this way corresponds to the above reasoning, and it essentially agrees with the naive estimator from Example (7.1). Figure 7.2 illustrates the situation.

For $n = 1$ (when only a single sample is taken) we have $T(x) = 0$ or N depending on whether $x = 0$ or 1. This 'all-or-nothing' estimator is obviously completely useless in practice. The reason is that it is based on almost no information. An estimation that relies on too few observations cannot be accurate indeed – in that case, one might as well make no observations at all and simply take a chance and guess.

We now turn to the general definition of the estimation rule above. Recall the notion of the standard model introduced on p. 194.

Definition. Let $(\mathcal{X}, \mathcal{F}, P_\vartheta : \vartheta \in \Theta)$ be a statistical standard model.

(a) The function $\rho : \mathcal{X} \times \Theta \to [0, \infty[$ defined by $\rho(x, \vartheta) = \rho_\vartheta(x)$ is called the *likelihood (or plausibility) function* of the model, and the mapping

$$\rho_x = \rho(x, \cdot) : \Theta \to [0, \infty[, \quad \vartheta \to \rho(x, \vartheta),$$

is called the *likelihood function for the outcome $x \in \mathcal{X}$.*

(b) An estimator $T : \mathcal{X} \to \Theta$ of ϑ is called a *maximum likelihood estimator*, if

$$\rho(x, T(x)) = \max_{\vartheta \in \Theta} \rho(x, \vartheta)$$

for each $x \in \mathcal{X}$, i.e., if the estimate $T(x)$ is a maximiser of the function ρ_x on Θ. This is also expressed by writing $T(x) = \arg\max \rho_x$. A common abbreviation for maximum likelihood estimator is MLE.

We will explore this notion in some further examples.

(7.5) Example. *Estimation of a fish stock.* A pond ('urn') contains an unknown number ϑ of carps ('balls'). To estimate ϑ, we first catch r fish, mark them ('paint them red') and then release them again ('return them to the urn'). Once the marked fish have mixed well with the others, we catch n fish and notice that x of them are marked.

The statistical model is clearly $\mathcal{X} = \{0, \ldots, n\}$, $\Theta = \{r, r+1, \ldots\}$, $P_\vartheta = \mathcal{H}_{n;r,\vartheta-r}$, and the likelihood function for an outcome $x \in \mathcal{X}$ is

$$\rho_x(\vartheta) = \frac{\binom{r}{x}\binom{\vartheta-r}{n-x}}{\binom{\vartheta}{n}}.$$

Which $T(x)$ maximises ρ_x? In analogy with the last example we find that, for $x \neq 0$, ρ_x is increasing on $\{r, \ldots, \lfloor nr/x \rfloor\}$ and decreasing for larger values of ϑ. Therefore $T(x) = \lfloor nr/x \rfloor$ is a maximum likelihood estimator of ϑ, in agreement with the intuition $r/\vartheta \approx x/n$.

For $x = 0$ we see that ρ_x is increasing on all of Θ; so we set $T(0) = \infty$. However, then $T(0) \notin \Theta$, but this formal problem can be fixed by adding ∞ to Θ and setting $P_\infty := \delta_0$. A more serious objection is that for small x, the estimate $T(x)$ depends heavily on whether we catch one more or one less of the marked fish. In this case the estimator T is hardly reliable, and it would be better to repeat the experiment and mark a larger number of fish.

(7.6) Example. *Estimation of the success probability.* A drawing pin can either land on its tip or on its head. Let us suppose it lands on the tip with probability ϑ. We want to find an estimator of the unknown ϑ when the pin is thrown n times. Of course, this is a toy application; but the same sort of problem appears whenever a number of independent observations is used to determine an unknown probability

ϑ of a 'success', for instance the medical effect of a drug, or the preference for a candidate in an election. The natural statistical model is the *binomial model*

$$\big(\{0,\dots,n\}, \mathscr{P}(\{0,\dots,n\}), \mathcal{B}_{n,\vartheta} : \vartheta \in [0,1]\big)$$

with the likelihood function $\rho_x(\vartheta) = \binom{n}{x} \vartheta^x (1-\vartheta)^{n-x}$. To determine a maximum likelihood estimator we consider the log-likelihood function $\log \rho_x$, which is simpler to handle. For $0 < \vartheta < 1$ we get

$$\frac{d}{d\vartheta} \log \rho_x(\vartheta) = \frac{d}{d\vartheta}\big[x \log \vartheta + (n-x)\log(1-\vartheta)\big] = \frac{x}{\vartheta} - \frac{n-x}{1-\vartheta}\,.$$

The last expression is decreasing in ϑ and vanishes exactly at the point $\vartheta = x/n$. Therefore, $T(x) = x/n$ is the (unique) maximum likelihood estimator of ϑ. This is another instance where the maximum likelihood principle gives the intuitively obvious estimator.

(7.7) Example. *Estimation of the content of an urn.* Suppose an urn contains a certain number of identical balls with distinct colours. Let E be the finite set of colours. (The previous Example (7.6) corresponds to the case $|E| = 2$.) We take a sample of size n with replacement. The aim is to estimate the number of balls of colour $a \in E$ contained in the urn, simultaneously for all colours. According to Section 2.2.1 we choose the sample space $\mathcal{X} = E^n$, the parameter set $\Theta = \{\vartheta \in [0,1]^E : \sum_{a\in E} \vartheta(a) = 1\}$ of all probability densities on E, as well as the family of product measures $P_\vartheta = \vartheta^{\otimes n}$, $\vartheta \in \Theta$. (Equivalently, one could also use the multinomial model from Section 2.2.2.) Then, the likelihood function is $\rho_x(\vartheta) = \prod_{a\in E} \vartheta(a)^{nL(a,x)}$; here $L(a,x) = |\{1 \le i \le n : x_i = a\}|/n$ is the relative frequency of colour a in sample x. The discrete density $L(x) = (L(a,x))_{a\in E} \in \Theta$ is called the *histogram* or the *empirical distribution* of x.

In analogy to Example (7.6), it is natural to guess that the mapping $L : \mathcal{X} \to \Theta$ is a maximum likelihood estimator. Instead of finding the maximum of the likelihood function by differentiation (which would require the method of Lagrange multipliers if $|E| > 2$, the only case of interest here), it is simpler to verify our conjecture directly. For arbitrary x and ϑ we get

$$\rho_x(\vartheta) = \rho_x(L(x)) \prod_a{}' \Big(\frac{\vartheta(a)}{L(a,x)}\Big)^{n\,L(a,x)},$$

where $\prod_a'$ runs over all $a \in E$ with $L(a,x) > 0$; since $s^0 = 1$ for $s \ge 0$, we can ignore the other factors. By the inequality $s \le e^{s-1}$, the primed product can be bounded by

$$\prod_a{}' \exp\big[n\,L(a,x)\big(\vartheta(a)/L(a,x) - 1\big)\big] \le \exp\big[n\,(1-1)\big] = 1\,.$$

This yields the required inequality $\rho_x(\vartheta) \leq \rho_x(L(x))$. Since $s < e^{s-1}$ for $s \neq 1$, we also see that the inequality is strict when $\vartheta \neq L(x)$. Therefore, L is the only maximum likelihood estimator.

(7.8) Example. *Range of random numbers, cf.* (7.3). Consider again the example of a random number generator on a TV show. The underlying statistical model is the product model

$$\left([0,\infty[^n, \mathscr{B}_{[0,\infty[}^{\otimes n}, \mathcal{U}_{[0,\vartheta]}^{\otimes n} : \vartheta > 0\right)$$

of scaled uniform distributions. Therefore, the likelihood function is

$$\rho_x(\vartheta) = \begin{cases} \vartheta^{-n} & \text{if } x_1, \ldots, x_n \leq \vartheta, \\ 0 & \text{otherwise,} \end{cases}$$

where $x = (x_1, \ldots, x_n) \in [0,\infty[^n$ and $\vartheta > 0$. The maximum likelihood estimator is therefore the estimator $\tilde{T}_n(x) = \max(x_1, \ldots, x_n)$ from Example (7.3).

(7.9) Example. *Measurements in physics.* We measure the current in an electric conductor under certain external conditions. The deflection of the pointer of the amperemeter does not only depend on the current, but is also perturbed by small inaccuracies introduced by the gauges and the test conditions. Just as in Example (7.2), we will therefore suppose that the deflection of the pointer is a normally distributed random variable with an unknown expectation m (which we are interested in) and a variance $v > 0$, which is also assumed to be unknown this time. We perform n independent experiments. Consequently, the statistical model we choose is the product model

$$(\mathcal{X}, \mathcal{F}, P_\vartheta : \vartheta \in \Theta) = \left(\mathbb{R}^n, \mathscr{B}^n, \mathcal{N}_{m,v}^{\otimes n} : m \in \mathbb{R}, v > 0\right).$$

We call this the *n-fold normal, or Gaussian, product model.* According to Example (3.30) the associated likelihood function takes the form

$$\rho_x(\vartheta) = \prod_{i=1}^n \phi_{m,v}(x_i) = (2\pi v)^{-n/2} \exp\left[-\sum_{i=1}^n \frac{(x_i - m)^2}{2v}\right],$$

where $x = (x_1, \ldots, x_n) \in \mathbb{R}^n$ and $\vartheta = (m, v) \in \Theta$. To maximise this expression, we first have to

▷ choose m such that the mean squared error $\frac{1}{n}\sum_{i=1}^n (x_i - m)^2$ is minimal, which is achieved for $m = M(x) := \frac{1}{n}\sum_{i=1}^n x_i$, the sample mean. This is an obvious consequence of the *shift formula*

$$(7.10) \qquad \frac{1}{n}\sum_{i=1}^n (x_i - m)^2 = \frac{1}{n}\sum_{i=1}^n (x_i - M(x))^2 + (M(x) - m)^2,$$

which follows from Pythagoras' theorem, see Figure 7.3. The reader should memorise the result that the average minimises the mean squared error.

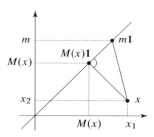

Figure 7.3. Writing $\mathbf{1} = (1, \ldots, 1)$ for the vector pointing in diagonal direction, equation (7.10) takes the form of Pythagoras' theorem $|x - m\mathbf{1}|^2 = |x - M(x)\mathbf{1}|^2 + |M(x)\mathbf{1} - m\mathbf{1}|^2$, which holds because $x - M(x)\mathbf{1} \perp \mathbf{1}$. In particular, $M(x)\mathbf{1}$ is the projection of x onto the diagonal. Here, the case $n = 2$ is shown.

Furthermore, we have to

▷ choose v such that $(2\pi v)^{-n/2} \exp\left[-\frac{1}{2v} \sum_{i=1}^{n}(x_i - M(x))^2\right]$ is maximal. Differentiating the logarithm of this expression with respect to v, we obtain

$$-\frac{d}{dv}\left(\frac{n}{2}\log v + \frac{1}{2v}\sum_{i=1}^{n}(x_i - M(x))^2\right) = -\frac{n}{2v} + \frac{1}{2v^2}\sum_{i=1}^{n}(x_i - M(x))^2.$$

The last term vanishes if and only if

$$v = V(x) := \frac{1}{n}\sum_{i=1}^{n}(x_i - M(x))^2,$$

and it is easy to check that $v = V(x)$ is indeed a maximiser.

We state the result as follows. (As always in product models, X_i denotes the ith projection.)

(7.11) Theorem. Maximum likelihood estimator in the Gaussian model. *The maximum likelihood estimator in the n-fold Gaussian product model is $T = (M, V)$. Here,*

$$M = \frac{1}{n}\sum_{i=1}^{n}X_i \quad and \quad V = \frac{1}{n}\sum_{i=1}^{n}(X_i - M)^2$$

are the sample mean and the sample variance, respectively.

Let us conclude this section with a natural generalisation of the maximum likelihood principle. Suppose it is not the parameter ϑ itself, but only a characteristic $\tau(\vartheta)$ that is to be estimated. Then, if T is a maximum likelihood estimator of ϑ, $\tau(T)$ is called a maximum likelihood estimator of $\tau(\vartheta)$.

(7.12) Example. *Failure times of appliances.* Consider the lifetime of a technical product. Suppose that the lifetime is exponentially distributed with unknown parameter $\vartheta > 0$, and that n appliances from different production series are tested. Hence, the statistical model is the n-fold product of the exponential distribution model $([0, \infty[, \mathscr{B}_{[0,\infty[}, \mathcal{E}_\vartheta : \vartheta > 0)$. It has the density $\rho_\vartheta = \vartheta^n \exp[-\vartheta \sum_{i=1}^n X_i]$. Solving the equation $\frac{d}{d\vartheta} \log \rho_\vartheta = 0$ for ϑ, one obtains the maximum likelihood estimator $T = 1/M$, where again $M = \frac{1}{n} \sum_{i=1}^n X_i$ denotes the sample mean. But we might be less interested in the parameter ϑ as such, but rather in the probability that an appliance fails before the end of the guarantee period t. If ϑ is the true parameter, this failure probability is $\tau(\vartheta) := 1 - e^{-\vartheta t}$ for each individual appliance. Therefore, the maximum likelihood estimator of the failure probability within the guarantee period is $\tau(T) = 1 - e^{-t/M}$.

7.4 Bias and Mean Squared Error

We will now investigate the quality of estimators. An elementary criterion is the following.

Definition. Let $(\mathcal{X}, \mathscr{F}, P_\vartheta : \vartheta \in \Theta)$ be a statistical model and $\tau : \Theta \to \mathbb{R}$ a real characteristic. An estimator $T : \mathcal{X} \to \mathbb{R}$ of τ is called *unbiased* if

$$\mathbb{E}_\vartheta(T) = \tau(\vartheta) \quad \text{for all } \vartheta \in \Theta.$$

Otherwise, $\mathscr{B}_T(\vartheta) = \mathbb{E}_\vartheta(T) - \tau(\vartheta)$ is called the *bias* of T at ϑ. Here, the existence of the expectations is understood.

In other words, an unbiased estimator avoids systematic errors. This is clearly a sensible criterion, but is not automatically compatible with the maximum likelihood principle. As we have seen in Examples (7.3) and (7.8) on guessing the range of random numbers, the estimator used by player B, the maximum likelihood estimator $\tilde{T}_n$, is not unbiased, but at least asymptotically so. There is a similar issue with the sample variance V from Theorem (7.11), as the following theorem shows.

(7.13) Theorem. Estimation of expectation and variance in real product models. *Let $n \geq 2$ and $(\mathbb{R}^n, \mathscr{B}^n, Q_\vartheta^{\otimes n} : \vartheta \in \Theta)$ be a real n-fold product model. Suppose that for each $\vartheta \in \Theta$ the expectation $m(\vartheta) = \mathbb{E}(Q_\vartheta)$ and the variance $v(\vartheta) = \mathbb{V}(Q_\vartheta)$ of Q_ϑ are defined. Then, the sample mean $M = \frac{1}{n} \sum_{i=1}^n X_i$ and the corrected sample variance $V^* = \frac{1}{n-1} \sum_{i=1}^n (X_i - M)^2$ are unbiased estimators of m and v, respectively.*

Proof. Fix $\vartheta \in \Theta$. By the linearity of the expectation, it is clear that $\mathbb{E}_\vartheta(M) = \frac{1}{n} \sum_{i=1}^n \mathbb{E}_\vartheta(X_i) = m(\vartheta)$. In particular, $\mathbb{E}_\vartheta(X_i - M) = 0$. Hence

$$(n-1)\,\mathbb{E}_\vartheta(V^*) = \sum_{i=1}^{n} \mathbb{V}_\vartheta(X_i - M) = n\,\mathbb{V}_\vartheta(X_1 - M)$$

$$= n\,\mathbb{V}_\vartheta\Big(\tfrac{n-1}{n}X_1 - \tfrac{1}{n}\sum_{j=2}^{n} X_j\Big)$$

$$= n\left(\big(\tfrac{n-1}{n}\big)^2 + (n-1)\tfrac{1}{n^2}\right)v(\vartheta) = (n-1)\,v(\vartheta)\,.$$

The second step follows by symmetry, and the fourth exploits Bienaymé's identity (4.23c), since the projections X_i are independent and thus uncorrelated with respect to the product measure $Q_\vartheta^{\otimes n}$. Dividing by $n-1$ yields $\mathbb{E}_\vartheta(V^*) = v(\vartheta)$. $\diamond$

Since V^* is unbiased, the maximum likelihood estimator $V = \frac{n-1}{n}V^*$ is biased, though its bias $\mathscr{B}_V(\vartheta) = -v(\vartheta)/n$ is negligible for large n. For this reason, one often finds V^* or $\sigma^* := \sqrt{V^*}$, but not V, among the statistical functions of calculators.

It is clearly desirable to avoid systematic errors and therefore to use unbiased estimators. However, this only becomes relevant if the values of the estimator T are typically close to the true value of τ. A convenient way of measuring the latter quality of an estimator T of τ is the *mean squared error*

$$\mathscr{E}_T(\vartheta) := \mathbb{E}_\vartheta\big((T - \tau(\vartheta))^2\big) = \mathbb{V}_\vartheta(T) + \mathscr{B}_T(\vartheta)^2\,;$$

the second equality is analogous to the shift formula (7.10). To keep $\mathscr{E}_T(\vartheta)$ as small as possible, both variance and bias must be minimised simultaneously. As the following example shows, it might be preferable to admit a bias in order to minimise the total error.

(7.14) Example. *A good estimator with bias.* Consider the binomial model, in which $\mathcal{X} = \{0, \ldots, n\}$, $\Theta = [0, 1]$, and $P_\vartheta = \mathcal{B}_{n,\vartheta}$. The maximum likelihood estimator of ϑ is $T(x) = x/n$, which is also unbiased. Furthermore, it even has the smallest variance among all unbiased estimators, as we will see soon. Its mean squared error is $\mathscr{E}_T(\vartheta) = n^{-2}\,\mathbb{V}(\mathcal{B}_{n,\vartheta}) = \vartheta(1-\vartheta)/n$. However, there is an estimator S of ϑ whose mean squared error is smaller for certain ϑ, namely

$$S(x) = \frac{x+1}{n+2}\,.$$

It is easily checked that $S(x) \geq T(x)$ if and only if $T(x) \leq 1/2$. That is, the values of S are always closer to the centre of $[0, 1]$ than those of T. Furthermore, S has the bias

$$\mathscr{B}_S(\vartheta) = \frac{n\vartheta + 1}{n+2} - \vartheta = \frac{1-2\vartheta}{n+2}$$

and the mean squared error

$$\mathscr{E}_S(\vartheta) = \mathbb{V}_\vartheta(S) + \mathscr{B}_S(\vartheta)^2 = \frac{n\vartheta(1-\vartheta) + (1-2\vartheta)^2}{(n+2)^2}\,.$$

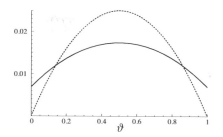

Figure 7.4. $\mathscr{E}_S(\vartheta)$ (solid) and $\mathscr{E}_T(\vartheta)$ (dashed) for $n = 10$.

As Figure 7.4 shows, the mean squared error of S is smaller than that of T when ϑ is close to $1/2$. Specifically, we have $\mathscr{E}_S(\vartheta) \le \mathscr{E}_T(\vartheta)$ if and only if

$$\frac{|\vartheta - \frac{1}{2}|^2}{\vartheta(1-\vartheta)} \le 1 + \frac{1}{n},$$

and the latter holds for arbitrary n as long as $|\vartheta - 1/2| \le 1/\sqrt{8} \approx 0.35$. So, if for some reason the circumstances make us believe that only central values of ϑ are possible, it is better to work with S instead of T.

The last example shows that unbiasedness should not be idolised. Nevertheless, in the following we will restrict our attention to unbiased estimators when minimising the mean squared error. Then, our aim becomes to minimise the variance.

7.5   Best Estimators

We are now looking for estimators that satisfy two of the performance criteria discussed before. On the one hand, we will require that the estimators are unbiased, so that the estimates are correct on average. In addition, we want to achieve that the estimators are best possible, in the sense that they vary no more than any other unbiased estimator. In particular, this will imply that their mean squared error is minimal.

Definition. Let $(\mathcal{X}, \mathscr{F}, P_\vartheta : \vartheta \in \Theta)$ be a statistical model. An unbiased estimator T of a real characteristic $\tau : \Theta \to \mathbb{R}$ is called a *minimum variance* or a *(uniformly) best estimator* if $\mathbb{V}_\vartheta(T) \le \mathbb{V}_\vartheta(S)$ for any other unbiased estimator S and all $\vartheta \in \Theta$. Here it is understood that T and the other S belong to the class $\bigcap_{\vartheta \in \Theta} \mathscr{L}^2(P_\vartheta)$ of estimators with existing variances.

To find best estimators, we consider only the case of one-parameter standard models with particularly nice properties; the multi-parameter case can be found, for example, in [41]. For these models, we will derive a lower bound for the variances of estimators. By investigating for which estimators this bound is attained, we will then find a convenient criterion for the existence of best estimators.

Definition. A one-parameter standard model $(\mathcal{X}, \mathcal{F}, P_\vartheta : \vartheta \in \Theta)$ is called *regular* if it satisfies the following conditions:

▷ Θ is an open interval in $\mathbb{R}$.

▷ The likelihood function on $\mathcal{X} \times \Theta$ is strictly positive and continuously differentiable in ϑ. So, in particular, there exists the so-called *score function*

$$U_\vartheta(x) := \frac{d}{d\vartheta} \log \rho(x, \vartheta) = \frac{\rho_x'(\vartheta)}{\rho_x(\vartheta)}.$$

▷ ρ allows to interchange the differentiation in ϑ and the integration over x, in that

$$(7.15) \qquad \int \frac{d}{d\vartheta} \rho(x, \vartheta)\, dx = \frac{d}{d\vartheta} \int \rho(x, \vartheta)\, dx .$$

(As always, for discrete $\mathcal{X}$ the integral is to be replaced by a sum.)

▷ For each $\vartheta \in \Theta$, the variance $I(\vartheta) := \mathbb{V}_\vartheta(U_\vartheta)$ exists and is non-zero.

The function $I : \vartheta \to I(\vartheta)$ is then called the *Fisher information* of the model, in honour of the British statistician Sir Ronald A. Fisher (1880–1962).

This definition requires a few comments. Consider first the interchange relation (7.15). In the continuous case, it certainly holds if every $\vartheta_0 \in \Theta$ admits a neighbourhood $N(\vartheta_0)$ such that

$$\int_{\mathcal{X}} \sup_{\vartheta \in N(\vartheta_0)} \left| \frac{d}{d\vartheta} \rho(x, \vartheta) \right| dx < \infty ;$$

this follows from the mean value theorem, which allows to bound the modulus of difference ratios by the maximal absolute derivative, and the dominated convergence theorem (for Lebesgue measure, in place of the probability measure in Problem 4.7; see also [57]). In the case of a countably infinite $\mathcal{X}$, one has a similar sufficient condition, and (7.15) holds trivially when $\mathcal{X}$ is finite. The significance of (7.15) comes from the relation

$$(7.16) \qquad \mathbb{E}_\vartheta(U_\vartheta) = \int \frac{d}{d\vartheta} \rho(x, \vartheta)\, dx = \frac{d}{d\vartheta} \int \rho(x, \vartheta)\, dx = \frac{d}{d\vartheta} 1 = 0 ,$$

which says that the score function of ϑ is centred with respect to P_ϑ. In particular, $I(\vartheta) = \mathbb{E}_\vartheta(U_\vartheta^2)$.

Next, consider the Fisher information. Why is it called information? As a justification we note two facts: First, I vanishes identically on an interval $\Theta_0 \subset \Theta$ if and only if $U_\vartheta(x) = 0$ for all $\vartheta \in \Theta_0$ and (almost) all $x \in \mathcal{X}$, that is, if ρ_x is constant on Θ_0 for (almost) all $x \in \mathcal{X}$. In that case no observation can distinguish between the parameters in Θ_0, which means that Θ_0 provides no statistical information. (This is why this case was excluded in the definition.) Secondly, the Fisher information adds up for independent observations, as stated in the remark below.

(7.17) Remark. *Additivity of Fisher information.* Let $\mathscr{M} = (\mathcal{X}, \mathscr{F}, P_\vartheta : \vartheta \in \Theta)$ be a regular statistical model with Fisher information I. Then the product model $\mathscr{M}^{\otimes n} = (\mathcal{X}^n, \mathscr{F}^{\otimes n}, P_\vartheta^{\otimes n} : \vartheta \in \Theta)$ has Fisher information $I^{\otimes n} := n\,I$.

Proof. $\mathscr{M}^{\otimes n}$ has the likelihood function $\rho_\vartheta^{\otimes n} = \prod_{k=1}^n \rho_\vartheta \circ X_k$ and thus the score function

$$U_\vartheta^{\otimes n} = \sum_{k=1}^n U_\vartheta \circ X_k \,,$$

where again X_k denotes the kth projection from $\mathcal{X}^n$ onto $\mathcal{X}$. Since the projections are independent under $P_\vartheta^{\otimes n}$, Bienaymé's identity (4.23c) implies that

$$I^{\otimes n}(\vartheta) = \mathbb{V}_\vartheta(U_\vartheta^{\otimes n}) = \sum_{k=1}^n \mathbb{V}_\vartheta(U_\vartheta \circ X_k) = n\,I(\vartheta)\,,$$

as claimed. $\diamond$

The following information inequality (which is also known as Cramér–Rao inequality) reveals the significance of the Fisher information. We call an unbiased estimator T *regular* if its expectation can be differentiated with respect to ϑ under the integral sign, in that

$$(7.18) \qquad \int T(x)\,\frac{d}{d\vartheta}\,\rho(x, \vartheta)\,dx = \frac{d}{d\vartheta}\int T(x)\,\rho(x, \vartheta)\,dx$$

for all ϑ.

(7.19) Theorem. *Information inequality; M. Fréchet 1943, C. R. Rao 1945, H. Cramér 1946. Suppose we are given a regular statistical model $(\mathcal{X}, \mathscr{F}, P_\vartheta : \vartheta \in \Theta)$, a continuously differentiable function $\tau : \Theta \to \mathbb{R}$ with $\tau' \neq 0$ that is to be estimated, and a regular unbiased estimator T of τ. Then,*

$$(7.20) \qquad \mathbb{V}_\vartheta(T) \geq \tau'(\vartheta)^2/I(\vartheta) \quad \text{for all } \vartheta \in \Theta\,.$$

Moreover, equality holds for all ϑ if and only if

$$T - \tau(\vartheta) = \tau'(\vartheta)\,U_\vartheta/I(\vartheta) \quad \text{for all } \vartheta \in \Theta\,,$$

i.e., if the model has the likelihood function

$$(7.21) \qquad \rho(x, \vartheta) = \exp[a(\vartheta)\,T(x) - b(\vartheta)]\,h(x)\,;$$

here $a : \Theta \to \mathbb{R}$ is a primitive function of I/τ', $h : \mathcal{X} \to \,]0, \infty[$ a measurable function, and $b(\vartheta) = \log \int_{\mathcal{X}} e^{a(\vartheta)\,T(x)} h(x)\,dx$ a normalising function.

Proof. Since U_ϑ is centred by (7.16) and T is regular and unbiased, we find that

(7.22) $\mathrm{Cov}_\vartheta (T, U_\vartheta) = \mathbb{E}_\vartheta (T U_\vartheta)$

$$= \int_\mathcal{X} T(x) \frac{d}{d\vartheta} \rho(x, \vartheta) \, dx$$

$$= \frac{d}{d\vartheta} \int_\mathcal{X} T(x) \rho(x, \vartheta) \, dx = \frac{d}{d\vartheta} \mathbb{E}_\vartheta (T) = \tau'(\vartheta)$$

for all $\vartheta \in \Theta$. Together with the Cauchy–Schwarz inequality from Theorem (4.23b), this already yields the required inequality.

To find out when equality holds, we have to argue slightly more carefully. Setting $c(\vartheta) = \tau'(\vartheta)/I(\vartheta)$ and using Theorem (4.23c), we can write

$$0 \le \mathbb{V}_\vartheta (T - c(\vartheta) U_\vartheta) = \mathbb{V}_\vartheta (T) + c(\vartheta)^2 \, \mathbb{V}_\vartheta (U_\vartheta) - 2c(\vartheta) \, \mathrm{Cov}_\vartheta (T, U_\vartheta)$$

$$= \mathbb{V}_\vartheta (T) - \tau'(\vartheta)^2 / I(\vartheta) \, .$$

This again yields the required inequality, and shows further that equality holds if and only if the random variable $T - c(\vartheta) U_\vartheta$ is constant P_ϑ-almost surely. Obviously, this constant must equal the expectation $\tau(\vartheta)$. Since P_ϑ has a positive density with respect to Lebesgue respectively the counting measure μ on $\mathcal{X}$, the latter is equivalent to the statement $\mu(T - \tau(\vartheta) \ne c(\vartheta) U_\vartheta) = 0$. Assume this holds for all ϑ. Then

$$\mu\big((T - \tau(\vartheta))/c(\vartheta) \ne U_\vartheta \text{ for some } \vartheta \in \Theta\big) = 0$$

because, for reasons of continuity, it is sufficient to consider rational ϑ, and a countable union of events of measure 0 has still measure 0. So, for μ-almost all x, (7.21) follows by indefinite integration with respect to ϑ. The converse direction is obvious.

$\diamond$

Theorem (7.19) admits two conclusions:

▷ If a regular experiment is repeated n times, then the variance of an unbiased estimator of τ is at least of order $1/n$. This follows from the information inequality (7.20) together with Remark (7.17). (This is no contradiction to the result of Example (7.3), where we found that the variances of the estimators T_n^* even decrease like n^{-2}. The reason is that the model there is not regular.)

▷ Suppose we can find a regular unbiased estimator T of τ such that equality holds in (7.20); such a T is called *Cramér–Rao efficient*. Then T is clearly a best estimator (if only in the class of all regular estimators of τ). Such Cramér–Rao efficient estimators, however, can exist only if (7.21) holds. Therefore, this last condition defines a particularly interesting class of statistical models, which we will now investigate further.

Definition. Let $\mathcal{M} = (\mathcal{X}, \mathcal{F}, P_\vartheta : \vartheta \in \Theta)$ be a one-parameter standard model whose parameter set Θ is an open interval. $\mathcal{M}$ is called an *exponential (statistical) model* and $\{P_\vartheta : \vartheta \in \Theta\}$ an *exponential family* with respect to the statistic $T : \mathcal{X} \to \mathbb{R}$, if the likelihood function takes the form (7.21) for a continuously differentiable function $a : \Theta \to \mathbb{R}$ with $a' \neq 0$ and a measurable function $h : \mathcal{X} \to]0, \infty[$.

(Note that the normalising function b is uniquely determined by a, h, and T. Also, it is assumed implicitly that T is not constant almost surely, because otherwise P_ϑ would not depend on ϑ.)

We will see some exponential families from Example (7.25) onwards. First we investigate their properties.

(7.23) Remark. *Properties of exponential families.* For every exponential model $\mathcal{M}$ the following holds.

(a) The normalising function b is continuously differentiable on Θ with derivative $b'(\vartheta) = a'(\vartheta)\, \mathbb{E}_\vartheta(T)$ for $\vartheta \in \Theta$.

(b) Every statistic $S : \mathcal{X} \to \mathbb{R}$ whose expectations $\mathbb{E}_\vartheta(S)$ exist is regular. In particular, $\mathcal{M}$ and T are regular, and $\tau(\vartheta) := \mathbb{E}_\vartheta(T)$ is continuously differentiable with derivative $\tau'(\vartheta) = a'(\vartheta)\, \mathbb{V}_\vartheta(T) \neq 0$, $\vartheta \in \Theta$.

(c) The Fisher information is given by $I(\vartheta) = a'(\vartheta)\, \tau'(\vartheta)$ for all $\vartheta \in \Theta$.

Proof. Without loss of generality, we can assume that $a(\vartheta) = \vartheta$ and so $a'(\vartheta) = 1$ for all $\vartheta \in \Theta$; the general case follows by re-parametrising and applying the chain rule to find the derivatives.

Now, let S be any real statistic for which $S \in \mathscr{L}^1(P_\vartheta)$ for all $\vartheta \in \Theta$. Then we can define the function

$$u_S(\vartheta) := e^{b(\vartheta)}\, \mathbb{E}_\vartheta(S) = \int_{\mathcal{X}} S(x)\, e^{\vartheta\, T(x)} h(x)\, dx$$

on Θ. We claim that u_S is differentiable arbitrarily often. Indeed, let $\vartheta \in \Theta$ and t be so small that $\vartheta \pm t \in \Theta$. Then, by Theorem (4.11c),

$$\sum_{k \geq 0} \frac{|t|^k}{k!} \int_{\mathcal{X}} |S(x)|\, |T(x)|^k e^{\vartheta\, T(x)}\, h(x)\, dx = \int_{\mathcal{X}} |S(x)|\, e^{\vartheta\, T(x) + |t\, T(x)|}\, h(x)\, dx$$

$$\leq \int_{\mathcal{X}} |S(x)|\, \big[e^{(\vartheta + t)\, T(x)} + e^{(\vartheta - t)T(x)} \big]\, h(x)\, dx < \infty.$$

Therefore, $ST^k \in \mathscr{L}^1(P_\vartheta)$ for all ϑ and k, and thus (for $S \equiv 1$) $T \in \mathscr{L}^2(P_\vartheta)$ for all ϑ. Furthermore, it follows that the series

$$\sum_{k \geq 0} \frac{t^k}{k!} \int_{\mathcal{X}} S(x)\, T(x)^k e^{\vartheta\, T(x)}\, h(x)\, dx$$

is absolutely convergent, and summation and integration can be interchanged. Hence, this series takes the value $u_S(\vartheta + t)$. As a consequence, u_S is in fact real analytic. In particular, we find that $u'_S(\vartheta) = e^{b(\vartheta)} \mathbb{E}_\vartheta(ST)$ and, in the special case $S \equiv 1$, $u'_1(\vartheta) = u_1(\vartheta) \mathbb{E}_\vartheta(T)$ as well as $u''_1(\vartheta) = u_1(\vartheta) \mathbb{E}_\vartheta(T^2)$. Thus, for $b = \log u_1$ we obtain $b'(\vartheta) = \mathbb{E}_\vartheta(T) =: \tau(\vartheta)$ and

$$\tau'(\vartheta) = b''(\vartheta) = u''_1(\vartheta)/u_1(\vartheta) - (u'_1(\vartheta)/u_1(\vartheta))^2 = \mathbb{V}_\vartheta(T).$$

This proves statement (a) and the second part of (b). Moreover, we can write

$$\frac{d}{d\vartheta} \mathbb{E}_\vartheta(S) = \frac{d}{d\vartheta} \left[u_S(\vartheta) e^{-b(\vartheta)} \right] = \left[u'_S(\vartheta) - u_S(\vartheta) b'(\vartheta) \right] e^{-b(\vartheta)}$$

$$= \mathbb{E}_\vartheta(ST) - \mathbb{E}_\vartheta(S) b'(\vartheta) = \mathbb{E}_\vartheta(SU_\vartheta) = \mathrm{Cov}_\vartheta(S, T),$$

since $U_\vartheta = T - b'(\vartheta)$ by (7.21). (Recall our assumption that $a(\vartheta) = \vartheta$.) So, if we replace T by S in the terms of equation (7.22), then the second and fifth term coincide, and so do the third and fourth. This proves the regularity of S, and in particular that of T. For $S \equiv 1$ we obtain (7.15). Hence, the first part of (b) has also been proved. Finally, since $U_\vartheta = T - b'(\vartheta)$ we see that $I(\vartheta) = \mathbb{V}_\vartheta(U_\vartheta) = \mathbb{V}_\vartheta(T) = \tau'(\vartheta)$, which implies (c) for our choice of $a(\vartheta)$. $\diamond$

Combining the above remark with Theorem (7.19), we arrive at the following convenient criterion for the existence of minimum variance estimators.

(7.24) Corollary. *Existence of best estimators. For every exponential model, the underlying statistic T is a best estimator of $\tau(\vartheta) := \mathbb{E}_\vartheta(T) = b'(\vartheta)/a'(\vartheta)$. Moreover, $I(\vartheta) = a'(\vartheta) \tau'(\vartheta)$ and $\mathbb{V}_\vartheta(T) = \tau'(\vartheta)/a'(\vartheta)$ for all $\vartheta \in \Theta$.*

Proof. By Remark (7.23b), the exponential model as well as any unbiased estimator S of τ are regular. Thus, Theorem (7.19) yields the required inequality $\mathbb{V}_\vartheta(S) \geq \mathbb{V}_\vartheta(T)$. The formulae for $I(\vartheta)$ and $\mathbb{V}_\vartheta(T)$ also follow from Remark (7.23). $\diamond$

Here are some standard examples of exponential families.

(7.25) Example. *Binomial distributions.* For fixed $n \in \mathbb{N}$, the binomial distributions $\{\mathcal{B}_{n,\vartheta} : 0 < \vartheta < 1\}$ form an exponential family on $\mathcal{X} = \{0, \ldots, n\}$. Indeed, the associated likelihood function $\rho(x, \vartheta) = \binom{n}{x} \vartheta^x (1 - \vartheta)^{n-x}$ takes the form (7.21) if we set $T(x) = \frac{x}{n}$, $a(\vartheta) = n \log \frac{\vartheta}{1-\vartheta}$, $b(\vartheta) = -n \log(1 - \vartheta)$, and $h(x) = \binom{n}{x}$. Hence, T is a best unbiased estimator of ϑ. We also find

$$a'(\vartheta) = n \left(\frac{1}{\vartheta} + \frac{1}{1-\vartheta} \right) = \frac{n}{\vartheta(1-\vartheta)}$$

and $\mathbb{V}_\vartheta(T) = \frac{1}{a'(\vartheta)} = \frac{\vartheta(1-\vartheta)}{n}$, which is already familiar from (4.27) and (4.34).

(7.26) Example. *Poisson distributions.* The Poisson distributions $\{\mathcal{P}_\vartheta : \vartheta > 0\}$ also form an exponential family, since the corresponding likelihood function is

$$\rho(x, \vartheta) = e^{-\vartheta} \frac{\vartheta^x}{x!} = \exp\big[(\log \vartheta)x - \vartheta\big] \frac{1}{x!}.$$

Hence, (7.21) holds with $T(x) = x$ and $a(\vartheta) = \log \vartheta$. We know that T is an unbiased estimator of ϑ, and so by Corollary (7.24) even a best estimator of ϑ. In particular, we once again obtain the equation $\mathbb{V}(\mathcal{P}_\vartheta) = \mathbb{V}_\vartheta(T) = 1/\frac{1}{\vartheta} = \vartheta$, cf. Example (4.36).

(7.27) Example. *Normal distributions.* (a) *Estimation of the expectation.* For a fixed variance $v > 0$, the associated family $\{\mathcal{N}_{\vartheta,v} : \vartheta \in \mathbb{R}\}$ of normal distributions on $\mathcal{X} = \mathbb{R}$ has the likelihood function

$$\rho(x, \vartheta) = (2\pi v)^{-1/2} \exp\big[-(x - \vartheta)^2/2v\big],$$

and so forms an exponential family with $T(x) = x$, $a(\vartheta) = \vartheta/v$, $b(\vartheta) = \vartheta^2/2v + \frac{1}{2}\log(2\pi v)$, and $h(x) = \exp[-x^2/2v]$. Therefore, $T(x) = x$ is a best estimator of ϑ. We also get $\mathbb{V}_\vartheta(T) = v$, as we already know from (4.28). Since the optimality inequality $\mathbb{V}_\vartheta(T) \leq \mathbb{V}_\vartheta(S)$ holds for arbitrary ϑ *and* v, T is even a best estimator of the expectation in the class of *all* normal distributions (with arbitrary v).

(b) *Estimation of the variance with known expectation.* For a fixed expectation $m \in \mathbb{R}$, the associated family $\{\mathcal{N}_{m,\vartheta} : \vartheta > 0\}$ of normal distributions has the likelihood function

$$\rho(x, \vartheta) = \exp\left[-\frac{1}{2\vartheta} T(x) - \frac{1}{2} \log(2\pi\vartheta)\right]$$

for $T(x) = (x - m)^2$. Since T is an unbiased estimator of the unknown variance ϑ, Corollary (7.24) shows that T is in fact a best estimator of ϑ. Differentiating the coefficient of T we get $\mathbb{V}_\vartheta(T) = 2\vartheta^2$, and so for the fourth moments of the centred normal distributions

$$\int x^4 \, \phi_{0,\vartheta}(x) \, dx = \mathbb{V}_\vartheta(T) + \mathbb{E}_\vartheta(T)^2 = 3\vartheta^2.$$

Obviously, this formula can also be obtained directly by partial integration; cf. Problem 4.17.

Our final remark shows that the previous examples can be extended to the corresponding product models describing independently repeated observations.

(7.28) Remark. *Exponential product models.* If $\mathcal{M} = (\mathcal{X}, \mathcal{F}, P_\vartheta : \vartheta \in \Theta)$ is an exponential model with respect to the statistic $T : \mathcal{X} \to \mathbb{R}$, then the n-fold product model $\mathcal{M}^{\otimes n} := (\mathcal{X}^n, \mathcal{F}^{\otimes n}, P_\vartheta^{\otimes n} : \vartheta \in \Theta)$ is also exponential with underlying statistic $T_n = \frac{1}{n}\sum_{i=1}^n T \circ X_i$. In particular, T_n is a best estimator of $\tau(\vartheta) = \mathbb{E}_\vartheta(T)$.

Proof. Suppose the likelihood function ρ of $\mathcal{M}$ takes the form (7.21). By Example (3.30), the likelihood function $\rho^{\otimes n}$ of $\mathcal{M}^{\otimes n}$ then has the product form

$$\rho^{\otimes n}(\,\cdot\,,\vartheta) = \prod_{i=1}^{n} \rho(X_i,\vartheta) = \exp\!\big[na(\vartheta)\,T_n - nb(\vartheta)\big]\prod_{i=1}^{n} h(X_i)\,.$$

This proves the claim. $\diamondsuit$

7.6   Consistent Estimators

We now turn to the third of the performance criteria for estimators introduced in Example (7.3), namely, consistency. It concerns the long-term behaviour when the observations can be repeated arbitrarily often. For each number n of observations, one constructs an estimator T_n and hopes that for large n its values are typically close to the true value of the quantity to be estimated. Let us make this idea more precise.

Let $(\mathcal{X}, \mathcal{F}, P_\vartheta : \vartheta \in \Theta)$ be a statistical model and $\tau : \Theta \to \mathbb{R}$ a real characteristic that shall be estimated. Further, let $(X_n)_{n \geq 1}$ be a sequence of random variables on $(\mathcal{X}, \mathcal{F})$ taking values in some event space $(E, \mathcal{E})$; this sequence is supposed to model the successive observations. For each $n \geq 1$, let $T_n : \mathcal{X} \to \mathbb{R}$ be an estimator of τ that is based on the first n observations, and thus takes the form $T_n = t_n(X_1, \ldots, X_n)$ for some appropriate statistic $t_n : E^n \to \mathbb{R}$.

Definition. The sequence $(T_n)_{n \geq 1}$ of estimators of τ is called (weakly) *consistent* if

$$P_\vartheta\big(|T_n - \tau(\vartheta)| \leq \varepsilon\big) \xrightarrow[n \to \infty]{} 1$$

for all $\varepsilon > 0$ and $\vartheta \in \Theta$, which means that $T_n \xrightarrow{P_\vartheta} \tau(\vartheta)$ as $n \to \infty$ for all $\vartheta \in \Theta$.

In the following, we restrict our attention to the standard case of *independent* observations. A suitable statistical model is then the infinite product model

$$(\mathcal{X}, \mathcal{F}, P_\vartheta : \vartheta \in \Theta) = \big(E^{\mathbb{N}}, \mathcal{E}^{\otimes \mathbb{N}}, Q_\vartheta^{\otimes \mathbb{N}} : \vartheta \in \Theta\big)$$

in the sense of Example (3.29). Here, $(E, \mathcal{E}, Q_\vartheta : \vartheta \in \Theta)$ is the statistical model of every individual observation, and $X_n : \mathcal{X} \to E$ is again the projection onto the nth coordinate (as always in product models). In this situation, it is natural to consider appropriate averages over the first n observations, because their consistency is a direct consequence of the (weak) law of large numbers. A first example is provided by the following theorem, which deals with the situation of Theorem (7.13), and states that both the sample mean and the sample variance are consistent.

(7.29) Theorem. Consistency of sample mean and variance. *Let $(E, \mathscr{E}) = (\mathbb{R}, \mathscr{B})$, and suppose that for every $\vartheta \in \Theta$ the expectation $m(\vartheta) = \mathbb{E}(Q_\vartheta)$ and the variance $v(\vartheta) = \mathbb{V}(Q_\vartheta)$ of Q_ϑ exist. In the infinite product model, let $M_n = \frac{1}{n} \sum_{i=1}^{n} X_i$ and $V_n^* = \frac{1}{n-1} \sum_{i=1}^{n} (X_i - M_n)^2$ be the unbiased estimators of m and v after n independent observations. Then, the sequences $(M_n)_{n \geq 1}$ and $(V_n^*)_{n \geq 2}$ are consistent.*

Proof. The consistency of the sequence (M_n) follows immediately from Theorem (5.7). (For this result to hold, only the existence of the expectations $m(\vartheta)$ is required.) To prove the consistency of (V_n^*), let ϑ be fixed and consider

$$\widetilde{V}_n = \frac{1}{n} \sum_{i=1}^{n} (X_i - m(\vartheta))^2$$

as well as $V_n = \frac{n-1}{n} V_n^*$. By the shift formula (7.10),

$$V_n = \widetilde{V}_n - (M_n - m(\vartheta))^2 \,,$$

and Theorem (5.7) implies that both

$$\widetilde{V}_n \xrightarrow{P_\vartheta} v(\vartheta) \quad \text{and} \quad (M_n - m(\vartheta))^2 \xrightarrow{P_\vartheta} 0.$$

Together with Lemma (5.8a), this shows that $V_n^* \xrightarrow{P_\vartheta} v(\vartheta)$. $\diamond$

As a rule, maximum likelihood estimators are also consistent. In the following theorem, we will make the simplifying assumption of unimodality.

(7.30) Theorem. Consistency of maximum likelihood estimators. *Consider a one-parameter standard model $(E, \mathscr{E}, Q_\vartheta : \vartheta \in \Theta)$ with likelihood function ρ. Suppose*

(a) *Θ is an open interval in $\mathbb{R}$, and $Q_\vartheta \neq Q_{\vartheta'}$ when $\vartheta \neq \vartheta'$,*

(b) *the n-fold product likelihood function $\rho^{\otimes n}(x, \vartheta) = \prod_{i=1}^{n} \rho(x_i, \vartheta)$ is unimodal in ϑ, for all $x \in E^{\mathbb{N}}$ and $n \geq 1$; that is, there exists a maximum likelihood estimator $T_n : E^{\mathbb{N}} \to \mathbb{R}$ such that the function $\vartheta \to \rho^{\otimes n}(x, \vartheta)$ is increasing for $\vartheta < T_n(x)$ and decreasing for $\vartheta > T_n(x)$.*

Then the sequence $(T_n)_{n \geq 1}$ of estimators of ϑ is consistent.

The assumption (b) of unimodality is certainly satisfied if $\log \rho(x_i, \cdot)$ is concave for each $x_i \in E$ with a slope that ranges from positive to negative values. For, this implies that $\log \rho^{\otimes n}(x, \cdot)$ exhibits the same property for arbitrary $x \in E^{\mathbb{N}}$, and the unimodality follows immediately. Figure 7.5 shows a typical unimodal function.

In the proof of Theorem (7.30), we will need the notion of relative entropy, which is of independent interest and will show up again in the context of test theory.

(7.31) Remark and Definition. *Relative entropy.* Let P and Q be two probability measures on a discrete or real event space $(E, \mathscr{E})$ with existing densities ρ and σ respectively. Define

$$H(P; Q) := \mathbb{E}_P \left(\log \frac{\rho}{\sigma} \right) = \int_E \rho(x) \log \frac{\rho(x)}{\sigma(x)} \, dx$$

if $P(\sigma = 0) = 0$, and $H(P; Q) := \infty$ otherwise. (Here, we set $0 \log 0 = 0$; in the discrete case the integral is to be replaced by a sum.) Then $H(P; Q)$ is well-defined, but possibly $= \infty$. Also, $H(P; Q) \geq 0$ with equality exactly for $P = Q$. $H(P; Q)$ is called the *relative entropy* of P with respect to Q; it was introduced by Kullback and Leibler (1951).

Proof. To prove that $H(P; Q)$ is well-defined, we need to show that the expectation $\mathbb{E}_P (\log \frac{\rho}{\sigma})$ is well-defined when $P(\sigma = 0) = 0$. For $x \in E$, let $f(x) = \rho(x)/\sigma(x)$ if $\sigma(x) > 0$, and $f(x) = 1$ otherwise. Since $P(\sigma > 0) = 1$, the product σf is also a density of P, and we can suppose without loss of generality that $\rho = \sigma f$.

For $s \geq 0$, set $\psi(s) = 1 - s + s \log s$. The function ψ is strictly convex and attains its minimum 0 exactly at the point 1; cf. Figure 11.3 on p. 298. In particular, $\psi \geq 0$, and this implies that the expectation $\mathbb{E}_Q (\psi \circ f) = \int_E \psi(f(x)) \sigma(x) \, dx \in [0, \infty]$ exists. By definition, we also know that the expectation $\mathbb{E}_Q (1 - f) = 1 - \mathbb{E}_P(1) = 0$ exists. By taking the difference, we obtain the existence of

$$\mathbb{E}_Q(f \log f) = \int_E \sigma(x) \, f(x) \, \log f(x) \, dx \in [0, \infty].$$

Since $\sigma f = \rho$, this shows the existence of $H(P; Q) \in [0, \infty]$.

If $H(P; Q) = 0$, then $\mathbb{E}_Q(\psi \circ f) = 0$ by the previous argument. Since $\psi \geq 0$, it follows (for example by the Markov inequality) that $Q(\psi \circ f = 0) = 1$. Since ψ has its only root at 1, we see that $Q(f = 1) = 1$, and this means that $Q = P$. $\diamond$

Proof of Theorem (7.30). Consider a fixed $\vartheta \in \Theta$ and let $\varepsilon > 0$ be so small that $\vartheta \pm \varepsilon \in \Theta$. By assumption (a) and Remark (7.31), we can find some $\delta > 0$ such that $\delta < H(Q_\vartheta; Q_{\vartheta \pm \varepsilon})$. Just as in Example (5.13), we will apply the law of large numbers to the log-likelihood functions $\log \rho_\vartheta^{\otimes n}$.

It suffices to show that

(7.32)
$$P_\vartheta \left(\frac{1}{n} \log \frac{\rho_\vartheta^{\otimes n}}{\rho_{\vartheta \pm \varepsilon}^{\otimes n}} > \delta \right) \to 1$$

as $n \to \infty$, because the assumption (b) of unimodality implies that

$$\bigcap_{\sigma = \pm 1} \left\{ \frac{1}{n} \log \frac{\rho_\vartheta^{\otimes n}}{\rho_{\vartheta + \sigma\varepsilon}^{\otimes n}} > \delta \right\} \subset \left\{ \rho_{\vartheta - \varepsilon}^{\otimes n} < \rho_\vartheta^{\otimes n} > \rho_{\vartheta + \varepsilon}^{\otimes n} \right\} \subset \left\{ \vartheta - \varepsilon < T_n < \vartheta + \varepsilon \right\};$$

see Figure 7.5.

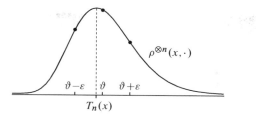

Figure 7.5. If the unimodal likelihood function $\rho^{\otimes n}(x,\cdot)$ for the outcome x is larger at the point ϑ than at the points $\vartheta \pm \varepsilon$, then its maximiser $T_n(x)$ must lie in the interval $]\vartheta - \varepsilon, \vartheta + \varepsilon[$.

To prove (7.32), we confine ourselves to the case when ε has positive sign. First suppose that $H(Q_\vartheta; Q_{\vartheta+\varepsilon}) < \infty$. As in the proof of (7.31), we then see that $f = \rho_\vartheta / \rho_{\vartheta+\varepsilon}$ is well-defined, and $\log f \in \mathcal{L}^1(Q_\vartheta)$. Theorem (5.7) therefore implies that

$$\frac{1}{n} \log \frac{\rho_\vartheta^{\otimes n}}{\rho_{\vartheta+\varepsilon}^{\otimes n}} = \frac{1}{n} \sum_{i=1}^{n} \log f(X_i) \xrightarrow{P_\vartheta} \mathbb{E}_\vartheta(\log f) = H(Q_\vartheta; Q_{\vartheta+\varepsilon})$$

and hence (7.32). Next, we consider the case that $H(Q_\vartheta; Q_{\vartheta+\varepsilon}) = \infty$, but still $Q_\vartheta(\rho_{\vartheta+\varepsilon} = 0) = 0$. Then f is still well-defined, and for every $c > 1$ we get that $h_c := \log \min(f, c) \in \mathcal{L}^1(Q_\vartheta)$. By Theorem (4.11c), $\mathbb{E}_\vartheta(h_c) \uparrow H(Q_\vartheta; Q_{\vartheta+\varepsilon}) = \infty$ as $c \uparrow \infty$. Therefore, there is a c such that $\mathbb{E}_\vartheta(h_c) > \delta$. As in the first case, Theorem (5.7) yields

$$P_\vartheta \left(\frac{1}{n} \log \frac{\rho_\vartheta^{\otimes n}}{\rho_{\vartheta+\varepsilon}^{\otimes n}} > \delta \right) \geq P_\vartheta \left(\frac{1}{n} \sum_{i=1}^{n} h_c(X_i) > \delta \right) \to 1$$

as $n \to \infty$. Finally, suppose that $Q_\vartheta(\rho_{\vartheta+\varepsilon} = 0) =: a > 0$. Since $Q_\vartheta(\rho_\vartheta > 0) = 1$, it follows that

$$P_\vartheta \left(\frac{1}{n} \log \frac{\rho_\vartheta^{\otimes n}}{\rho_{\vartheta+\varepsilon}^{\otimes n}} = \infty \right) = P_\vartheta \left(\rho_{\vartheta+\varepsilon}^{\otimes n} = 0 \right) = 1 - (1-a)^n \to 1,$$

which again implies (7.32). $\diamond$

Here are a few examples of the above consistency theorem.

(7.33) Example. *Estimating the Poisson parameter.* How large is the average number of insurance claims that a car insurance receives per year, or what is the average number of people that use a saver ticket on the train? Since these quantities can be assumed to be Poisson distributed, we consider the Poisson model $E = \mathbb{Z}_+$, $\Theta =]0, \infty[$, $Q_\vartheta = \mathcal{P}_\vartheta$ with the likelihood function $\rho(x, \vartheta) = e^{-\vartheta} \vartheta^x / x!$. Clearly, $\log \rho(x, \vartheta) = x \log \vartheta - \vartheta - \log x!$ is concave in ϑ. So, assumption (b) of the previous theorem holds with $T_n = M_n := \frac{1}{n} \sum_{i=1}^{n} X_i$, the sample mean. This shows that the sequence (T_n) is consistent, as follows also from Theorem (7.29).

(7.34) Example. *Estimating the parameter of an exponential distribution.* How long does it typically take until the first customer arrives at a counter? Or how long until the first call arrives at a telephone switchboard? In a first approximation, these waiting times can be assumed to be exponentially distributed; so we have to find an estimator of the parameter of the exponential distribution. The associated likelihood function is $\rho(x, \vartheta) = \vartheta e^{-\vartheta x}$, where $x \in E = [0, \infty[$ and $\vartheta \in \Theta =]0, \infty[$. Clearly, the function $\log \rho(x, \vartheta) = -\vartheta x + \log \vartheta$ is concave in ϑ, so the assumption (b) of unimodality in the previous theorem is satisfied. As noticed in Example (7.12), the maximum likelihood estimator based on n independent observations is $T_n := 1/M_n$. Hence, both Theorem (7.29) as well as Theorem (7.30) yield its consistency.

(7.35) Example. *Estimating the range of a uniform distribution.* Recall the situation of Example (7.3), where on a TV show the candidates have to guess the range of a random number generator. According to Example (7.8), $\tilde{T}_n = \max(X_1, \ldots, X_n)$ is the maximum likelihood estimator after n independent observations. In fact, the n-fold product likelihood function can be written in the form $\vartheta^{-n} 1_{[\tilde{T}_n, \infty[}(\vartheta)$, which reveals its unimodal structure. Hence, Theorem (7.30) once again yields the consistency of the sequence $(\tilde{T}_n)$, which was already shown directly in Example (7.3).

7.7   Bayes Estimators

We will close this chapter with an excursion into so-called Bayesian statistics, an alternative approach to the problem of estimation. Instead of minimising the mean squared error uniformly for all ϑ, the aim is here to minimise a mean squared error that is averaged over ϑ. The following example provides some motivation.

(7.36) Example. *Car insurance.* A customer approaches an insurance company to get a quote for a new car insurance. He has a particular driving style that leads, with probability $\vartheta \in [0, 1]$, to at least one claim per year. Obviously, the insurance agent does not know ϑ, but he has statistics about the claim frequencies for the total population. These provide him with an a priori risk assessment, i.e., a probability measure α on $\Theta = [0, 1]$. Let us assume for simplicity that α is 'smooth', in that it has a Lebesgue density α. The agent's presumed (and thus subjective) probability that the customer will have a claim in k out of n years is then given by

$$\int_0^1 d\vartheta \, \alpha(\vartheta) \, \mathcal{B}_{n,\vartheta}(\{k\}) \,.$$

(This corresponds to the claim frequency in a two-step model, where first a claim probability ϑ is chosen at random according to α, and then n Bernoulli experiments are performed with parameter ϑ.)

Suppose now that the customer would like a new contract after n years, and the agent needs to re-assess the risk. He then knows that there was a claim in $x \in \{0, \ldots, n\}$ years, say. How is he going to adapt his a priori assessment α to this information? In analogy to Bayes' formula (3.3b), it is plausible to replace the prior density α by the posterior density

$$\pi_x(\vartheta) = \frac{\alpha(\vartheta)\, \mathcal{B}_{n,\vartheta}(\{x\})}{\int_0^1 dp\, \alpha(p)\, \mathcal{B}_{n,p}(\{x\})},$$

namely the conditional distribution density of ϑ given x years with claims.

Definition. Let $(\mathcal{X}, \mathscr{F}, P_\vartheta : \vartheta \in \Theta)$ be a parametric standard model with likelihood function $\rho : \mathcal{X} \times \Theta \to [0, \infty[$. Then, every probability density α on $(\Theta, \mathscr{B}_\Theta^d)$ is called a *prior density* and the associated probability measure $\boldsymbol{\alpha}$ on $(\Theta, \mathscr{B}_\Theta^d)$ a *prior distribution*. Furthermore, for every $x \in \mathcal{X}$, the density

$$\pi_x(\vartheta) = \frac{\alpha(\vartheta)\, \rho(x, \vartheta)}{\int_\Theta dt\, \alpha(t)\, \rho(x, t)}$$

on Θ is called the *posterior density*, and the associated probability measure $\boldsymbol{\pi}_x$ the *posterior distribution*, relative to the observed value x and the prior distribution $\boldsymbol{\alpha}$.

(In order that $\pi_x(\vartheta)$ is well-defined, we need to assume that ρ is strictly positive and measurable jointly in both variables, i.e., relative to $\mathscr{F} \otimes \mathscr{B}_\Theta^d$. If Θ is discrete, and so $\mathscr{B}_\Theta^d = \mathscr{P}(\Theta)$, the latter condition is automatically satisfied, and the integral must be replaced by a sum over Θ; α and π_x are then discrete densities.)

In statistics, the so-called Bayesian or subjective school of thought interprets the prior distribution $\boldsymbol{\alpha}$ as a subjective assessment of the situation. This certainly makes sense when the probability measures P_ϑ are interpreted according to the subjective rather than the frequentist interpretation. Note, however, that 'subjective' does not mean 'arbitrary'. The prior distribution of the insurance agent in Example (7.36) depends on the claim statistics, which provide an 'objective' information.

What does this mean for Example (7.36)? Let us assume that nothing is known about ϑ, and thus choose the prior distribution $\boldsymbol{\alpha} = \mathcal{U}_{[0,1]}$. Then, $\alpha(\vartheta) = 1$ for all $\vartheta \in [0, 1]$, and therefore

$$\pi_x(\vartheta) = \frac{\binom{n}{x} \vartheta^x (1-\vartheta)^{n-x}}{\int_0^1 \binom{n}{x} s^x (1-s)^{n-x}\, ds} = \frac{\vartheta^x (1-\vartheta)^{n-x}}{\mathrm{B}(x+1, n-x+1)}$$

for all $x \in \{0, \ldots, n\}$ and $\vartheta \in [0, 1]$; recall the definition (2.21) of the beta function. Hence, $\pi_x(\vartheta)$ is the density of the beta distribution with parameters $x+1$ and $n-x+1$, which means that

$$\boldsymbol{\pi}_x = \boldsymbol{\beta}_{x+1, n-x+1}.$$

What can one say about π_x when the number n of observations is large? To indicate the n-dependence of π_x, we now write $\pi_x^{(n)}$. Then

$$\mathbb{E}(\pi_x^{(n)}) = S_n(x) := \frac{x+1}{n+2} \quad \text{and} \quad \mathbb{V}(\pi_x^{(n)}) = \frac{(x+1)(n-x+1)}{(n+2)^2(n+3)} \leq \frac{1}{n}$$

by Example (4.29). That is, the expectation of $\pi_x^{(n)}$ is the biased estimator of ϑ we considered in Example (7.14), and the variances converge to 0 as $n \to \infty$. In view of Chebyshev's inequality (5.5), the latter means that $\pi_x^{(n)}$ concentrates more and more around $S_n(x)$ as the number n of observations increases. So, with growing information, the initial uncertainty disappears, and deviations from the average become less and less likely. Taking also into account that $S_n \to \vartheta$ in $\mathcal{B}_{n,\vartheta}$-probability, we arrive at the consistency statement

$$(7.37) \qquad \mathcal{B}_{n,\vartheta}\left(x : \pi_x^{(n)}([\vartheta - \varepsilon, \vartheta + \varepsilon]) \geq 1 - \varepsilon\right) \xrightarrow[n \to \infty]{} 1$$

for all $\varepsilon > 0$ and $\vartheta \in [0, 1]$; compare Figure 7.6.

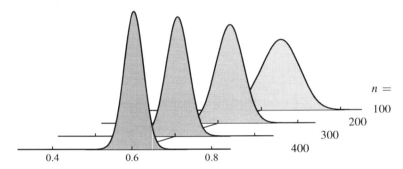

Figure 7.6. The posterior distributions $\pi_{x_n}^{(n)}$ in (7.37) for x_n chosen randomly according to $\mathcal{B}_{n,0.6}$ plotted for various n.

Let us now return to our initial question: How will the insurance agent estimate the unknown claim probability ϑ when he knows that a claim happened to occur in x years? An obvious strategy is again to minimise a suitable mean squared error.

Definition. Let $(\mathcal{X}, \mathcal{F}, P_\vartheta : \vartheta \in \Theta)$ be a parametric standard model with a likelihood function $\rho > 0$ that is measurable jointly in both variables, and let $\boldsymbol{\alpha}$ be a prior distribution on Θ with density α. Furthermore, let $\tau : \Theta \to \mathbb{R}$ be a measurable real characteristic with $\mathbb{E}_\alpha(\tau^2) < \infty$. An estimator $T : \mathcal{X} \to \mathbb{R}$ of τ is called a *Bayes estimator* relative to the prior distribution $\boldsymbol{\alpha}$, if the mean squared error

$$\mathscr{E}_T(\alpha) := \mathbb{E}_\alpha(\mathscr{E}_T) = \int_\Theta d\vartheta\, \alpha(\vartheta) \int_\mathcal{X} dx\, \rho(x, \vartheta)\, (T(x) - \tau(\vartheta))^2$$

(where the mean is taken over both x and ϑ) is minimal among all estimators of τ.

Instead of the mean squared error, one could use any other sensible risk function, but we will not pursue this possibility further here.

The following theorem shows that Bayes estimators can be obtained directly from the posterior distribution. We set $P_\alpha = \int d\vartheta\, \alpha(\vartheta)\, P_\vartheta$, which means that P_α is the probability measure on $\mathcal{X}$ with density $\rho_\alpha(x) := \int_\Theta \rho(x,\vartheta)\, \alpha(\vartheta)\, d\vartheta$.

(7.38) Theorem. *Bayes estimator and posterior distribution. Under the conditions of the definition, the Bayes estimator of τ corresponding to the prior density α is P_α-almost surely unique, and is given by*

$$T(x) := \mathbb{E}_{\pi_x}(\tau) = \int_\Theta \pi_x(\vartheta)\, \tau(\vartheta)\, d\vartheta\,, \quad x \in \mathcal{X}.$$

Here, π_x is the posterior distribution corresponding to α and x. (For discrete Θ, the integral has to be replaced by a sum.)

Proof. By definition, $\alpha(\vartheta)\rho_\vartheta(x) = \rho_\alpha(x)\pi_x(\vartheta)$. So, if T is defined as above and S is an arbitrary estimator, we obtain by interchanging the order of integration (Fubini's theorem, see e.g. [56]):

$$\mathscr{E}_S(\alpha) - \mathscr{E}_T(\alpha)$$
$$= \int_\mathcal{X} dx\, \rho_\alpha(x) \int_\Theta d\vartheta\, \pi_x(\vartheta)\big[S(x)^2 - 2\,S(x)\,\tau(\vartheta) - T(x)^2 + 2\,T(x)\,\tau(\vartheta)\big]$$
$$= \int_\mathcal{X} dx\, \rho_\alpha(x)\big[S(x)^2 - 2\,S(x)\,T(x) - T(x)^2 + 2\,T(x)^2\big]$$
$$= \int_\mathcal{X} dx\, \rho_\alpha(x)\big[S(x) - T(x)\big]^2 \geq 0.$$

This yields the claim. $\diamond$

For Example (7.36), we therefore obtain: The estimator $S(x) = (x+1)/(n+2)$ known from Example (7.14) is the unique Bayes estimator for the prior distribution $\alpha = \mathcal{U}_{[0,1]}$. Let us give a second example.

(7.39) Example. *Bayes estimate of the expectation of a normal distribution when the variance is known.* Let $(\mathbb{R}^n, \mathscr{B}^n, \mathcal{N}_{\vartheta,v}{}^{\otimes n} : \vartheta \in \mathbb{R})$ be the n-fold Gaussian product model with fixed variance $v > 0$, which has the likelihood function

$$\rho(x,\vartheta) = (2\pi v)^{-n/2} \exp\Big[-\frac{1}{2v} \sum_{i=1}^n (x_i - \vartheta)^2\Big].$$

We choose a prior distribution which is also normal, namely $\alpha = \mathcal{N}_{m,u}$ for $m \in \mathbb{R}$ and $u > 0$. Using the maximum likelihood estimator $M(x) = \frac{1}{n} \sum_{i=1}^n x_i$ and appropriate constants c_x, $c_x' > 0$, we can then write

$$\pi_x(\vartheta) = c_x \, \exp\left[-\frac{1}{2u}(\vartheta - m)^2 - \frac{1}{2v}\sum_{i=1}^{n}(x_i - \vartheta)^2\right]$$

$$= c_x' \, \exp\left[-\frac{\vartheta^2}{2}\left(\frac{1}{u} + \frac{n}{v}\right) + \vartheta\left(\frac{m}{u} + \frac{n}{v}M(x)\right)\right]$$

$$= \phi_{T(x),u^*}(\vartheta) \, ;$$

here $u^* = 1/\left(\frac{1}{u} + \frac{n}{v}\right)$ and

$$T(x) = \frac{\frac{1}{u}m + \frac{n}{v}M(x)}{\frac{1}{u} + \frac{n}{v}} \, .$$

(Since π_x and $\phi_{T(x),u^*}$ are both probability densities, the factor c_x'' that appears in the last step is necessarily equal to 1.) Hence, we obtain $\pi_x = \mathcal{N}_{T(x),u^*}$ and, in particular, $T(x) = \mathbb{E}(\pi_x)$. Theorem (7.38) thus tells us that T is the Bayes estimator corresponding to the prior distribution $\alpha = \mathcal{N}_{m,u}$. Note that T is a convex combination of m and M, which gives M more and more weight as either the number n of observations or the prior uncertainty u increases. In the limit as $n \to \infty$, we obtain an analogue of the consistency statement (7.37); see also Problem 7.29.

Problems

7.1   Forest mushrooms are examined in order to determine their radiation burden. For this purpose, n independent samples are taken, and, for each, the number of decays in a time unit is registered by means of a Geiger counter. Set up a suitable statistical model and find an unbiased estimator of the radiation burden.

7.2   *Shifted uniform distributions.* Consider the product model $(\mathbb{R}^n, \mathcal{B}^n, \mathcal{U}_\vartheta^{\otimes n} : \vartheta \in \mathbb{R})$, where $\mathcal{U}_\vartheta$ is the uniform distribution on the interval $[\vartheta - \frac{1}{2}, \vartheta + \frac{1}{2}]$. Show that

$$M = \frac{1}{n}\sum_{i=1}^{n}X_i \quad \text{and} \quad T = \frac{1}{2}\left(\max_{1 \leq i \leq n} X_i + \min_{1 \leq i \leq n} X_i\right)$$

are unbiased estimators of ϑ. *Hint:* Use the symmetry of $\mathcal{U}_\vartheta$ for $\vartheta = 0$.

7.3^{S} *Discrete uniform distribution model.* A lottery drum contains N lots labelled with the numbers $1, 2, \ldots, N$. Little Bill, who is curious about the total number N of lots, uses an unwatched moment to take a lot at random, read off its number, and return it to the drum. He repeats this n times.

(a) Find a maximum likelihood estimator T of N that is based on the observed numbers $x_1, \ldots, x_n$. Is it unbiased?

(b) Find an approximation to the relative expectation $\mathbb{E}_N(T)/N$ for large N. *Hint:* Treat a suitable expression as a Riemann sum.

7.4 Consider again the setting of Problem 7.3. This time little Bill draws n lots *without* replacing them. Find the maximum likelihood estimator T of N, calculate $\mathbb{E}_N(T)$, and give an unbiased estimator of N.

7.5 Determine a maximum likelihood estimator

(a) in the situation of Problem 7.1,

(b) in the product model $(]0, 1[^n, \mathscr{B}_{]0,1[}^{\otimes n}, Q_\vartheta^{\otimes n} : \vartheta > 0)$, where $Q_\vartheta = \beta_{\vartheta,1}$ is the probability measure on $]0, 1[$ with density $\rho_\vartheta(x) = \vartheta x^{\vartheta - 1}$,

and check whether it is unique.

7.6 [S] *Phylogeny.* When did the most recent common ancestor V of two organisms A and B live? In the 'infinite sites mutation model', it is assumed that the mutations of a fixed gene occur along the lines of descent from V to A and V to B at the times of independent Poisson processes with known intensity (i.e., mutation rate) $\mu > 0$. It is also assumed that each mutation changes a different nucleotide in the gene sequence. Let x be the observed number of nucleotides that differ in the sequences of A and B. What is your maximum likelihood estimate of the age of V? First specify the statistical model!

7.7 A certain butterfly species is split into three types 1, 2 and 3, which occur in the geno-typical proportions $p_1(\vartheta) = \vartheta^2$, $p_2(\vartheta) = 2\vartheta(1 - \vartheta)$ and $p_3(\vartheta) = (1 - \vartheta)^2$, $0 \leq \vartheta \leq 1$. Among n butterflies of this species you have caught, you find n_i specimens of type i. Determine a maximum likelihood estimator T of ϑ. (Do not forget to consider the extreme cases $n_1 = n$ and $n_3 = n$.)

7.8 [S] At the summer party of the rabbit breeders' association, there is a prize draw for K rabbits. The organisers print $N \geq K$ tickets, of which K are winning, the remaining ones are blanks. Much to his mum's dismay, little Bill brings x rabbits home, $1 \leq x \leq K$. How many tickets did he probably buy? Give an estimate using the maximum likelihood method.

7.9 Consider the geometric model $(\mathbb{Z}_+, \mathscr{P}(\mathbb{Z}_+), \mathcal{G}_\vartheta : \vartheta \in]0, 1])$. Determine a maximum likelihood estimator of the unknown parameter ϑ. Is it unbiased?

7.10 Consider the statistical product model $(\mathbb{R}^n, \mathscr{B}^n, Q_\vartheta^{\otimes n} : \vartheta \in \mathbb{R})$, where Q_ϑ is the so-called *two-sided exponential distribution* or *Laplace distribution* centred at ϑ, namely the probability measure on $(\mathbb{R}, \mathscr{B})$ with density

$$\rho_\vartheta(x) = \tfrac{1}{2} e^{-|x - \vartheta|}, \quad x \in \mathbb{R}.$$

Find a maximum likelihood estimator of ϑ and show that it is unique for even n only. *Hint:* Use Problem 4.15.

7.11 *Estimate of a transition matrix.* Let $X_0, \ldots, X_n$ be a Markov chain with finite state space E, known initial distribution α and unknown transition matrix Π. For $a, b \in E$, let $L^{(2)}(a, b) = |\{1 \leq i \leq n : X_{i-1} = a, X_i = b\}|/n$ be the relative frequency of the letter pair (a, b) in the 'random word' $(X_0, \ldots, X_n)$. The random matrix $L^{(2)} = (L^{(2)}(a, b))_{a,b \in E}$ is called the *empirical pair distribution*. Define the empirical transition matrix T on E by

$$T(a, b) = L^{(2)}(a, b)/L(a) \quad \text{if } L(a) := \sum_{c \in E} L^{(2)}(a, c) > 0,$$

and arbitrarily otherwise. Specify the statistical model and show that T is a maximum likelihood estimator of Π. *Hint:* You can argue as in Example (7.7).

7.12 Consider the binomial model of Example (7.14). For any given n, find an estimator of ϑ for which the mean squared error does not depend on ϑ.

7.13 *Unbiased estimators can be bad.* Consider the model $(\mathbb{N}, \mathscr{P}(\mathbb{N}), P_\vartheta : \vartheta > 0)$ of the conditional Poisson distributions

$$P_\vartheta(\{n\}) = \mathcal{P}_\vartheta(\{n\}|\mathbb{N}) = \frac{\vartheta^n}{n!\,(e^\vartheta - 1)}\,, \quad n \in \mathbb{N}.$$

Show that the only unbiased estimator of $\tau(\vartheta) = 1 - e^{-\vartheta}$ is the (useless) estimator $T(n) = 1 + (-1)^n$, $n \in \mathbb{N}$.

7.14 ^{S} *Uniqueness of best estimators.* In a statistical model $(\mathcal{X}, \mathscr{F}, P_\vartheta : \vartheta \in \Theta)$, let S, T be two best unbiased estimators of a real characteristic $\tau(\vartheta)$. Show that $P_\vartheta(S = T) = 1$ for all ϑ. *Hint:* Consider the estimator $(S + T)/2$.

7.15 Consider the negative binomial model $(\mathbb{Z}_+, \mathscr{P}(\mathbb{Z}_+), \overline{B}_{r,\vartheta} : 0 < \vartheta < 1)$ for given $r > 0$. Determine a best estimator of $\tau(\vartheta) = 1/\vartheta$ and determine its variance explicitly for each ϑ.

7.16 Consider the n-fold Gaussian product model $(\mathbb{R}^n, \mathscr{B}^n, \mathcal{N}_{m,\vartheta}{}^{\otimes n} : \vartheta > 0)$ with known expectation $m \in \mathbb{R}$ and unknown variance. Show that the statistic

$$T = \sqrt{\frac{\pi}{2}}\,\frac{1}{n}\sum_{i=1}^{n}|X_i - m|$$

on $\mathbb{R}^n$ is an unbiased estimator of $\tau(\vartheta) = \sqrt{\vartheta}$, but that there is no ϑ at which $\mathbb{V}_\vartheta(T)$ reaches the Cramér–Rao bound $\tau'(\vartheta)^2/I(\vartheta)$.

7.17 ^{S} *Randomised response.* In a survey on a delicate topic ('Do you take hard drugs?') it is difficult to protect the privacy of the people questioned and at the same time to get reliable answers. That is why the following 'unrelated question method' was suggested. A deck of cards is prepared such that half of the cards contain the delicate question A and the other half a harmless question B, which is unrelated to question A ('Did you go to the cinema last week?'). The interviewer asks the candidate to shuffle the cards, then to choose a card without showing it to anyone, and to answer the question found on this card. The group of people questioned contains a known proportion p_B of people affirming question B (cinema-goers). Let $\vartheta = p_A$ be the unknown probability that the sensitive question A is answered positively. Suppose n people are questioned independently. Specify the statistical model, find a best estimator of ϑ, and determine its variance.

7.18 ^{S} *Shifted uniform distributions.* Consider the situation of Problem 7.2. Compute the variances $\mathbb{V}_\vartheta(M)$ and $\mathbb{V}_\vartheta(T)$ of the estimators M and V, and decide which of them you would recommend for practical use. *Hint:* For $n \geq 3$ and $\vartheta = 1/2$, determine first the joint distribution density of $\min_{1 \leq i \leq n} X_i$ and $\max_{1 \leq i \leq n} X_i$, and then the distribution density of T. Use also (2.23).

7.19 ^{S} *Sufficiency and completeness.* Let $(\mathcal{X}, \mathscr{F}, P_\vartheta : \vartheta \in \Theta)$ be a statistical model and $T : \mathcal{X} \to \Sigma$ a statistic with (for simplicity) countable range Σ. T is called *sufficient* if there exists a family $\{Q_s : s \in \Sigma\}$ of probability measures on $(\mathcal{X}, \mathscr{F})$ that do not depend on ϑ and satisfy $P_\vartheta(\cdot | T = s) = Q_s$ whenever $P_\vartheta(T = s) > 0$. T is called *complete* if $g \equiv 0$ is the only function $g : \Sigma \to \mathbb{R}$ such that $\mathbb{E}_\vartheta(g \circ T) = 0$ for all $\vartheta \in \Theta$. Let τ be a real characteristic. Show the following.

(a) *Rao–Blackwell 1945/47.* If T is sufficient, then every unbiased estimator S of τ can be improved as follows: Let $g_S(s) := \mathbb{E}_{Q_s}(S)$ for $s \in \Sigma$; then the estimator $g_S \circ T$ is unbiased and satisfies $\mathbb{V}_\vartheta(g_S \circ T) \leq \mathbb{V}_\vartheta(S)$ for all $\vartheta \in \Theta$.

(b) *Lehmann–Scheffé 1950.* If T is sufficient and complete and $g \circ T$ is an unbiased estimator of τ, then $g \circ T$ is in fact a best estimator of τ.

7.20 Let $(\mathcal{X}, \mathscr{F}, P_\vartheta : \vartheta \in \Theta)$ be an exponential model relative to a statistic T, and suppose for simplicity that T takes values in $\Sigma := \mathbb{Z}_+$. Show that T is sufficient and complete.

7.21 Recall the situation of Problem 7.3 and show that the maximum likelihood estimator T to be determined there is sufficient and complete.

7.22 *Relative entropy and Fisher information.* Let $(\mathcal{X}, \mathscr{F}, P_\vartheta : \vartheta \in \Theta)$ be a regular statistical model with finite sample space $\mathcal{X}$. Show that

$$\lim_{\varepsilon \to 0} \varepsilon^{-2} H(P_{\vartheta+\varepsilon}; P_\vartheta) = I(\vartheta)/2 \quad \text{for all } \vartheta \in \Theta.$$

7.23 [S] *Exponential families and maximisation of entropy.* Let $\{P_\vartheta : \vartheta \in \mathbb{R}\}$ be an exponential family relative to a statistic T on a finite set $\mathcal{X}$. According to physical convention, we assume that $a(\vartheta) = -\vartheta$ and $h \equiv 1$. That is, the likelihood function takes the form

(7.40) $$\rho_\vartheta(x) = \exp[-\vartheta\, T(x)]/Z(\vartheta), \quad x \in \mathcal{X},$$

with $Z(\vartheta) = \exp b(\vartheta)$. Let $t_- = \min_{x \in \mathcal{X}} T(x)$, $t_+ = \max_{x \in \mathcal{X}} T(x)$ and $t \in\,]t_-, t_+[$. Show:

(a) There exists a unique $\vartheta = \vartheta(t) \in \mathbb{R}$ with $\mathbb{E}_\vartheta(T) = t$.

(b) Consider an arbitrary probability measure P on $\mathcal{X}$ with $\mathbb{E}_P(T) = t$ and, in analogy to (5.14), its entropy $H(P) := -\sum_{x \in \mathcal{X}} P(\{x\}) \log P(\{x\})$. Also, let $\vartheta = \vartheta(t)$. Then

$$H(P) \leq H(P_\vartheta) = \vartheta t + \log Z(\vartheta)$$

with equality for $P = P_\vartheta$ only. *Hint:* Compute the relative entropy $H(P; P_\vartheta)$.

Note: In view of this maximum entropy property, distributions of the form (7.40) are taken as equilibrium distributions in statistical physics. They are called *Boltzmann–Gibbs distributions* in honour of Ludwig Boltzmann (1844–1906) and J. Willard Gibbs (1839–1903). In the physical setting, $T(x)$ is the internal energy of a state x, and $1/\vartheta(t)$ is proportional to the temperature that is associated to the mean energy t.

7.24 [S] *Estimation of the mutation rate in the infinite alleles model.* For given $n \geq 1$, consider Ewens' sampling distribution $\rho_{n,\vartheta}$ with unknown mutation rate $\vartheta > 0$, as defined in Problem 6.5a. Show the following.

(a) $\{\rho_{n,\vartheta} : \vartheta > 0\}$ is an exponential family and $K_n(x) := \sum_{i=1}^n x_i$ (the number of different clans in the sample) is a best unbiased estimator of $\tau_n(\vartheta) := \sum_{i=0}^{n-1} \frac{\vartheta}{\vartheta+i}$.

(b) The maximum likelihood estimator of ϑ is $T_n := \tau_n^{-1} \circ K_n$. (Note that τ_n is strictly increasing.)

(c) The sequence $(K_n / \log n)_{n \geq 1}$ of estimators of ϑ is asymptotically unbiased and consistent. But the squared error of $K_n / \log n$ is of order $1/\log n$, so it converges to 0 very slowly (in spite of its optimality).

7.25 *Estimation by the method of moments.* Let $(\mathbb{R}, \mathscr{B}, Q_\vartheta : \vartheta \in \Theta)$ be a real-valued statistical model, and let $r \in \mathbb{N}$ be given. Suppose that for each $\vartheta \in \Theta$ and every $k \in \{1, \ldots, r\}$, the kth moment $m_k(\vartheta) := \mathbb{E}_\vartheta(\mathrm{Id}_{\mathbb{R}}^k)$ of Q_ϑ exists. Furthermore, let $g : \mathbb{R}^r \to \mathbb{R}$ be continuous, and consider the real characteristic $\tau(\vartheta) := g(m_1(\vartheta), \ldots, m_r(\vartheta))$. In the associated infinite product model $(\mathbb{R}^{\mathbb{N}}, \mathscr{B}^{\otimes \mathbb{N}}, Q_\vartheta^{\otimes \mathbb{N}} : \vartheta \in \Theta)$, one can then define the estimator

$$T_n := g\Big(\frac{1}{n}\sum_{i=1}^{n} X_i, \frac{1}{n}\sum_{i=1}^{n} X_i^2, \ldots, \frac{1}{n}\sum_{i=1}^{n} X_i^r\Big)$$

of τ, which is based on the first n observations. Show that the sequence (T_n) is consistent.

7.26 [S] Consider the two-sided exponential model of Problem 7.10. For each $n \geq 1$, let T_n be a maximum likelihood estimator based on n independent observations. Show that the sequence (T_n) is consistent.

7.27 Verify the consistency statement (7.37) for the posterior distributions in the binomial model of Example (7.36).

7.28 *Dirichlet and multinomial distributions.* As a generalisation of Example (7.36), consider an urn model in which each ball has one of a finite number s of colours (instead of only two). Let Θ be the set of all probability densities on $\{1, \ldots, s\}$. Suppose that the prior distribution α on Θ is the *Dirichlet distribution* $\mathcal{D}_\rho$ for some parameter $\rho \in \,]0, \infty[^s$, which is defined by the equation

$$\mathcal{D}_\rho(A) = \frac{\Gamma\big(\sum_{i=1}^{s} \rho(i)\big)}{\prod_{i=1}^{s} \Gamma(\rho(i))} \int 1_A(\vartheta) \prod_{i=1}^{s} \vartheta_i^{\rho(i)-1} d\vartheta_1 \ldots d\vartheta_{s-1}, \quad A \in \mathscr{B}_\Theta.$$

Here, the integral runs over all $(\vartheta_1, \ldots, \vartheta_{s-1})$ for which $\vartheta := (\vartheta_1, \ldots, \vartheta_{s-1}, 1 - \sum_{i=1}^{s-1} \vartheta_i)$ belongs to Θ. The fact that $\mathcal{D}_\rho$ is indeed a probability measure will follow for instance from Problem 9.8. (In the case $\rho \equiv 1$, $\mathcal{D}_\rho$ is the uniform distribution on Θ.) We take a sample of size n with replacement. For each colour composition $\vartheta \in \Theta$, the colour histogram then has the multinomial distribution $\mathcal{M}_{n,\vartheta}$. Determine the associated posterior distribution.

7.29 [S] *Asymptotics of the residual uncertainty as the information grows.* Consider Example (7.39) in the limit as $n \to \infty$. Let $x = (x_1, x_2, \ldots)$ be a sequence of observed values in $\mathbb{R}$ such that the sequence of averages $M_n(x) = \frac{1}{n}\sum_{i=1}^{n} x_i$ remains bounded. Let $\pi_x^{(n)}$ be the posterior density corresponding to the outcomes $(x_1, \ldots, x_n)$ and the prior distribution $\mathcal{N}_{m,u}$. Let $\theta_{n,x}$ be a random variable with distribution $\pi_x^{(n)}$. Show that the rescaled random variables $\sqrt{n/v}\,(\theta_{n,x} - M_n(x))$ converge in distribution to $\mathcal{N}_{0,1}$.

7.30 *Gamma and Poisson distribution.* Let $(\mathbb{Z}_+^n, \mathscr{P}(\mathbb{Z}_+^n), \mathcal{P}_\vartheta^{\otimes n} : \vartheta > 0)$ be the n-fold Poisson product model. Suppose the prior distribution is given by $\alpha = \Gamma_{a,r}$, the gamma distribution with parameters $a, r > 0$. Find the posterior density π_x for each $x \in \mathbb{Z}_+^n$, and determine the Bayes estimator of ϑ.

Chapter 8

Confidence Regions

Estimators are useful for getting a first idea of the true value of an unknown quantity, but have the drawback that one does not know how reliable the estimates are. To take care of the unpredictability of chance, one should not specify a single value, but rather a whole region that depends on the observed outcome and can be guaranteed to contain the true value with sufficiently large certainty. Such so-called confidence regions are the subject of this chapter.

8.1 Definition and Construction

To begin with, consider again Example (7.6) about tossing a drawing pin. (This is our toy model of an independently repeated experiment that involves two alternatives with unknown probability of success.) What is the probability $\vartheta \in [0, 1]$ that the pin lands on its tip? Suppose Alice and Bob each throw it $n = 100$ times. Alice finds that the drawing pin lands on its tip $x = 40$ times, say, whereas Bob observes only $x' = 30$ 'successes'. So Alice estimates $\vartheta = x/n = 0.4$, but Bob insists on $\vartheta = x'/n = 0.3$. Who is right? Obviously neither of them! Actually, even if it should turn out that indeed $\vartheta = 0.4$, this would only mean that Alice made a lucky guess. To obtain more reliable conclusions, the random fluctuations of the estimates must be taken into account, and error probabilities must be specified. This can be done by making statements of the form 'With a certainty of 95%, ϑ lies in the (random) interval $]T - \varepsilon, T + \varepsilon[$'. Here T is an appropriate estimator and ε a suitable error bound.

Definition. Let $(\mathcal{X}, \mathcal{F}, P_\vartheta : \vartheta \in \Theta)$ be a statistical model and Σ an arbitrary set. Suppose $\tau : \Theta \to \Sigma$ is a characteristic of the parameter that is to be determined, and $0 < \alpha < 1$. A *confidence region for τ with error level α (or confidence level $1-\alpha$)* is a mapping $C : \mathcal{X} \to \mathcal{P}(\Sigma)$ that assigns to each possible outcome $x \in \mathcal{X}$ a set $C(x) \subset \Sigma$ so that

$$(8.1) \qquad \inf_{\vartheta \in \Theta} P_\vartheta \big(x \in \mathcal{X} : C(x) \ni \tau(\vartheta) \big) \geq 1 - \alpha .$$

To make sure that these probabilities are well-defined, it is also assumed that all sets $\{C(\cdot) \ni s\} := \{x \in \mathcal{X} : C(x) \ni s\}$ with $s \in \Sigma$ belong to the σ-algebra $\mathcal{F}$. If $\Sigma = \mathbb{R}$ and each $C(x)$ is an interval, one speaks of a *confidence interval*. C is then also called an *interval estimator*, as opposed to the usual estimators, which, for the sake of distinction, are also named *point estimators*.

This definition requires two comments.

▷ Condition (8.1) is certainly satisfied when $C(x) = \Sigma$ for all x. Such a large confidence region, however, is completely useless! After all, we would like to obtain as much information about $\tau(\vartheta)$ as possible, and so the sets $C(x)$ should be chosen *as small as possible*. On the other hand, the error probability α should also be as small as possible. These two demands are in conflict: It is impossible to keep α small and, at the same time, all sets $C(\cdot)$. In fact, by condition (8.1), the smaller we choose α, the larger the sets $C(\cdot)$ must be. The point is therefore to find the right balance between the error probability α and the size of the regions $C(\cdot)$. This can only be done as the case arises.

▷ To avoid a possible misunderstanding, consider again the experiment with the drawing pin. If Alice obtains the estimate $T(x) = 0.4$ for $\tau(\vartheta) := \vartheta$, and so (by some rule) specifies the confidence interval $]0.3, 0.5[$ with confidence level 90%, then this does *not* mean that ϑ belongs to $]0.3, 0.5[$ in 90% of all cases. Such an interpretation would tacitly assume that ϑ is random. However, although ϑ is unknown, it has a specific value (which characterises the actual behaviour of the drawing pin and shall be determined by the experiment), and is definitely not random. Rather, the observed value x and the resulting $C(x)$ are random. So the correct formulation reads: In 90% of all observations, the random interval $C(\cdot)$ contains the true value ϑ (regardless of what this value is).

Figure 8.1 illustrates the basic principles of confidence regions.

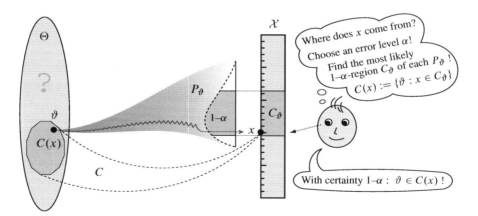

Figure 8.1. Estimation with error bounds: An unknown quantity ϑ is to be identified. The true ϑ determines a probability measure P_ϑ, which leads to the observed outcome x, say. The statistician must first analyse P_ϑ for each possible ϑ and then, given an error level α, find a smallest possible set $C_\vartheta \subset \mathcal{X}$ such that $P_\vartheta(C_\vartheta) \geq 1-\alpha$. Then he can specify a set $C(x) \subset \Theta$ that contains the true ϑ with certainty at least $1 - \alpha$, regardless of the true value of ϑ.

How can one construct a confidence region? Fortunately, there is a general principle, which we will now describe. For simplicity we confine ourselves to the most important case that the parameter ϑ itself has to be identified, which means that $\Sigma = \Theta$
and τ is the identity; the general case will be sketched in Example (8.4).

So suppose that $(\mathcal{X}, \mathcal{F}, P_\vartheta : \vartheta \in \Theta)$ is a statistical model. Each confidence region
$C : x \to C(x)$ for ϑ can be represented uniquely by its graph

$$C = \{(x, \vartheta) \in \mathcal{X} \times \Theta : \vartheta \in C(x)\}$$

(again denoted by C); see Figure 8.2. For each $x \in \mathcal{X}$, $C(x)$ then corresponds to the
vertical section through C at x, and

$$C_\vartheta := \{C(\cdot) \ni \vartheta\} = \{x \in \mathcal{X} : (x, \vartheta) \in C\} \in \mathcal{F}$$

is the horizontal section through C at ϑ. Condition (8.1) thus means that, for all
$\vartheta \in \Theta$, the horizontal section at ϑ has P_ϑ-probability at least $1 - \alpha$:

$$\inf_{\vartheta \in \Theta} P_\vartheta(C_\vartheta) \geq 1 - \alpha \,.$$

This observation leads us to the following construction principle.

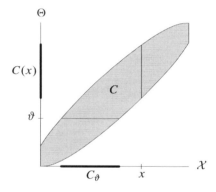

Figure 8.2. The construction of confidence regions.

(8.2) Method of constructing confidence regions. Choose an error level $0 < \alpha < 1$
and proceed as follows.

▷ For each $\vartheta \in \Theta$, determine a smallest possible set $C_\vartheta \in \mathcal{F}$ that is still large
   enough to satisfy $P_\vartheta(C_\vartheta) \geq 1 - \alpha$. For example, in the case of a standard model
   one can invoke the maximum likelihood principle to set

$$C_\vartheta = \{x \in \mathcal{X} : \rho_\vartheta(x) \geq c_\vartheta\} \,,$$

   where $c_\vartheta > 0$ is such that (8.1) holds as tightly as possible. In other words, one
   collects into C_ϑ as many x with the largest likelihood as are necessary to exceed
   the confidence level $1 - \alpha$.

▷ Then, as in Figure 8.2, set $C = \{(x, \vartheta) \in \mathcal{X} \times \Theta : x \in C_\vartheta\}$ and

$$C(x) = \{\vartheta \in \Theta : C_\vartheta \ni x\}.$$

The mapping C so constructed is then a confidence region for ϑ with error level α.

Here is an example.

(8.3) Example. *Emission control.* Because of high inspection costs, out of $N = 10$ power stations only $n = 4$ are selected at random and their emissions are checked. One needs a confidence region for the number ϑ of power stations whose emission values are too high. This situation corresponds to a sample of size 4 without replacement from an urn containing 10 balls, of which an unknown number ϑ is black. Hence, the appropriate model is $\mathcal{X} = \{0, \ldots, 4\}$, $\Theta = \{0, \ldots, 10\}$, $P_\vartheta = \mathcal{H}_{4;\vartheta,10-\vartheta}$. The corresponding likelihood function is listed in Table 8.1.

Table 8.1. Construction of a confidence region (underlined) in the hypergeometric model. The values for $\vartheta > 5$ follow by symmetry.

ϑ	$\binom{10}{4} \mathcal{H}_{4;\vartheta,10-\vartheta}(\{x\})$				
5	5	<u>50</u>	<u>100</u>	<u>50</u>	5
4	15	<u>80</u>	<u>90</u>	24	1
3	35	<u>105</u>	<u>63</u>	7	0
2	<u>70</u>	<u>112</u>	28	0	0
1	<u>126</u>	<u>84</u>	0	0	0
0	<u>210</u>	0	0	0	0
	0	1	2	3	4	x

Suppose we choose $\alpha = 0.2$. To apply the method (8.2), we select from each row of Table 8.1 the largest entries until their sum is at least $168 = 0.8 \binom{10}{4}$. These are underlined and define a confidence region C with

$$C(0) = \{0, 1, 2\}, \quad C(1) = \{1, \ldots, 5\}, \quad C(2) = \{3, \ldots, 7\},$$
$$C(3) = \{5, \ldots, 9\}, \quad C(4) = \{8, 9, 10\}.$$

Due to the small sample size, the sets $C(x)$ are rather large, even though the error level α was also chosen quite large. Therefore, despite the additional costs, one should examine more power stations!

As we have emphasised above, the main point in constructing confidence regions is to find a set C_ϑ of 'most likely outcomes' for every $\vartheta \in \Theta$. So suppose for instance that $\mathcal{X} = \mathbb{R}$, and P_ϑ has a Lebesgue density ρ_ϑ, which is unimodal in the sense that it first increases up to a peak and then decreases again. Then one would choose C_ϑ as an 'interval around the peak' which is just large enough to ensure that the 'tails' to the left and right of C_ϑ together have only probability α. In this context, the notion of a quantile becomes useful.

Definition. Let Q be a probability measure on $(\mathbb{R}, \mathscr{B})$ and $0 < \alpha < 1$. A number $q \in \mathbb{R}$ with $Q(]-\infty, q]) \geq \alpha$ and $Q([q, \infty[) \geq 1 - \alpha$ is called an α-*quantile of Q*. An $1/2$-quantile of Q is simply a *median*, and a $(1-\alpha)$-quantile is also called an α-*fractile* of Q. The $1/4$- and $3/4$-quantiles are known as the lower and upper *quartiles*. The quantiles of a real-valued random variable X are defined as the quantiles of its distribution.

This definition explains the name of the quantile transformations in Proposition (1.30). As is obvious from Figure 8.3, an α-quantile is a point q at which the distribution function F_Q of Q crosses (or jumps over) the level α. The existence of an α-quantile thus follows directly from (1.29). In general, the α-quantile is not unique, but it is if F_Q is strictly increasing in a neighbourhood of q, so in particular if Q admits a Lebesgue density ρ whose support $\{\rho > 0\}$ is an interval. The notion of a quantile will be used mainly in this latter case, in which the α-quantile is the unique value q such that $\int_{-\infty}^{q} \rho(x)\, dx = \alpha$, cf. Figure 8.3.

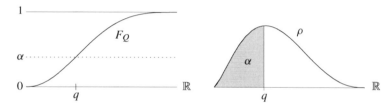

Figure 8.3. Definition of the α-quantile q as the (or some) point at which the distribution function F_Q crosses the level α (left), or for which the area under the Lebesgue density ρ of Q to the left of q takes the value α (right).

The quantiles of some familiar distributions can be either found in the tables in the appendix or in other collections of statistical tables such as [46, 53], for example. Alternatively, one can use appropriate software as, for instance, `Mathematica` (where the command `Quantile[dist,a]` provides the a-quantile of a distribution `dist`).

The following example demonstrates the natural role of quantiles in the construction of confidence intervals. In addition, it shows how to generalise the construction method (8.2) if only a function of the parameter ϑ is to be identified.

(8.4) Example. *Confidence interval for the expectation in the Gaussian product model.* Consider the n-fold Gaussian product model

$$(\mathscr{X}, \mathscr{F}, P_{\vartheta} : \vartheta \in \Theta) = \left(\mathbb{R}^n, \mathscr{B}^n, \mathcal{N}_{m,v}^{\otimes n} : m \in \mathbb{R}, v > 0\right).$$

Let us look for a confidence interval for the expectation coordinate $m(\vartheta) = m$ of the parameter $\vartheta = (m, v)$. Generalising the method (8.2), we need to find for every $m \in \mathbb{R}$ a smallest possible set $C_m \in \mathscr{B}$ such that $P_{\vartheta}(C_{m(\vartheta)}) \geq 1 - \alpha$ for all ϑ; the required confidence region for $m(\vartheta)$ then takes the form

$$C(x) = \{m \in \mathbb{R} : C_m \ni x\}.$$

This task is greatly simplified by the scaling properties of normal distributions, as follows. For each $m \in \mathbb{R}$ consider the statistic

$$T_m = (M - m) \sqrt{n/V^*}$$

from $\mathbb{R}^n$ to $\mathbb{R}$; as before, we write $M = \frac{1}{n} \sum_{i=1}^n X_i$ for the sample mean, and $V^* = \frac{1}{n-1} \sum_{i=1}^n (X_i - M)^2$ for the corrected sample variance. The point is now that the distribution $Q := P_\vartheta \circ T_{m(\vartheta)}^{-1}$ does not depend on the parameter ϑ; this property is expressed by calling $(Q; T_m : m \in \mathbb{R})$ a *pivot* for $m(\vartheta)$. For the proof, let $\vartheta = (m, v)$ be given and consider the standardisation mapping

$$S_\vartheta = \left(\frac{X_i - m}{\sqrt{v}} \right)_{1 \le i \le n}$$

from $\mathbb{R}^n$ to $\mathbb{R}^n$. By the scaling properties of the normal distributions stated in Problem 2.15, it then follows that $P_\vartheta \circ S_\vartheta^{-1} = \mathcal{N}_{0,1}^{\otimes n}$. On the other hand, we have $M \circ S_\vartheta = (M - m)/\sqrt{v}$ and

$$V^* \circ S_\vartheta = \frac{1}{n-1} \sum_{i=1}^n \left(\frac{X_i - m}{\sqrt{v}} - \frac{M - m}{\sqrt{v}} \right)^2 = \frac{V^*}{v},$$

and therefore $T_{m(\vartheta)} = T_0 \circ S_\vartheta$. Hence we obtain

$$P_\vartheta \circ T_{m(\vartheta)}^{-1} = P_\vartheta \circ S_\vartheta^{-1} \circ T_0^{-1} = \mathcal{N}_{0,1}^{\otimes n} \circ T_0^{-1} =: Q.$$

The distribution Q will be calculated explicitly in Theorem (9.17): Q is the so-called *t-distribution with $n - 1$ degrees of freedom*, or t_{n-1}*-distribution* for short. By the symmetry of $\mathcal{N}_{0,1}$ and T_0, Q is symmetric. Moreover, we will see that Q has a Lebesgue density, which is decreasing on $[0, \infty[$.

  In analogy to (8.2), we now proceed as follows. For a given error level α, let I be the shortest interval such that $Q(I) \ge 1 - \alpha$. In view of the properties of Q just discussed, this interval I is centred around the origin, and is in fact equal to $I =]-t, t[$ for the $\alpha/2$-fractile $t > 0$ of Q. So, setting $C_m = T_m^{-1}]-t, t[$ we can conclude that $P_\vartheta(C_{m(\vartheta)}) = 1 - \alpha$ for all $\vartheta = (m, v)$, and we obtain the following theorem.

(8.5) Theorem. Confidence interval for the expectation in the Gaussian product model. *Consider the n-fold Gaussian product model for n independent, normally distributed experiments with unknown expectation m and unknown variance. For $0 < \alpha < 1$ let $t := F_Q^{-1}(1 - \frac{\alpha}{2})$ be the $\alpha/2$-fractile of the t_{n-1}-distribution. Then,*

$$C(\cdot) =]M - t \sqrt{V^*/n}, \; M + t \sqrt{V^*/n}[$$

is a confidence interval for m with error level α.

A substantial generalisation of this theorem will follow later in Corollary (12.19). A typical application is the following.

(8.6) Example. *Comparison of two sleeping pills.* The effect of two sleeping pills *A* and *B* is to be compared. They are given to $n = 10$ patients in two consecutive nights, and the duration of sleep is measured. (This is the experimental design of *matched pairs*, in the special case when each patient is matched with himself.) A classical experiment provided the data of Table 8.2 for the difference in sleep duration.

Table 8.2. Differences in sleep duration for sleeping pills *A* or *B*, according to [5, p. 215].

Patient	1	2	3	4	5	6	7	8	9	10
Difference	1.2	2.4	1.3	1.3	0.0	1.0	1.8	0.8	4.6	1.4

Since this difference is positive for all patients, it is clear that drug *B* has greater effect. More precisely, one finds that the outcome *x* has mean $M(x) = 1.58$ and corrected variance $V^*(x) = 1.513$. Accepting that sleep duration is the result of many small independent influences that add up, we can invoke the central limit theorem to assume that the sleep duration is (approximately) normally distributed with unknown parameters. Theorem (8.5) then applies, where *m* stands for the mean difference in sleep duration. To be specific, let us take error level $\alpha = 0.025$. Table C in the appendix then yields the $\alpha/2$-fractile $t = 2.72$ (by linear interpolation), and the resulting confidence interval for *m* is $C(x) = {]}0.52, 2.64{[}$.

8.2 Confidence Intervals in the Binomial Model

We now take a closer look at the binomial model

$$\mathcal{X} = \{0, \dots, n\}, \quad \Theta = {]}0, 1{[}, \quad P_\vartheta = \mathcal{B}_{n,\vartheta} \,.$$

For example, one can again think of throwing a drawing pin *n* times, or of a survey of *n* randomly chosen people concerning two presidential candidates, or of any other independently repeated experiment dealing with two alternatives. Our objective is to find a confidence interval for the unknown 'success probability' ϑ. We will present three different approaches to this problem.

Method 1: Application of Chebyshev's inequality. The best estimator of ϑ is $T(x) = x/n$. It is therefore natural to set

(8.7) $$C(x) = {]}\tfrac{x}{n} - \varepsilon, \, \tfrac{x}{n} + \varepsilon{[} \,,$$

where $\varepsilon > 0$ is to be determined appropriately. For this choice of *C*, condition (8.1) takes the form

$$\mathcal{B}_{n,\vartheta}\left(x : \left|\tfrac{x}{n} - \vartheta\right| \geq \varepsilon\right) \leq \alpha \,.$$

By Chebyshev's inequality (5.5), these probabilities are at most

$$\frac{\mathbb{V}(\mathcal{B}_{n,\vartheta})}{n^2 \varepsilon^2} = \frac{\vartheta(1-\vartheta)}{n\varepsilon^2} \, .$$

Since ϑ is unknown, we bound this further by $1/(4n\varepsilon^2)$. So condition (8.1) certainly holds if $4n\alpha\varepsilon^2 \geq 1$. This inequality shows how the number of samples n, the error level α and the precision ε of the estimate relate to each other. For example, if $n = 1000$ and $\alpha = 0.025$, then the best possible precision we can obtain is $\varepsilon = 1/\sqrt{100} = 0.1$. Conversely, for the same level $\alpha = 0.025$, we already need $n = 4000$ samples to achieve a precision of $\varepsilon = 0.05$. (Finding the minimal number of cases required to achieve a given certainty and precision is an important issue in practice, for example in medical studies.)

The advantage of this method is that the calculations are easy, and the error bounds do not rely on any approximations. The disadvantage is that Chebyshev's inequality is not adapted to the binomial distribution, and is thus rather crude. The quantities ε or n so obtained are therefore unnecessarily large.

Method 2: Using the normal approximation. Again we define C as in (8.7), but now we suppose that n is so large that the central limit theorem (5.24) can be applied. We write

$$\mathcal{B}_{n,\vartheta}\left(x : \left|\tfrac{x}{n} - \vartheta\right| < \varepsilon\right) = \mathcal{B}_{n,\vartheta}\left(x : \left|\tfrac{x-n\vartheta}{\sqrt{n\vartheta(1-\vartheta)}}\right| < \varepsilon\sqrt{\tfrac{n}{\vartheta(1-\vartheta)}}\right)$$

$$\approx \Phi\left(\varepsilon\sqrt{\tfrac{n}{\vartheta(1-\vartheta)}}\right) - \Phi\left(-\varepsilon\sqrt{\tfrac{n}{\vartheta(1-\vartheta)}}\right)$$

$$= 2\,\Phi\left(\varepsilon\sqrt{\tfrac{n}{\vartheta(1-\vartheta)}}\right) - 1$$

with Φ as in (5.21). Specifically, if we take $\alpha = 0.025$ and introduce a safety margin of 0.02 for the approximation error, then condition (8.1) will hold as soon as $2\,\Phi(\varepsilon\sqrt{\tfrac{n}{\vartheta(1-\vartheta)}}) - 1 \geq 0.975 + 0.02$, so if

$$\varepsilon\,\sqrt{\tfrac{n}{\vartheta(1-\vartheta)}} \geq \Phi^{-1}(0.9975) = 2.82 \, .$$

The last value is taken from Table A in the appendix. Since $\vartheta(1-\vartheta) \leq 1/4$, we therefore obtain, for $n = 1000$ and $\alpha = 0.025$, the sufficient condition

$$\varepsilon \geq 2.82/\sqrt{4000} \approx 0.0446 \, .$$

As compared to the first method, the confidence interval has shrunk to less than half its size, but one must be aware of the approximation error (which was treated above in a rather crude way).

If the resulting confidence interval is still too large for the required precision of the estimate, one needs more observations (i.e., the drawing pin must be thrown more often). As in the case of the first method, one can easily modify the above calculation to determine the number of experiments that are needed to achieve the required precision and level of confidence.

In the third and most precise method, we will take advantage of some particular properties of the binomial distribution.

(8.8) Lemma. Monotonicity properties of the binomial distribution. *Let $n \geq 1$ and $\mathcal{X} = \{0, \ldots, n\}$. Then the following is true.*

(a) *For every $0 < \vartheta < 1$, the function $\mathcal{X} \ni x \to \mathcal{B}_{n,\vartheta}(\{x\})$ is strictly increasing on $\{0, \ldots, \lceil (n+1)\vartheta - 1 \rceil\}$ and strictly decreasing on $\{\lfloor (n+1)\vartheta \rfloor, \ldots, n\}$. So it is maximal for $x = \lfloor (n+1)\vartheta \rfloor$ (and also for $(n+1)\vartheta - 1$ if this is an integer).*

(b) *For every $0 \neq x \in \mathcal{X}$, the function $\vartheta \to \mathcal{B}_{n,\vartheta}(\{x, \ldots, n\})$ is continuous and strictly increasing on $[0, 1]$. In fact, it is linked to a beta distribution via the relation $\mathcal{B}_{n,\vartheta}(\{x, \ldots, n\}) = \boldsymbol{\beta}_{x,n-x+1}([0, \vartheta])$.*

Proof. (a) For each $x \geq 1$ one easily finds that

$$\frac{\mathcal{B}_{n,\vartheta}(\{x\})}{\mathcal{B}_{n,\vartheta}(\{x-1\})} = \frac{(n-x+1)\vartheta}{x(1-\vartheta)} \, .$$

This ratio is greater than 1 if and only if $x < (n+1)\vartheta$.

(b) Let $U_1, \ldots, U_n$ be i.i.d. random variables that are uniformly distributed on $[0, 1]$. Using Theorem (3.24), we see that $1_{[0,\vartheta]} \circ U_1, \ldots, 1_{[0,\vartheta]} \circ U_n$ is a Bernoulli sequence with parameter ϑ. Theorem (2.9) thus implies that the sum $S_\vartheta = \sum_{i=1}^n 1_{[0,\vartheta]} \circ U_i$ has the binomial distribution $\mathcal{B}_{n,\vartheta}$. Hence

$$\mathcal{B}_{n,\vartheta}(\{x, \ldots, n\}) = P(S_\vartheta \geq x) \, .$$

Now, the condition $S_\vartheta \geq x$ means that at least x of the random numbers $U_1, \ldots, U_n$ are at most ϑ. In terms of the xth order statistic $U_{x:n}$ (i.e., the xth smallest of the values of $U_1, \ldots, U_n$), this can be expressed by writing $U_{x:n} \leq \vartheta$. As we have seen in Section 2.5.3, the last event has probability $\boldsymbol{\beta}_{x,n-x+1}([0, \vartheta])$. This proves the required relation between binomial and beta distributions. Since the beta distribution has a strictly positive density, we obtain, in particular, the required continuity and strict monotonicity in ϑ. (For an alternative, purely formal, proof see Problem 8.8.) $\diamond$

Now we are sufficiently prepared for

Method 3: Application of binomial and beta quantiles. In contrast to the previous two methods, we will not use the symmetric approach (8.7) here, which is centred at the relative frequency x/n, but instead make direct use of the construction (8.2). For every $\vartheta \in \,]0, 1[$, we then need to find a C_ϑ such that $\mathcal{B}_{n,\vartheta}(C_\vartheta) \geq 1 - \alpha$. Lemma (8.8a) shows that C_ϑ should be an appropriate section in the middle of $\mathcal{X} = \{0, \ldots, n\}$. Since deviations both to the right and to the left are equally unwelcome, we cut off the same probability $\alpha/2$ from both sides. That is, we set

$$C_\vartheta = \{x_-(\vartheta), \ldots, x_+(\vartheta)\},$$

where

$$x_-(\vartheta) = \max\{x \in \mathcal{X} : \mathcal{B}_{n,\vartheta}(\{0, \ldots, x-1\}) \le \alpha/2\},$$
$$x_+(\vartheta) = \min\{x \in \mathcal{X} : \mathcal{B}_{n,\vartheta}(\{x+1, \ldots, n\}) \le \alpha/2\}.$$

In other words, $x_-(\vartheta)$ is the largest $\alpha/2$-quantile of $\mathcal{B}_{n,\vartheta}$ and $x_+(\vartheta)$ the smallest $\alpha/2$-fractile. To find the confidence interval $C(x)$ corresponding to the outcome x, we must solve the condition $x \in C_\vartheta$ for ϑ. This can be done with the help of Lemma (8.8b). For $x \ne 0$, this gives us the equivalence

$$x \le x_+(\vartheta) \Leftrightarrow \boldsymbol{\beta}_{x,n-x+1}([0, \vartheta]) = \mathcal{B}_{n,\vartheta}(\{x, \ldots, n\}) > \alpha/2$$
(8.9) $$\Leftrightarrow \vartheta > p_-(x) := \text{the } \alpha/2\text{-quantile of } \boldsymbol{\beta}_{x,n-x+1}.$$

If we set $p_-(0) = 0$, then the relation $x \le x_+(\vartheta) \Leftrightarrow \vartheta > p_-(x)$ also holds for $x = 0$. Similarly, we obtain $x \ge x_-(\vartheta) \Leftrightarrow \vartheta < p_+(x)$, where for $x < n$

(8.10) $$p_+(x) := \text{the } \alpha/2\text{-fractile of } \boldsymbol{\beta}_{x+1,n-x} = 1 - p_-(n-x),$$

and $p_+(x) = 1$ for $x = n$. Therefore, the condition $x \in C_\vartheta$ is equivalent to the condition $p_-(x) < \vartheta < p_+(x)$. The method (8.2) thus yields the following result.

(8.11) Theorem. Confidence intervals in the binomial model. *Consider the binomial model* $(\{0, \ldots, n\}, \mathcal{B}_{n,\vartheta} : 0 < \vartheta < 1)$, *let* $0 < \alpha < 1$, *and suppose that the functions* p_- *and* p_+ *are defined by* (8.9) *and* (8.10). *Then the mapping* $x \to]p_-(x), p_+(x)[$ *is a confidence interval for* ϑ *with error level* α.

As a matter of fact, the functions p_- and p_+ are also defined for non-integer x. Figure 8.2 shows these continuous interpolations for the parameter values $n = 20$, $\alpha = 0.1$.

How can one determine p_- and p_+? For small n, one can use the tables of the binomial distribution (as for example in [46, 53]). One may also use Table D for the F-distributions that will be introduced in the next chapter, since these are closely related to the beta distributions; see Remark (9.14). Alternatively, one can use the Mathematica command `Quantile[BetaDistribution[a,b],q]` for the q-quantile of $\boldsymbol{\beta}_{a,b}$. The values listed in Table 8.3 are obtained in this way. If we compare these with the results of the first two methods, we see that the confidence intervals found by the last method are the smallest ones. This is most evident if x/n is close to 0 or 1, and the asymmetry with respect to the estimate x/n is then particularly obvious.

Table 8.3. Various confidence intervals for the binomial model, and the corresponding quantities obtained from the normal approximation (in italics). For reasons of symmetry, we only need to consider the case $x \leq n/2$.

x/n	.05	.1	.15	.2	.25	.3	.35	.4	.45	.5
				$n = 20$,	$\alpha = .2$					
$p_-(x)$	.0053	.0269	.0564	.0902	.1269	.1659	.2067	.2491	.2929	.3382
$p_+(x)$	.1810	.2448	.3042	.3607	.4149	.4673	.5180	.5673	.6152	.6618
$\tilde{p}_-(x)$	.0049	.0279	.0580	.0921	.1290	.1679	.2086	.2508	.2944	.3393
$\tilde{p}_+(x)$	.1867	.2489	.3072	.3628	.4163	.4680	.5182	.5670	.6145	.6607

x/n	.05	.1	.15	.2	.25	.3	.35	.4	.45	.5
				$n = 100$,	$\alpha = .1$					
$p_-(x)$	.0199	.0553	.0948	.1367	.1802	.2249	.2708	.3175	.3652	.4136
$p_+(x)$	.1023	.1637	.2215	.2772	.3313	.3842	.4361	.4870	.5371	.5864
$\tilde{p}_-(x)$	.0213	.0569	.0964	.1382	.1816	.2262	.2718	.3184	.3658	.4140
$\tilde{p}_+(x)$	.1055	.1662	.2235	.2788	.3325	.3850	.4366	.4872	.5370	.5860

x/n	.05	.1	.15	.2	.25	.3	.35	.4	.45	.5
				$n = 1000$,	$\alpha = .02$					
$p_-(x)$	.0353	.0791	.1247	.1713	.2187	.2666	.3151	.3639	.4132	.4628
$p_+(x)$	.0684	.1242	.1782	.2311	.2833	.3350	.3861	.4369	.4872	.5372
$\tilde{p}_-(x)$	.0358	.0796	.1252	.1718	.2191	.2670	.3153	.3641	.4133	.4628
$\tilde{p}_+(x)$	.0692	.1248	.1787	.2315	.2837	.3352	.3863	.4370	.4873	.5372

For large n, one can also use the normal approximation

$$\mathcal{B}_{n,\vartheta}(\{0, \ldots, x\}) \approx \Phi\left(\frac{x + \frac{1}{2} - n\vartheta}{\sqrt{n\vartheta(1-\vartheta)}}\right).$$

That is, one can approximate $p_+(x)$ by the solution $\tilde{p}_+(x)$ of the equation

$$\Phi\left(\frac{x + \frac{1}{2} - n\vartheta}{\sqrt{n\vartheta(1-\vartheta)}}\right) = \frac{\alpha}{2}$$

(in the variable ϑ), which by applying Φ^{-1} and squaring is turned into

(8.12) $$\left(x + \tfrac{1}{2} - n\vartheta\right)^2 = n\vartheta(1-\vartheta)\,\Phi^{-1}(\alpha/2)^2.$$

The $\alpha/2$-quantile $\Phi^{-1}(\alpha/2)$ of the standard normal distribution $\mathcal{N}_{0,1}$ can be found in Table A, and then one can solve the quadratic equation (8.12) for ϑ. Since $\Phi^{-1}(\alpha/2) < 0$, one finds that $\tilde{p}_+(x)$ is the larger of the two solutions of (8.12). Similarly, $p_-(x)$ can be approximated by the solution $\tilde{p}_-(x)$ of the equation

$$1 - \Phi\left(\frac{x - \frac{1}{2} - n\vartheta}{\sqrt{n\vartheta(1-\vartheta)}}\right) = \frac{\alpha}{2},$$

which, due to the anti-symmetry of Φ, leads to equation (8.12) with $-\frac{1}{2}$ instead of $\frac{1}{2}$, and $\tilde{p}_-(x)$ is the smaller of its two solutions. (If we ignore the discreteness corrections $\pm\frac{1}{2}$, $\tilde{p}_-(x)$ and $\tilde{p}_+(x)$ are the two solutions of the same quadratic equation.) Table 8.3 compares the approximate confidence intervals thus obtained with the exact intervals.

(8.13) Example. *Public acceptance of evolution.* In 2005 a survey of the acceptance of the theory of evolution was conducted in 34 countries. National samples of adults were asked whether or not they agreed with the statement 'Human beings, as we know them, developed from earlier species of animals' or whether they were not sure or did not know. Table 8.4 shows the results from a selection of countries.

Table 8.4. Acceptance of evolution in 14 countries: Percentage of agreement and size n of the national sample (according to [49]).

Country	IS	F	J	GB	N	D	I	NL	CZ	PL	A	GR	USA	TR
% pro	85	80	78	75	74	72	69	68	66	58	57	54	40	27
n	500	1021	2146	1308	976	1507	1006	1005	1037	999	1034	1000	1484	1005

Is the large diversity of the results only due to statistical outliers? To deal with this question, one should specify a confidence interval for each country. In reality, the candidates for the survey are chosen by sampling *without* replacement, so the number of supporters in each national sample is hypergeometrically distributed. However, because of the large total populations, Theorem (2.14) allows to replace the hypergeometric distributions by binomial distributions, so that we can apply Theorem (8.11). For error level $\alpha = 0.001$, we obtain the confidence intervals shown in Figure 8.4.

Notice that this small error level only applies to each individual country. But the statement that in *each* of the 14 countries the proportion of supporters lies in the given interval can still be made with a certainty of $0.999^{14} \approx 0.986$. Indeed, the

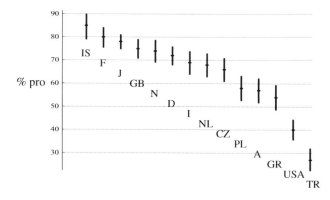

Figure 8.4. Confidence intervals for the data of Table 8.4.

event that this statement holds is the intersection of 14 events that are independent (by the independence of the choice of candidates in the different countries) and have probability at least 0.999 each. Further data and a more precise analysis of the social and religious correlations can be found in [49].

8.3 Order Intervals

In many applications it is difficult to figure out a natural class $\{P_\vartheta : \vartheta \in \Theta\}$ of probability measures to set up a model. Recall, for instance, Example (8.6) where two sleeping pills were compared. One can doubt whether the assumption of a normal distribution is really justified in this case. This stimulates the interest in methods that require only mild assumptions on the probability measures and, instead of making use of specific properties of a certain model, are fairly universal. Such methods are called *non-parametric*, because then the class of probability measures cannot be indexed by a finite dimensional parameter set. We will now discuss such a method. The key idea is to take advantage of the natural ordering of the real numbers, but to ignore their linear structure.

Suppose we are given n independent observations $X_1, \ldots, X_n$ with values in $\mathbb{R}$ and an unknown distribution Q. Which information about Q can be extracted from these observations? To get an overview, one will plot the observed values $x_1, \ldots, x_n$ on the real line. Unless two of these happen to coincide, the values are then automatically ordered according to size. This ordering is made precise by the so-called order statistics, which we already encountered in Section 2.5.3 and which we will now investigate in more detail.

To avoid that two different observations take the same value, we stipulate that the distribution Q of the individual observations X_i satisfies the condition

$$(8.14) \qquad\qquad Q(\{x\}) = 0 \quad \text{for all } x \in \mathbb{R}.$$

A probability measure Q on $(\mathbb{R}, \mathscr{B})$ with property (8.14) is called *continuous* or *diffuse*. Note that (8.14) is equivalent to the assumption that the distribution function F_Q is continuous, and so holds in particular when Q has a Lebesgue density ρ. It implies that

$$(8.15) \qquad\qquad P\left(X_i \neq X_j \text{ for all } i \neq j\right) = 1.$$

Namely, let $i \neq j$, $\ell \geq 2$, $t_0 = -\infty$, $t_\ell = \infty$, and t_k be a k/ℓ-quantile of Q for $0 < k < \ell$. Then $Q(]t_{k-1}, t_k]) = 1/\ell$ for $1 \leq k \leq \ell$, and thus

$$P(X_i = X_j) \leq P\left(\bigcup_{k=1}^{\ell} \{X_i, X_j \in]t_{k-1}, t_k]\}\right) \leq \sum_{k=1}^{\ell} Q(]t_{k-1}, t_k])^2 = 1/\ell.$$

Letting $\ell \to \infty$, it follows that $P(X_i = X_j) = 0$. Hence, in the following we need not worry about possible 'ties' $X_i = X_j$.

Definition. The *order statistics* $X_{1:n}, \ldots, X_{n:n}$ of the random variables $X_1, \ldots, X_n$ are defined via the recursion

$$X_{1:n} = \min_{1 \le i \le n} X_i, \quad X_{j:n} = \min\{X_i : X_i > X_{j-1:n}\} \quad \text{for } 1 < j \le n.$$

In view of (8.15), this means that

$$X_{1:n} < X_{2:n} < \cdots < X_{n:n}, \quad \{X_{1:n}, \ldots, X_{n:n}\} = \{X_1, \ldots, X_n\}$$

almost surely; cf. p. 44. In short, for each realisation of $X_1, \ldots, X_n$, the value of $X_{j:n}$ is the jth smallest among the realised values. The situation is illustrated in Figure 8.5.

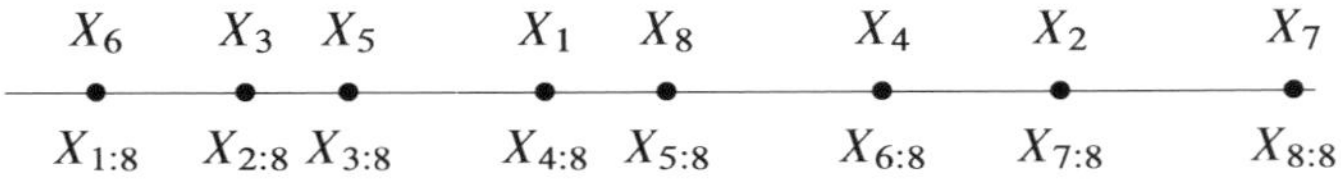

Figure 8.5. The relation between the observations $X_1, \ldots, X_8$ and the corresponding order statistics $X_{1:8}, \ldots, X_{8:8}$.

The following remark gives another characterisation of the order statistics.

(8.16) Remark. *Order statistics and empirical distribution.* The preceding definition of order statistics relies on property (8.15), the absence of ties. A more general definition is

$$X_{j:n} = \min \left\{ c \in \mathbb{R} : \sum_{i=1}^{n} 1_{\{X_i \le c\}} \ge j \right\},$$

which coincides with the preceding definition in the absence of ties, but also works when ties are possible. In words, $X_{j:n}$ is the smallest number that exceeds at least j of the outcomes. In particular,

(8.17) $$\left\{X_{j:n} \le c\right\} = \left\{ \sum_{i=1}^{n} 1_{\{X_i \le c\}} \ge j \right\}$$

for all $c \in \mathbb{R}$ and $1 \le j \le n$. Together with Remark (1.26), this implies that the order statistics are really statistics, in that they are measurable. Next, consider the random function $F_L : c \to \frac{1}{n} \sum_{i=1}^{n} 1_{\{X_i \le c\}}$. By definition, $X_{j:n}$ is then precisely the value at which F_L reaches, or jumps over, the level j/n. F_L is called the *empirical distribution function*, for it is the distribution function of the *empirical distribution* $L = \frac{1}{n} \sum_{i=1}^{n} \delta_{X_i}$ of $X_1, \ldots, X_n$, that is, of the (random) probability measure on $\mathbb{R}$ that assigns the weight $1/n$ to each observation X_i. So, $X_{j:n}$ is the smallest j/n-quantile of L. Conversely, it is clear that $L = \frac{1}{n} \sum_{j=1}^{n} \delta_{X_{j:n}}$. This shows that the collection of all order statistics is in a one-to-one correspondence with the empirical distribution.

How much information do the order statistics contain about the true distribution Q of the individual observations? One object of interest is the 'representative value' of Q. In the present context it is not sensible to identify 'representative value' with 'expectation'. One reason is that the class of all continuous Q includes the heavy-tailed probability measures whose expectation does not exist. Another point is the fact that the sample mean M is strongly influenced by the outcomes of large modulus, the so-called outliers. Both problems derive from the fact that both the expectation and the sample mean make use of the linear structure of $\mathbb{R}$. On the other hand, they ignore the total ordering of $\mathbb{R}$, which forms the basis of the order statistics. Now, the most natural 'representative value' of Q in the sense of ordering is certainly the median; cf. p. 100. To see that the median only depends on the ordering, one can observe the following. If μ is a median of Q and $T : \mathbb{R} \to \mathbb{R}$ is an order-preserving (i.e., strictly increasing) mapping, then $T(\mu)$ is a median of $Q \circ T^{-1}$. We will now show that the order statistics can indeed be used to construct confidence intervals for the median.

Given a continuous probability measure Q we write $\mu(Q)$ for an arbitrary median of Q. Condition (8.14) then implies that

(8.18) $Q(]-\infty, \mu(Q)]) = Q([\mu(Q), \infty[) = 1/2 \, .$

Let us also introduce the abbreviation

(8.19) $b_n(\alpha) = \max\{1 \leq m \leq n : \mathcal{B}_{n,1/2}(\{0, \dots, m-1\}) \leq \alpha\}$

for the largest α-quantile of the binomial distribution $\mathcal{B}_{n,1/2}$.

(8.20) Theorem. *Order intervals for the median. Let $X_1, \dots, X_n$ be independent, real-valued random variables with an unknown distribution Q, which is assumed to be continuous. Also, let $0 < \alpha < 1$ and $k = b_n(\alpha/2)$. Then $[X_{k:n}, X_{n-k+1:n}]$ is a confidence interval for $\mu(Q)$ with error level α.*

Proof. Without loss of generality, we can work with the canonical statistical model $(\mathbb{R}^n, \mathscr{B}^n, Q^{\otimes n} : Q$ continuous$)$, and identify the observations $X_1, \dots, X_n$ with the projections from $\mathbb{R}^n$ onto $\mathbb{R}$. Then, for each continuous Q,

$$Q^{\otimes n}(X_{k:n} > \mu(Q)) = Q^{\otimes n}\left(\sum_{i=1}^{n} 1_{\{X_i \leq \mu(Q)\}} < k \right)$$

$$= \mathcal{B}_{n,1/2}(\{0, \dots, k-1\}) \leq \alpha/2 \, .$$

Here, the first equality follows from (8.17), and the second from the fact that, by (8.18), the indicator functions $1_{\{X_i \leq \mu(Q)\}}$ form a Bernoulli sequence with parameter $1/2$. In the same way we find

$$Q^{\otimes n}(X_{n-k+1:n} < \mu(Q)) = \mathcal{B}_{n,1/2}(\{0, \dots, k-1\}) \leq \alpha/2 \, .$$

Both inequalities together yield the inequality

$$Q^{\otimes n}(\mu(Q) \in [X_{k:n}, X_{n-k+1:n}]) \geq 1 - \alpha,$$

which was to be proved. $\diamond$

As an application, we return to Example (8.6).

(8.21) Example. *Comparison of two sleeping pills.* How can one quantify the effect of the two sleeping pills, if one considers the assumption of a normal distribution in (8.6) to be hardly plausible? The difference in the sleep duration of a single patient has some distribution Q on $\mathbb{R}$, which we can assume to be continuous, even if, in practice, only values rounded to the next minute can be measured. The median of Q is a plausible measure of the average difference in sleep. (Note that $\mu(Q) = m$ when we stick to the normal assumption $Q = \mathcal{N}_{m,v}$.) Choosing the same error level $\alpha = 0.025$ as in Example (8.6), we find the value $k = b_{10}(0.0125) = 2$ (for instance from the binomial table in [46, 53]). For the data from Table 8.2, Theorem (8.20) thus gives the confidence interval $[0.8, 2.4]$. It may come as a surprise that, for our data set, the order interval is shorter than the confidence interval derived in Example (8.6) under the stronger hypothesis of a normal distribution.

Finally, let us come back to Remark (8.16). There we saw that the jth order statistic $X_{j:n}$ is nothing but the smallest j/n-quantile (or the unique $(j - \frac{1}{2})/n$-quantile) of the empirical distribution $L = \frac{1}{n} \sum_{j=1}^{n} \delta_{X_j}$. In general, the quantiles of L are called the *sample quantiles*. Of special interest is the median

$$(8.22) \qquad \mu(L) = \begin{cases} X_{k+1:n} & \text{if } n = 2k+1, \\ (X_{k:n} + X_{k+1:n})/2 & \text{if } n = 2k \end{cases}$$

of L, the so-called *sample median*. (This is clearly the unique median if n is odd, and otherwise it is the central one.) Similarly, one can define the sample quartiles.

It is well known that graphical representations are a useful tool for getting a first impression of measured data. One possibility, which uses the sample quantiles, is the so-called *box-and-whisker plot*, or simply *box plot*, of Figure 8.6. According to

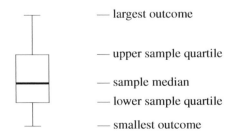

Figure 8.6. Structure of a box plot.

Theorem (8.20), the box between the sample quartiles defines a confidence interval for the true median with error level $\alpha = 2\mathcal{B}_{n,1/2}(\{0, \ldots, \lfloor n/4 \rfloor\})$, which for large n coincides approximately with $2\Phi(-\sqrt{n}/2)$. (There are variations of the box plot in which the outliers, namely the atypically large or small observations, are defined in a certain way and depicted separately.)

Problems

8.1 *Shifted exponential distributions.* Consider the statistical model $(\mathbb{R}, \mathcal{B}, P_\vartheta : \vartheta \in \mathbb{R})$, where P_ϑ is the probability measure with Lebesgue density $\rho_\vartheta(x) = e^{-(x-\vartheta)} 1_{[\vartheta,\infty[}(x)$, $x \in \mathbb{R}$. Construct a minimal confidence interval for ϑ with error level α.

8.2 Return to Problem 7.3 of estimating an unknown number N of lots in a lottery drum, and let T be the maximum likelihood estimator determined there. For a given error level α, find the smallest possible confidence region for N that has the form $C(x) = \{T(x), \ldots, c(T(x))\}$.

8.3 *Combination of confidence intervals.* Let $(\mathcal{X}, \mathcal{F}, P_\vartheta : \vartheta \in \Theta)$ be a statistical model and $\tau_1, \tau_2 : \Theta \to \mathbb{R}$ two real characteristics of ϑ. You are given two individual confidence intervals $C_1(\cdot)$ and $C_2(\cdot)$ for τ_1 and τ_2 with arbitrary error levels α_1 and α_2, respectively. Use these to construct a confidence rectangle for $\tau = (\tau_1, \tau_2)$ with a prescribed error level α.

8.4 ^{S} *Scaled uniform distributions.* Consider the product model $(\mathbb{R}^n, \mathcal{B}^n, \mathcal{U}_{[\vartheta,2\vartheta]}^{\otimes n} : \vartheta > 0)$, where $\mathcal{U}_{[\vartheta,2\vartheta]}$ is the uniform distribution on $[\vartheta, 2\vartheta]$. For $\vartheta > 0$, let $T_\vartheta = \max_{1 \le i \le n} X_i/\vartheta$.

(a) For which probability measure Q on $(\mathbb{R}, \mathcal{B})$ is $(Q; T_\vartheta : \vartheta > 0)$ a pivot?

(b) Use this pivot to construct a confidence interval for ϑ of minimal length for a given error level α.

8.5 Let $(\mathbb{R}^n, \mathcal{B}^n, Q_\vartheta^{\otimes n} : \vartheta \in \Theta)$ be a real n-fold product model with continuous distribution functions $F_\vartheta = F_{Q_\vartheta}$. Furthermore, let $T_\vartheta = -\sum_{i=1}^n \log F_\vartheta(X_i)$, $\vartheta \in \Theta$. For which probability measure Q on $(\mathbb{R}, \mathcal{B})$ is $(Q; T_\vartheta : \vartheta \in \Theta)$ a pivot? *Hint:* Problem 1.18, Corollary (3.36).

8.6 *Consequences of choosing the wrong model.* An experimenter takes n independent and normally distributed measurements with unknown expectation m. He is certain he knows the variance $v > 0$.

(a) Which confidence interval for m will he specify for a given error level α?

(b) Which error level does this confidence interval have if the true variance can take an arbitrary positive value?

Hint: Use Example (3.32).

8.7 *Determining the speed of light.* In 1879 the American physicist (and Nobel prize laureate of 1907) Albert Abraham Michelson made five series of 20 measurements each to determine the speed of light; you can find his results under http://lib.stat.cmu.edu/DASL/ Datafiles/Michelson.html. Suppose that the measurements are normally distributed with unknown m and v, and determine confidence intervals for the speed of light with error level 0.02, individually for each series as well as jointly for all measurements.

8.8 *Beta representation of the binomial distribution.* Show by differentiating with respect to p that

$$\mathcal{B}_{n,p}(\{k+1,k+2,\ldots,n\}) = \binom{n}{k}(n-k)\int_0^p t^k(1-t)^{n-k-1}dt = \beta_{k+1,n-k}([0,p])$$

for all $0 < p < 1$, $n \in \mathbb{N}$ and $k \in \{0,1,\ldots,n-1\}$.

8.9 [S] *Confidence points.* Consider the Gaussian product model $(\mathbb{R}^n, \mathcal{B}^n, P_\vartheta : \vartheta \in \mathbb{Z})$ with $P_\vartheta = \mathcal{N}_{\vartheta,v}^{\otimes n}$ for a known variance $v > 0$ and unknown *integer* mean. Let $\mathrm{ni} : \mathbb{R} \to \mathbb{Z}$ be the 'nearest integer function', that is, for $x \in \mathbb{R}$ let $\mathrm{ni}(x) \in \mathbb{Z}$ be the integer closest to x, with the convention $\mathrm{ni}(z - \frac{1}{2}) = z$ for $z \in \mathbb{Z}$. Show the following.

(a) $\widetilde{M} := \mathrm{ni}(M)$ is a maximum likelihood estimator of ϑ.

(b) Under P_ϑ, $\widetilde{M}$ has the discrete distribution $P_\vartheta(\widetilde{M} = k) = \Phi(a_+(k)) - \Phi(a_-(k))$ with $a_\pm(k) := (k - \vartheta \pm \frac{1}{2})\sqrt{n/v}$. Furthermore, $\widetilde{M}$ is unbiased.

(c) For each $\alpha > 0$ it is the case that $\inf_{\vartheta \in \mathbb{Z}} P_\vartheta(\widetilde{M} = \vartheta) \geq 1 - \alpha$ for all sufficiently large $n \in \mathbb{N}$.

8.10 *History of the European Union.* On June 23, 1992, the German newspaper Frankfurter Allgemeine Zeitung (FAZ) reported that 26% of all German people agreed with a single European currency; furthermore, 50% were in favour of an enlargement of the European Union towards the East. The percentages were based on a poll by the Allensbach institute among approximately 2200 people. Would it make sense to state more precise percentages (e.g., with one decimal digit)? Use the normal approximation to construct an approximate confidence interval with error level 0.05, and consider its length.

8.11 [S] In a chemical plant, n fish are kept in the sewage system. Their survival probability ϑ serves as an indicator for the degree of pollution. How large must n be to ensure that ϑ can be deduced from the number of dead fish with 95% confidence up to a deviation of ± 0.05? Use (a) Chebyshev's inequality and (b) the normal approximation.

8.12 Consider the binomial model $(\{0,\ldots,n\}, \mathcal{P}(\{0,\ldots,n\}), \mathcal{B}_{n,\vartheta} : 0 < \vartheta < 1)$. Use the method discussed after Theorem (8.11) to determine an approximate confidence interval for ϑ with error level $\alpha = 0.02$ and $n = 1000$, and verify some of the corresponding entries of Table 8.3.

8.13 *Is the drawing of the lottery numbers fair?* Using the method of the previous problem, determine, for some of the lottery numbers between 1 and 49, an approximate confidence interval for the probability that the respective number is drawn. First choose an error level and obtain the up-to-date lottery statistics, for instance under www.dielottozahlen.de/lotto/ 6aus49/statistiken.html (for Germany) or www.lottery.co.uk/statistics/ (for the UK). Does the result correspond to your expectations?

8.14 [S] *Non-uniqueness of quantiles.* Let Q be a probability measure on $(\mathbb{R}, \mathcal{B})$, $0 < \alpha < 1$, q an α-quantile of Q, and $q' > q$. Show that q' is another α-quantile of Q if and only if $Q(]-\infty, q]) = \alpha$ and $Q(]q, q'[) = 0$.

8.15 [S] *Median-unbiased estimators.* Consider the non-parametric product model $(\mathbb{R}^n, \mathcal{B}^n, Q^{\otimes n} : Q$ continuous$)$. An estimator T of a real characteristic $\tau(Q)$ is called median-unbiased if, for every continuous Q, $\tau(Q)$ is a median of $Q^{\otimes n} \circ T^{-1}$. Show the following.

(a) If T is a median-unbiased estimator of $\tau(Q)$ and $f : \mathbb{R} \to \mathbb{R}$ is monotone, then $f \circ T$ is a median-unbiased estimator of $f \circ \tau(Q)$. Under which assumption on f does the analogous statement hold for (expectation-)unbiased estimators?

(b) If n is odd and T is the sample median, then the median $\mu(\mathcal{U}_{]0,1[}) = 1/2$ of the uniform distribution on $]0, 1[$ is also a median of $\mathcal{U}_{]0,1[}^{\otimes n} \circ T^{-1}$.

(c) For odd n, the sample median is a median-unbiased estimator of the median $\mu(Q)$. *Hint:* Use (1.30).

8.16 Take the data from Example (8.6) to plot the empirical distribution function, i.e., the distribution function F_L of the empirical distribution L. Determine the sample median and the sample quartiles and draw the corresponding box plot.

8.17 *Sensitivity of sample mean, sample median and trimmed mean.* Let $T_n : \mathbb{R}^n \to \mathbb{R}$ be a statistic that assigns an 'average' to n real outcomes. The sensitivity function

$$S_n(x) = n\big(T_n(x_1, \ldots, x_{n-1}, x) - T_{n-1}(x_1, \ldots, x_{n-1})\big)$$

then describes how much T_n depends on a single outcome $x \in \mathbb{R}$ when the other outcomes $x_1, \ldots, x_{n-1} \in \mathbb{R}$ are fixed. Determine and plot S_n if T_n is (a) the sample mean, (b) the sample median, and (c) the α-trimmed mean

$$(X_{\lfloor n\alpha \rfloor + 1:n} + \cdots + X_{n-\lfloor n\alpha \rfloor:n})/(n - 2\lfloor n\alpha \rfloor)$$

for $0 \le \alpha < 1/2$.

8.18 [S] *Distribution density of order statistics.* Let $X_1, \ldots, X_n$ be i.i.d. real random variables with continuous distribution density ρ, and suppose that the set $\mathcal{X} := \{\rho > 0\}$ is an interval. Determine the distribution density of the kth order statistic $X_{k:n}$. *Hint:* Use either Problem 1.18 and Section 2.5.3, or (8.17) and (8.8b).

8.19 [S] *Law of large numbers for order statistics.* Consider a sequence $(X_i)_{i \ge 1}$ of i.i.d. random variables taking values in an interval $\mathcal{X}$. Let Q be the common distribution of the X_i, and suppose that the distribution function $F = F_Q$ of Q is strictly increasing on $\mathcal{X}$. Also, let $0 < \alpha < 1$ and (j_n) be any sequence in $\mathbb{N}$ with $j_n/n \to \alpha$. Show the following. As $n \to \infty$, the order statistics $X_{j_n:n}$ converge in probability to the α-quantile of Q.

8.20 [S] *Normal approximation of order statistics.* Let $(X_i)_{i \ge 1}$ be a sequence of i.i.d. random variables with common distribution Q on an interval $\mathcal{X}$. Assume that the distribution function $F = F_Q$ is differentiable on $\mathcal{X}$ with continuous derivative $\rho = F' > 0$. Let $0 < \alpha < 1$, $q \in \mathcal{X}$ the associated α-quantile of Q, and (j_n) a sequence in $\mathbb{N}$ with $|j_n - \alpha n|/\sqrt{n} \to 0$. Show that

$$\sqrt{n}(X_{j_n:n} - q) \xrightarrow{d} \mathcal{N}_{0,v}$$

as $n \to \infty$, where $v = \alpha(1-\alpha)/\rho(q)^2$. *Hint:* Use (8.17) and the fact that Corollary (5.24) still holds when the underlying success probability depends on n but converges to a limit.

8.21 *Determining the speed of light.* Consider the data collected by Michelson in order to determine the speed of light, see Problem 8.7. Which confidence intervals for the speed of light can you specify, individually for each series as well as jointly for all measurements, if you want to avoid any assumptions about the distribution of the measurements? Fix the same error level $\alpha = 0.02$ as in Problem 8.7, and compare your results.

Chapter 9

Around the Normal Distributions

This short chapter deals with some basic distributions of statistics that arise as distributions of certain transforms of independent standard normal random variables. Linear transformations produce the multivariate normal (or Gaussian) distributions, and certain quadratics or fractions of quadratics lead to the chi-square, F- and t-distributions. These play an important role in the construction of confidence intervals and tests for statistical experiments with normal, or asymptotically normal, distributions.

9.1 The Multivariate Normal Distributions

A basic tool for this section is the following analytical result, which describes the behaviour of probability densities under smooth transformations.

(9.1) Proposition. Transformation of densities. *Let $\mathcal{X} \subset \mathbb{R}^n$ be open and P a probability measure on $(\mathcal{X}, \mathscr{B}^n_{\mathcal{X}})$ with Lebesgue density ρ. Furthermore, suppose $\mathcal{Y} \subset \mathbb{R}^n$ is open and $T : \mathcal{X} \to \mathcal{Y}$ a diffeomorphism, i.e., a continuously differentiable bijection with Jacobian determinant $\det DT(x) \neq 0$ for all $x \in \mathcal{X}$. Then the distribution $P \circ T^{-1}$ of T on $\mathcal{Y}$ has the Lebesgue density*

$$\rho_T(y) = \rho(T^{-1}(y))\,|\det DT^{-1}(y)|, \quad y \in \mathcal{Y}.$$

Proof. The change-of-variables theorem for multidimensional Lebesgue integrals (see for instance [58]) implies that

$$P \circ T^{-1}(A) = \int_{T^{-1}A} \rho(x)\,dx = \int_A \rho(T^{-1}(y))\,|\det DT^{-1}(y)|\,dy$$

for every open $A \subset \mathcal{Y}$. By the uniqueness theorem (1.12), this equation even holds for all $A \in \mathscr{B}^n_{\mathcal{Y}}$, as claimed. $\diamond$

Our first application of this result concerns affine transformations of random vectors with independent, standard normally distributed coordinates. For an arbitrary matrix B we write B_{ij} for the entry in the ith row and jth column, and $\mathsf{B}^\top$ for the corresponding transposed matrix. Moreover, $\mathsf{E} = (\delta_{ij})_{1 \leq i,j \leq n}$ stands for the identity matrix of the underlying dimension n.

(9.2) Theorem. *Multivariate normal distributions. Let $X_1, \ldots, X_n$ be i.i.d. random variables with standard normal distribution $\mathcal{N}_{0,1}$, and let $X = (X_1, \ldots, X_n)^\top$ be the corresponding random column vector. Also, let $B \in \mathbb{R}^{n \times n}$ be an invertible real $n \times n$ matrix, and $m \in \mathbb{R}^n$ a fixed column vector. Then $Y := BX + m$ has the distribution density*

$$(9.3) \qquad \phi_{m,C}(y) = (2\pi)^{-n/2} |\det C|^{-1/2} \exp\left[-\tfrac{1}{2}(y-m)^\top C^{-1}(y-m)\right],$$

$y \in \mathbb{R}^n$. Here, $C = BB^\top$, and the coordinates Y_i of Y satisfy $\mathbb{E}(Y_i) = m_i$ and $\mathrm{Cov}(Y_i, Y_j) = C_{ij}$, $1 \le i, j \le n$.

Proof. Since the coordinates X_i of X are independent, Example (3.30) shows that the random vector X has the product distribution density

$$\prod_{i=1}^{n} \phi_{0,1}(x_i) = (2\pi)^{-n/2} \exp\left[-\tfrac{1}{2}x^\top x\right] = \phi_{0,E}(x),$$

$x \in \mathbb{R}^n$. By Proposition (9.1), it follows that Y has the distribution density

$$\phi_{0,E}(B^{-1}(y-m))\,|\det B^{-1}| = (2\pi)^{-n/2}|\det B|^{-1}\exp\left[-\tfrac{1}{2}(y-m)^\top C^{-1}(y-m)\right].$$

Since $\det C = |\det B|^2$, the last expression is equal to $\phi_{m,C}(y)$, as required. Furthermore,

$$\mathbb{E}(Y_i) = \mathbb{E}\left(\sum_{j=1}^{n} B_{ij} X_j + m_i\right) = \sum_{j=1}^{n} B_{ij}\,\mathbb{E}(X_j) + m_i = m_i$$

and

$$\mathrm{Cov}(Y_i, Y_j) = \mathrm{Cov}\left(\sum_{k=1}^{n} B_{ik} X_k, \sum_{l=1}^{n} B_{jl} X_l\right)$$

$$= \sum_{k,l=1}^{n} B_{ik} B_{jl}\,\mathrm{Cov}(X_k, X_l) = \sum_{k=1}^{n} B_{ik} B_{jk} = C_{ij},$$

since $\mathrm{Cov}(X_k, X_l) = \delta_{kl}$ by the independence of the X_i. ◇

Now let $C \in \mathbb{R}^{n \times n}$ be an arbitrary positive definite symmetric $n \times n$ matrix. Then, by the diagonalisation theorem of linear algebra (see [40], Section VIII-4, for instance), there exist an orthogonal matrix O and a diagonal matrix D with diagonal entries $D_{ii} > 0$ such that $C = ODO^\top$. In particular, writing $D^{1/2}$ for the diagonal matrix with diagonal entries $\sqrt{D_{ii}}$, and setting $B = OD^{1/2}$, we obtain that $C = BB^\top$ for an invertible matrix B. The function $\phi_{m,C}$ in (9.3) is therefore the distribution density of a random vector, and hence a probability density. It defines a probability measure $\mathcal{N}_n(m, C)$ on $\mathbb{R}^n$. By the preceding theorem, this probability measure has expectation vector m and covariance matrix C.

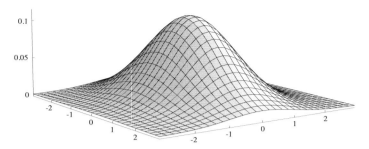

Figure 9.1. Density of the centred bivariate normal distribution with covariance matrix $\left(\begin{smallmatrix} 2 & 0 \\ 0 & 1 \end{smallmatrix}\right)$. The section lines are proportional to Gaussian bell curves. The lines of constant height (not shown) are ellipses.

Definition. For every positive definite symmetric matrix $\mathsf{C} \in \mathbb{R}^{n \times n}$ and every vector $m \in \mathbb{R}^n$, the probability measure $\mathcal{N}_n(m, \mathsf{C})$ on $(\mathbb{R}^n, \mathscr{B}^n)$ with the density $\phi_{m,\mathsf{C}}$ of (9.3) is called the *n-dimensional*, or *multivariate*, *normal* (or *Gaussian*) *distribution* with mean vector m and covariance matrix C. In particular, $\mathcal{N}_n(0, \mathsf{E}) = \mathcal{N}_{0,1}^{\otimes n}$ is the *multivariate standard normal distribution*.

The shapes of bivariate normal densities for two choices of C are shown in Figures 9.1 and 9.4. For $\mathsf{C} = \mathsf{E}$, $\mathcal{N}_n(0, \mathsf{E})$ exhibits the following invariance property, which will be basic for the statistical analysis of i.i.d. experiments with normal distribution.

(9.4) Corollary. *Rotational invariance of the multivariate standard normal distribution. The distribution $\mathcal{N}_n(0, \mathsf{E}) = \mathcal{N}_{0,1}^{\otimes n}$ is invariant under all orthogonal transformations (i.e., the rotoreflections) of $\mathbb{R}^n$.*

Proof. Let O be an orthogonal $n \times n$ matrix; we identify O with the linear transformation $x \to \mathsf{O}x$ of $\mathbb{R}^n$. If X is a random vector with distribution $\mathcal{N}_n(0, \mathsf{E})$, then $\mathcal{N}_n(0, \mathsf{E}) \circ \mathsf{O}^{-1}$ is the distribution of $\mathsf{O}X$. By Theorem (9.2), the latter is given by $\mathcal{N}_n(0, \mathsf{C})$, where $\mathsf{C} = \mathsf{O}\mathsf{O}^\top = \mathsf{E}$. Hence $\mathcal{N}_n(0, \mathsf{E}) \circ \mathsf{O}^{-1} = \mathcal{N}_n(0, \mathsf{E})$. $\diamond$

How do the multivariate normal distributions behave under general affine transformations?

(9.5) Theorem. *Transformation of multivariate normal distributions. Let Y be an $\mathcal{N}_n(m, \mathsf{C})$-distributed random vector in $\mathbb{R}^n$. For any $k \leq n$, let $\mathsf{A} \in \mathbb{R}^{k \times n}$ be a real $k \times n$ matrix of full rank, and $a \in \mathbb{R}^k$. Then, the random vector $Z = \mathsf{A}Y + a$ has the k-dimensional normal distribution $\mathcal{N}_k(\mathsf{A}m + a, \mathsf{A}\mathsf{C}\mathsf{A}^\top)$.*

Proof. Without loss of generality, we can set $a = 0$ and $m = 0$; the general case then follows by shifting the coordinates. By Theorem (9.2), we can also assume that $Y = \mathsf{B}X$ for an $\mathcal{N}_n(0, \mathsf{E})$-distributed random vector X and an invertible matrix B with $\mathsf{B}\mathsf{B}^\top = \mathsf{C}$. Let L be the subspace of $\mathbb{R}^n$ that is spanned by the row vectors of the $k \times n$

matrix AB. Since A has full rank, $\boldsymbol{L}$ has dimension k. By the Gram–Schmidt or-thonormalisation process [40], there exists an orthonormal basis $u_1, \ldots, u_k$ of $\boldsymbol{L}$ that can be extended to an orthonormal basis $u_1, \ldots, u_n$ of $\mathbb{R}^n$. Let O be the orthogonal matrix with row vectors $u_1, \ldots, u_n$. It is then easy to check that AB = (R|0) O, where R is the invertible $k \times k$ matrix that describes the change of basis of $\boldsymbol{L}$; (R|0) denotes the $k \times n$ matrix that has been augmented by zeros.

By Corollary (9.4), the random vector $\tilde{X} = OX$ still has the distribution $\mathcal{N}_n(0, \mathsf{E})$. Its coordinates $\tilde{X}_1, \ldots, \tilde{X}_n$ are thus independent with distribution $\mathcal{N}_{0,1}$. A fortiori, the coordinates of the shortened random vector $\hat{X} = (\tilde{X}_1, \ldots, \tilde{X}_k)^\top$ are independent with distribution $\mathcal{N}_{0,1}$. By Theorem (9.2) we thus find that $AY = ABX = (\mathsf{R}|0)\tilde{X} = \mathsf{R}\hat{X}$ has the distribution $\mathcal{N}_k(0, \mathsf{RR}^\top)$. Since

$$\mathsf{RR}^\top = (\mathsf{R}|0)(\mathsf{R}|0)^\top = (\mathsf{R}|0)\,\mathsf{OO}^\top(\mathsf{R}|0)^\top = \mathsf{ACA}^\top,$$

this is the desired result. $\diamond$

In the special case of $n = 2$, $k = 1$, $\mathsf{C} = \left(\begin{smallmatrix} v_1 & 0 \\ 0 & v_2 \end{smallmatrix}\right)$ and $\mathsf{A} = (1, 1)$, the theorem again yields the convolution statement of Example (3.32).

It is useful to extend the definition of the multivariate normal distribution to sym-metric matrices C that are only positive semidefinite, rather than (strictly) positive definite. (So, the eigenvalue zero is allowed.) This extension again uses the spectral decomposition $\mathsf{C} = \mathsf{ODO}^\top$ into an orthogonal matrix O and a non-negative diagonal matrix D; this representation is unique up to reflection and permutation of the coordi-nates.

Definition. Let $\mathsf{C} \in \mathbb{R}^{n \times n}$ be a positive semidefinite symmetric matrix, and suppose O and D are as above. Then, for every $m \in \mathbb{R}^n$, the multivariate normal distribution $\mathcal{N}_n(m, \mathsf{C})$ on $(\mathbb{R}^n, \mathscr{B}^n)$ is defined as the image of $\bigotimes_{i=1}^n \mathcal{N}_{0, \mathsf{D}_{ii}}$ under the affine trans-formation $x \to \mathsf{O}x + m$. Here we set $\mathcal{N}_{0,0} = \delta_0$, the Dirac distribution at the point $0 \in \mathbb{R}$.

In the positive definite case, Theorem (9.5) shows that this definition coincides with the previous one. In the general case, if 0 is an eigenvalue of C with multiplicity k, $\mathcal{N}_n(m, \mathsf{C})$ does not have a Lebesgue density, but 'lives' on an $(n - k)$-dimensional affine subspace of $\mathbb{R}^n$, namely the image of the linear subspace $\mathsf{D}(\mathbb{R}^n) = \{x \in \mathbb{R}^n : \mathsf{D}_{ii} = 0 \Rightarrow x_i = 0\}$ under the Euclidean motion $x \to \mathsf{O}x + m$.

9.2 The χ^2-, F- and t-Distributions

To control the sample variances in Gaussian models, one needs to know the distri-bution of the sum of squares of independent centred normal random variables. Such sums of squares turn out to have a gamma distribution. The basic observation is the following.

(9.6) Remark. *The square of a standard normal variable.* If X is an $\mathcal{N}_{0,1}$-distributed random variable, then X^2 has the gamma distribution $\Gamma_{1/2,1/2}$.

Proof. For reasons of symmetry, $|X|$ has the distribution density $2\,\phi_{0,1}$ on $\mathcal{X} = \,]0, \infty[$. (We can ignore the case $X = 0$, since it only occurs with probability 0.) Furthermore, $T : x \to x^2$ is a diffeomorphism from $\mathcal{X}$ onto itself with inverse function $T^{-1}(y) = \sqrt{y}$. Proposition (9.1) thus tells us that $X^2 = T(|X|)$ has the distribution density

$$\rho_T(y) = 2\,\phi_{0,1}(\sqrt{y})\,\frac{1}{2}\,y^{-1/2} = \frac{1}{\sqrt{2\pi}}\,e^{-y/2}y^{-1/2} = \frac{\Gamma(1/2)}{\sqrt{\pi}}\,\gamma_{1/2,1/2}(y)\,;$$

the last equation follows from the Definition (2.20) of the gamma densities. Since both ρ_T and $\gamma_{1/2,1/2}$ integrate to 1, we necessarily get $\Gamma(1/2) = \sqrt{\pi}$, and the claim follows. $\diamond$

The preceding remark suggests a further study of gamma distributions.

(9.7) Proposition. Relation between beta and gamma distributions. *Let $\alpha, r, s > 0$ and X, Y be independent random variables with gamma distributions $\Gamma_{\alpha,r}$ and $\Gamma_{\alpha,s}$, respectively. Then $X + Y$ and $X/(X + Y)$ are independent with distributions $\Gamma_{\alpha,r+s}$ and $\beta_{r,s}$, respectively.*

Proof. We know from (3.28) and (3.30) that the joint distribution of (X, Y) is the product measure $\Gamma_{\alpha,r} \otimes \Gamma_{\alpha,s}$ on $\mathcal{X} = \,]0, \infty[^2$, which has the density

$$\rho(x, y) = \gamma_{\alpha,r}(x)\,\gamma_{\alpha,s}(y) = \frac{\alpha^{r+s}}{\Gamma(r)\,\Gamma(s)}\,x^{r-1}y^{s-1}e^{-\alpha(x+y)}\,, \quad (x, y) \in \mathcal{X}.$$

We consider the diffeomorphism

$$T(x, y) = \left(x + y,\ \frac{x}{x + y}\right)$$

from $\mathcal{X}$ to $\mathcal{Y} := \,]0, \infty[\,\times\,]0, 1[$. Its inverse function is $T^{-1}(u, v) = (uv, u(1 - v))$ with Jacobian matrix

$$DT^{-1}(u, v) = \begin{pmatrix} v & u \\ 1-v & -u \end{pmatrix}.$$

Hence $|\det DT^{-1}(u, v)| = u$. Applying Proposition (9.1), we thus find that the random vector

$$\left(X + Y,\ \frac{X}{X + Y}\right) = T(X, Y)$$

has the distribution density

$$\rho_T(u,v) = \rho(uv, u(1-v))\, u$$

$$= \frac{\alpha^{r+s}}{\Gamma(r)\,\Gamma(s)}\, u^{r+s-1}\, e^{-\alpha u}\, v^{r-1}(1-v)^{s-1}$$

$$= \frac{\Gamma(r+s)}{\Gamma(r)\,\Gamma(s)}\, B(r,s)\, \gamma_{\alpha,r+s}(u)\, \beta_{r,s}(v), \quad (u,v) \in \mathcal{Y};$$

in the last step we have used the formulas (2.20) and (2.22) for the gamma and beta densities. Since ρ_T as well as $\gamma_{\alpha,r+s}$ and $\beta_{r,s}$ integrate to 1, the coefficient is necessarily equal to 1. This means that

(9.8) $$B(r,s) = \frac{\Gamma(r)\,\Gamma(s)}{\Gamma(r+s)} \quad \text{for } r,s > 0,$$

and ρ_T is a product density with the required factors. $\diamond$

The relation (9.8), which is well-known from analysis, is a nice by-product of the above proof. The statement about the distribution of $X + Y$ is equivalent to the following convolution property, which generalises the earlier Corollary (3.36).

(9.9) Corollary. Convolution of gamma distributions. *For all $\alpha, r, s > 0$, one has the identity*

$$\Gamma_{\alpha,r} * \Gamma_{\alpha,s} = \Gamma_{\alpha,r+s};$$

that is, if the scale parameter α is fixed, the gamma distributions form a convolution semigroup.

Combining this corollary with Remark (9.6), we obtain the following result about the distribution of the sum of squares of independent standard normal random variables.

(9.10) Theorem. The chi-square distributions. *Let $X_1, \ldots, X_n$ be i.i.d. random variables with standard normal distribution $\mathcal{N}_{0,1}$. Then $\sum_{i=1}^{n} X_i^2$ has the gamma distribution $\Gamma_{1/2,n/2}$.*

Definition. For each $n \geq 1$, the gamma distribution $\chi_n^2 := \Gamma_{1/2,n/2}$ with parameters $1/2, n/2$ and density

(9.11) $$\chi_n^2(x) := \gamma_{1/2,n/2}(x) = \frac{x^{n/2-1}}{\Gamma(n/2)\, 2^{n/2}}\, e^{-x/2}, \quad x > 0,$$

is called the *chi-square distribution with n degrees of freedom*, or the χ_n^2-distribution for short.

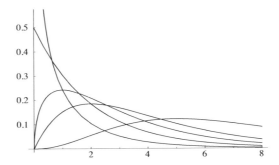

Figure 9.2. Densities of the χ_n^2-distributions for $n = 1, 2, 3, 4, 7$.

Figure 9.2 shows the densities χ_n^2 for various n. As it will turn out, the chi-square distributions play a central role in test theory. For example, the unbiased sample variance V^* in the Gaussian product model has a (rescaled) chi-square distribution, see Theorem (9.17) below. Next, we consider ratios of sums of squares of standard normal random variables.

(9.12) Theorem. The Fisher distributions. *Let $X_1, \ldots, X_m, Y_1, \ldots, Y_n$ be i.i.d. random variables with standard normal distribution $\mathcal{N}_{0,1}$. Then the quotient*

$$F_{m,n} := \frac{1}{m} \sum_{i=1}^{m} X_i^2 \Big/ \frac{1}{n} \sum_{j=1}^{n} Y_j^2$$

has the distribution density

$$(9.13) \qquad f_{m,n}(x) = \frac{m^{m/2} n^{n/2}}{B(m/2, n/2)} \frac{x^{m/2-1}}{(n+mx)^{(m+n)/2}}, \quad x > 0.$$

Proof. We know from Theorem (9.10) that $X := \sum_{i=1}^{m} X_i^2$ has the gamma distribution $\chi_m^2 = \Gamma_{1/2, m/2}$, and $Y := \sum_{j=1}^{n} Y_j^2$ the gamma distribution $\chi_n^2 = \Gamma_{1/2, n/2}$. By Theorem (3.24), X and Y are independent. Proposition (9.7) thus shows that $Z := X/(X + Y)$ has the beta distribution $\beta_{m/2, n/2}$. But now we can write

$$F_{m,n} = \frac{n}{m} \frac{X}{Y} = \frac{n}{m} \frac{Z}{1-Z} = T(Z)$$

for the diffeomorphism $T(x) = \frac{n}{m} \frac{x}{1-x}$ from $]0, 1[$ to $]0, \infty[$. Its inverse function is $T^{-1}(y) = \frac{my}{n+my}$. Hence, by Proposition (9.1), $F_{m,n}$ has the distribution density

$$\beta_{m/2, n/2}\left(\frac{my}{n+my}\right) \frac{mn}{(n+my)^2} = f_{m,n}(y),$$

and the proof is complete. $\diamond$

Definition. In honour of R. A. Fisher, the distribution $\mathcal{F}_{m,n}$ on $]0, \infty[$ with density $f_{m,n}$ of (9.13) is called the *Fisher distribution* with m and n degrees of freedom, or shorter the $F_{m,n}$-*distribution*.

The F-distributions play an important role for the comparison of various sample variances in Gaussian models, as we will see, in particular, in Chapter 12. Their relation to the beta distributions that came up in the preceding proof is spelled out again in the following remark.

(9.14) Remark. *Relation between Fisher and beta distributions.* For given $m, n \in \mathbb{N}$, let $T(x) = \frac{n}{m} \frac{x}{1-x}$. Then the identity $\mathcal{F}_{m,n} = \beta_{m/2, n/2} \circ T^{-1}$ is true. That is,

$$\mathcal{F}_{m,n}(]0, c]) = \beta_{m/2, n/2}\left(\left[0, \tfrac{mc}{n+mc}\right]\right)$$

for all $c > 0$. So if c is the α-quantile of $\mathcal{F}_{m,n}$, then $\frac{mc}{n+mc}$ is the α-quantile of $\beta_{m/2, n/2}$. Hence, one can obtain the quantiles of the beta distributions with integer or half-integer parameters from those of the F-distributions (and vice versa).

As we will see soon, the 'symmetrically signed root of $\mathcal{F}_{1,n}$' plays a special role.

(9.15) Corollary. *The Student distributions. Let $X, Y_1, \ldots, Y_n$ be i.i.d. random variables with standard normal distribution $\mathcal{N}_{0,1}$. Then*

$$T = X \Big/ \sqrt{\tfrac{1}{n} \sum_{j=1}^{n} Y_j^2}$$

has the distribution density

(9.16) $\qquad \tau_n(x) = \left(1 + \dfrac{x^2}{n}\right)^{-\frac{n+1}{2}} \Big/ B(1/2, n/2)\, \sqrt{n}\,, \quad x \in \mathbb{R}.$

Proof. Theorem (9.12) states that T^2 has the distribution $\mathcal{F}_{1,n}$. Proposition (9.1) thus shows that $|T| = \sqrt{T^2}$ has the density $f_{1,n}(y^2)\, 2y$, $y > 0$. But the symmetry of $\mathcal{N}_{0,1}$ implies that T is symmetrically distributed, which means that T and $-T$ have the same distribution. Therefore, T has the distribution density $f_{1,n}(y^2)\,|y| = \tau_n(y)$. $\diamond$

Definition. The probability measure t_n on $(\mathbb{R}, \mathscr{B})$ with the density τ_n defined in (9.16) is called *Student's t-distribution with n degrees of freedom*, or t_n-*distribution* for short.

For $n = 1$, one obtains the *Cauchy distribution* with density $\tau_1(x) = \frac{1}{\pi} \frac{1}{1+x^2}$, which already appeared in Problem 2.5.

The t-distribution was published in 1908 by the statistician W. S. Gosset, who, at that time, was working for the Guinness brewery. He had to publish it under the pseudonym 'Student', because the brewery did not allow its employees to publish scientific work.

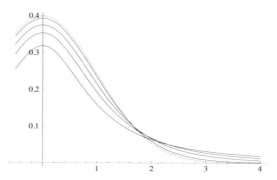

Figure 9.3. Densities of t_n-distributions for $n = 1, 2, 4, 16$; and density of the standard normal distribution (dotted) for comparison.

The densities τ_n of the t_n-distributions are illustrated in Figure 9.3 for several values of n. As is directly apparent from (9.16), the τ_n tend, in the limit as $n \to \infty$, to $\phi_{0,1}$, the density of the standard normal distribution; see also Problem 9.13. For fixed n, however, $\tau_n(x)$ decays in x only algebraically, in contrast to the superexponential decay of $\phi_{0,1}(x)$.

The significance of the t-distributions comes from the following theorem that exploits the rotational invariance of the multivariate standard normal distribution. In particular, the theorem answers the question about the distribution Q in Example (8.4), which we used to determine a confidence interval for the expectation of a normal distribution with unknown variance. As in this example, we consider the Gaussian product model together with the unbiased estimators

$$M = \frac{1}{n} \sum_{i=1}^{n} X_i, \quad V^* = \frac{1}{n-1} \sum_{i=1}^{n} (X_i - M)^2$$

for expectation and variance.

(9.17) Theorem. Student 1908. *Consider the n-fold Gaussian product model*

$$(\mathbb{R}^n, \mathscr{B}^n, \mathcal{N}_{m,v}^{\otimes n} : m \in \mathbb{R}, v > 0),$$

and let $\vartheta = (m, v) \in \mathbb{R} \times]0, \infty[$ be arbitrary. Then the following statements hold under $P_\vartheta = \mathcal{N}_{m,v}^{\otimes n}$.

(a) *M and V^* are independent.*

(b) *M has the distribution $\mathcal{N}_{m,v/n}$, and $\frac{n-1}{v} V^*$ has the distribution χ_{n-1}^2.*

(c) *$T_m := \frac{\sqrt{n}(M-m)}{\sqrt{V^*}}$ has the distribution t_{n-1}.*

At first glance, the independence statement (a) may cause some surprise, since M appears in the definition of V^*. Recall, however, that independence does not mean causal independence, but rather describes a proportional overlap of probabilities. In the present case, the independence is a consequence of the rotational invariance of the multivariate standard normal distribution; the statement fails for other probability measures. Statement (c) says that the distribution Q in Example (8.4) equals t_{n-1}.

Proof. We again write $X = (X_1, \ldots, X_n)^\top$ for the identity mapping on $\mathbb{R}^n$, and $\mathbf{1} = (1, \ldots, 1)^\top$ for the diagonal vector in $\mathbb{R}^n$. Moreover, we will assume without loss of generality that $m = 0$ and $v = 1$; otherwise we can replace X_i by $(X_i - m)/\sqrt{v}$, noticing that m and $\sqrt{v}$ cancel in the definition of V^* and T_m, respectively.

Let O be an orthogonal $n \times n$ matrix of the form

$$\mathsf{O} = \begin{pmatrix} \frac{1}{\sqrt{n}} & \cdots & \frac{1}{\sqrt{n}} \\ \text{arbitrary, but} \\ \text{orthogonal} \end{pmatrix};$$

such a matrix can be constructed by extending the unit vector $\frac{1}{\sqrt{n}}\mathbf{1}$ to an orthonormal basis. The matrix O maps the diagonal onto the 1-axis; indeed, since the row vectors of the second up to the nth row of O are orthogonal to $\mathbf{1}$, we find that $\mathsf{O}\mathbf{1} = (\sqrt{n}, 0, \ldots, 0)^\top$. Figure 9.4 illustrates the situation.

Now, let $Y = \mathsf{O}X$ and Y_i be the ith coordinate of Y. By Corollary (9.4) we know that Y has the same distribution as X, namely $\mathcal{N}_n(0, \mathsf{E}) = \mathcal{N}_{0,1}^{\otimes n}$. By Remark (3.28), the product form of this distribution implies that the $Y_1, \ldots, Y_n$ are indepen-

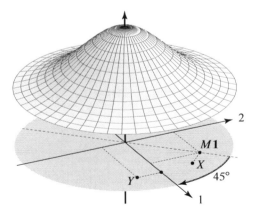

Figure 9.4. The case $n = 2$: A point $X \in \mathbb{R}^2$ is randomly chosen according to the Gaussian bell distribution $\mathcal{N}_2(0, \mathsf{E})$ (lifted upwards to expose the plane $\mathbb{R}^2$). $M\mathbf{1}$ is the projection of X onto the diagonal. By a rotation of $45°$, $M\mathbf{1}$ is mapped onto the 1-axis, and X moves to the point Y, which, by the rotational symmetry of the Gaussian bell, still has the distribution $\mathcal{N}_2(0, \mathsf{E})$. In particular, $Y_1 = M|\mathbf{1}| = M\sqrt{2}$, and $|Y_2|$ is the distance between X and $M\mathbf{1}$, viz. $\sqrt{V^*}$.

dent. The point is now that $M = \frac{1}{\sqrt{n}} \sum_{i=1}^{n} \frac{1}{\sqrt{n}} X_i = \frac{1}{\sqrt{n}} Y_1$ and, on the other hand,

$$(n-1) V^* = \sum_{i=1}^{n} (X_i - M)^2 = \sum_{i=1}^{n} X_i^2 - n M^2 = |Y|^2 - Y_1^2 = \sum_{i=2}^{n} Y_i^2$$

because $|Y| = |X|$. Statement (a) is thus an immediate consequence of the independence of the Y_i. Claim (b) follows from the scaling property of the normal distribution (cf. Problem 2.15) and Theorem (9.10). Finally, statement (c) follows from (a), (b), and Corollary (9.15). $\diamond$

As we have seen in Example (8.4), the quantiles and fractiles of the t_n-distributions are needed for the construction of confidence intervals in Gaussian models. They are also required in various test procedures, which will be discussed in the next chapters. Likewise, the quantiles of the χ^2 and F-distributions are needed in this context. A selection of these quantiles can be found in the tables on pp. 386 ff. Missing quantiles can either be obtained by interpolation, or by approximation as in Problems 9.11 and 9.13. Alternatively, one can use suitable computer programs such as `Maple`, `Mathematica`, `R`, or `SciLab`.

Problems

9.1 Find the distribution function and the distribution density of $Y = e^X$ when X has distribution (a) $\mathcal{N}_{m,v}$, or (b) $\mathcal{E}_\alpha$. In case (a), the distribution of Y is called the *lognormal distribution* for m and v, and in case (b) the *Pareto distribution* for α.

9.2 ^{S} *Best linear prediction.* Suppose that the joint distribution of the random variables $X_1, \ldots, X_n$ is an n-dimensional normal distribution. Show the following.

   (a) $X_1, \ldots, X_n$ are independent if and only if they are pairwise uncorrelated.
   (b) There exists a linear combination $\hat{X}_n := \sum_{i=0}^{n-1} a_i X_i$ of $X_1, \ldots, X_{n-1}$ and the constant $X_0 := 1$ such that $\hat{X}_n - X_n$ is independent of $X_1, \ldots, X_{n-1}$ and has expectation 0. *Hint:* Minimise the mean squared deviation $\mathbb{E}((\hat{X}_n - X_n)^2)$ and use (a).

9.3 *Geometry of the bivariate normal distribution.* Suppose $\mathsf{C} = \begin{pmatrix} v_1 & c \\ c & v_2 \end{pmatrix}$ is positive definite, and let $\phi_{0,\mathsf{C}}$ be the density of the corresponding bivariate centred normal distribution. Show the following.

   (a) The contour lines $\{x \in \mathbb{R}^2 : \phi_{0,\mathsf{C}}(x) = h\}$ for $0 < h < (2\pi \sqrt{\det \mathsf{C}})^{-1}$ are ellipses. Determine the principal axes.
   (b) The sections $\mathbb{R} \ni t \to \phi_{0,\mathsf{C}}(a + tb)$ with $a, b \in \mathbb{R}^2$, $b \neq 0$ are proportional to one-dimensional normal densities $\phi_{m,v}$ with suitable m, v.

9.4 Let X be an $\mathcal{N}_n(0, \mathsf{E})$-distributed n-dimensional random vector, and let A and B be $k \times n$ and $l \times n$ matrices of rank k and l, respectively. Show that AX and BX are independent if and only if $AB^\top = 0$. *Hint:* Assume without loss of generality that $k + l \leq n$. In the proof of the 'only if' direction, verify first that the $(k+l) \times n$ matrix $\mathsf{C} := \begin{pmatrix} A \\ B \end{pmatrix}$ has rank $k + l$, and conclude that $\mathsf{C}X$ has distribution $\mathcal{N}_{k+l}(0, \mathsf{C}\mathsf{C}^\top)$.

9.5[S] *Normal distributions as maximum-entropy distributions.* Let C be a positive definite symmetric $n \times n$ matrix, and consider the class $\mathscr{W}_C$ of all probability measures P on $(\mathbb{R}^n, \mathscr{B}^n)$ with the properties

> P is centred with covariance matrix C, that is, the projections $X_i : \mathbb{R}^n \to \mathbb{R}$ satisfy $\mathbb{E}(X_i) = 0$ and $\mathrm{Cov}(X_i, X_j) = C_{ij}$ for all $1 \leq i, j \leq n$, and

> P has a Lebesgue density ρ, and there exists the *differential entropy*

$$H(P) = -\int_{\mathbb{R}^n} dx \, \rho(x) \log \rho(x).$$

Show that

$$H(\mathcal{N}_n(0, C)) = \frac{n}{2} \log\left[2\pi e (\det C)^{1/n}\right] = \max_{P \in \mathscr{W}_C} H(P).$$

Hint: Consider the relative entropy $H(P; \mathcal{N}_n(0, C))$; cf. Remark (7.31).

9.6[S] *Balls of maximal $\mathcal{N}_n(0, E)$-probability.* Show that

$$\mathcal{N}_n(0, E)(|X - m| < r) \leq \mathcal{N}_n(0, E)(|X| < r)$$

for all $n \in \mathbb{N}$, $r > 0$ and $m \in \mathbb{R}^n$; here, X stands for the identity mapping on $\mathbb{R}^n$. That is, balls have maximal $\mathcal{N}_n(0, E)$-probability when they are centred. *Hint:* You can use induction on n and Fubini's theorem (Problem 4.8).

9.7 *Moments of the chi-square and t-distributions.* Let Y and Z be real random variables with distributions χ_n^2 and t_n, respectively.

(a) Show that $\mathbb{E}(Y^{-k/2}) = \Gamma((n-k)/2)/[\Gamma(n/2) \, 2^{k/2}]$ for $k < n$.

(b) Determine the moments of Z up to order $n-1$ and show that the nth moment of Z does not exist. *Hint:* Use Problem 4.17.

9.8 *Gamma and Dirichlet distributions.* Let $s \geq 2$, $\alpha > 0$, $\rho \in \,]0, \infty[^s$ and suppose that $X_1, \ldots, X_s$ are independent random variables with gamma distributions $\Gamma_{\alpha, \rho(1)}, \ldots, \Gamma_{\alpha, \rho(s)}$. Also, let $X = \sum_{i=1}^s X_i$. Show that the random vector $(X_i/X)_{1 \leq i \leq s}$ is independent of X and has the Dirichlet distribution $\mathcal{D}_\rho$ that was defined in Problem 7.28.

9.9 *Non-central chi-square distributions.* Let $a \in \mathbb{R}^n$ and X be an $\mathcal{N}_n(0, E)$-distributed n-dimensional random vector. Show that the distribution of $|X + a|^2$ depends only on (n and) the Euclidean norm $|a|$ of a. This distribution is called the non-central χ_n^2-distribution with non-centrality parameter $\lambda = |a|^2$. *Hint:* Corollary (9.4).

9.10[S] *Representation of non-central chi-square distributions.* Let $X, X_1, X_2, \ldots$ be i.i.d. random variables with standard normal distribution $\mathcal{N}_{0,1}$, and let $a, a_1, a_2, \ldots \in \mathbb{R}$. Show the following.

(a) $Y = (X + a)^2$ has the distribution density

$$\rho_a(y) = (2\pi y)^{-1/2} \, e^{-(y + a^2)/2} \, \cosh(a\sqrt{y}), \quad y > 0.$$

(b) The probability density ρ_a is also the distribution density of $\sum_{i=1}^{2Z+1} X_i^2$, where Z is a $\mathcal{P}_{a^2/2}$-distributed random variable that is independent of the X_i. *Hint:* Use the series expansion of cosh.

(c) Let $n \geq 1$ and define $\lambda = \sum_{i=1}^{n} a_i^2$. Then the sum $\sum_{i=1}^{n}(X_i + a_i)^2$ has the same distribution as $\sum_{i=1}^{2Z+n} X_i^2$, where Z is a $\mathcal{P}_{\lambda/2}$-distributed random variable that is independent of the X_i.

9.11 *Approximation of chi-square quantiles.* For $0 < \alpha < 1$ and $n \in \mathbb{N}$, let $\chi_{n;\alpha}^2$ be the α-quantile of the χ_n^2-distribution and $\Phi^{-1}(\alpha)$ the α-quantile of the standard normal distribution. Show that $(\chi_{n;\alpha}^2 - n)/\sqrt{2n} \to \Phi^{-1}(\alpha)$ as $n \to \infty$. *Hint:* Use Problem 4.17 or Example (7.27b).

9.12^S *Fisher's approximation.* Suppose that S_n has the chi-square distribution χ_n^2. Use Problem 5.21 to show that $\sqrt{2S_n} - \sqrt{2n} \xrightarrow{d} \mathcal{N}_{0,1}$. Deduce that $\sqrt{2\chi_{n;\alpha}^2} - \sqrt{2n} \to \Phi^{-1}(\alpha)$ as $n \to \infty$. Use the tables in the appendix to compare this approximation with the one from the previous problem for the levels $\alpha = 0.6, 0.9$ and 0.99 as well as $n = 10, 25, 50$.

9.13 *Approximation of t- and F-distributions.* Let $c \in \mathbb{R}$, $0 < \alpha < 1$ and $m \in \mathbb{N}$ be given. Show that

(a) $t_n(]-\infty, c]) \to \Phi(c)$ and $t_{n;\alpha} \to \Phi^{-1}(\alpha)$,

(b) $\mathcal{F}_{m,n}([0,c]) \to \chi_m^2([0, mc])$ and $f_{m,n;\alpha} \to \chi_{m;\alpha}^2/m$

as $n \to \infty$. Here, $t_{n;\alpha}$, $f_{m,n;\alpha}$, $\chi_{m;\alpha}^2$, and $\Phi^{-1}(\alpha)$ are the α-quantiles of t_n, $\mathcal{F}_{m,n}$, χ_m^2, and $\mathcal{N}_{0,1}$, respectively.

9.14 *Non-central F-distributions.* Let $m, n \in \mathbb{N}$, $a_1, \ldots, a_m \in \mathbb{R}$, $\lambda = \sum_{i=1}^{m} a_i^2$, and $X_1, \ldots, X_m, Y_1, \ldots, Y_n$ be i.i.d. standard normal random variables. Show that the random variable

$$F_{m,n;a_1,\ldots,a_m} := \frac{1}{m} \sum_{i=1}^{m}(X_i + a_i)^2 \Big/ \frac{1}{n} \sum_{j=1}^{n} Y_j^2$$

has the distribution density

$$f_{m,n,\lambda}(y) := \sum_{k \geq 0} \mathcal{P}_\lambda(\{k\}) \, f_{m+2k,n}\left(\frac{my}{m+2k}\right) \frac{m}{m+2k}$$

$$= e^{-\lambda} \sum_{k \geq 0} \frac{\lambda^k n^{n/2} m^{m/2+k}}{k! \, B(m/2+k, n/2)} \frac{y^{m/2+k-1}}{(n+my)^{(m+n)/2+k}}$$

for $y \geq 0$. The corresponding probability measure is called the non-central $F_{m,n}$-distribution with non-centrality parameter λ. *Hint:* Problem 9.10c.

9.15 *Non-central Student distributions.* The non-central t-distribution $t_{n,\lambda}$ with n degrees of freedom and non-centrality parameter $\lambda > 0$ is defined as the distribution of the random variables $T = Z/\sqrt{S/n}$ for independent random variables Z and S with distribution $\mathcal{N}_{\lambda,1}$ and χ_n^2, respectively. Show that $t_{n,\lambda}$ has the distribution density

$$\tau_{n,\lambda}(x) = \frac{1}{\sqrt{\pi n} \, \Gamma(n/2) \, 2^{(n+1)/2}} \int_0^\infty ds \; s^{(n-1)/2} \exp\left[-s/2 - (x\sqrt{s/n} - \lambda)^2/2\right].$$

Hint: First determine the joint distribution of Z and S.

9.16 *Confidence region in the Gaussian product model.* Consider the n-fold Gaussian product model with unknown parameter $\vartheta = (m, v) \in \mathbb{R} \times]0, \infty[$. Let $\alpha \in]0, 1[$ be given and define $\beta_\pm = (1 \pm \sqrt{1 - \alpha})/2$. Also, let $u = \Phi^{-1}(\beta_+)$ be the β_+-quantile of $\mathcal{N}_{0,1}$ and $c_\pm = \chi^2_{n-1;\beta_\pm}$ the $\beta_\pm$-quantile of χ^2_{n-1}. Show that

$$C(\cdot) = \left\{ (m, v) : |m - M| \le u \sqrt{v/n}, \; (n-1)\, V^*/c_+ \le v \le (n-1)\, V^*/c_- \right\}$$

is a confidence region for ϑ with error level α. Make a sketch of $C(\cdot)$.

9.17 ^{S} *Two-sample problem in Gaussian product models with known variance.* Consider independent random variables $X_1, \ldots, X_n, Y_1, \ldots, Y_n$, where each X_i has distribution $\mathcal{N}_{m,v}$ and each Y_j distribution $\mathcal{N}_{m',v}$. The expectations m, m' are assumed to be unknown, but the common variance $v > 0$ is known. Construct a confidence circle for (m, m') with error level α.

9.18 (a) Let $X_1, \ldots, X_n$ be independent, $\mathcal{N}_{m,v}$-distributed random variables with known m and unknown v. Determine a confidence interval for v with error level α.

(b) *Two-sample problem with known expectations.* Let $X_1, \ldots, X_k, Y_1, \ldots, Y_l$ be independent random variables, where the X_i are $\mathcal{N}_{m,v}$-distributed with known m, and the Y_j are $\mathcal{N}_{m',v'}$-distributed with known m'. The variances $v, v' > 0$ are supposed to be unknown. Determine a confidence interval for v/v' with error level α.

9.19 ^{S} *Sequential confidence intervals of given maximal length, Charles M. Stein 1945.* Let $X_1, X_2, \ldots$ be independent, $\mathcal{N}_{m,v}$-distributed random variables with unknown parameters m and v. Furthermore, let $n \ge 2$ and $0 < \alpha < 1$ be given and $t = t_{n-1;1-\alpha/2}$ the $\alpha/2$-fractile of the t_{n-1}-distribution. Consider the unbiased estimators

$$M_n = \frac{1}{n} \sum_{i=1}^{n} X_i \quad \text{and} \quad V_n^* = \frac{1}{n-1} \sum_{i=1}^{n} (X_i - M_n)^2$$

of m and v after n experiments. Finally, let $\varepsilon > 0$ and consider the random sample size $N = \max\{n, \lceil (t/\varepsilon)^2 V_n^* \rceil\}$ (where $\lceil x \rceil$ denotes the smallest integer $\ge x$) as well as the associated estimator $M_N = \sum_{i=1}^{N} X_i/N$ of m after N experiments. Show the following.

(a) $(M_N - m)\sqrt{N/v}$ is independent of V_n^* and has standard normal distribution $\mathcal{N}_{0,1}$.

(b) $(M_N - m)\sqrt{N/V_n^*}$ is t_{n-1}-distributed, and $]M_N - t\sqrt{V_n^*/N}, \; M_N + t\sqrt{V_n^*/N}[$ is a confidence interval for m with error level α and total length at most 2ε.

Chapter 10

Hypothesis Testing

In the theory of estimation, the observations are used to make a sensible guess of the unknown quantity. By way of contrast, the objective of test theory is to develop rules for the rational behaviour in random situations that require a clear-cut decision (with possibly far-reaching consequences). The decision problem is stated in terms of a hypothesis about the true random mechanism that governs the observations. The observation results are then used to decide whether or not the hypothesis can be accepted. As the observations are random, such a decision can obviously be erroneous. The goal is therefore to find decision rules that keep the probability of error as small as possible – irrespectively of what is actually the case.

10.1 Decision Problems

As a motivation we recall Example (7.1).

(10.1) Example. *Quality checking.* An orange importer receives a delivery of $N = 10\,000$ oranges. He only has to pay the agreed price if at most 5% have gone bad. To check whether this is the case, he takes a sample of size $n = 50$ and decides on a threshold value c, namely the maximal number of bad oranges in the sample he is willing to tolerate. He then uses the following decision rule:

$$\text{at most } c \text{ bad oranges in the sample} \Rightarrow \text{accept the delivery,}$$

$$\text{more than } c \text{ bad oranges in the sample} \Rightarrow \text{demand price reduction.}$$

Clearly, the point is to make a sensible choice of the threshold c. How can this be done? In general, one proceeds as follows.

Statistical decision procedure.

Step 1: Formulation of the statistical model. As always, the first step is to specify the appropriate statistical model. In the present example, this is the hypergeometric model $\mathcal{X} = \{0, \dots, n\}$, $\Theta = \{0, \dots, N\}$, and $P_\vartheta = \mathcal{H}_{n;\vartheta, N-\vartheta}$ for $\vartheta \in \Theta$.

Step 2: Formulation of null hypothesis and alternative. The parameter set Θ is decomposed into two subsets Θ_0 and Θ_1 according to the following principle:

$\vartheta \in \Theta_0 \quad \Leftrightarrow \quad \vartheta$ is satisfactory, that is, ϑ is considered a normal case that requires no particular action.

$\vartheta \in \Theta_1 \quad \Leftrightarrow \quad \vartheta$ is problematic, i.e., ϑ represents a deviation from the normal case, which should be detected whenever it occurs.

This decomposition being made, one says that the *null hypothesis* $H_0 : \vartheta \in \Theta_0$ is to be tested against the *alternative* $H_1 : \vartheta \in \Theta_1$. (It is advisable to speak of the *null* hypothesis instead of simply the hypothesis. Otherwise, there is danger of confusion because the alternative describes the 'suspicious case' and so corresponds to what is colloquially called the hypothesis.) In our example, from the importer's point of view, things are

$\triangleright$ satisfactory   if $\vartheta \in \Theta_0 = \{0, \ldots, 500\}$ (quality is fine),
$\triangleright$ problematic   if $\vartheta \in \Theta_1 = \{501, \ldots, 10\,000\}$ (quality is insufficient).

An unscrupulous supplier would possibly pursue the opposite interests, and thus interchange the indices 0 and 1.

Step 3: Choice of a significance level. There are two possible sources of error: a false rejection of the null hypothesis ('type I error'), and a false acceptance of the null hypothesis ('type II error'). The first is considered as so embarrassing that it should be avoided as much as possible. Therefore, one specifies a so-called *significance level* $0 < \alpha < 1$, for example $\alpha = 0.05$, and requires of the decision rule (still to be found) that the probability of a false rejection of the null hypothesis should not exceed α. (Note that this requirement introduces an asymmetry between null hypothesis and alternative.)

Step 4: Choice of a decision rule. One constructs a statistic $\varphi : \mathcal{X} \rightarrow [0, 1]$ as follows. If $x \in \mathcal{X}$ was observed, then $\varphi(x)$ indicates the tendency of deciding in favour of the alternative. Specifically, this means:

$\varphi(x) = 0$ $\Leftrightarrow$ the null hypothesis is accepted, i.e., the suspicion that the alternative might be true is not justified by the observed x.

$\varphi(x) = 1$ $\Leftrightarrow$ on the basis of x, the null hypothesis is rejected, and the alternative is supposed to be true.

$0 < \varphi(x) < 1$ $\Leftrightarrow$ x does not allow a clear-cut decision, and so one should decide at random, with a probability $\varphi(x)$ for rejection of the null hypothesis.

In our example, the importer can use a decision rule like

$$\varphi(x) = \begin{cases} 1 & \text{if } x > c, \\ 1/2 & \text{if } x = c, \\ 0 & \text{if } x < c, \end{cases}$$

and the above question about the choice of c can be answered. Namely, c should be the smallest number such that the requirement of Step 3 is still satisfied.

Step 5: Performing the experiment. This should be done only now, at the end of the procedure. Why not earlier? Because otherwise deception and self-deception are almost inevitable! Suppose you have a suspicion that you would like to verify, and you make the respective observations right at the beginning. Then you can steal a glance at the data and

> ▷ adjust null hypothesis and alternative to the data,

> ▷ choose the level and the decision rule appropriately, and

> ▷ if necessary eliminate troublesome 'outliers',

until the decision rule leads to the desired result. If you proceed in this fashion (and, unfortunately, human nature has a tendency to do so), then there are no probabilities involved, and the result is predetermined. The test then only serves the purpose of giving a preconceived opinion a pseudo-scientific justification!

Figure 10.1 illustrates the main ideas of the statistical decision procedure.

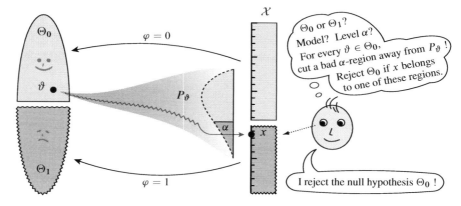

Figure 10.1. The principle of testing. Θ is partitioned into the 'normal cases' Θ_0 and the 'problematic cases' Θ_1; using the observed x, one tries to find out which part the true ϑ actually belongs to. The statistician analyses the model as usual, specifies a significance level α and, in the non-randomised case, decomposes the sample space $\mathcal{X}$ into an acceptance region $\{\varphi = 0\}$ and a rejection region $\{\varphi = 1\}$. This decomposition should be such that the 'embarrassing error' (namely $\vartheta \in \Theta_0$, but $\varphi(x) = 1$, as illustrated here) has at most probability α, and such that the reversed error is also as unlikely as possible. This can be achieved by cutting an atypical α-region away from every P_ϑ with $\vartheta \in \Theta_0$.

What is the mathematical core of the above decision process?

Definition. Let $(\mathcal{X}, \mathcal{F}, P_\vartheta : \vartheta \in \Theta)$ be a statistical model, and assume $\Theta = \Theta_0 \cup \Theta_1$ is a decomposition of Θ into null hypothesis and alternative.

(a) Every statistic $\varphi : \mathcal{X} \to [0, 1]$ (if interpreted as a decision rule) is called a *test* of Θ_0 against Θ_1. A test φ is called *non-randomised* if $\varphi(x) = 0$ or 1 for all $x \in \mathcal{X}$, otherwise it is termed *randomised*. In the first case, the set $\{x \in \mathcal{X} : \varphi(x) = 1\}$ is called the *rejection region* or *critical region*, and $\{x \in \mathcal{X} : \varphi(x) = 0\}$ the *acceptance region* of the test φ.

(b) The worst case probability of a type I error is $\sup_{\vartheta \in \Theta_0} \mathbb{E}_{\vartheta}(\varphi)$; it is called the *size*, or *effective level*, of φ. A test φ is called a *test of (significance, or error) level α*, if $\sup_{\vartheta \in \Theta_0} \mathbb{E}_{\vartheta}(\varphi) \leq \alpha$.

(c) The function $G_{\varphi} : \Theta \to [0, 1]$, $G_{\varphi}(\vartheta) = \mathbb{E}_{\vartheta}(\varphi)$, is called the *power function* of the test φ. For $\vartheta \in \Theta_1$, $G_{\varphi}(\vartheta)$ is called the *power* of φ at ϑ. So the power is the probability that the alternative is recognised when it is true, and $\beta_{\varphi}(\vartheta) = 1 - G_{\varphi}(\vartheta)$ is the probability of a type II error, namely that the alternative is not recognised as true, so that the null hypothesis is falsely accepted.

The preceding discussion leads to the following two

Requirements of a test φ:

▷ $G_{\varphi}(\vartheta) \leq \alpha$ for all $\vartheta \in \Theta_0$; that is, φ must respect the level α, in that the error probability of type I is at most α.

▷ $G_{\varphi}(\vartheta) \overset{!}{=}$ max for all $\vartheta \in \Theta_1$; that is, the power should be as large as possible, or equivalently, a type II error should be as unlikely as possible.

These requirements lead to the following notion.

Definition. A test φ of Θ_0 against Θ_1 is called a *(uniformly) most powerful test of level α*, if its size is at most α, and its power exceeds that of any other test ψ of level α; the latter means that

$$G_{\varphi}(\vartheta) \geq G_{\psi}(\vartheta) \quad \text{for all } \vartheta \in \Theta_1 .$$

(A common abbreviation in the literature is 'UMP test'.)

Our aim is now to find most powerful tests. Note, however, that a most powerful test is not necessarily good, that is, good enough for the specific decision problem at hand. In each application, one faces the problem of finding the right balance between the significance level and the power of a test. The smaller the level, the smaller, in general, the power as well. In other words, the harder one tries to avoid a type I error, the smaller are the chances of discovering the alternative when it is true, and the more likely is a type II error. If level and power do not allow a well-grounded decision, then the only (possibly inconvenient) way out is to increase the available information by making more or better observations. The following example will illustrate this and further problems.

(10.2) Example. *Extrasensory perception, binomial test.* A medium claims that using her extrasensory powers she can identify playing cards that are lying face down on a table. This shall be verified in the following experiment: the queen of hearts and the king of hearts from a brand new deck are placed face down in a random arrangement in front of the medium, and the medium is supposed to identify the queen of hearts. This shall be repeated $n = 20$ times. Following the textbook closely, the experimenter actually proceeds as follows:

▷ A suitable model is clearly the binomial model with $\mathcal{X} = \{0, \ldots, n\}$, $P_\vartheta = \mathcal{B}_{n,\vartheta}$ and $\Theta = [\frac{1}{2}, 1]$ (because simply by guessing the medium can already achieve the success probability $\frac{1}{2}$).

▷ The decision problem to be tested consists of the null hypothesis $\Theta_0 = \{\frac{1}{2}\}$ and the alternative $\Theta_1 =]\frac{1}{2}, 1]$. Indeed, the embarrassing error would be to acknowledge extrasensory powers of the medium, even though she purely relies on guessing.

▷ A reasonable level of significance is $\alpha = 0.05$; this is small enough to defend a positive test result against critics.

▷ As in Example (10.1), it is natural to use a test of the form $\varphi = 1_{\{c,\ldots,n\}}$ for some appropriate threshold value $c \in \mathcal{X}$; a detailed justification follows below. (Right after Theorem (10.10), we will even see that such a φ is optimal for its size.) By looking up a table of binomial quantiles, the experimenter finds that he has to choose $c = 15$ in order to adhere to the level α. It then follows that $G_\varphi(\frac{1}{2}) = \mathcal{B}_{n,1/2}(\{15, \ldots, n\}) \approx 0.0207$. 'Even better', reasons the experimenter, 'the size is actually smaller, and therefore the test result more convincing.'

▷ The test is carried out, and the medium achieves $x = 14$ successes. Therefore, $\varphi(x) = 0$ and the experimenter has to tell the medium (and the public) that the test could not confirm the extrasensory powers.

Thinking it over, the experimenter finds this result rather disappointing. He is impressed by the number of successes (and the aura of the medium) and argues as follows:

'It is only the set-up of my experiment that is to blame for the result. If I had chosen the test $\psi = 1_{\{14,\ldots,n\}}$, then I would have attested that the medium has extrasensory powers, and ψ still has level $\mathcal{B}_{n,1/2}(\{14, \ldots, n\}) \approx 0.0577$ – hardly more than what I originally intended to achieve. Also, if the medium really has a success probability of 0.7, then my test gives her only a chance of 41% to recognises these powers. Indeed, one has $G_\varphi(0.7) = \mathcal{B}_{n,0.7}(\{15, \ldots, n\}) \approx 0.4164$, whereas $G_\psi(0.7) \approx 0.6080$.'

These statements about ψ would be correct, if the threshold level $c = 14$ was chosen *before* the experiment. Now, afterwards, it is in fact not the number 14 that is chosen as the threshold, but the observed x. This means that ψ is actually defined by the equation $\psi(x) = 1_{\{x,\ldots,n\}}(x)$, and is thus identically 1. In other words, the decision rule ψ does not rely on the observation, but rather accepts the alternative as true; in particular, its size $G_\psi(\frac{1}{2})$ is equal to 1, and so is its power $G_\psi(\vartheta)$ for all $\vartheta \in \Theta_1$.

It is true that the power of φ is rather poor, but this was already known when the experiment was set up. If this power did not seem to be sufficient, one could have

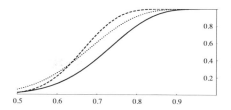

Figure 10.2. Power functions of the tests with rejection region $\{15, \ldots, 20\}$ (solid line) and $\{14, \ldots, 20\}$ (dotted line) for $n = 20$, as well as $\{27, \ldots, 40\}$ for $n = 40$ (dashed line). The values at the point $1/2$ correspond to the respective test sizes; they are $0.0207, 0.0577, 0.0192$.

increased the number n of experiments. For the given level and the corresponding choice of c, this would have improved the power of φ. Indeed, as Figure 10.2 shows, one obtains a significantly better power function if the medium has to achieve 27 successes in 40 attempts, even though $27/40 < 15/20$.

Why should the test take the form $\varphi = 1_{\{c,\ldots,n\}}$? Could we not have chosen a test of the form $\chi = 1_{\{c,\ldots,d\}}$ with $d < n$? As the size of χ is obviously smaller than that of φ, this might appear attractive. However, if the medium is in good shape and achieves more than d successes, then the hypothesis of extrasensory powers has to be rejected for χ! In other words, even though the power of χ increases for moderate abilities, it decreases again for strong abilities. In particular, we get $G_\chi(1) = 0 < G_\chi(\frac{1}{2})$. Thus, if we use χ and the medium performs well, then the extrasensory powers are only accepted with a smaller probability than if the medium were purely guessing. To exclude such absurdities, one introduces the following notion.

Definition. A test φ is called *unbiased of level α*, if

$$G_\varphi(\vartheta_0) \leq \alpha \leq G_\varphi(\vartheta_1) \quad \text{for all } \vartheta_0 \in \Theta_0 \text{ and } \vartheta_1 \in \Theta_1,$$

which means that one decides for the alternative with a higher probability when it is correct than when it is false.

In the following, we will mainly deal with the problems of existence and construction of most powerful tests. But unbiasedness will sometimes also come into play.

10.2 Neyman–Pearson Tests

In this section, we consider the particularly simple case of a binary decision problem, in which one has to choose between just two probability measures P_0 and P_1. So we consider a statistical model of the form $(\mathcal{X}, \mathscr{F}; P_0, P_1)$ with $\Theta = \{0, 1\}$, and the null hypothesis $\Theta_0 = \{0\}$ and the alternative $\Theta_1 = \{1\}$ are *simple* in that they are singleton sets. Furthermore, we suppose that the model is a standard model, so that P_0 and P_1 are given by suitable probability densities ρ_0 and ρ_1 on $\mathcal{X}$.

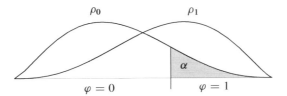

Figure 10.3. The construction of Neyman–Pearson tests.

We are in search of a most powerful test φ of P_0 against P_1 for a given level α. To get an intuition, we look at the densities ρ_0 and ρ_1, cf. Figure 10.3. According to the maximum likelihood principle, it is natural to decide for the alternative as soon as $\rho_1(x)$, the likelihood of the observed x under the alternative, dominates sufficiently strongly over $\rho_0(x)$. The degree of dominance of ρ_1 over ρ_0 can be described by the *likelihood ratio*

$$R(x) = \begin{cases} \rho_1(x)/\rho_0(x) & \text{if } \rho_0(x) > 0, \\ \infty & \text{if } \rho_0(x) = 0. \end{cases}$$

'Sufficiently strong dominance' therefore means that the likelihood ratio $R(x)$ exceeds an appropriately chosen threshold value c. We are thus led to tests of the form

$$\varphi(x) = \begin{cases} 1 & \text{if } R(x) > c, \\ 0 & \text{if } R(x) < c. \end{cases}$$

(The case $R(x) = c$ is deliberately excluded.) Such a φ is called a *Neyman–Pearson test* for the threshold value c (after the Ukrainian statistician Jerzy Neyman, 1894–1981, who later worked in Poland and the USA, and the Briton Egon Sharpe Pearson, 1895–1980). The following theorem, which is fundamental in test theory, shows that tests of this type are in fact optimal.

(10.3) Theorem. Neyman–Pearson lemma, 1932. *Let $(\mathcal{X}, \mathcal{F}; P_0, P_1)$ be a standard model with simple hypothesis and simple alternative, and let $0 < \alpha < 1$ be a given significance level. Then the following holds.*

(a) *There exists a Neyman–Pearson test φ such that $\mathbb{E}_0(\varphi) = \alpha$ (which therefore exhausts the level α completely).*

(b) *Each Neyman–Pearson test φ with $\mathbb{E}_0(\varphi) = \alpha$ is a most powerful test of level α, and each most powerful test ψ of level α is indistinguishable from a Neyman–Pearson test.*

Proof. (a) Let c be an arbitrary α-fractile of $P_0 \circ R^{-1}$. Such a fractile exists since R takes the value ∞ only on the set $\{\rho_0 = 0\}$, which has probability 0 under P_0; hence $P_0(R < \infty) = 1$, and $P_0 \circ R^{-1}$ is a probability measure on $\mathbb{R}$. By definition, we thus get that $P_0(R \geq c) \geq \alpha$ and $P_0(R > c) \leq \alpha$, and therefore

$$\alpha - P_0(R > c) \leq P_0(R \geq c) - P_0(R > c) = P_0(R = c).$$

We now distinguish two cases. If $P_0(R = c) = 0$, then $P_0(R > c) = \alpha$ by the above, and so $\varphi = 1_{\{R > c\}}$ is a Neyman–Pearson test with $\mathbb{E}_0(\varphi) = \alpha$. If in contrast $P_0(R = c) > 0$, then

$$\gamma := \frac{\alpha - P_0(R > c)}{P_0(R = c)}$$

is a well-defined number in $[0, 1]$, and

$$\varphi(x) = \begin{cases} 1 & \text{if } R(x) > c, \\ \gamma & \text{if } R(x) = c, \\ 0 & \text{if } R(x) < c \end{cases}$$

is a Neyman–Pearson test with $\mathbb{E}_0(\varphi) = P_0(R > c) + \gamma \, P_0(R = c) = \alpha$.

(b) Let φ be a Neyman–Pearson test with $\mathbb{E}_0(\varphi) = \alpha$ and threshold value c, and suppose ψ is an arbitrary test of level α. We write

$$\mathbb{E}_1(\varphi) - \mathbb{E}_1(\psi) = \int_{\mathcal{X}} (\varphi(x) - \psi(x)) \, \rho_1(x) \, dx \, ;$$

in the discrete case, the integral is to be replaced by a sum. Now, if $\varphi(x) > \psi(x)$, then $\varphi(x) > 0$ and so $R(x) \geq c$, which means that $\rho_1(x) \geq c \, \rho_0(x)$. If, on the other hand, $\varphi(x) < \psi(x)$, then $\varphi(x) < 1$ and thus $\rho_1(x) \leq c \, \rho_0(x)$. In either case we get

$$f_1(x) := (\varphi(x) - \psi(x)) \, \rho_1(x) \geq c \, (\varphi(x) - \psi(x)) \, \rho_0(x) =: c \, f_0(x) \, .$$

Hence, integrating (or summing) over x yields

$$\mathbb{E}_1(\varphi) - \mathbb{E}_1(\psi) = \int_{\mathcal{X}} f_1(x) \, dx \geq c \int_{\mathcal{X}} f_0(x) \, dx = c \, (\alpha - \mathbb{E}_0(\psi)) \geq 0 \, .$$

Therefore, as claimed, φ is a most powerful test for α.

Conversely, suppose that ψ is an arbitrary most powerful test for α and φ is defined as above. Since φ is also a most powerful test for α, the powers must coincide, in that $\mathbb{E}_1(\varphi) = \mathbb{E}_1(\psi)$. So, the above inequality holds with equality everywhere. This is only possible if $f_1(x) = c \, f_0(x)$ for Lebesgue-almost all x (and in the discrete case even for all x). We thus conclude that $\psi = \varphi$ (almost) everywhere on $\{R \neq c\}$. Thus, ψ is also a Neyman–Pearson test, at least outside an exceptional set N of Lebesgue measure 0. Since $P_0(N) = P_1(N) = 0$, observed values in N do not appear and can thus be ignored. $\diamond$

So we have seen that Neyman–Pearson tests are optimal, at least in the case of a simple hypothesis and a simple alternative. But are they also useful? This depends on how large their power is. The power is obviously the larger the more information is available. Thus we can ask: How quickly does the power increase for independently repeated observations? To answer this, it is convenient, as in Section 7.6, to work in the framework of an infinite product model. Therefore, let $(E, \mathscr{E}; Q_0, Q_1)$ be a statistical standard model with simple hypothesis $\Theta_0 = \{0\}$ and simple alternative $\Theta_1 = \{1\}$, and let

$$(\mathcal{X}, \mathscr{F}, P_\vartheta : \vartheta \in \{0,1\}) = \left(E^{\mathbb{N}}, \mathscr{E}^{\otimes \mathbb{N}}, Q_\vartheta^{\otimes \mathbb{N}} : \vartheta \in \{0,1\}\right)$$

be the associated infinite product model. For simplicity, we require that the densities ρ_0 and ρ_1 of Q_0 and Q_1 are both strictly positive. As usually we write $X_i : \mathcal{X} \to E$ for the ith projection, and we recall the definition of the relative entropy in (7.31). We will, of course, assume that Q_0 and Q_1 are different, so that $H(Q_0; Q_1) > 0$.

(10.4) Theorem. Gaining power by information, Charles M. Stein 1956. *In the above situation, let $(\varphi_n)_{n \geq 1}$ be a sequence of Neyman–Pearson tests with $\mathbb{E}_0(\varphi_n) = \alpha$, where φ_n uses the observations $X_1, \ldots, X_n$ only. Then, as $n \to \infty$, the power $\mathbb{E}_1(\varphi_n)$ tends to 1 with exponential speed. Specifically,*

$$\lim_{n \to \infty} \frac{1}{n} \log[1 - \mathbb{E}_1(\varphi_n)] = -H(Q_0; Q_1),$$

which roughly means that $\mathbb{E}_1(\varphi_n) \approx 1 - e^{-n\,H(Q_0;Q_1)}$ for large n.

Proof. For $n \geq 1$ and $\vartheta \in \{0,1\}$, let $\rho_\vartheta^{\otimes n} = \prod_{i=1}^n \rho_\vartheta(X_i)$ be the n-fold product density, and let $R_n = \rho_1^{\otimes n} / \rho_0^{\otimes n}$ be the likelihood ratio after the first n observations. Set $h = \log(\rho_0/\rho_1)$ and

$$h_n = -\frac{1}{n} \log R_n = \frac{1}{n} \sum_{i=1}^n h(X_i).$$

Then $\mathbb{E}_0(h) = H(Q_0; Q_1)$ by definition, and the tests φ_n take the form

$$\varphi_n = \begin{cases} 1 & \text{if } h_n < a_n, \\ 0 & \text{if } h_n > a_n \end{cases}$$

for suitable constants $a_n \in \mathbb{R}$.

We will first show that $\limsup_{n \to \infty} \frac{1}{n} \log[1 - \mathbb{E}_1(\varphi_n)] \leq -\mathbb{E}_0(h)$. If $1 - \varphi_n > 0$, then $h_n \geq a_n$ by the nature of φ_n, and thus $\rho_0^{\otimes n} \geq e^{n\,a_n} \rho_1^{\otimes n}$. We conclude that

$$1 \geq \mathbb{E}_0(1-\varphi_n) \geq e^{n\,a_n}\, \mathbb{E}_1(1-\varphi_n).$$

It is therefore sufficient to pick an arbitrary $a < \mathbb{E}_0(h)$ and to show that $a_n > a$ for all sufficiently large n. Since $P_0(h_n \leq a_n) \geq \mathbb{E}_0(\varphi_n) = \alpha > 0$, this will follow once we have shown that $P_0(h_n \leq a) \to 0$ as $n \to \infty$. As we have seen in the proof of (7.32), this is a direct consequence of the weak law of large numbers, irrespective of whether $h \in \mathscr{L}^1(P_0)$ or $\mathbb{E}_0(h) = H(Q_0; Q_1) = \infty$.

Conversely, let us verify that $\liminf_{n \to \infty} \frac{1}{n} \log[1 - \mathbb{E}_1(\varphi_n)] \geq -\mathbb{E}_0(h)$. For this we can assume that $\mathbb{E}_0(h) < \infty$, hence $h \in \mathscr{L}^1(P_0)$. Fix any $a > \mathbb{E}_0(h)$. Using again the weak law of large numbers, Theorem (5.7), we find that

$$
\begin{aligned}
P_0\big(\rho_1^{\otimes n} \geq e^{-na}\rho_0^{\otimes n}\big) &= P_0(h_n \leq a) \\
&\geq P_0\big(|h_n - \mathbb{E}_0(h)| \leq a - \mathbb{E}_0(h)\big) \geq \tfrac{1+\alpha}{2}
\end{aligned}
$$

for all sufficiently large n, and therefore

$$
\begin{aligned}
\mathbb{E}_1(1-\varphi_n) &= \mathbb{E}_0\big((1-\varphi_n)\rho_1^{\otimes n}/\rho_0^{\otimes n}\big) \\
&\geq \mathbb{E}_0\big(e^{-na}(1-\varphi_n)1_{\{h_n \leq a\}}\big) \\
&\geq e^{-na}\,\mathbb{E}_0\big(1_{\{h_n \leq a\}} - \varphi_n\big) \\
&\geq e^{-na}\left(\tfrac{1+\alpha}{2} - \alpha\right) = e^{-na}\,\tfrac{1-\alpha}{2}
\end{aligned}
$$

eventually. Taking logarithms, dividing by n, and letting $n \to \infty$, the claim follows.

$\diamond$

Stein's lemma discloses the statistical significance of relative entropy. The larger the relative entropy between two probability measures Q_0 and Q_1, the faster the power of the optimal tests of Q_0 against Q_1 grows with the number of observations, and the easier it is to distinguish Q_0 from Q_1 in an experiment. The relative entropy thus measures how well two probability measures can be distinguished by statistical means.

(10.5) Example. *Test for the expectation of two normal distributions.* Consider the probability measures $Q_0 = \mathcal{N}_{m_0,v}$ and $Q_1 = \mathcal{N}_{m_1,v}$ on $E = \mathbb{R}$ for fixed $m_0 < m_1$ and $v > 0$. We want to test the null hypothesis $H_0 : m = m_0$ against the alternative $H_1 : m = m_1$ on the basis of n observations.

As an application, one can imagine that a satellite component shall be checked whether or not it is working. To find out, a test signal is sent to the satellite. If the component works fine, this signal activates a response signal that is emitted for n seconds. Due to noise, the average signal intensity arriving on earth in each second is random. It can be assumed to have a normal distribution with either expectation $m_0 = 0$ (if the component fails) or $m_1 > 0$ (if everything is fine). Obviously, the embarrassing type I error would be to infer that the component is working fine, even though it has actually failed.

As above, we consider the corresponding infinite product model. The likelihood ratio for the first n observations is given by

$$(10.6) \qquad R_n = \exp\left[-\frac{1}{2v}\sum_{i=1}^{n}\left((X_i - m_1)^2 - (X_i - m_0)^2\right)\right]$$

$$= \exp\left[\frac{n}{2v}\left(2(m_1 - m_0)\,M_n - m_1^2 + m_0^2\right)\right];$$

here, $M_n = \frac{1}{n}\sum_{i=1}^{n} X_i$ again denotes the sample mean. Using the notation of the last proof, we conclude that

$$h_n = -\frac{m_1 - m_0}{v}\,M_n + \frac{m_1^2 - m_0^2}{2v}\,.$$

For a given size α, the Neyman–Pearson test of m_0 against m_1 after n observations is thus $\varphi_n = 1_{\{M_n > b_n\}}$, where the constant b_n is determined by the condition

$$\alpha = P_0(M_n > b_n) = \mathcal{N}_{m_0, v/n}(]b_n, \infty[) = 1 - \Phi\left((b_n - m_0)\sqrt{n/v}\right).$$

Solving for b_n, we find

$$(10.7) \qquad\qquad b_n = m_0 + \sqrt{v/n}\,\Phi^{-1}(1 - \alpha)\,.$$

What can be said about the power of φ_n? We calculate

$$H(P_0; P_1) = \mathbb{E}_0(h_n) = -\frac{m_1 - m_0}{v}\,m_0 + \frac{m_1^2 - m_0^2}{2v} = \frac{(m_1 - m_0)^2}{2v}\,;$$

in words, the relative entropy of two normal distributions with the same variance is simply the squared difference of their expected values (up to a factor of $1/2v$). Theorem (10.4) thus shows that $\mathbb{E}_1(1 - \varphi_n) \approx \exp[-n\,(m_0 - m_1)^2/2v]$.

This result can actually be sharpened. Using (10.7) and the definition of φ_n, one finds that

$$\mathbb{E}_1(1 - \varphi_n) = P_1(M_n \le b_n) = \mathcal{N}_{m_1, v/n}(]-\infty, b_n])$$
$$= \Phi\left((b_n - m_1)\sqrt{n/v}\right) = \Phi\left(u_\alpha - \eta\sqrt{n}\right);$$

here we set $u_\alpha = \Phi^{-1}(1 - \alpha)$ and $\eta = (m_1 - m_0)/\sqrt{v}$ for brevity. Now, Problem 5.15 implies that $\Phi(c) \sim \phi(c)/|c|$ as $c \to -\infty$. Hence,

$$\mathbb{E}_1(1 - \varphi_n) \underset{n \to \infty}{\sim} \frac{1}{\eta\sqrt{2\pi n}}\,\exp\left[-n\,\eta^2/2 + \sqrt{n}\,\eta u_\alpha - u_\alpha^2/2\right].$$

This formula describes the exact asymptotic behaviour of the power of φ_n.

10.3 Most Powerful One-Sided Tests

In the binary test problem of a simple null hypothesis and a simple alternative, the Neyman–Pearson lemma tells us how optimal tests can be constructed. Based on this experience, we now want to find most powerful tests for composite null hypotheses and alternatives. This turns out to be quite easy when the underlying likelihood function exhibits an appropriate monotonicity property. We explain this first for our standard example (10.1).

(10.8) Example. *Quality checking.* Consider again the familiar situation of the orange importer, which is described by the hypergeometric model $\mathcal{X} = \{0, \ldots, n\}$, $\Theta = \{0, \ldots, N\}$, and $P_\vartheta = \mathcal{H}_{n;\vartheta, N-\vartheta}$ for $\vartheta \in \Theta$, where $n < N$. We want to test the null hypothesis $\Theta_0 = \{0, \ldots, \vartheta_0\}$ against the alternative $\Theta_1 = \{\vartheta_0 + 1, \ldots, N\}$ at a given level $0 < \alpha < 1$. (Previously, we have considered the specific values $n = 50$, $N = 10\,000$, $\vartheta_0 = 500$.) It is natural to consider a test φ of the form

$$\varphi(x) = \begin{cases} 1 & \text{if } x > c, \\ \gamma & \text{if } x = c, \\ 0 & \text{if } x < c. \end{cases}$$

We determine the constants c and γ so that the power function of φ takes the value α at the parameter ϑ_0 that separates Θ_0 from Θ_1. This can be achieved in the same way as in the proof of Theorem (10.3a). Namely, we first choose c as an α-fractile of P_{ϑ_0}, and then let γ be such that

$$G_\varphi(\vartheta_0) = P_{\vartheta_0}(\{c+1, \ldots, n\}) + \gamma\, P_{\vartheta_0}(\{c\}) = \alpha.$$

Note that c and γ depend only on ϑ_0.

We will now show that the test φ so constructed is actually a uniformly most powerful level-α test of Θ_0 against Θ_1. The proof relies on a monotonicity property of the densities ρ_ϑ of $P_\vartheta = \mathcal{H}_{n;\vartheta, N-\vartheta}$. Namely, for $\vartheta' > \vartheta$, the likelihood ratio $R_{\vartheta':\vartheta}(x) := \rho_{\vartheta'}(x)/\rho_\vartheta(x)$ is increasing in x. Indeed, for $x \leq \vartheta$ we can write

$$R_{\vartheta':\vartheta}(x) = \prod_{k=\vartheta}^{\vartheta'-1} \frac{\rho_{k+1}(x)}{\rho_k(x)} = \prod_{k=\vartheta}^{\vartheta'-1} \frac{(k+1)(N-k-n+x)}{(N-k)(k+1-x)},$$

and the last expression is clearly increasing in x; for $x > \vartheta$ we have $R_{\vartheta':\vartheta}(x) = \infty$. To exploit this monotonicity, let $\tilde{c} = R_{\vartheta':\vartheta}(c)$. The inequality $R_{\vartheta':\vartheta}(x) > \tilde{c}$ then implies that $x > c$ and thus $\varphi(x) = 1$; likewise, in the case $R_{\vartheta':\vartheta}(x) < \tilde{c}$ we obtain $\varphi(x) = 0$. Therefore, φ is a Neyman–Pearson test of the (simple) null hypothesis $\{\vartheta\}$ against the (simple) alternative $\{\vartheta'\}$. In particular, for $\vartheta = \vartheta_0$ and arbitrary $\vartheta' > \vartheta_0$, Theorem (10.3b) shows that φ is a most powerful level-α test of ϑ_0 against *every* $\vartheta' \in \Theta_1$. It is thus a uniformly most powerful test of ϑ_0 against the entire alternative Θ_1.

It remains to be shown that φ has level α even as a test of the entire Θ_0 against Θ_1, i.e., that $G_\varphi(\vartheta) \leq \alpha$ for all $\vartheta \in \Theta_0$. Since $G_\varphi(\vartheta_0) = \alpha$, it suffices to show that the power function G_φ is increasing. So let $\vartheta < \vartheta'$. As we have just shown, φ is a Neyman–Pearson test of ϑ against ϑ', so by Theorem (10.3b) it is a most powerful test of level $\beta := G_\varphi(\vartheta)$. In particular, it is more powerful than the constant test $\psi \equiv \beta$. It follows that $G_\varphi(\vartheta') \geq G_\psi(\vartheta') = \beta = G_\varphi(\vartheta)$, as claimed. Altogether, we see that in the hypergeometric model the intuitively obvious test procedure is indeed optimal; so it is unnecessary to search for better procedures.

The only thing that remains to be done for the orange importer is to determine the constants c and γ for the level he has chosen. For $\alpha = 0.025$ and the specific values N, n, ϑ_0 given previously, one obtains (for instance by using Mathematica) the values $c = 6$ and $\gamma = 0.52$. Since N is large, one can also approximate the hypergeometric distribution by the binomial distribution and the latter by the normal distribution (or the Poisson distribution). Then one obtains again $c = 6$ and a slightly modified γ.

The essential ingredient of the above optimality proof was the monotonicity of the likelihood ratios. This property will now be defined in general.

Definition. A statistical standard model $(\mathcal{X}, \mathscr{F}, P_\vartheta : \vartheta \in \Theta)$ with $\Theta \subset \mathbb{R}$ is said to have *increasing likelihood ratios* (or *increasing density ratios*) relative to a statistic $T : \mathcal{X} \to \mathbb{R}$, if for all $\vartheta < \vartheta'$ the density ratio $R_{\vartheta':\vartheta} := \rho_{\vartheta'}/\rho_\vartheta$ is an increasing function of T, in that $R_{\vartheta':\vartheta} = f_{\vartheta':\vartheta} \circ T$ for an increasing function $f_{\vartheta':\vartheta}$.

(10.9) Example. *Exponential models.* Every (one-parameter) exponential model has increasing likelihood ratios. Indeed, inserting the expression (7.21) into the likelihood ratio we find that

$$R_{\vartheta':\vartheta} = \exp\big[\big(a(\vartheta') - a(\vartheta)\big)\,T + \big(b(\vartheta) - b(\vartheta')\big)\big]$$

whenever $\vartheta < \vartheta'$. By assumption, the coefficient function $\vartheta \to a(\vartheta)$ is either strictly increasing or strictly decreasing. In the first case we have $a(\vartheta') - a(\vartheta) > 0$, so that $R_{\vartheta':\vartheta}$ is an increasing function of T; in the second case, $R_{\vartheta':\vartheta}$ is an increasing function of the statistic $-T$.

The results of Example (10.8) can be directly extended to all models with increasing likelihood ratios.

(10.10) Theorem. One-sided tests for models with increasing likelihood ratios. *Suppose $(\mathcal{X}, \mathscr{F}, P_\vartheta : \vartheta \in \Theta)$ with $\Theta \subset \mathbb{R}$ is a statistical standard model with increasing likelihood ratios relative to a statistic T. Also, let $\vartheta_0 \in \Theta$ and $0 < \alpha < 1$. Then* *there exists a uniformly most powerful test φ of level α for the one-sided test problem $H_0 : \vartheta \leq \vartheta_0$ against $H_1 : \vartheta > \vartheta_0$. This takes the form*

$$\varphi(x) = \begin{cases} 1 & \text{if } T(x) > c, \\ \gamma & \text{if } T(x) = c, \\ 0 & \text{if } T(x) < c, \end{cases}$$

where c and γ are uniquely determined by the condition $G_\varphi(\vartheta_0) = \alpha$. Moreover, the power function G_φ is increasing.

Proof. The arguments of Example (10.8) immediately carry over to the general case; it is enough to replace x by $T(x)$. For example, the equation that determines c and γ now reads $G_\varphi(\vartheta_0) = P_{\vartheta_0}(T > c) + \gamma\, P_{\vartheta_0}(T = c) = \alpha$. $\diamond$

Theorem (10.10) also covers the case of a right-sided hypothesis $H_0 : \vartheta \geq \vartheta_0$ against a left-sided alternative $H_1 : \vartheta < \vartheta_0$; it is sufficient to multiply both ϑ and T by -1. The most powerful test then takes an analogous form, with '$<$' and '$>$' interchanged.

As we have seen in Section 7.5, many of the classical statistical models belong to the class of exponential models and thus have monotone likelihood ratios. One instance is the binomial model, see Example (7.25). In particular, the test φ in Example (10.2) about extrasensory perception is a uniformly most powerful test for the test problem $H_0 : \vartheta = 1/2$ against $H_1 : \vartheta > 1/2$. Another prominent example is the Gaussian model, which we now discuss in two variants.

(10.11) Example. *One-sided Gauss test, known variance.* Let us return to Example (7.2). On the basis of n independent measurements, it is to be tested whether the brittleness of a cooling pipe falls below a prescribed threshold m_0. The model is the n-fold Gaussian product model $(\mathbb{R}^n, \mathscr{B}^n, \mathcal{N}_{m,v}{}^{\otimes n} : m \in \mathbb{R})$ with known variance $v > 0$. According to Example (7.27a) and Remark (7.28), this is an exponential model; the underlying statistic is the sample mean M, and the associated coefficient $a(\vartheta) = n\vartheta/v$ is increasing. By Example (10.9), Theorem (10.10) and equation (10.7), the test with rejection region

$$\{M > m_0 + \sqrt{v/n}\ \Phi^{-1}(1-\alpha)\}$$

is therefore a most powerful level-α test of $H_0 : m \leq m_0$ against $H_1 : m > m_0$. This test is called the *one-sided Gauss test*.

(10.12) Example. *One-sided chi-square variance test, known expectation.* Suppose the genetic variability of a certain type of cereal grains is to be determined. Using n observations, it shall be decided whether the variance of some characteristic quantity, for example the culm length, exceeds some minimum value. Let us make the modelling assumption that the logarithm of the culm length of each individual plant is normally distributed with known expectation m (the average logarithmic culm length) and an unknown variance $v > 0$. (Indeed, it is plausible to assume that the genetic factors affect the culm length multiplicatively, and so the logarithmic culm length is affected additively. By the central limit theorem, one can thus suppose that the logarithmic culm lengths are approximately normally distributed.) The model is then the n-fold Gaussian product model $(\mathbb{R}^n, \mathscr{B}^n, \mathcal{N}_{m,v}^{\otimes n} : v > 0)$ with known expectation m, and one has to test a null hypothesis of the form $H_0 : v \geq v_0$ against the alternative $H_1 : v < v_0$. Now, we know from Example (7.27b) and Remark (7.28) that the product normal distributions with fixed mean m form an exponential family relative to the statistic $T = \sum_{i=1}^n (X_i - m)^2$. Theorem (10.10) therefore applies, and the most powerful test φ of given level α has the rejection region

$$\left\{ \sum_{i=1}^n (X_i - m)^2 < v_0 \, \chi_{n;\alpha}^2 \right\}.$$

Here, $\chi_{n;\alpha}^2$ is the α-quantile of the χ_n^2-distribution; this is the correct threshold because $\mathbb{E}_{v_0}(\varphi) = \chi_n^2([0, \chi_{n;\alpha}^2]) = \alpha$ by Theorem (9.10). The test φ is therefore called a *one-sided χ^2-test*.

10.4 Parameter Tests in the Gaussian Product Model

In the previous two examples, we have assumed that one of the parameters in the Gaussian product model is known, and we have derived most powerful tests for the free parameter. In this section we will consider the two-parameter case, in which both expectation and variance of the normal distributions are unknown. So we consider the two-parameter Gaussian product model

$$(\mathcal{X}, \mathscr{F}, P_\vartheta : \vartheta \in \Theta) = \left(\mathbb{R}^n, \mathscr{B}^n, \mathcal{N}_{m,v}^{\otimes n} : m \in \mathbb{R}, v > 0 \right).$$

In this situation, it seems natural to modify the tests in Examples (10.11) and (10.12) in such a way that the unknown nuisance parameter, which is not supposed to be tested, is simply replaced by its unbiased estimator. But are the resulting tests optimal? Well, we will find out as we proceed. We will first consider tests for the variance and then for the expectation (where the latter are of greater importance).

10.4.1 Chi-Square Tests for the Variance

We start with the *left-sided test problem*

(V−) $H_0 : v \leq v_0$ against $H_1 : v > v_0$

for the variance; the threshold $v_0 > 0$ and a level α are given. In other words, we have $\Theta_0 = \mathbb{R} \times \,]0, v_0]$ and $\Theta_1 = \mathbb{R} \times \,]v_0, \infty[$.

As an application, we can imagine that the quality of a gauge is to be tested. As before, it is then quite natural to assume that the measurements taken are i.i.d. and normally distributed. A high-quality gauge should achieve a variance below a tolerance threshold v_0.

If m were known, the most powerful test would have the rejection region

$$\Big\{ \sum_{i=1}^{n} (X_i - m)^2 > v_0 \, \chi^2_{n;1-\alpha} \Big\},$$

in analogy to Example (10.12); $\chi^2_{n;1-\alpha}$ is the α-fractile of the χ^2_n-distribution. Now, as m is unknown, it suggests itself to replace m by its best estimator M. If v_0 is the true variance, Theorem (9.17) tells us that the resulting test statistic $(n-1)V^*/v_0$ still has a χ^2-distribution, but with only $(n-1)$ degrees of freedom. Therefore, the fractile $\chi^2_{n;1-\alpha}$ must be replaced by $\chi^2_{n-1;1-\alpha}$. Hence we can guess that the test with rejection region

(10.13) $$\{(n-1)\,V^* > v_0\,\chi^2_{n-1;1-\alpha}\}$$

is optimal. But is this really true?

Before addressing this question, we want to develop an alternative heuristics based on the maximum likelihood principle. Consider again Figure 10.3, but suppose now that both null hypothesis and alternative are composite. Then, given the outcome x, one would certainly decide for the alternative as soon as the maximal likelihood of the alternative, namely $\sup_{\vartheta \in \Theta_1} \rho_\vartheta(x)$, dominates sufficiently strongly over the maximal likelihood $\sup_{\vartheta \in \Theta_0} \rho_\vartheta(x)$ of the null hypothesis. This can be rephrased by saying that one decides for the alternative whenever the *(generalised) likelihood ratio*

(10.14) $$R(x) = \frac{\sup_{\vartheta \in \Theta_1} \rho_\vartheta(x)}{\sup_{\vartheta \in \Theta_0} \rho_\vartheta(x)}$$

exceeds a threshold a. One is thus led to tests of the form

(10.15) $$\varphi = \begin{cases} 1 & \text{if } R > a, \\ 0 & \text{if } R < a, \end{cases}$$

which are called *likelihood ratio tests*. Encouraged by the Neyman–Pearson lemma, one can hope that this recipe leads to good optimality properties even in fairly general situations.

How does a likelihood ratio test look like for the test problem (V−)? The likelihood function in the n-fold Gaussian product model is $\rho_{m,v} = \phi_{m,v}^{\otimes n}$. As in Example (7.9), we therefore obtain in the case $V > v_0$

$$R = \frac{\sup_{m \in \mathbb{R},\, v > v_0} \phi_{m,v}^{\otimes n}}{\sup_{m \in \mathbb{R},\, v \le v_0} \phi_{m,v}^{\otimes n}} = \frac{\sup_{v > v_0} v^{-n/2} \exp\left[-n\,V/2v\right]}{\sup_{v \le v_0} v^{-n/2} \exp\left[-n\,V/2v\right]}$$

$$= \exp\left[\frac{n}{2}\Big(\frac{V}{v_0} - \log \frac{V}{v_0} - 1 \Big) \right];$$

in the alternative case one obtains the reciprocal of the last expression. Since the function $s \to s - \log s$ is increasing for $s > 1$, we can conclude that R is a strictly increasing function of V, and thus also of V^*. A likelihood ratio test for the test problem (V−) therefore has the rejection region (10.13). The punch-line is that such a test is indeed optimal. Namely, the following holds.

(10.16) Theorem. Left-sided χ^2-test for the variance of a normal distribution. *In the n-fold Gaussian product model, the test with the rejection region*

$$\left\{ \sum_{i=1}^{n} (X_i - M)^2 > v_0 \, \chi^2_{n-1;1-\alpha} \right\}$$

is a uniformly most powerful level-α test of the null hypothesis $H_0 : v \leq v_0$ against the alternative $H_1 : v > v_0$. Here, M is the sample mean and $\chi^2_{n-1;1-\alpha}$ the α-fractile of the χ^2_{n-1}-distribution.

In terms of the random vector $X = (X_1, \ldots, X_n)^\top$ and the diagonal vector $\mathbf{1} = (1, \ldots, 1)^\top$ used before, the test statistic in the preceding theorem can be written in the suggestive form $\sum_{i=1}^{n} (X_i - M)^2 = |X - M\mathbf{1}|^2$. The null hypothesis will therefore be *accepted* if and only if X is close enough to its projection onto the diagonal; cf. Figure 7.3 on p. 204. So the acceptance region is a cylinder oriented in the direction of $\mathbf{1}$. The following proof should be skipped at first reading.

Proof. The main idea of the proof is to reduce the two-parameter problem at hand to a one-parameter problem, by averaging over the nuisance parameter m with the help of a suitably chosen prior distribution.

We fix a parameter $\vartheta_1 = (m_1, v_1) \in \Theta_1$ in the alternative and consider a family of probability measures of the form

$$\bar{P}_v = \int \mathsf{w}_v(dm) \, P_{m,v} , \quad 0 < v \leq v_1 .$$

We want to choose the probability weights w_v on $(\mathbb{R}, \mathscr{B})$ in such a way that $\bar{P}_v$ is as close as possible to P_{ϑ_1}, so that it is as hard as possible to distinguish $\bar{P}_v$ from P_{ϑ_1}. Each such w_v is called a *least favourable prior distribution*. Since we are dealing with normal distributions, it seems natural to choose w_v as a normal distribution as well. Specifically, we set $\mathsf{w}_v = \mathcal{N}_{m_1,(v_1-v)/n}$ for $v < v_1$ and $\mathsf{w}_{v_1} = \delta_{m_1}$. (This is indeed an unfavourable case, since

$$\bar{P}_v \circ M^{-1} = \int \mathcal{N}_{m_1,(v_1-v)/n}(dm) \, \mathcal{N}_{m,v/n}$$

$$= \mathcal{N}_{m_1,(v_1-v)/n} \star \mathcal{N}_{0,v/n} = \mathcal{N}_{m_1,v_1/n} = P_{\vartheta_1} \circ M^{-1}$$

by Example (3.32); so one cannot distinguish $\bar{P}_v$ from P_{ϑ_1} by only observing the sample mean.)

The density $\bar{\rho}_v$ of $\bar{P}_v$ is obtained by integrating the density of $P_{m,v}$ with respect to $\mathbf{w}_v$. We thus find for $v < v_1$

$$\bar{\rho}_v = \int dm \, \phi_{m_1,(v_1-v)/n}(m) \prod_{i=1}^{n} \phi_{m,v}(X_i)$$

$$= c_1(v) \int dm \, \exp\left[-\frac{(m-m_1)^2}{2(v_1-v)/n} - \sum_{i=1}^{n} \frac{(X_i-m)^2}{2v}\right],$$

where $c_1(v)$ is a suitable constant. Together with the shift formula (7.10), this yields

$$\bar{\rho}_v = c_1(v) \exp\left[-\frac{n-1}{2v} V^*\right] \int dm \, \exp\left[-\frac{(m_1-m)^2}{2(v_1-v)/n} - \frac{(m-M)^2}{2v/n}\right].$$

Up to a constant factor, the integral in the last expression coincides with the convoluted density $\phi_{0,(v_1-v)/n} \star \phi_{M,v/n}(m_1)$, which equals $\phi_{M,v_1/n}(m_1)$ by Example (3.32). So

$$\bar{\rho}_v = c_2(v) \exp\left[-\frac{n-1}{2v} V^* - \frac{(m_1-M)^2}{2v_1/n}\right]$$

for some constant $c_2(v)$. This holds also for $v = v_1$ if we set $\bar{P}_{v_1} := P_{\vartheta_1} = \mathcal{N}_{m_1,v_1}^{\otimes n}$. As a result, the probability measures $\{\bar{P}_v : 0 < v \le v_1\}$ form an exponential family with respect to the statistic $T = V^*$ with increasing coefficient function $a(v) = -\frac{n-1}{2v}$. Theorem (10.10) thus provides us with a uniformly most powerful level-α test φ of the null hypothesis $\{\bar{P}_v : v \le v_0\}$ against the alternative $\{\bar{P}_{v_1}\}$. It takes the form $\varphi = 1_{\{V^*>c\}}$, where the constant c is determined by the condition $\alpha = \bar{G}_\varphi(v_0) = \bar{P}_{v_0}(V^* > c)$. In particular, c depends only on v_0 (and n). More precisely, using Theorem (9.17b) we obtain the relation

$$\bar{P}_v(V^* > c) = \int \mathcal{N}_{m_1,(v_1-v)/n}(dm) \, P_{m,v}(V^* > c) = \chi_{n-1}^2(]\tfrac{n-1}{v}c, \infty[)$$

for all $v \le v_1$. Setting $v = v_0$, we find that $c = \frac{v_0}{n-1} \chi_{n-1;1-\alpha}^2$. For this c, Theorem (9.17b) further shows that

$$G_\varphi(\vartheta) = \chi_{n-1}^2([\tfrac{n-1}{v} c, \infty[) \le \alpha$$

for arbitrary $\vartheta = (m, v) \in \Theta_0$. Hence, φ respects the level α even as a test of Θ_0 against ϑ_1.

It remains to be shown that φ is a uniformly most powerful level-α test of Θ_0 against the entire alternative Θ_1. Suppose ψ is an arbitrary test of Θ_0 against Θ_1 of level α. Then

$$\bar{G}_\psi(v) = \int \mathbf{w}_v(dm) \, G_\psi(m, v) \le \alpha$$

for all $v \le v_0$; that is, ψ respects the level α also as a test of $\{\bar{P}_v : v \le v_0\}$ against $\{\bar{P}_{v_1}\} = \{P_{\vartheta_1}\}$. But φ is optimal for this test problem, so that $G_\psi(\vartheta_1) \le G_\varphi(\vartheta_1)$. Since $\vartheta_1 \in \Theta_1$ was chosen arbitrarily, the optimality of φ follows. $\diamond$

What about the (reversed) *right-sided test problem*

(V+) $H_0 : v \geq v_0$ against $H_1 : v < v_0$

for the variance? Is it possible to obtain a most powerful test by swapping the relations '>' and '<' in the above theorem (and replacing $\chi^2_{n-1;1-\alpha}$ by $\chi^2_{n-1;\alpha}$)? Unfortunately, the answer is no.

First of all, it is no longer possible to take advantage of least favourable prior distributions. Namely, if one chooses a fixed parameter (m, v) in the alternative (so
that $v < v_0$), then the corresponding normal distribution $P_{m,v} = \mathcal{N}_n(m\mathbf{1}, v\mathsf{E})$ has a
sharper peak than the normal distributions in the null hypothesis, and the peaks of the
latter will only become flatter by averaging. Therefore, it is impossible to approach
$P_{m,v}$ by mixtures of distributions in the null hypothesis.

But it is not only the argument that breaks down, but also the statement itself!
Indeed, let $m \in \mathbb{R}$ be fixed and $c = \chi^2_{n;\alpha}$, and consider the test φ_m with rejection
region $\{|X - m\mathbf{1}|^2 < v_0\, c\}$. This φ_m respects the level α on the entire null hypothesis
$\Theta_0 = \mathbb{R} \times [v_0, \infty[$; this is because, by Problem 9.6,

$$G_{\varphi_m}(m', v) = \mathcal{N}_n(m'\mathbf{1}, v\mathsf{E})\big(|X - m\mathbf{1}|^2 < v_0\, c\big)$$

$$\leq \mathcal{N}_n(m\mathbf{1}, v\mathsf{E})\big(|X - m\mathbf{1}|^2 < v_0\, c\big)$$

$$= \chi^2_n([0, v_0\, c/v]) \leq \alpha$$

for arbitrary $(m', v) \in \Theta_0$. On the other hand, let (m, v) with $v < v_0$ be any point in
the alternative. Example (10.12) then tells us that φ_m has the largest power at (m, v)
among all tests ψ such that $\mathbb{E}_{m,v_0}(\psi) \leq \alpha$. This means that different tests of level
α are most powerful at different points. In other words, a *uniformly* most powerful
level-α test does not exist!

However, the tests φ_m that are most powerful for a given m have the disadvantage
of being biased. Namely, for arbitrary $m, m' \in \mathbb{R}$ and $v > 0$ we get

$$G_{\varphi_m}(m', v) = \mathcal{N}_n(0, v\mathsf{E})\big(|X - (m - m')\mathbf{1}|^2 < v_0\, c\big) \to 0 \quad \text{as } |m'| \to \infty,$$

so that $G_{\varphi_m} < \alpha$ on a large part of the alternative. In contrast, the test φ with rejection
region $\{|X - M\mathbf{1}|^2 < v_0\, \chi^2_{n-1;\alpha}\}$, which is analogous to Theorem (10.16), is unbiased
of level α. This is because, for $m \in \mathbb{R}$ and $v < v_0$,

$$G_{\varphi}(m, v) = \chi^2_{n-1}\big([0, \tfrac{v_0}{v} \chi^2_{n-1;\alpha}]\big) > \alpha$$

by Student's theorem (9.17). So maybe φ is most powerful among all *unbiased* tests
of level α? Indeed, this is the case.

(10.17) Theorem. Right-sided χ^2-test for the variance of a normal distribution. *In the n-fold Gaussian product model, the test with rejection region*

$$\left\{ \sum_{i=1}^{n} (X_i - M)^2 < v_0 \, \chi^2_{n-1;\alpha} \right\}$$

is uniformly most powerful within the class of all unbiased *level-α tests of $H_0 : v \geq v_0$ against $H_1 : v < v_0$. Here, $\chi^2_{n-1;\alpha}$ is the α-quantile of χ^2_{n-1}.*

We omit the proof because we will present a very similar argument in the next theorem. Problem 10.18 will deal with the two-sided test problem for the variance.

10.4.2   *t*-Tests for the Expectation

We now turn to tests for the expectation parameter m in the two-parameter Gaussian product model. We consider first the *one-sided test problem*

$$(\text{M}-) \qquad\qquad H_0 : m \leq m_0 \text{ against } H_1 : m > m_0 .$$

(In the case of the analogous right-sided problem, one merely has to interchange the '<' and '>' signs; contrary to the variance tests in the preceding subsection, there are no principal differences here.) In contrast to the Gauss test in Example (10.11), the variance is now supposed to be unknown. This means that $\Theta_0 = \,]-\infty, m_0] \times \,]0, \infty[$ and $\Theta_1 = \,]m_0, \infty[\, \times \,]0, \infty[$.

Which test procedure is suggested by the maximum-likelihood principle? Since the maximum of $\phi^{\otimes n}_{m,v}$ over v is attained at $\widetilde{V}_m := |X - m\mathbf{1}|^2/n$, the likelihood ratio (10.14) takes the form

$$R = \frac{\sup_{m>m_0, v>0} \phi^{\otimes n}_{m,v}}{\sup_{m\leq m_0, v>0} \phi^{\otimes n}_{m,v}} = \frac{\sup_{m>m_0} \widetilde{V}_m^{-n/2}}{\sup_{m\leq m_0} \widetilde{V}_m^{-n/2}} = \begin{cases} (V/\widetilde{V}_{m_0})^{n/2} & \text{if } M \leq m_0, \\ (\widetilde{V}_{m_0}/V)^{n/2} & \text{if } M \geq m_0. \end{cases}$$

The last identity follows from the shift formula (7.10), which further implies that $\widetilde{V}_{m_0}/V = 1 + T^2_{m_0}/(n-1)$ for

$$T_{m_0} = (M - m_0)\sqrt{n/V^*} .$$

Hence, R is a strictly increasing function of T_{m_0}. A likelihood ratio test for the test problem (M$-$) thus has a rejection region of the form $\{T_{m_0} > t\}$. By Theorem (9.17), T_{m_0} has the t-distribution t_{n-1} under each $P_{m_0,v}$. A given level α is therefore exhausted if we set $t = t_{n-1;1-\alpha}$, the α-fractile of the t_{n-1}-distribution. The test obtained in this way is called *one-sided Student's t-test*.

As in the case of the right-sided test problem (V+) for the variance, it follows from Example (10.11) that a uniformly most powerful test does not exist. The Gauss tests, however, which are optimal when the variance v is fixed, are biased when v is unknown. In contrast, the t-test turns out to be uniformly most powerful among all unbiased tests.

(10.18) Theorem. One-sided t-test for the expectation. *In the n-fold Gaussian product model, the test φ with the rejection region*

$$\left\{(M - m_0)\sqrt{n/V^*} > t_{n-1;1-\alpha}\right\}$$

is uniformly most powerful within the class of all unbiased level-α tests of the null hypothesis $H_0 : m \le m_0$ against the alternative $H_1 : m > m_0$. Here, $t_{n-1;1-\alpha}$ is the α-fractile of the t_{n-1}-distribution.

Proof. Step 1: Preparations. Shifting coordinates, we can achieve that $m_0 = 0$. Further, we write the likelihood function in the form

$$\rho_{\mu,\eta} = (2\pi v)^{-n/2}\exp\left[-\sum_{i=1}^{n}(X_i - m)^2/2v\right] = c(\mu,\eta)\exp\left[\mu\,\widetilde{M} - \eta\,S\right]$$

with $\mu = m\sqrt{n}/v$, $\eta = 1/2v$, $\widetilde{M} = \sqrt{n}\,M$, $S = |X|^2 = \sum_{i=1}^{n} X_i^2$, and an appropriate normalising constant $c(\mu,\eta)$. In the variables (μ,η), the test problem (M−) takes the form $H_0 : \mu \le 0$ against $H_1 : \mu > 0$, and the test statistic T_0 of the t-test can be written as

$$T_0 = \sqrt{n-1}\,\widetilde{M}\Big/\sqrt{S - \widetilde{M}^2}\,.$$

The t-test φ therefore has the rejection region

$$\left\{\widetilde{M} > r\sqrt{S - \widetilde{M}^2}\right\} = \left\{\widetilde{M} > f(S)\right\},$$

where $r = t_{n-1;1-\alpha}/\sqrt{n-1}$ and $f(S) = r\sqrt{S/(1+r^2)}$.

Step 2: Behaviour at the boundary $\mu = 0$. Let ψ be an arbitrary unbiased test of level α. By continuity, it then follows that $\mathbb{E}_{0,\eta}(\psi) = \alpha$ for $\mu = 0$ and every $\eta > 0$. The same holds for φ because

$$\mathbb{E}_{0,\eta}(\varphi) = P_{0,\eta}\big(T_0 > t_{n-1;1-\alpha}\big) = \alpha$$

by Student's theorem (9.17). Hence $\mathbb{E}_{0,\eta}(\varphi - \psi) = 0$ for all $\eta > 0$. This result can be sharpened considerably, as follows. Let $\eta = \gamma + k$ for $\gamma > 0$ and $k \in \mathbb{Z}_+$. Since $\mathbb{E}_{0,\gamma+k}(\varphi - \psi)$ differs from $\mathbb{E}_{0,\gamma}(e^{-kS}\,[\varphi - \psi])$ only by a normalising factor, we find that

(10.19) $$\mathbb{E}_{0,\gamma}\big(g(e^{-S})\,[\varphi - \psi]\big) = 0$$

for each monomial $g(p) = p^k$. By linearity, this statement extends to arbitrary polynomials g and, by the Weierstrass approximation theorem (proved in Example (5.10)), also to arbitrary continuous functions $g : [0,1] \to \mathbb{R}$. This implies further that

(10.20) $$\mathbb{E}_{0,\eta}\big(h(S)\,[\varphi - \psi]\big) = 0$$

for all $\eta > 0$ and all continuous functions $h : [0, \infty[\to \mathbb{R}$ that satisfy $h(u)e^{-\delta u} \to 0$ as $u \to \infty$ for any $\delta > 0$. Indeed, if $0 < \delta < \eta$ is fixed, $\gamma = \eta - \delta$, and $g : [0, 1] \to \mathbb{R}$ is defined by $g(p) = h(\log \frac{1}{p}) p^{\delta}$ for $0 < p \le 1$ and $g(0) = 0$, then g is continuous, and $g(e^{-S}) = h(S)e^{-\delta S}$. Inserting this into (10.19) we obtain (10.20).

Step 3: The Neyman–Pearson argument. Let $(\mu, \eta) \in \Theta_1 =]0, \infty[^2$ be arbitrarily chosen. The likelihood ratio

$$R_{\mu:0,\eta} := \rho_{\mu,\eta}/\rho_{0,\eta} = c \exp[\mu \, \widetilde{M}]$$

with $c = c(\mu, \eta)/c(0, \eta)$ is then a strictly increasing function of $\widetilde{M}$. The rejection region of φ can therefore be rewritten as $\{R_{\mu:0,\eta} > h(S)\}$, where $h = c \exp[\mu f]$. Together with (10.20), this shows that

$$\mathbb{E}_{\mu,\eta}(\varphi - \psi) = \mathbb{E}_{0,\eta}\big([R_{\mu:0,\eta} - h(S)] \, [\varphi - \psi]\big) \, .$$

By our choice of $h(S)$, the two square brackets in the last expression have the same sign, so their product is non-negative. We thus arrive at the inequality $\mathbb{E}_{\mu,\eta}(\varphi) \ge \mathbb{E}_{\mu,\eta}(\psi)$, which means that φ has at least the same power as ψ. $\diamond$

We now turn to the *two-sided test problem*

(M=) $\qquad\qquad\qquad H_0 : m = m_0 \text{ against } H_1 : m \ne m_0$

for the expectation m. Hence $\Theta_0 = \{m_0\} \times]0, \infty[$.

For example, suppose a physical theory predicts the result $m_0 \in \mathbb{R}$ in a certain experiment, and this is to be checked by n independent measurements. In many cases, the outcomes can be considered as realisations of normally distributed random variables, for which both the expectation (the true result of measurement) and the variance (the precision of the experimental set-up) are unknown. The experimenter is then faced with the present test problem.

As before, the likelihood ratio gives an indication for a plausible procedure. Mimicking the calculation in the one-sided test problem (M−), one finds the equation

$$R = \left(1 + \frac{|T_{m_0}|^2}{n-1}\right)^{n/2},$$

which says that R is a strictly increasing function of $|T_{m_0}|$. Hence, a likelihood ratio test φ for the two-sided test problem (M=) has a rejection region of the form $\{|T_{m_0}| > t\}$. By Theorem (9.17), φ exhausts a given level α if t is chosen as the $\alpha/2$-fractile of t_{n-1}. This so-called *two-sided Student's t-test* proves to be a most powerful unbiased test.

(10.21) Theorem. Two-sided t-test for the expectation. *In the n-fold Gaussian product model, the test φ with rejection region*

$$\{|M - m_0|\sqrt{n/V^*} > t_{n-1;1-\alpha/2}\}$$

is uniformly most powerful among all unbiased level-α tests of $H_0 : m = m_0$ against $H_1 : m \neq m_0$. Here, $t_{n-1;1-\alpha/2}$ is the $\alpha/2$-fractile of the t_{n-1}-distribution.

Proof. We proceed exactly as in the proof of Theorem (10.18) and use the same notation.

Step 1. We again introduce the new variables μ, η. The test problem then reduces to $H_0 : \mu = 0$ against $H_1 : \mu \neq 0$, and the rejection region of φ takes the form

$$\left\{|\widetilde{M}| > r\sqrt{S - \widetilde{M}^2}\right\} = \left\{|\widetilde{M}| > f(S)\right\}.$$

Step 2. Let ψ be an arbitrary unbiased level-α test. Then, as in the one-sided case, the power function of ψ is identically α on the null hypothesis $H_0 : \mu = 0$. By construction, the same applies to φ. This leads again to equation (10.20). Additionally, we note the following.

For all $\eta > 0$, the function $\mu \to \mathbb{E}_{\mu,\eta}(\psi)$ has a global minimum at $\mu = 0$. Hence, its derivative $\frac{\partial}{\partial \mu}\mathbb{E}_{\mu,\eta}(\psi)|_{\mu=0} = \mathbb{E}_{0,\eta}(\widetilde{M}\,\psi)$ vanishes; regarding the existence of the derivative, see Remark (7.23) and the exponential form of $\rho_{\mu,\eta}$. Likewise, φ satisfies the identity

$$\mathbb{E}_{0,\eta}(\widetilde{M}\,\varphi) = c(0, \eta)\int_{\mathbb{R}^n} dx\, e^{-\eta\,S(x)}\,\widetilde{M}(x)\,\varphi(x) = 0,$$

because φ and S are symmetric under the reflection $x \to -x$, whereas $\widetilde{M}$ is skew-symmetric. The same reasoning as for (10.20) thus allows us to conclude that

(10.22) $$\mathbb{E}_{0,\eta}\big(h(S)\,\widetilde{M}\,[\varphi - \psi]\big) = 0$$

for all $\eta > 0$ and all continuous and at most subexponentially growing functions h.

Step 3. Let $\mu \neq 0$ and $\eta > 0$ be chosen arbitrarily. The likelihood ratio $R_{\mu:0,\eta} = c\exp[\mu\,\widetilde{M}]$ is then a strictly convex function of $\widetilde{M}$. As can be seen in Figure 10.4, we can thus write the rejection region $\{|\widetilde{M}| > f(S)\}$ in the form

$$\{R_{\mu:0,\eta} > a(S) + b(S)\,\widetilde{M}\};$$

here $a(S)$ and $b(S)$ are chosen such that the straight line $u \to a(S) + b(S)\,u$ intersects the exponential function $u \to c\exp[\mu\,u]$ precisely at the points $u = \pm f(S)$, i.e., $a = c\cosh(\mu f)$ and $b = c\sinh(\mu f)/f$. But (10.20) and (10.22) can be combined into

$$\mathbb{E}_{0,\eta}\big([a(S) + b(S)\widetilde{M}]\,[\varphi - \psi]\big) = 0,$$

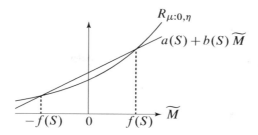

Figure 10.4. Characterisation of an interval via the secant of a convex function.

and thus

$$\mathbb{E}_{\mu,\eta}(\varphi - \psi) = \mathbb{E}_{0,\eta}\big([R_{\mu:0,\eta} - a(S) - b(S)\widetilde{M}]\,[\varphi - \psi]\big) \geq 0\,.$$

The last inequality comes from the fact that, by construction, the two square brackets always have the same sign. Setting $\psi \equiv \alpha$ we see, in particular, that φ is unbiased of level α. $\diamond$

Typical power functions of one- and two-sided t-tests are shown in Figure 10.5. To calculate them one can exploit the fact that, on the alternative Θ_1, the test statistic T_{m_0} has a non-central t_{n-1}-distribution, as introduced in Problem 9.15. For large n, one can also use a normal approximation; cf. Problems 10.21 and 10.22. Here is an application of the t-test in the experimental setting of matched pairs.

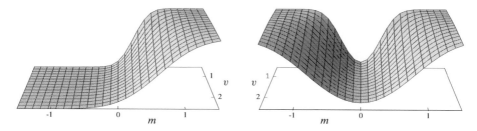

Figure 10.5. Power functions of the one-sided (left) and the two-sided (right) t-tests for $m_0 = 0, n = 12$ and $\alpha = 0.1$.

(10.23) Example. *Comparison of two sleeping pills.* Recall the situation of Example (8.6). Two sleeping pills A and B are administered to $n = 10$ patients, and for each patient the difference in sleep duration is measured; the latter is assumed to be normally distributed with unknown parameters m and v. We test the null hypothesis $H_0 : m = 0$, which says that both sleeping pills are equally efficient, at the level $\alpha = 0.01$. For the data vector x of Example (8.6), we obtain that $T_0(x) = 1.58\sqrt{10/1.513} = 4.06$, and this value is larger than the quantile $t_{9;0.995} = 3.25$. The null hypothesis is therefore rejected on the basis of x; in other words, the effect of the two sleeping pills is not the same. Since $T_0(x) > 0$, it is plausible to guess that B is more efficient.

Reading the preceding example the reader will wonder: Is it sensible to consider the sleeping pills as equally efficient only in the case when $m = 0$ precisely? Would it not be sufficient to require that $|m| \le 0.1$, say? This leads us to the following general comment on 'sharp' test problems.

(10.24) Remark. *Sharp null hypotheses.* In the test problem (M=) above and many test problems considered later, the null hypothesis is sharp, in that Θ_0 has lower dimension than Θ, so that Θ_1 is dense in Θ. But then it is rather naive to expect that 'nature' could ever choose the null hypothesis! The resolution of the real numbers is too fine for the real world of chance. So, such test problems should actually be considered as idealisations of what one really has in mind. Namely, (M=) is an idealised version of the 'true' test problem $H_0 : m \approx m_0$ against $H_1 : m \not\approx m_0$. In words, one wants to test whether or not m is reasonably close to the prescribed value m_0 or, in general, whether or not ϑ deviates no more from Θ_0 than one is willing to tolerate. The vague terms 'reasonably close' and 'willing to tolerate' are then specified by the shape of the power function G_φ: Its 'region of increase' separates an 'effective null hypothesis' from an 'effective alternative'; cf. Figure 10.5. The size α of the test and the number n of experiments should be tuned in such a way that the increase of G_φ is steep enough to make the resulting 'effective' test problem appropriate for the situation at hand. Of course, a similar remark applies also to one-sided tests, for which the boundary of Θ_0 is 'blurred' by the power function of the respective test.

As a final comment to this Section 10.4, we emphasise once more that the optimality of the χ^2- and t-tests for variance and expectation depends crucially on the assumption of normal distributions. If this assumption is not valid, these tests can produce misleading results.

Problems

10.1 *Relation between confidence regions and tests.* Let $(\mathcal{X}, \mathcal{F}, P_\vartheta : \vartheta \in \Theta)$ be a statistical model. Show the following:

(a) If $C : \mathcal{X} \to \mathscr{P}(\Theta)$ is a confidence region with error level α and $\vartheta_0 \in \Theta$ is chosen arbitrarily, then $\{\vartheta_0 \notin C(\cdot)\}$ is the rejection region of a test of $H_0 : \vartheta = \vartheta_0$ against $H_1 : \vartheta \ne \vartheta_0$ of level α.

(b) Conversely, if for each $\vartheta_0 \in \Theta$ a non-randomised level-α test of $H_0 : \vartheta = \vartheta_0$ against $H_1 : \vartheta \ne \vartheta_0$ is given, then these tests can be combined to construct a confidence region with error level α.

10.2 *Testing the range of a uniform distribution.* In the model $(\mathbb{R}^n, \mathscr{B}^n, \mathcal{U}_{[0,\vartheta]}^{\otimes n} : \vartheta > 0)$ of Example (7.3), consider the test problem $H_0 : \vartheta = 1$ against $H_1 : \vartheta \ne 1$. Determine the power function of the test with acceptance region $\{\frac{1}{2} < \max\{X_1, \ldots, X_n\} \le 1\}$.

10.3   A delivery of 10 appliances contains an unknown number of defective items. Unfortunately, a defect can only be discovered by a very expensive quality check. A buyer who is only
interested in a completely flawless delivery performs the following inspection upon delivery.
He checks 5 appliances. If they are all faultless, he accepts the delivery, otherwise he returns
it. Describe this procedure from the viewpoint of test theory and determine the size of the
test. How many appliances must be checked to keep the probability for a false acceptance at
or below 0.1?

10.4 [S] Specify a most powerful test of $H_0 : P = P_0$ against $H_1 : P = P_1$ of level $\alpha \in
]0, 1/2[$ in each of the following cases:

(a) $P_0 = \mathcal{U}_{]0,2[}$, $P_1 = \mathcal{U}_{]1,3[}$.

(b) $P_0 = \mathcal{U}_{]0,2[}$, P_1 has the density $\rho_1(x) = x \, 1_{]0,1]}(x) + \frac{1}{2} 1_{[1,2[}(x)$.

10.5   During a raid, the police searches a gambler and discovers a coin, of which another
player claims that 'heads' appears with a probability of $p = 0.75$ instead of $p = 0.5$. Because
of time restrictions, the coin can only be checked $n = 10$ times. Choose a null hypothesis
and an alternative according to the principle 'Innocent until proven guilty' and specify a most
powerful test of significance level $\alpha = 0.01$. (Use a calculator!)

10.6   On the basis of n draws in the lottery '6 out of 49', it is to be tested whether '13' is an
unlucky number, because it appears less frequently than expected. Formulate the test problem
and specify a most powerful test of approximate level $\alpha = 0.1$ (use the normal approximation
of binomial distributions). How would you decide for the 2944 draws of the German 'Saturday
lottery' between 9 Oct 1955 and 24 Mar 2012, where the '13' was drawn 300 times only, and
thus ended up right at the lower end of the frequency chart?

10.7 [S] *Neyman–Pearson geometry.* In the situation of the Neyman–Pearson lemma (10.3),
let $G^*(\alpha) := \sup\{\mathbb{E}_1(\psi) : \psi$ a test with $\mathbb{E}_0(\psi) \le \alpha\}$ be the greatest possible power that can
be achieved for the level $0 < \alpha < 1$. Show the following:

(a) G^* is increasing and concave.

(b) If φ is a Neyman–Pearson test with threshold value c and size $\alpha := \mathbb{E}_0(\varphi) \in]0, 1[$, then
c is the slope of a tangent to G^* at the point α. *Hint:* Use the fact that $\mathbb{E}_1(\varphi) - c\, \mathbb{E}_0(\varphi) \ge
\mathbb{E}_1(\psi) - c\, \mathbb{E}_0(\psi)$ for *every* test ψ.

10.8   *Bayes tests.* Let φ be a test of P_0 against P_1 in a binary standard model $(\mathcal{X}, \mathscr{F}; P_0, P_1)$,
and let $\alpha_0, \alpha_1 > 0$. Show that φ minimises the weighted error probability $\alpha_0\, \mathbb{E}_0(\varphi) +
\alpha_1\, \mathbb{E}_1(1 - \varphi)$ if and only if φ is a Neyman–Pearson test with threshold value $c = \alpha_0/\alpha_1$.
Such a φ is called a Bayes test with a priori assessment (α_0, α_1).

10.9 [S] *Minimax-Tests.* Consider a binary standard model $(\mathcal{X}, \mathscr{F}; P_0, P_1)$. A test φ of P_0
against P_1 is called a minimax test, if the maximum of the type I and type II error probabilities
is minimal. Establish the existence of a Neyman–Pearson test φ with $\mathbb{E}_0(\varphi) = \mathbb{E}_1(1 - \varphi)$,
and show that it is a minimax test.

10.10 Among 3000 children born in a hospital, 1578 were boys. On the basis of this result,
would you, with 95% certainty, insist on the null hypothesis that newborns are male with
probability 1/2?

10.11 Consider the situation of Example (10.5), where the proper functioning of a satellite
was tested. The satellite manufacturer has the choice between two systems A and B. The

signal-to-noise ratio of system A is $m_1^{(A)}/\sqrt{v} = 2$, at a price of € 10^5. System B, with a ratio of $m_1^{(B)}/\sqrt{v} = 1$, costs only € 10^4. For both systems, each second of transmitting a signal costs € 10^2, and the satellite is to be tested a total of 100 times. For each individual test, the transmission must be long enough to ensure that the error probabilities of type I and II are both ≤ 0.025. Which system should the manufacturer use?

10.12^{S} *Normal approximation for Neyman–Pearson tests.* Let $(E, \mathcal{E}; Q_0, Q_1)$ be a statistical standard model with simple null hypothesis and alternative and strictly positive densities ρ_0, ρ_1. Consider the negative log-likelihood ratio $h = \log(\rho_0/\rho_1)$ and assume that its variance $v_0 = \mathbb{V}_0(h)$ exists. In the associated infinite product model, let R_n be the likelihood ratio after n observations. Show that the Neyman–Pearson test of a given level $0 < \alpha < 1$ has a rejection region of the form

$$\left\{\log R_n > -n\, H(Q_0; Q_1) + \sqrt{n v_0}\, \Phi^{-1}(1 - \alpha + \eta_n)\right\}$$

with $\eta_n \to 0$ as $n \to \infty$. *Hint:* Determine the asymptotic size of these tests when $\eta_n = \eta \neq 0$ does not depend on n.

10.13 In the situation of Problem 7.1, determine a uniformly most powerful test of level $\alpha = 0.05$ for the null hypothesis that the radiation burden is at most 1, based on $n = 20$ independent observations. Plot the power function (using a suitable program).

10.14^{S} *Optimality of the power function on the null hypothesis.* Let $\Theta \subset \mathbb{R}$ and $(\mathcal{X}, \mathcal{F}, P_\vartheta : \vartheta \in \Theta)$ be a statistical model with increasing likelihood ratios relative to a statistic T. For given $\vartheta_0 \in \Theta$, let φ be a uniformly most powerful level-α test of the null hypothesis $H_0 : \vartheta \leq \vartheta_0$ against the alternative $H_1 : \vartheta > \vartheta_0$. Show that the power function of φ is minimal on the null hypothesis, in that $G_\varphi(\vartheta) \leq G_\psi(\vartheta)$ for all $\vartheta \leq \vartheta_0$ and every test ψ with $\mathbb{E}_{\vartheta_0}(\psi) = \alpha$.

10.15 *Testing the life span of appliances.* Consider the n-fold product of the statistical model $(]0, \infty[, \mathcal{B}_{]0,\infty[}, Q_\vartheta : \vartheta > 0)$, where Q_ϑ is the Weibull distribution with known exponent $\beta > 0$ and unknown scale parameter $\vartheta > 0$; recall Problem 3.27. That is, Q_ϑ has the Lebesgue density

$$\rho_\vartheta(x) = \vartheta \beta\, x^{\beta - 1} \exp\left[-\vartheta x^\beta\right], \quad x > 0.$$

(a) Show that, under $Q_\vartheta^{\otimes n}$, $T = \vartheta \sum_{i=1}^n X_i^\beta$ has the gamma distribution $\Gamma_{1,n}$. *Hint:* Corollary (9.9).

(b) Determine a uniformly most powerful level-α test φ for the null hypothesis $H_0 : \vartheta \leq \vartheta_0$ (stating that the mean life span exceeds a prescribed minimum) against $H_1 : \vartheta > \vartheta_0$.

(c) Let $\vartheta_0 = 1$ and $\alpha = 0.01$. How large must n be to ensure that $G_\varphi(2) \geq 0.95$? Use the central limit theorem.

10.16 A newspaper chooses the winner of its weekly competition by drawing (with replacement) from the post cards received, until a correct answer is found. Last week, 7 cards had to be checked; so the responsible editor suspects that the proportion p of correct solutions was less than 50%, implying that the question was too hard. Is he right with this conclusion? Set up a suitable statistical model and use the given outcome to perform a test of level α to decide between $H_0 : p \geq 0.5$ and $H_1 : p < 0.5$.

10.17 *Two-sided binomial test.* Construct a two-sided binomial test of level α, that is, a test in the binomial model of the null hypothesis $H_0 : \vartheta = \vartheta_0$ against the alternative $H_1 : \vartheta \neq \vartheta_0$, where $0 < \vartheta_0 < 1$. Also deduce an asymptotic version of this test by using the de Moivre–Laplace theorem (5.22).

10.18 [S] *Two-sided chi-square variance test.* In the two-parameter Gaussian product model, consider the two-sided test problem $H_0 : v = v_0$ against $H_1 : v \neq v_0$ for the variance. It is natural to consider a decision rule that accepts H_0 if $c_1 \leq \frac{n-1}{v_0} V^* \leq c_2$ for suitably chosen c_1 and c_2.

(a) Calculate the power function G of this test and show that

$$\frac{\partial G}{\partial v}(m, v_0) \gtreqqless 0 \text{ if and only if } v \gtreqqless v_0 \frac{c_2 - c_1}{(n-1) \log(c_2/c_1)} .$$

(b) Naively, one would choose c_1, c_2 so that

$$P_{m,v_0}\left(\tfrac{n-1}{v_0} V^* < c_1\right) = P_{m,v_0}\left(\tfrac{n-1}{v_0} V^* > c_2\right) = \alpha/2 .$$

Select the values $\alpha = 0.02$ and $n = 3$, show that the corresponding test is biased, and sketch G.

(c) How can one construct an unbiased test of the above type?

(d) How does the corresponding likelihood ratio test look like?

10.19 Consider the two-parameter Gaussian product model and show that a uniformly most powerful test for the one-sided test problem $H_0 : m \leq 0$ against $H_1 : m > 0$ does not exist.

10.20 [S] Let φ be the one-sided t-test of Theorem (10.18). Show that its power function is monotone in m, so that φ is indeed an unbiased test of size α. *Hint:* Problem 2.15.

10.21 Let φ be a one-sided or two-sided t-test for the expectation in the two-parameter Gaussian product model. Express the power function $G_\varphi(m, v)$ of φ in terms of the non-central t-distributions defined in Problem 9.15.

10.22 [S] *Approximate power function of the t-test.* Let φ_n be the t-test for the one-sided test problem $H_0 : m \leq 0$ against $H_1 : m > 0$ at a given level α in the two-parameter n-fold Gaussian product model. Show the following: For large n, the power function of φ_n admits the normal approximation

$$G_{\varphi_n}(m, v) \approx \Phi\left(\Phi^{-1}(\alpha) + m \sqrt{n/v}\right) .$$

Hint: Combine Theorems (7.29) and (9.17b).

10.23 A supplier of teaching materials delivers a set of electric resistors and claims that their resistances are normally distributed with mean 55 and standard deviation 5 (each measured in Ohm). Specify a test of level $\alpha = 0.05$ for each of the two test problems

(a) $H_0 : m \leq 55$ against $H_1 : m > 55$,

(b) $H_0 : v \leq 25$ against $H_1 : v > 25$,

on the basis of 10 independent measurements (which are assumed to be normal with unknown mean and variance). How would you decide when the following values are measured for 10 resistors: 45.9   68.5   56.8   60.0   57.7   63.0   48.2   59.0   55.2   50.6.

10.24 ^{S} *Two-sample problem in the Gaussian product model.* Let $X_1, \ldots, X_k, X'_1, \ldots, X'_l$ be independent random variables with distribution $\mathcal{N}_{m,v}$ and $\mathcal{N}_{m',v}$, respectively; the parameters m, m' and v are assumed to be unknown. Show that each likelihood ratio test for the test problem $H_0 : m \le m'$ against $H_1 : m > m'$ has a rejection region of the form

$$\left\{ M - M' > c\sqrt{\left(\tfrac{1}{k} + \tfrac{1}{l}\right)V^*} \right\}$$

with $M = \frac{1}{k}\sum_{i=1}^k X_i$, $M' = \frac{1}{l}\sum_{j=1}^l X'_j$, and

$$V^* = \tfrac{1}{k+l-2}\left(\sum_{i=1}^k (X_i - M)^2 + \sum_{j=1}^l (X'_j - M')^2 \right).$$

10.25 ^{S} *p-value and combination of tests.* Consider all tests with a rejection region of the form $\{T > c\}$ for a given real-valued statistic T. Suppose that the distribution function F of T does not depend on ϑ on the null hypothesis Θ_0, in that $P_\vartheta(T \le c) = F(c)$ for all $\vartheta \in \Theta_0$ and $c \in \mathbb{R}$. In particular, this implies that the test with rejection region $\{T > c\}$ has the size $1 - F(c)$. The *p-value* $p(x)$ of a sample $x \in \mathcal{X}$ is then defined as the largest size α that, given x, still leads to the acceptance of the null hypothesis, i.e., $p(x) = 1 - F \circ T(x)$. Assume that F is continuous and strictly monotone on the interval $\{0 < F < 1\}$ and show the following:

(a) Under the null hypothesis, $p(\cdot)$ has the distribution $\mathcal{U}_{]0,1[}$. *Hint:* Problem 1.18.

(b) The test with rejection region $\{p(\cdot) < \alpha\}$ is equivalent to the size-α test with rejection region of the form $\{T > c\}$.

(c) If $p_1(\cdot), \ldots, p_n(\cdot)$ are the *p*-values for n independent studies using the test statistic T, then $S = -2\sum_{i=1}^n \log p_i(\cdot)$ is χ^2_{2n}-distributed on the null hypothesis, and the rejection region $\{S > \chi^2_{2n;1-\alpha}\}$ defines a size-α test that combines the different studies.

Chapter 11

Asymptotic Tests and Rank Tests

How can one verify whether or not a dice is fair? And how can one check whether, for example, the milk production of cows depends on the nutrition? The chi-square tests for discrete data used in this context are asymptotic tests, in the sense that their size can only be determined in the limit as the number of observations tends to infinity. These limiting results are based on the central limit theorem. Quite different decision rules are needed when one wants to decide whether one drug is more efficient than another one. The so-called order and rank tests that serve this purpose will be discussed in the last section.

11.1   Normal Approximation of Multinomial Distributions

In contrast to the last Section 10.4, where the random observations were simply assumed to be normally distributed, we will now turn to so-called asymptotic tests, in which the normal distribution appears only approximately when the number of observations gets large. These are the object of the next two sections; here we will lay the theoretical foundations. Namely, we will prove a central limit theorem for multinomially distributed random vectors. But first we have to extend the notion of convergence in distribution to include vector-valued random variables.

Definition. Let $s \in \mathbb{N}$, $(Y_n)_{n \geq 1}$ be a sequence of $\mathbb{R}^s$-valued random vectors, and Y a random vector with distribution Q on $\mathbb{R}^s$. One says that Y_n *converges in distribution to Y respectively Q*, written as $Y_n \xrightarrow{d} Y$ or $Y_n \xrightarrow{d} Q$, if $P(Y_n \in A) \to Q(A)$ as $n \to \infty$ for all $A \in \mathscr{B}^s$ with $Q(\partial A) = 0$. Here, ∂A denotes the topological boundary of A.

> In principle, Y and all the Y_n could be defined on different probability spaces. However, by choosing the model appropriately, one can always achieve that they are all defined on the same probability space; we will assume this from now on for simplicity. One should also note that the preceding definition can be extended immediately to the case of random variables with values in an arbitrary topological space; see also Problem 11.1. The following remark, however, makes use of the special structure of $\mathbb{R}^s$.

(11.1) Remark. *Characterisation of convergence in distribution.* In the situation of the definition, the convergence $Y_n \xrightarrow{d} Q$ already holds if only $P(Y_n \in A) \to Q(A)$ for all orthants A of the form $A = \prod_{i=1}^{s}]-\infty, a_i]$, where the $a_i \in]-\infty, \infty]$ are chosen such that $Q(\partial A) = 0$.

Sketch of proof. In view of the additivity of probability measures, the system of all A with $P(Y_n \in A) \to Q(A)$ is closed under taking proper differences and finite disjoint unions. By taking differences successively, the convergence statement can therefore be extended from orthants to arbitrary half-open rectangular boxes (which can be half-sided infinite because $a_i = \infty$ was allowed), and further to finite unions of such boxes.

Given an arbitrary $A \in \mathscr{B}^s$, one can find a sequence of sets B_k with $Q(\partial B_k) = 0$, which can be represented as the disjoint union of finitely many half-open boxes, and which decrease to the closure $\bar{A}$ of A. (Indeed, let $W_\varepsilon(a) \subset \mathbb{R}^s$ be the half-open cube with centre $a \in \mathbb{R}^s$ and edge length $\varepsilon > 0$. The function $\varepsilon \to Q(W_\varepsilon(a))$ is then increasing and bounded and thus has at most countably many discontinuities. But $Q(\partial W_\varepsilon(a)) = 0$ when ε is a point of continuity. Therefore, one can choose B_k as an appropriate union of such $W_\varepsilon(a)$.) It follows that

$$\limsup_{n \to \infty} P(Y_n \in A) \le \inf_{k \ge 1} \lim_{n \to \infty} P(Y_n \in B_k) = \inf_{k \ge 1} Q(B_k) = Q(\bar{A}).$$

Conversely, applying this result to the complement A^c in place of A, we find that $\liminf_{n \to \infty} P(Y_n \in A) \ge Q(A^o)$, where A^o denotes the interior of A. Therefore, if $Q(\partial A) = 0$ then $Q(\bar{A}) = Q(A^o) = Q(A)$, and the convergence $P(Y_n \in A) \to Q(A)$ follows. $\diamond$

In particular, the remark shows that, for $s = 1$, the above definition of convergence in distribution coincides with the definition given previously. Namely, $Y_n \xrightarrow{d} Q$ if and only if $P(Y_n \le a) \to Q(]-\infty, a])$ for all $a \in \mathbb{R}$ with $Q(\{a\}) = 0$, i.e., if the distribution function of Y_n converges to the one of Q at all points where the latter is continuous.

Next we provide some general properties of convergence in distribution needed later. Statement (a) is known as the 'continuous mapping theorem' and (b) and (c) as the Cramér–Slutsky theorem.

(11.2) Proposition. Stability properties of convergence in distribution. *Let $s, r \in \mathbb{N}$ and suppose X and X_n, $n \ge 1$, are $\mathbb{R}^s$-valued random vectors such that $X_n \xrightarrow{d} X$. Then the following statements hold.*

(a) *If $f : \mathbb{R}^s \to \mathbb{R}^r$ is continuous, then $f(X_n) \xrightarrow{d} f(X)$.*

(b) *If $(Y_n)_{n \ge 1}$ is a sequence of random vectors in $\mathbb{R}^s$ with $|Y_n| \xrightarrow{P} 0$, then $X_n + Y_n \xrightarrow{d} X$.*

(c) *Let M be a fixed $r \times s$ matrix and $(\mathsf{M}_n)_{n \ge 1}$ a sequence of random matrices in $\mathbb{R}^{r \times s}$. If $\|\mathsf{M}_n - \mathsf{M}\| \xrightarrow{P} 0$, then $\mathsf{M}_n X_n \xrightarrow{d} \mathsf{M} X$.*

Proof. (a) Take any $A \in \mathscr{B}^r$ with $P(f(X) \in \partial A) = 0$. By the continuity of f, $\partial(f^{-1}A) \subset f^{-1}(\partial A)$, and so $P(X \in \partial(f^{-1}A)) = 0$. The convergence $X_n \xrightarrow{d} X$

therefore implies that

$$P(f(X_n) \in A) = P(X_n \in f^{-1}A) \xrightarrow[n\to\infty]{} P(X \in f^{-1}A) = P(f(X) \in A).$$

(b) Let $A \in \mathscr{B}^s$ be such that $P(X \in \partial A) = 0$. For $\varepsilon > 0$, let A^ε be the open ε-neighbourhood of A (i.e., the union of all open ε-balls with centre in A). The function $\varepsilon \to P(X \in A^\varepsilon)$ is increasing and thus has at most countably many discontinuity points. If ε is a point of continuity, then $P(X \in \partial A^\varepsilon) = 0$ and therefore

$$P(X_n + Y_n \in A) \le P(|Y_n| \ge \varepsilon) + P(X_n \in A^\varepsilon) \to P(X \in A^\varepsilon),$$

so in the limit $\varepsilon \to 0$

$$\limsup_{n\to\infty} P(X_n + Y_n \in A) \le P(X \in \bar{A}).$$

Applying this result to A^c we obtain

$$\liminf_{n\to\infty} P(X_n + Y_n \in A) \ge P(X \in A^o).$$

Since by assumption $P(X \in A^o) = P(X \in \bar{A}) = P(X \in A)$, this implies our claim.

(c) Statement (a) implies that $\mathsf{M}X_n \xrightarrow{d} \mathsf{M}X$. In view of (b), it is therefore sufficient to show that $|(\mathsf{M}_n - \mathsf{M})X_n| \xrightarrow{P} 0$. So let $\varepsilon > 0$ be arbitrary. Then we can write

$$P(|(\mathsf{M}_n - \mathsf{M})X_n| \ge \varepsilon) \le P(\|\mathsf{M}_n - \mathsf{M}\| \ge \delta) + P(|X_n| \ge \varepsilon/\delta),$$

where $\|\cdot\|$ is the operator norm and $\delta > 0$ is chosen arbitrarily. The same argument as in (b) shows that $P(|X| = \varepsilon/\delta) = 0$ for all but at most countably many $\delta > 0$. For these δ we obtain

$$\limsup_{n\to\infty} P(|(\mathsf{M}_n - \mathsf{M})X_n| \ge \varepsilon) \le P(|X| \ge \varepsilon/\delta).$$

Letting $\delta \to 0$, our claim follows. $\diamond$

After these preparations we can now address the approximation problem. Let $E = \{1,\dots,s\}$ be a finite set with $s \ge 2$, ρ a probability density on E, and $X_1, X_2, \dots$ a sequence of i.i.d. random variables with values in E and distribution density ρ, i.e., $P(X_k = i) = \rho(i)$ for all $i \in E$ and $k \in \mathbb{N}$. (For example, one can think of sampling infinitely often with replacement from an urn; E is the set of colours, $\rho(i)$ the proportions of balls with colour $i \in E$, and X_k the colour of the kth ball.) We consider the absolute frequencies

$$h_n(i) = |\{1 \le k \le n : X_k = i\}|$$

of the individual outcomes $i \in E$ up to time n. Theorem (2.9) then states that the random vector $h_n = (h_n(i))_{i\in E}$ has the multinomial distribution $\mathcal{M}_{n,\rho}$ on $\mathbb{Z}_+^s$.

To derive a central limit theorem for the h_n, we must standardise the random vectors h_n appropriately. By Example (4.27), every $h_n(i)$ has expectation $n\rho(i)$ and variance $n\rho(i)(1-\rho(i))$. The resulting standardisation $(h_n(i)-n\rho(i))/\sqrt{n\rho(i)(1-\rho(i))}$, however, is not suitable, because the $h_n(i)$ with $i \in E$ sum up to n and are, therefore, not independent of each other. It is rather the random vector

(11.3)
$$h_{n,\rho}^* = \left(\frac{h_n(i)-n\rho(i)}{\sqrt{n\rho(i)}}\right)_{1\le i\le s}$$

which has the correct standardisation, as we will see.

Note first that $h_{n,\rho}^*$ always belongs to the hyperplane

$$\boldsymbol{H}_\rho = \left\{x \in \mathbb{R}^s : \sum_{i=1}^s \sqrt{\rho(i)}\, x_i = 0\right\}.$$

The following theorem will show that $h_{n,\rho}^*$ converges in distribution to the 'multivariate standard normal distribution on $\boldsymbol{H}_\rho$'. More precisely, let O_ρ be an orthogonal matrix whose last column is the unit vector $u_\rho = (\sqrt{\rho(i)})_{1\le i\le s}$. So O_ρ describes a rotation that maps the hyperplane $\{x \in \mathbb{R}^s : x_s = 0\}$ onto the hyperplane $\boldsymbol{H}_\rho$. Further, let E_{s-1} be the diagonal matrix with a 1 in the first $s-1$ diagonal entries and 0 otherwise. The matrix $\Pi_\rho = \mathsf{O}_\rho \mathsf{E}_{s-1} \mathsf{O}_\rho^\top$ then describes the projection onto the hyperplane $\boldsymbol{H}_\rho$, and we get $\Pi_\rho^\top = \Pi_\rho = \Pi_\rho \Pi_\rho$. Now we define

(11.4) $\mathcal{N}_\rho := \mathcal{N}_s(0, \Pi_\rho) = \mathcal{N}_s(0, \mathsf{E}) \circ \Pi_\rho^{-1}$.

So, $\mathcal{N}_\rho$ is the image of the s-dimensional standard normal distribution $\mathcal{N}_s(0, \mathsf{E})$ under the projection of $\mathbb{R}^s$ onto $\boldsymbol{H}_\rho$. Equivalently, $\mathcal{N}_\rho$ is the image of $\mathcal{N}_s(0, \mathsf{E}_{s-1}) = \mathcal{N}_{0,1}^{\otimes(s-1)} \otimes \delta_0$ under the rotation O_ρ, as follows from the form of Π_ρ and the definition after Theorem (9.5); see also Figure 11.1. Now, the following central limit theorem holds.

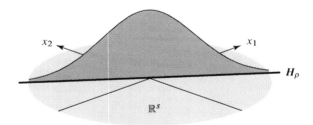

Figure 11.1. Illustration of $\mathcal{N}_\rho$, the standard normal distribution on the hyperplane $\boldsymbol{H}_\rho$ in $\mathbb{R}^s$, for $s = 2$ and $\rho = (0.4, 0.6)$.

(11.5) Theorem. Normal approximation of multinomial distributions. *Let $(X_i)_{i \geq 1}$ be independent random variables with values in E and identical distribution density ρ, and let $h^*_{n,\rho}$, as in (11.3), be the corresponding standardised histogram after n observations. Then*

$$h^*_{n,\rho} \xrightarrow{d} \mathcal{N}_\rho$$

as $n \to \infty$.

The proof is preceded by a lemma, which is also of independent interest. It will allow to replace the dependent random variables $h^*_{n,\rho}(i)$, $i \in E$, by independent random variables, so that we can apply the central limit theorem.

(11.6) Lemma. Poisson representation of multinomial distributions. *Consider a family $(Z_k(i))_{k \geq 1, 1 \leq i \leq s}$ of independent random variables for which $Z_k(i)$ has the Poisson distribution $\mathcal{P}_{\rho(i)}$. Further, let $S_n(i) = \sum_{k=1}^{n} Z_k(i)$, $S_n = (S_n(i))_{1 \leq i \leq s}$, and $N_n = \sum_{i=1}^{s} S_n(i)$. Then*

$$P(S_n = \ell | N_n = m) = \mathcal{M}_{m,\rho}(\{\ell\}) = P(h_m = \ell)$$

for all $m, n \in \mathbb{N}$ and all $\ell = (\ell(i))_{1 \leq i \leq s} \in \mathbb{Z}_+^s$ with $\sum_{i=1}^{s} \ell(i) = m$.

Figure 11.2 illustrates this result (see Problem 3.22 for an alternative formulation).

Proof. According to the convolution formula (4.41) for Poisson distributions, $S_n(i)$ and N_n have the Poisson distributions $\mathcal{P}_{n\rho(i)}$ and $\mathcal{P}_n$, respectively. Using the multinomial coefficient $\binom{m}{\ell}$ from (2.8), we can therefore write

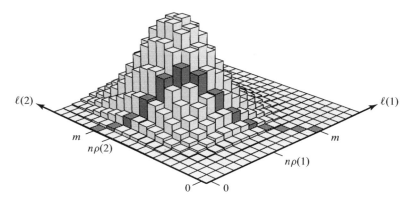

Figure 11.2. The section through the histogram of $P \circ S_n^{-1} = \bigotimes_{i=1}^{s} \mathcal{P}_{n\rho(i)}$ along the hyperplane $\sum_{i=1}^{s} \ell(i) = m$ is proportional to the histogram of $\mathcal{M}_{m,\rho}$. Here we illustrate the case $s = 2$, $\rho = (0.4, 0.6)$, $n = 18$, $m = 13$.

$$P(S_n = \ell \,|\, N_n = m) = \left(\prod_{i=1}^{s} e^{-n\rho(i)} \frac{(n\rho(i))^{\ell(i)}}{\ell(i)!} \right) \Big/ e^{-n} \frac{n^m}{m!}$$

$$= \binom{m}{\ell} \prod_{i=1}^{s} \rho(i)^{\ell(i)} = \mathcal{M}_{m,\rho}(\{\ell\}),$$

as claimed. $\diamond$

The lemma suggests the following *idea for the proof of Theorem* (11.5). Consider the standardised random variables

$$S_n^*(i) = \frac{S_n(i) - n\rho(i)}{\sqrt{n\rho(i)}}, \quad N_n^* = \frac{N_n - n}{\sqrt{n}} = \sum_{i=1}^{s} \sqrt{\rho(i)}\, S_n^*(i),$$

as well as the vectors $S_n^* = (S_n^*(i))_{1 \le i \le s}$ and $Y_n^* = \mathsf{O}_\rho^\top S_n^*$. By the definition of the matrix O_ρ, we then obtain $Y_n^*(s) = N_n^*$ and therefore

$$P(h_n^* \in A) = P(S_n^* \in A \,|\, N_n^* = 0) = P(Y_n^* \in \mathsf{O}_\rho^{-1} A \,|\, Y_n^*(s) = 0).$$

It follows from the central limit theorem (5.29) that

(11.7) $$S_n^* \xrightarrow{d} \mathcal{N}_{0,1}^{\otimes s} = \mathcal{N}_s(0, \mathsf{E}).$$

Jointly with Proposition (11.2a) and Corollary (9.4), this shows that

(11.8) $$Y_n^* \xrightarrow{d} \mathcal{N}_s(0, \mathsf{E}) \circ \mathsf{O}_\rho = \mathcal{N}_s(0, \mathsf{E}).$$

It is therefore plausible to suspect that

$$P(Y_n^* \in \mathsf{O}_\rho^{-1} A \,|\, Y_n^*(s) = 0) \xrightarrow[n \to \infty]{} \mathcal{N}_s(0, \mathsf{E})(\mathsf{O}_\rho^{-1} A \,|\, \{x \in \mathbb{R}^s : x_s = 0\}).$$

Even though this last conditional probability is not well-defined, since the condition has probability 0, it can be identified with $\mathcal{N}_s(0, \mathsf{E}_{s-1})(\mathsf{O}_\rho^{-1} A) = \mathcal{N}_\rho(A)$ in a natural way. Hence we obtain the claim of the theorem. The only point that needs some effort is the *conditional* central limit theorem required. This can be handled by certain monotonicity arguments.

Proof of Theorem (11.5). We will continue to use the random variables as introduced above.

Step 1: Weakening the condition on N_n^*. Let A be an orthant as in Remark (11.1). The decisive property of orthants we will use here is that they are decreasing with respect to the natural partial ordering on $\mathbb{R}^s$. Namely, if $x \in A$ and $y \le x$ coordinate-wise, then also $y \in A$. For arbitrary $m, n \ge 1$, we now consider

$$h_{m,n}^* := \big((h_m(i) - n\rho(i)) / \sqrt{n\rho(i)} \big)_{1 \le i \le s}$$

and

$$q_{m,n} := P(h^*_{m,n} \in A) = P(S^*_n \in A | N_n = m);$$

the last equality follows from Lemma (11.6). Since $h^*_{n,n} = h^*_n$, we are actually only interested in the case $m = n$, but to weaken the 'sharp' condition $m = n$ we will now allow m to deviate slightly from n. Since $h_m \leq h_{m+1}$ coordinate-wise, it follows that $\{h^*_{m,n} \in A\} \supset \{h^*_{m+1,n} \in A\}$, which is to say that $q_{m,n}$ is decreasing in m. Together with the case-distinction formula (3.3a), we thus obtain the inequality

$$q_{n,n} \geq \sum_{m=n}^{\lfloor n+\varepsilon\sqrt{n}\rfloor} q_{m,n} \, P(N_n = m | N^*_n \in [0, \varepsilon]) = P(S^*_n \in A | N^*_n \in [0, \varepsilon])$$

for every $\varepsilon > 0$, and similarly $q_{n,n} \leq P(S^*_n \in A | N^*_n \in [-\varepsilon, 0])$.

Step 2: Application of the central limit theorem. Using the independence of the coordinates of S^*_n and the central limit theorem (5.29), we find that

$$P(S^*_n \in A) = \prod_{i=1}^{s} P(S^*_n(i) \leq a_i) \xrightarrow[n\to\infty]{} \prod_{i=1}^{s} \Phi(a_i) = \mathcal{N}_s(0, \mathsf{E})(A)$$

for each orthant $A = \prod_{i=1}^{s}]\infty, a_i]$. In view of Remark (11.1), this is equivalent to the convergence result (11.7), and this in turn implies statement (11.8). Hence, fixing again an orthant A we obtain for every slab $U_\varepsilon = \{x \in \mathbb{R}^s : x_s \in [0, \varepsilon]\}$

$$P(S^*_n \in A | N^*_n \in [0, \varepsilon]) = \frac{P(Y^*_n \in O^{-1}_\rho A \cap U_\varepsilon)}{P(Y^*_n \in U_\varepsilon)} \to \mathcal{N}_s(0, \mathsf{E})(O^{-1}_\rho A | U_\varepsilon) =: q_\varepsilon.$$

To analyse the last expression, we use Example (3.30) and Fubini's theorem to write

$$q_\varepsilon = \mathcal{N}_{0,1}([0, \varepsilon])^{-1} \int_0^\varepsilon dt \, \phi(t) \, f_{A,\rho}(t),$$

where

$$f_{A,\rho}(t) = \mathcal{N}_{s-1}(0, \mathsf{E})\big(x \in \mathbb{R}^{s-1} : (x, t) \in O^{-1}_\rho A\big).$$

Now, if $t \to 0$, then $1_{O^{-1}_\rho A}(x, t) \to 1_{O^{-1}_\rho A}(x, 0)$ for Lebesgue-almost all x, and therefore $f_{A,\rho}(t) \to f_{A,\rho}(0) = \mathcal{N}_\rho(A)$ by the dominated convergence theorem; cf. Problem 4.7b. It follows that $q_\varepsilon \to \mathcal{N}_\rho(A)$ as $\varepsilon \to 0$. Together with the first step, this implies

$$\liminf_{n\to\infty} q_{n,n} \geq \sup_{\varepsilon>0} q_\varepsilon \geq \mathcal{N}_\rho(A).$$

Similarly, using the upper bound of the first step, we arrive at the reversed inequality $\limsup_{n\to\infty} q_{n,n} \leq \mathcal{N}_\rho(A)$. By Remark (11.1), this proves the theorem. $\diamond$

11.2   The Chi-Square Test of Goodness of Fit

How can one check whether a dice is fair? Or whether the pseudo-random numbers produced by some algorithm are sufficiently random? Or whether an inheritance theory is correct in predicting certain frequencies for specific variations of a certain trait? To answer any of these questions, one will perform n independent experiments with results in a finite set $E = \{1, \ldots, s\}$, and observe the histogram of relative frequencies of the individual outcomes $i \in E$. If this histogram is close enough to the anticipated probability distribution (for instance to the equidistribution for a dice), then one will accept this distribution, otherwise reject it. But what does 'close enough' mean? To answer this question, we first need to formalise the statistical model.

As the number n of independent observations is supposed to tend to infinity later on, it is again convenient to work in an infinite product model. Since each single observation takes values in $E = \{1, \ldots, s\}$, the sample space is thus $\mathcal{X} = E^{\mathbb{N}}$, which is equipped with the product σ-algebra $\mathscr{F} = \mathscr{P}(E)^{\otimes \mathbb{N}}$. Each single experiment can have any conceivable distribution, but we suppose that each individual outcome has a non-vanishing probability. Therefore, we take Θ to be the set of all strictly positive probability densities on $\{1, \ldots, s\}$, viz.

$$\Theta = \left\{ \vartheta = (\vartheta(i))_{1 \le i \le s} \in \,]0, 1[^s \,:\, \sum_{i=1}^{s} \vartheta(i) = 1 \right\}.$$

Each $\vartheta \in \Theta$ will also be interpreted as a probability measure on $E = \{1, \ldots, s\}$, so we will not distinguish between probability measures and their densities. For $\vartheta \in \Theta$, we let $P_\vartheta = \vartheta^{\otimes \mathbb{N}}$ be the infinite product of ϑ on $E^{\mathbb{N}}$; this ensures that the individual experiments are i.i.d. Our statistical model is therefore the infinite product model

$$(\mathcal{X}, \mathscr{F}, P_\vartheta : \vartheta \in \Theta) = \left(E^{\mathbb{N}}, \mathscr{P}(E)^{\otimes \mathbb{N}}, \vartheta^{\otimes \mathbb{N}} : \vartheta \in \Theta \right).$$

As usual, the kth observation is described by the kth projection $X_k : \mathcal{X} \to E$.

Now, the questions raised at the beginning can be rephrased in general terms as follows: *Do the observations X_k have the specific distribution $\rho = (\rho(i))_{1 \le i \le s}$ that is anticipated on theoretical grounds?* (For the dice and the pseudo-random numbers, ρ is the uniform distribution, and for the genetic problem, it is the theoretical frequency distribution of the different variations of the trait.) So, our aim is to test the null hypothesis $H_0 : \vartheta = \rho$ against the alternative $H_1 : \vartheta \ne \rho$; in other words, we set $\Theta_0 = \{\rho\}$ and $\Theta_1 = \Theta \setminus \{\rho\}$. (Note that the null hypothesis is sharp, so that the comments of Remark (10.24) should be kept in mind.)

For this purpose, we consider for each $n \ge 1$ the (absolute) frequencies

$$h_n(i) = |\{1 \le k \le n : X_k = i\}|$$

of the individual outcomes $1 \leq i \leq s$ up to time n. The corresponding (random) vector of the relative frequencies

$$L_n = \left(\frac{h_n(1)}{n}, \ldots, \frac{h_n(s)}{n} \right)$$

is called the *(empirical) histogram*, or the *empirical distribution*, after n observations. How can one decide whether L_n is close enough to ρ, so that the null hypothesis can be accepted?

Taking again the maximum likelihood principle as a source of inspiration, we wonder about the form of a likelihood ratio test for our problem. The likelihood ratio after n observations reads

$$R_n = \frac{\sup_{\vartheta \in \Theta_1} \prod_{i=1}^{s} \vartheta(i)^{h_n(i)}}{\prod_{i=1}^{s} \rho(i)^{h_n(i)}}.$$

Since Θ_1 is dense in Θ, the supremum in the numerator coincides, by continuity, with the supremum over the entire Θ. The latter is attained for $\vartheta = L_n$ because L_n is the maximum likelihood estimator of ϑ; recall Example (7.7). Therefore,

$$(11.9) \qquad \log R_n = n \sum_{i=1}^{s} L_n(i) \log \frac{L_n(i)}{\rho(i)} = n\, H(L_n; \rho),$$

where $H(\,\cdot\,;\,\cdot\,)$ is the relative entropy introduced in (7.31). In words, up to a factor of n, the logarithmic likelihood ratio is nothing but the relative entropy of the empirical distribution L_n relative to the null hypothesis ρ. A likelihood ratio test based on n independent observations thus takes the form

$$\varphi_n = \begin{cases} 1 & \text{if } n\, H(L_n; \rho) > c, \\ 0 & \text{if } n\, H(L_n; \rho) \leq c \end{cases}$$

for a suitable constant c (which possibly depends on n). Of course, c must be chosen to match the level α. As a first step in this direction we observe that, under the null hypothesis P_ρ and in the limit as $n \to \infty$, the relative entropy can be replaced by a quadratic Taylor approximation.

(11.10) Proposition. *Quadratic approximation of relative entropy. Let*

$$(11.11) \qquad D_{n,\rho} = \sum_{i=1}^{s} \frac{(h_n(i) - n\rho(i))^2}{n\rho(i)} = n \sum_{i=1}^{s} \rho(i) \left(\frac{L_n(i)}{\rho(i)} - 1 \right)^2.$$

Then the convergence in probability

$$n\, H(L_n; \rho) - D_{n,\rho}/2 \xrightarrow{P_\rho} 0$$

holds in the limit as $n \to \infty$.

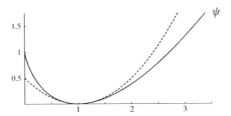

Figure 11.3. The function ψ and its Taylor parabola at the point 1 (dashed line).

Proof. We first assume that $D_{n,\rho} \leq c$ for some given $c > 0$. With the abbreviation $a(i) = \frac{L_n(i)}{\rho(i)} - 1$, this means that

$$n \sum_{i=1}^{s} \rho(i)\, a(i)^2 = D_{n,\rho} \leq c \,.$$

Hence $a(i)^2 \leq c/n\rho(i)$ or, in terms of the Landau symbol, $a(i) = O(n^{-1/2})$ for all i. On the other hand, we know from Remark (7.31) that

$$n\, H(L_n; \rho) = n \sum_{i=1}^{s} \rho(i)\, \psi\big(1 + a(i)\big),$$

where $\psi(u) = 1 - u + u \log u$. The function ψ attains its minimum 0 at the point $u = 1$, and its Taylor approximation at this point is $\psi(u) = (u-1)^2/2 + O(|u-1|^3)$, see Figure 11.3. So we obtain

$$n\, H(L_n; \rho) = n \sum_{i=1}^{s} \rho(i)\big(a(i)^2/2 + O(n^{-3/2})\big)$$

$$= D_{n,\rho}/2 + O(n^{-1/2}) \,.$$

Turning to the proof of the stated convergence in probability, we pick an arbitrary $\varepsilon > 0$. It then follows from the above that

$$A_n := \{\, |n\, H(L_n; \rho) - D_{n,\rho}/2| > \varepsilon \,\} \subset \{D_{n,\rho} > c\}$$

for every $c > 0$ and all sufficiently large n. We also know from Theorem (2.9) that each $h_n(i)$ has a binomial distribution, so that, by Example (4.27),

$$\mathbb{E}_\rho(D_{n,\rho}) = \sum_{i=1}^{s} \frac{\mathbb{V}_\rho(h_n(i))}{n\rho(i)} = \sum_{i=1}^{s}(1 - \rho(i)) = s - 1 \,.$$

The Markov inequality (5.4) therefore implies that $P(A_n) \leq (s-1)/c$ for all sufficiently large n and for arbitrary c. This means that $P(A_n) \to 0$, as desired. $\diamond$

The proposition states that, on the null hypothesis, a likelihood ratio test is essentially equivalent to the following test.

Definition. Let $D_{n,\rho}$ be defined by (11.10). A test of the null hypothesis $H_0 : \vartheta = \rho$ against the alternative $H_1 : \vartheta \neq \rho$ with a rejection region of the form $\{D_{n,\rho} > c\}$ for some $c > 0$ is called a *chi-square test of goodness of fit* after n observations, or simply a χ^2-goodness-of-fit test.

So, for a χ^2-test of goodness of fit, the function $D_{n,\rho}$ is used to measure the deviation of the observed histogram L_n from the hypothetical distribution ρ. For practical calculations of $D_{n,\rho}$, the formula

$$D_{n,\rho} = n \sum_{i=1}^{s} \frac{L_n(i)^2}{\rho(i)} - n$$

is particularly convenient. It is immediately obtained from (11.11) by expanding the square. This formula shows, in particular, that the null hypothesis is accepted if and only if L_n falls into the centred ellipsoid with semi-axis $\sqrt{\rho(i)(1 + \frac{c}{n})}$ in the ith direction, or more precisely, into its intersection with the simplex Θ; see Figure 11.4.

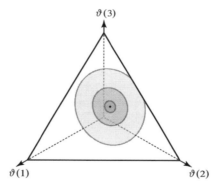

Figure 11.4. Acceptance ellipses of the χ^2-goodness-of-fit test in the simplex Θ for $s = 3$, $\rho = (3, 4, 5)/12$ (the point at the centre), $\alpha = 1\%$, and $n = 25, 100, 1000$. The null hypothesis is accepted if L_n falls into the corresponding ellipse.

Why is this called a χ^2-test? This is explained by the following theorem, which states that the asymptotic distribution of $D_{n,\rho}$ under the null hypothesis is a χ^2-distribution. Its discoverer Karl Pearson (1857–1936) is the father of Egon S. Pearson, who is referred to in the Neyman–Pearson lemma.

(11.12) Theorem. K. Pearson 1900. *Under the null hypothesis P_ρ, in the limit as $n \to \infty$, the statistic $D_{n,\rho}$ converges in distribution to the chi-square distribution with $s - 1$ degrees of freedom. That is,*

$$\lim_{n \to \infty} P_\rho(D_{n,\rho} \leq c) = \chi^2_{s-1}([0, c])$$

for all $c > 0$.

Combining the theorem with Propositions (11.10) and (11.2b), we also obtain that $2n\,H(L_n;\rho) \xrightarrow{\ d\ } \chi^2_{s-1}$ under the null hypothesis.

Proof. We consider the standardised frequency vector $h^*_{n,\rho}$ from (11.3). By definition, $D_{n,\rho} = |h^*_{n,\rho}|^2$, and Theorem (11.5) states that, under P_ρ, $h^*_{n,\rho}$ converges in distribution to the multivariate Gauss distribution $\mathcal{N}_\rho$. The latter is the image of $\mathcal{N}_{0,1}^{\otimes(s-1)} \otimes \delta_0$ under the rotation O_ρ. Specifically, for the rotation invariant set $A = \{|\cdot|^2 \le c\}$ we obtain

$$P_\rho(D_{n,\rho} \le c) = P_\rho(h^*_{n,\rho} \in A) \xrightarrow[n\to\infty]{} \mathcal{N}_\rho(A)$$
$$= \mathcal{N}_{0,1}^{\otimes(s-1)}(x \in \mathbb{R}^{s-1} : |x|^2 \le c) = \chi^2_{s-1}([0,c]) \,.$$

In the last step, we have used Theorem (9.10), which also implies that $\mathcal{N}_\rho(\partial A) = \chi^2_{s-1}(\{c\}) = 0$, as required for the second step. The proof is thus complete. $\diamond$

Pearson's theorem allows to choose the threshold c of the χ^2-goodness-of-fit test such that a given size α is reached at least approximately for large n. Namely, one simply takes $c = \chi^2_{s-1;1-\alpha}$, the α-fractile of the χ^2-distribution with $s-1$ degrees of freedom. Then $\chi^2_{s-1}(]c,\infty[) = \alpha$ by definition, and Theorem (11.12) shows that the χ^2-goodness-of-fit test with threshold level c has approximately size α when n is large enough. A rule of thumb claims that the approximation is satisfactory if $n \ge 5/\min_{1\le i\le s}\rho(i)$. For smaller n one should use the exact distribution of $D_{n,\rho}$, which can be derived from the multinomial distribution – as we have seen in the proof of Theorem (11.12). We will now present two applications.

(11.13) Example. *Mendel's peas.* One of the classical experiments of genetics, published by Gregor Mendel in 1865 to support his theory, is the following. He observed in peas the two traits 'shape' and 'colour' with the respective variants 'round' (A) or 'wrinkled' (a) and 'yellow' (B) or 'green' (b). The variant 'round' is dominant, i.e., the three genotypes AA, Aa and aA all lead to a round phenotype, whereas only the peas of genotype aa are wrinkled. Similarly, 'yellow' is dominant. So if one considers the offspring of a plant of the so-called double heterozygote genotype AaBb, then according to Mendel's theory, the four possible phenotypes should appear in the frequency ratio 9 : 3 : 3 : 1. Table 11.1 lists Mendel's experimental data.

Table 11.1. Mendel's counts for $n = 556$ peas.

	yellow	green
round	315	108
wrinkled	101	32

Do these results support the theory? In other words, can one confirm the null hypothesis that the data follow the theoretical frequency distribution $\rho = (9,3,3,1)/16$?

Using the χ^2-test of goodness of fit, the question can be answered in the following way. Let us choose the relatively large level $\alpha = 0.1$ to achieve a reasonably large power, so that we can be quite certain that a deviation from the theory will be detected. We have $s = 4$, so we need the 0.9-quantile of the χ^2-distribution with 3 degrees of freedom. Table B in the appendix yields $\chi^2_{3;0.9} = 6.3$. For the data set x of Table 11.1 we then find

$$D_{n,\rho}(x) = \frac{16}{556}\left(\frac{315^2}{9} + \frac{108^2}{3} + \frac{101^2}{3} + \frac{32^2}{1}\right) - 556 = 0.470 < 6.3\,.$$

Hence, this sample really does confirm Mendel's theory. But it comes as a surprise that the value of $D_{n,\rho}(x)$ is actually that small. Table B shows that $\chi^2_3([0, 0.5]) < 0.1$; the probability of such a small deviation from the theoretical prediction is therefore less than 10%. This gave reason to suspect that Mendel manipulated his numbers to make his theory more convincing. Be this as it may, another 35 years had to pass before Pearson proved his theorem, so Mendel himself could not have estimated the typical order of magnitude of stochastic fluctuations.

(11.14) Example. *Munich council elections.* On May 10, 2001, the German newspaper 'Süddeutsche Zeitung' published the result of an opinion poll among $n = 783$ Munich citizens that had been asked the so-called Sunday question: 'If there were a local council election next Sunday, which party would you vote for?'. The newspaper emphasised that, for the first time since the previous election, the social democrats (SPD) were leading. Does the result of the poll really confirm a shift in the voters' preferences? This is not evident from the histograms in Figure 11.5.

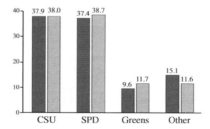

Figure 11.5. Results of the council election in Munich on 10/03/1996 (dark) and the opinion poll on 10/05/2001(light), in percentages.

But let ρ denote the probability density that holds the percentages in the last council election. Our question then reads: Is the sample of the opinion poll still based on this ρ? This is precisely the null hypothesis of a χ^2-test of goodness of fit. For the significance level $\alpha = 0.02$, Table B yields the fractile $\chi^2_{3;0.98} = 9.837$; the sample x from the opinion poll gives the value $D_{n,\rho}(x) = 10.3$. The χ^2-test thus leads to a rejection of the null hypothesis. So the poll indicates a change in the voters' opinions.

However, the decision is rather close. Since $\chi^2_{3;0.99} = 11.34$, the null hypothesis had to be accepted for the level $\alpha = 0.01$.

The χ^2-test of goodness of fit is constructed in a way that makes its size converge to the chosen value $\alpha > 0$ in the limit as $n \to \infty$. A version of Stein's theorem (10.4) (cf. Problem 11.4) therefore implies that, at each point $\vartheta \neq \rho$, the power tends to 1 at exponential rate, i.e., the probability of a type II error vanishes exponentially fast. It is thus natural to wonder if also the probability of a type I error can be made to vanish with exponential rate. To achieve this, the relative entropy may not be replaced by its quadratic approximation, but must rather be used in its original form to measure the deviation of the histogram L_n from the null hypothesis ρ. For a suitable choice of the threshold, the error probabilities of both type I and type II then converge to 0 exponentially fast, as is made precise below.

(11.15) Theorem. W. Hoeffding's entropy test, 1965. *For given $a > 0$, let φ_n be the test with rejection region $\{H(L_n; \rho) > a\}$. Then the following holds.*

(a) *The size of φ_n decays exponentially fast with rate a, i.e.,*

$$\lim_{n \to \infty} \frac{1}{n} \log \mathbb{E}_\rho(\varphi_n) = -a \, .$$

(b) *For each $\vartheta \neq \rho$, the power of φ_n satisfies the estimate*

$$\mathbb{E}_\vartheta(\varphi_n) \geq 1 - \exp\Big[-n \min_{\nu : H(\nu;\rho) \leq a} H(\nu; \vartheta)\Big] \, ,$$

and there is no sequence of tests satisfying (a) whose power tends to 1 at a faster rate.

For $H(\vartheta; \rho) > a$, the minimum in the last exponent is non-zero by Remark (7.31); so, in this case, the power converges to 1 at exponential rate. No conclusions can be drawn in the alternative case. So, a must be chosen rather small if one wants to obtain an exponential decay of the type II error probabilities on a large part of the alternative.

Proof. We will only prove a part of the theorem, namely the exponential decay; for the optimality of the bounds we refer, for instance, to [13, Problem 2.13]. The crucial inequality is the following 'large deviation estimate'. If $\rho \in \Theta$ and A is an arbitrary subset of $\bar{\Theta}$, the set of *all* (not necessarily strictly positive) probability densities on E, then

(11.16) $\displaystyle \frac{1}{n} \log P_\rho(L_n \in A) \leq - \inf_{\nu \in A} H(\nu; \rho) + \delta_n \, ,$

where $\delta_n = \frac{s}{n} \log(n+1) \to 0$ as $n \to \infty$. Indeed, let $\bar{\Theta}_n$ be the set of all $\nu \in \bar{\Theta}$ such that $n\, \nu(i) \in \mathbb{Z}_+$ for all $1 \leq i \leq s$. Since each such $\nu(i)$ can take at most $n+1$

different values, we have the bound $|\bar{\Theta}_n| \leq (n+1)^s = e^{n\delta_n}$. Next, for $v \in \bar{\Theta}_n$ we can write, replacing P_ρ by P_v,

$$P_\rho(L_n = v) = \mathbb{E}_v\left(1_{\{L_n=v\}} \prod_{i \in E : v(i) > 0} \left(\frac{\rho(i)}{v(i)}\right)^{h_n(i)}\right)$$

$$= \mathbb{E}_v\left(1_{\{L_n=v\}} e^{-n H(v;\rho)}\right) \leq e^{-n H(v;\rho)} .$$

Summing over all $v \in A$, we thus find that

$$P_\rho(L_n \in A) = P_\rho(L_n \in A \cap \bar{\Theta}_n) \leq |\bar{\Theta}_n| \exp\left[-n \inf_{v \in A} H(v; \rho)\right],$$

and (11.16) follows.

Specifically, we now set $A = \{v \in \bar{\Theta} : H(v; \rho) > a\}$. The rejection region of φ_n is then equal to $\{L_n \in A\}$, and (11.16) yields the inequality

$$\frac{1}{n} \log \mathbb{E}_\rho(\varphi_n) \leq -a + \delta_n ,$$

which is the part of statement (a) that is important in applications. Likewise, replacing ρ by ϑ and A by A^c, we obtain from (11.16) that

$$1 - \mathbb{E}_\vartheta(\varphi_n) = P_\vartheta(L_n \in A^c) \leq \exp\left[n\left(-\min_{v \in A^c} H(v; \vartheta) + \delta_n\right)\right].$$

Up to the error term δ_n, this is the first claim of (b). As A^c turns out to be a convex set, a refined estimate shows that the error term δ_n is actually superfluous; see [13]. $\diamond$

11.3 The Chi-Square Test of Independence

We begin again with an illustrative example.

(11.17) Example. *Environmental awareness and educational background.* Does education have an influence on environmental awareness? In a survey by the German polling organisation EMNID, $n = 2004$ randomly chosen persons were asked how much they felt affected by pollutants (with $a = 4$ possible answers ranging from 'not at all' to 'severely'). At the same time, the interviewee's educational background was registered (in $b = 5$ steps from 'unskilled' to 'university degree'). The results are summarised in Table 11.2, a so-called *contingency table* that lists for each pair $(i, j) \in \{1, \ldots, a\} \times \{1, \ldots, b\}$ the number of people with the corresponding pair of features (together with the respective row and column sums).

It is plain that the answer 'not at all' strongly dominates in columns 1 to 3, whereas, in columns 4 and 5, the favourite reply is 'slightly'. But are these differences significant enough to support the claim that there is a correlation between environmental awareness and educational background?

Table 11.2. Contingency table for the environment question; according to [60].

Effect	Education 1	2	3	4	5	Σ
not at all	212	434	169	79	45	939
slightly	85	245	146	93	69	638
moderately	38	85	74	56	48	301
severely	20	35	30	21	20	126
Σ	355	799	419	249	182	2004

Such questions about the correlation of two characteristics appear in many different contexts. For instance, does the effect of a drug depend on its pharmaceutical form? Is a driver's reaction rate related to his/her gender or a certain medication? Is there any connection between musicality and mathematical ability? Does the error frequency in the production of an item depend on the day of the week, and the milk production of cows on the nutrition? And so on. The general test problem can be formulated as follows.

Let $A = \{1, \ldots, a\}$ and $B = \{1, \ldots, b\}$ be the sets of the possible specifications of the characteristics that are tested for independence, where $a, b \geq 2$. Then, each individual observation takes values in $E = A \times B$; the elements of E will, for brevity, be denoted by ij, with $i \in A$ and $j \in B$. The unknown parameter is the probability distribution of the individual observations; as before, we will assume that each element of E occurs with non-zero probability. So we define Θ as the set of all strictly positive probability densities on E, viz.

$$\Theta = \Theta_E := \left\{ \vartheta = (\vartheta(ij))_{ij \in E} \in]0, 1[^E : \sum_{ij \in E} \vartheta(ij) = 1 \right\}.$$

As in the last section, we identify each $\vartheta \in \Theta$ with the corresponding probability measure on E. Our statistical model is then the infinite product model

$$(\mathcal{X}, \mathcal{F}, P_\vartheta : \vartheta \in \Theta) = \left(E^{\mathbb{N}}, \mathscr{P}(E)^{\otimes \mathbb{N}}, \vartheta^{\otimes \mathbb{N}} : \vartheta \in \Theta \right).$$

For each $\vartheta \in \Theta$, we denote by

$$\vartheta^A = (\vartheta^A(i))_{i \in A}, \qquad \vartheta^A(i) = \sum_{j \in B} \vartheta(ij),$$

$$\vartheta^B = (\vartheta^B(j))_{j \in B}, \qquad \vartheta^B(j) = \sum_{i \in A} \vartheta(ij),$$

the two marginal distributions of ϑ on A and B, respectively.

What are the null hypothesis and the alternative? By Remark (3.28), the two characteristics we consider are independent if and only if the true ϑ has the form of a product, which means that $\vartheta = \alpha \otimes \beta$ for two probability densities α and β on A

and B, respectively. It then follows that $\alpha = \vartheta^A$ and $\beta = \vartheta^B$. The null hypothesis can therefore be written in the form

$$H_0 : \vartheta = \vartheta^A \otimes \vartheta^B .$$

Correspondingly, we choose

$$\Theta_0 = \left\{ \alpha \otimes \beta = (\alpha(i)\,\beta(j))_{ij \in E} : \alpha \in \Theta_A,\ \beta \in \Theta_B \right\}$$

and $\Theta_1 = \Theta \setminus \Theta_0$.

What is a sensible way of testing Θ_0 against Θ_1? After n observations $X_1, \ldots, X_n$, we obtain the *contingency table*

$$h_n(ij) = |\{1 \le k \le n : X_k = ij\}| , \quad ij \in E .$$

The random matrix

$$L_n = \big(L_n(ij)\big)_{ij \in E} := \big(h_n(ij)/n\big)_{ij \in E}$$

describes the empirical joint distribution of the two characteristics, and its marginal distributions L_n^A and L_n^B describe the empirical relative frequencies of the individual characteristics; the corresponding absolute frequencies are

$$h_n^A(i) := \sum_{j \in B} h_n(ij) = n\, L_n^A(i) \quad \text{and} \quad h_n^B(j) := \sum_{i \in A} h_n(ij) = n\, L_n^B(j) .$$

In terms of this notation, it is natural to accept the null hypothesis H_0 if L_n is sufficiently close to the product distribution $L_n^A \otimes L_n^B$. But what does 'sufficiently close' mean in this context?

In order to find a reasonable measure of distance, we again consider the likelihood ratio R_n after n observations. In the following, we will omit the index n for brevity. Since Θ_1 is dense in Θ, we can argue as in the proof of (11.9) to obtain

$$R = \frac{\max_{\vartheta \in \Theta} \prod_{ij \in E} \vartheta(ij)^{h(ij)}}{\max_{\alpha \otimes \beta \in \Theta_0} \prod_{ij \in E} (\alpha(i)\beta(j))^{h(ij)}}$$

$$= \frac{\max_{\vartheta \in \Theta} \prod_{ij \in E} \vartheta(ij)^{h(ij)}}{\big(\max_{\alpha \in \Theta_A} \prod_{i \in A} \alpha(i)^{h^A(i)}\big)\big(\max_{\beta \in \Theta_B} \prod_{j \in B} \beta(j)^{h^B(j)}\big)}$$

$$= \prod_{ij \in E} L(ij)^{h(ij)} \Big/ \Big[\Big(\prod_{i \in A} L^A(i)^{h^A(i)}\Big)\Big(\prod_{j \in B} L^B(j)^{h^B(j)}\Big)\Big]$$

$$= \prod_{ij \in E} \Big(\frac{L(ij)}{L^A(i)\,L^B(j)}\Big)^{n\, L(ij)} = \exp\big[n\, H(L;\, L^A \otimes L^B)\big] .$$

The relative entropy $H(L; L^A \otimes L^B)$ is also called the *mutual information* of L^A and L^B. One can then prove a counterpart of Proposition (11.10). Namely, under the null hypothesis, the expression $n\,H(L; L^A \otimes L^B)$ admits the quadratic approximation

$$\widetilde{D}_n = n \sum_{ij \in E} L^A(i)\, L^B(j) \Big(\frac{L(ij)}{L^A(i)\, L^B(j)} - 1\Big)^2,$$

which can also be written in the variants

$$\widetilde{D}_n = \sum_{ij \in E} \frac{(h(ij) - h^A(i)\, h^B(j)/n)^2}{h^A(i)\, h^B(j)/n} = n \Big(\sum_{ij \in E} \frac{L(ij)^2}{L^A(i)\, L^B(j)} - 1 \Big);$$

see Problem 11.9. The asymptotic distribution of $\widetilde{D}_n$ can be derived from Theorem (11.5), as follows.

(11.18) Theorem. Extended theorem of Pearson. *For every $\rho = \alpha \otimes \beta$ in the null hypothesis Θ_0, the sequence $\widetilde{D}_n$, under P_ρ, converges in distribution to the chi-square distribution $\chi^2_{(a-1)(b-1)}$. That is,*

$$\lim_{n \to \infty} P_{\alpha \otimes \beta}(\widetilde{D}_n \le c) = \chi^2_{(a-1)(b-1)}([0, c])$$

for all $c > 0$.

Why is the number of degrees of freedom equal to $(a-1)(b-1)$ rather than $ab - 1$, as one might expect from Theorem (11.12)? As we have already seen in Student's theorem (9.17), the number of degrees of freedom is reduced by 1 for each unknown parameter that must be estimated. Estimating α by L^A costs $a - 1$ degrees of freedom (since $\sum_{i \in A} \alpha(i) = 1$), and estimating β requires $b - 1$ degrees of freedom. The total number $ab - 1$ of degrees of freedom is therefore diminished by $(a-1) + (b-1)$, so that only $(a-1)(b-1)$ degrees of freedom remain. The following proof will show this in more detail. But first we will discuss the practical implications of Theorem (11.18), namely the following test procedure.

Chi-square test of independence. For given $0 < \alpha < 1$ let $c = \chi^2_{(a-1)(b-1);1-\alpha}$ be the α-fractile of the chi-square distribution with $(a-1)(b-1)$ degrees of freedom. For sufficiently large n, the test with rejection region $\{\widetilde{D}_n > c\}$ for the null hypothesis $H_0 : \vartheta = \vartheta^A \otimes \vartheta^B$ against the alternative $H_1 : \vartheta \ne \vartheta^A \otimes \vartheta^B$ then has approximately size α.

What is the result of this test in the case of Example (11.17), the question about environmental awareness? There we have $a = 4$ and $b = 5$, so the number of degrees of freedom is 12. Let us choose the significance level 1%. The appropriate fractile is then $\chi^2_{12;0.99} = 26.22$, by Table B. The corresponding χ^2-test therefore has the rejection region $\{\widetilde{D}_n > 26.22\}$. For the data x with the histogram $h_n(x)$ of Table 11.2, one finds the value $\widetilde{D}_n(x) = 125.01$. Hence, the null hypothesis of independence is clearly rejected. (However, one should not rush to the conclusion that there is a causal relationship between education and environmental awareness. For instance, both could be influenced by a third factor that we have ignored here. Furthermore, one should recall the remarks after Example (3.16).)

The following proof of Theorem (11.18) might become more transparent in the special case $a = b = 2$ of 2×2 contingency tables; see Problem 11.10.

Proof of Theorem (11.18). We set $s = ab$ and $r = (a-1)(b-1)$ for brevity. Let $\rho = \alpha \otimes \beta \in \Theta_0$ be a fixed product distribution on E. We must show that $\widetilde{D}_n \xrightarrow{d} \chi_r^2$ with respect to P_ρ. Just as in Theorem (11.5), we consider the standardised random matrix

$$h_{n,\alpha\beta}^* = \left(\frac{h_n(ij) - n\,\alpha(i)\beta(j)}{\sqrt{n\,\alpha(i)\beta(j)}} \right)_{ij \in E}.$$

It is convenient to interpret $h_{n,\alpha\beta}^*$ as a random vector in $\mathbb{R}^s$. Theorem (11.5) gives us the asymptotic distribution of $h_{n,\alpha\beta}^*$, but unfortunately this is not the vector that enters the definition of $\widetilde{D}_n$. Instead, we have $\widetilde{D}_n = |\tilde{h}_n|^2$, where

$$\tilde{h}_n = \left(\frac{h_n(ij) - n\,L_n^A(i)L_n^B(j)}{\sqrt{n\,L_n^A(i)L_n^B(j)}} \right)_{ij \in E}.$$

So it is this vector that is relevant to us. But, as we will see, it is related to $h_{n,\alpha\beta}^*$. Namely, if n is large, $\tilde{h}_n$ comes close to the projection of $h_{n,\alpha\beta}^*$ onto a suitable subspace $L \subset \mathbb{R}^s$. We proceed in three steps.

Step 1: Definition of the subspace L. Consider the random vector

$$h_n^\circ = \left(\frac{h_n(ij) - n\,L_n^A(i)L_n^B(j)}{\sqrt{n\,\alpha(i)\beta(j)}} \right)_{ij \in E},$$

which is 'intermediate' between $h_{n,\alpha\beta}^*$ and $\tilde{h}_n$. The numerators of its coordinates add to zero when we sum over $i \in A$ or $j \in B$. Formally this means that the vector h_n° in $\mathbb{R}^s$ is orthogonal to the vectors

$$a_\ell = \left(\sqrt{\alpha(i)}\,\delta_{j\ell} \right)_{ij \in E} \quad \text{and} \quad b_k = \left(\delta_{ki}\,\sqrt{\beta(j)} \right)_{ij \in E}$$

with $k \in A$ and $\ell \in B$; here δ_{ki} is the Kronecker-delta (which is equal to 1 for $k = i$, and 0 otherwise). Namely, we find, for instance, that

$$h_n^\circ \cdot a_\ell := \sum_{ij \in E} h_n^\circ(ij)\,a_\ell(ij) = \sum_{i \in A} h_n^\circ(i\ell)\,\sqrt{\alpha(i)} = 0.$$

Therefore, let

$$L^{\perp} = \mathrm{span}\big(a_{\ell}, b_k : k \in A,\, \ell \in B\big)$$

be the subspace of $\mathbb{R}^s$ spanned by these vectors, and let L be the orthogonal complement of $L^{\perp}$. Then $h_n^{\circ} \in L$ by definition. We note that $L^{\perp}$ has dimension $a+b-1$, so that L has dimension r. Indeed, the identity

$$(11.19) \qquad \sum_{\ell \in B} \sqrt{\beta(\ell)}\, a_{\ell} = \sum_{k \in A} \sqrt{\alpha(k)}\, b_k = \big(\sqrt{\alpha(i)\,\beta(j)}\,\big)_{ij \in E}$$

shows that $\dim L^{\perp} \leq a+b-1$. On the other hand, it is easy to verify the orthogonality relations $a_{\ell} \cdot a_{\ell'} = \delta_{\ell\ell'}$, $b_k \cdot b_{k'} = \delta_{kk'}$ and $a_{\ell} \cdot b_k = \sqrt{\alpha(k)\beta(\ell)}$, which imply, after a short calculation, that $a+b-1$ of the vectors a_{ℓ} and b_k are linearly independent.

Now, let u_s be the vector in $L^{\perp}$ defined by the right-hand side of (11.19). By the Gram–Schmidt procedure (cf., for instance, [40]) we can extend u_s to an orthonormal basis $u_{r+1}, \ldots, u_s$ of $L^{\perp}$, and the latter to an orthonormal basis $u_1, \ldots, u_s$ of $\mathbb{R}^s$. It is then clear that $L = \mathrm{span}(u_1, \ldots, u_r)$. Let $\mathsf{O}_{\alpha\beta}$ denote the orthogonal matrix with columns $u_1, \ldots, u_s$ and E_r the diagonal matrix whose first r diagonal entries consist of a 1 and which is 0 everywhere else. The matrix $\Pi_{\alpha\beta} = \mathsf{O}_{\alpha\beta}\, \mathsf{E}_r\, \mathsf{O}_{\alpha\beta}^{\top}$ then describes the orthogonal projection onto the subspace L.

Step 2: The deviation from the projection. We will now show that h_n° is approximately equal to $\Pi_{\alpha\beta}\, h_{n,\alpha\beta}^{*}$. To this end, we first claim that

$$\Pi_{\alpha\beta}\, h_{n,\alpha\beta}^{*}(ij) = \frac{h_n(ij) + n\,\alpha(i)\beta(j) - \alpha(i)h_n^B(j) - h_n^A(i)\beta(j)}{\sqrt{n\,\alpha(i)\beta(j)}}$$

for $ij \in E$. To see this, one only has to check that the vector defined by the right-hand side is orthogonal to all a_{ℓ} and b_k and thus belongs to L, and further that its difference relative to $h_{n,\alpha\beta}^{*}$ is an element of $L^{\perp}$; this is an easy exercise. So we can write

$$h_n^{\circ}(ij) = \Pi_{\alpha\beta}\, h_{n,\alpha\beta}^{*}(ij) + \eta_n^A(i)\, \eta_n^B(j)$$

with

$$\eta_n^A(i) = \frac{n^{1/4}}{\sqrt{\alpha(i)}}\,\big(L_n^A(i) - \alpha(i)\big), \quad \eta_n^B(j) = \frac{n^{1/4}}{\sqrt{\beta(j)}}\,\big(L_n^B(j) - \beta(j)\big).$$

Now, Chebyshev's inequality shows that, for each $i \in A$ and $\varepsilon > 0$, one has

$$P_{\alpha\otimes\beta}\big(|\eta_n^A(i)| \geq \varepsilon\big) \leq \frac{1-\alpha(i)}{\sqrt{n}\,\varepsilon^2},$$

because $h_n^A(i)$ has the distribution $\mathcal{B}_{n,\alpha(i)}$ under $P_{\alpha\otimes\beta}$. This implies the convergence in probability $|\eta_n^A| \xrightarrow{P_{\alpha\otimes\beta}} 0$. Similarly, it follows that $|\eta_n^B| \xrightarrow{P_{\alpha\otimes\beta}} 0$ and thus

$$|h_n^\circ - \Pi_{\alpha\beta} h_{n,\alpha\beta}^*| = |\eta_n^A| |\eta_n^B| \xrightarrow{P_{\alpha\otimes\beta}} 0 \,.$$

Step 3: Application of the central limit theorem. Theorem (11.5) tells us that

$$h_{n,\alpha\beta}^* \xrightarrow{d} \mathcal{N}_{\alpha\otimes\beta} := \mathcal{N}_s(0, \mathsf{E}_{s-1}) \circ \mathsf{O}_{\alpha\beta}^{-1} \,.$$

Together with Proposition (11.2a) and the equation $\Pi_{\alpha\beta}\mathsf{O}_{\alpha\beta} = \mathsf{O}_{\alpha\beta}\mathsf{E}_r$, it follows that

$$\begin{aligned}
\Pi_{\alpha\beta} h_{n,\alpha\beta}^* \xrightarrow{d} \widetilde{\mathcal{N}}_{\alpha\otimes\beta} &:= \mathcal{N}_{\alpha\otimes\beta} \circ \Pi_{\alpha\beta}^{-1} \\
&= \mathcal{N}_s(0, \mathsf{E}_{s-1}) \circ (\mathsf{O}_{\alpha\beta}\mathsf{E}_r)^{-1} = \mathcal{N}_s(0, \mathsf{E}_r) \circ \mathsf{O}_{\alpha\beta}^{-1} \,.
\end{aligned}$$

Using Proposition (11.2bc) and Step 2 above, we can therefore conclude that also $h_n^\circ \xrightarrow{d} \widetilde{\mathcal{N}}_{\alpha\otimes\beta}$, and further that $\tilde{h}_n \xrightarrow{d} \widetilde{\mathcal{N}}_{\alpha\otimes\beta}$. In particular, we obtain for the sphere $A = \{x \in \mathbb{R}^s : |x|^2 \le c\}$ that

$$\begin{aligned}
P_{\alpha\otimes\beta}(\widetilde{D}_n \le c) &= P_{\alpha\otimes\beta}(\tilde{h}_n \in A) \\
&\xrightarrow[n\to\infty]{} \widetilde{\mathcal{N}}_{\alpha\otimes\beta}(A) = \mathcal{N}_s(0, \mathsf{E}_r)(A) = \chi_r^2([0, c]) \,,
\end{aligned}$$

as stated by the theorem. The last equality follows from Theorem (9.10) because $\mathcal{N}_s(0, \mathsf{E}_r) = \mathcal{N}_{0,1}^{\otimes r} \otimes \delta_0^{\otimes(s-r)}$, and the next to last from the rotational invariance of A. The same argument also implies that $\widetilde{\mathcal{N}}_{\alpha\otimes\beta}(\partial A) = \chi_r^2(\{c\}) = 0$, as needed for the convergence. $\diamond$

11.4 Order and Rank Tests

This section is devoted to certain non-parametric decision rules. As stated before, it is the general aim of non-parametric methods to avoid any use of particular properties of a specific model; instead, one exploits quite general structural properties only. In the case of real-valued observations, this can be achieved by exploiting only the order structure of the real line, rather than its linear structure. This idea was already implemented in Section 8.3; here we will follow the same approach.

As in Section 8.3, we suppose we are given a sequence $X_1, \ldots, X_n$ of i.i.d. observations with an unknown distribution Q on $\mathbb{R}$. Once more we will only require that Q has the continuity property

$$(11.20) \qquad\qquad Q(\{x\}) = 0 \text{ for all } x \in \mathbb{R} \,,$$

which allows a convenient use of the order statistics $X_{1:n}, \ldots, X_{n:n}$. These are the building blocks of one-sided and two-sided tests for the median $\mu(Q)$. Moreover, we will introduce the corresponding rank statistics and use them to construct tests for the comparison of two distributions P and Q.

11.4.1 Median Tests

In this subsection we consider a non-parametric analogue of the t-test; the role of the expectation parameter m is taken over by the median. Here is an example for motivation.

(11.21) Example. *Tyre profiles.* A tyre company intends to compare a newly developed profile (A) with an approved profile (B). The company thus equips n vehicles with tyres of type A, and in a second stage with tyres of type B. For each car and each profile, an emergency stop is performed at a certain speed, and the stopping distances are measured. (As in Example (10.23), this is a case of matched pairs.) In a specific instance, the data of Table 11.3 were measured.

Table 11.3. Measurements of stopping distances (std), according to [43].

vehicle	1	2	3	4	5	6	7	8	9	10
std(B) – std(A)	0.4	−0.2	3.1	5.0	10.3	1.6	0.9	−1.4	1.7	1.5

Can one infer from these data whether profile A behaves differently from profile B? Or whether profile A is actually better than profile B? Denote by Q the distribution of the difference between the stopping distances of profiles B and A. If both profiles work equally well, Q is symmetric with respect to 0 and so has the median $\mu(Q) = 0$. If, however, A is better than B (with typically shorter stopping distances), then $\mu(Q) > 0$. The first question thus leads to the two-sided test problem $H_0 : \mu(Q) = 0$ against $H_1 : \mu(Q) \neq 0$, whereas the second question yields the one-sided test problem $H_0 : \mu(Q) \leq 0$ against $H_1 : \mu(Q) > 0$.

In Theorem (8.20), we have already seen how order statistics can be used to define a confidence interval. To deal with the test problems above, it is therefore sufficient to recall the general relation between confidence intervals and tests that was stated in Problem 10.1. The resulting theorem is the following. As in (8.19), we write $b_n(\alpha)$ for the largest α-quantile of the binomial distribution $\mathcal{B}_{n,1/2}$.

(11.22) Theorem. *Sign test for the median. Let α be a given level, $\mu_0 \in \mathbb{R}$, and $X_1, \ldots, X_n$ be a sequence of i.i.d. random variables with an unknown continuous distribution Q on $\mathbb{R}$. Then the following holds.*

(a) *For the two-sided test problem $H_0 : \mu(Q) = \mu_0$ against $H_1 : \mu(Q) \neq \mu_0$, the acceptance region $\{X_{k:n} \leq \mu_0 \leq X_{n-k+1:n}\}$ with $k := b_n(\alpha/2)$ determines a test of level α.*

(b) *For the one-sided test problem $H_0 : \mu(Q) \leq \mu_0$ against $H_1 : \mu(Q) > \mu_0$, the acceptance region $\{X_{k:n} \leq \mu_0\}$ with $k := b_n(\alpha)$ defines a test of level α. Its power function is a strictly increasing function of $p(Q) := Q(]\mu_0, \infty[)$.*

Proof. Statement (a) follows immediately from Theorem (8.20) and Problem 10.1, and the proof of (b) is very similar. Indeed, as we have seen in (8.17), the rejection event $X_{k:n} > \mu_0$ occurs if and only if at most $k - 1$ observations fall into the interval $]-\infty, \mu_0]$. Writing $S_n^- = \sum_{i=1}^n 1_{\{X_i \leq \mu_0\}}$ for the number of observations in this interval, we obtain from Theorem (2.9) that

$$Q^{\otimes n}(X_{k:n} > \mu_0) = Q^{\otimes n}(S_n^- < k) = \mathcal{B}_{n,1-p(Q)}(\{0, \ldots, k-1\}).$$

By Lemma (8.8b), the last probability is an increasing function of $p(Q)$. Now, the null hypothesis $\mu(Q) \leq \mu_0$ implies that $p(Q) \leq 1/2$. So, according to the choice of k, the last probability is at most α, as claimed. $\diamond$

The name sign test comes from the fact that the test statistic S_n^- counts how often the difference $X_i - \mu_0$ has a negative sign. Let us also note that, in the above, we have confined ourselves to the non-randomised case for simplicity. However, since S_n^- is discrete, a given size can only be reached if one allows for randomised tests. For the one-sided test problem, it then turns out that the randomised sign test is optimal for its size, see Problem 11.12.

What is the result of the sign test in the case of Example (11.21)? Let us choose the level $\alpha = 0.025$. Since $n = 10$, the appropriate binomial quantile is then $b_{10}(0.0125) = 2$; cf. Example (8.21). The data vector x of Table 11.3 yields the order statistics $X_{2:10}(x) = -0.2 < 0$ and $X_{9:10}(x) = 5.0 > 0$. The null hypothesis $H_0 : \mu(Q) = 0$ in the two-sided test problem can therefore not be rejected (even though the opposite appears to be true). For the one-sided test problem, see Problem 11.13.

The sign tests for the median have the advantage that their level α can be determined without detailed knowledge of Q. Their obvious disadvantage is that they simply count the number of outcomes above respectively below μ_0, but completely ignore how much the outcomes differ from μ_0. To facilitate the discrimination between null hypothesis and alternative, one should exploit the information about the relative positions of the outcomes. This information is captured by the rank statistics, which we will introduce now. (A corresponding modification of the sign tests is the subject of Problem 11.18.)

11.4.2 Rank Statistics and the Two-Sample Problem

Do non-smokers, on average, have a higher temperature in their finger tips than smokers? Do patients live longer, on average, when they receive the newly developed treatment 1 rather than the old treatment 2? Is the petrol consumption, on average, higher for brand 1 than for brand 2? Is the weight increase for calves, on average, greater for feeding method 1 than for feeding method 2?

As we will see, such questions can be answered without making any assumptions about the underlying distributions. (In the above examples, it would be difficult to justify the popular assumption of a normal distribution.) But first we need to clarify what it means to be 'higher, longer, or greater on average'. The appropriate concept is the partial ordering of probability measures called *stochastic domination*.

Definition. Let P, Q be two probability measures on $(\mathbb{R}, \mathscr{B})$. P is said to be *stochastically smaller* than Q, or P is *stochastically dominated* by Q, written as $P \preceq Q$, if $P(]c, \infty[) \leq Q(]c, \infty[)$ for all $c \in \mathbb{R}$. If also $P \neq Q$, we write $P \prec Q$.

The relation $P \preceq Q$ indicates that realisations of Q are typically larger than realisations of P; see Problem 11.14 for an equivalent way of expressing this fact.

(11.23) Example. *Stochastic ordering in exponential families.* For binomial distributions, one has $\mathcal{B}_{n,p} \prec \mathcal{B}_{n,p'}$ when $p < p'$; this follows immediately from Lemma (8.8b). Next, a simple translation argument shows that $\mathcal{N}_{m,v} \prec \mathcal{N}_{m',v}$ for $m < m'$. Further, it is clear from (2.19) that $\Gamma_{\alpha,r} \prec \Gamma_{\alpha,r+1}$ for $r \in \mathbb{N}$, and Corollary (9.9) shows that, in fact, $\Gamma_{\alpha,r} \prec \Gamma_{\alpha,r'}$ for all real $0 < r < r'$. All these facts are special cases of the following general result.

Suppose $\{P_\vartheta : \vartheta \in \Theta\}$ is a (one-parameter) exponential family on $\mathbb{R}$ for which both the underlying statistic T and the coefficient function $a(\vartheta)$ are strictly increasing. Then $P_\vartheta \prec P_{\vartheta'}$ for $\vartheta < \vartheta'$. Indeed, in this case $P_\vartheta(]c, \infty[)$ is precisely the power function of the test with rejection region $\{T > T(c)\}$, which is increasing in ϑ by Example (10.9) and Theorem (10.10).

We now consider the following *two-sample problem*: Suppose we are given $n = k + l$ independent observations $X_1, \ldots, X_n$. We assume that $X_1, \ldots, X_k$ is a sample from an unknown continuous distribution P, and $X_{k+1}, \ldots, X_{k+l}$ a sample from a second continuous distribution Q that is also unknown. In the canonical statistical model

$$\left(\mathbb{R}^{k+l}, \mathscr{B}^{k+l}, P^{\otimes k} \otimes Q^{\otimes l} : P, Q \text{ continuous} \right),$$

X_i is simply the projection of $\mathbb{R}^{k+l}$ onto the ith coordinate. We will investigate the test problem

$$H_0 : P = Q \quad \text{against} \quad H_1 : P \prec Q.$$

Does this mean that we know in advance that $P \preceq Q$? No, for otherwise it would be pointless to perform a test! It rather means that the null hypothesis $P = Q$ should only be rejected if $P \prec Q$. Strictly speaking, as the null hypothesis is sharp, it should only be rejected if Q is 'significantly larger' than P; cf. Remark (10.24). (A typical application follows in (11.27).) Consequently, the basic idea of the test will be to reject the hypothesis H_0 whenever the majority of the observations $X_1, \ldots, X_k$ with distribution P is smaller than the observations $X_{k+1}, \ldots, X_{k+l}$ sampled from Q. How can this be made precise?

Consider the order statistics $X_{1:n}, \ldots, X_{n:n}$ associated with the total sample $X_1, \ldots, X_k, X_{k+1}, \ldots, X_n$. At which position does the ith observation X_i appear in the order statistics? This is described by the following rank statistics.

Definition. The *rank statistics* of a sample sequence $X_1, \ldots, X_n$ are defined as the random variables $R_1, \ldots, R_n$ given by $R_i = |\{1 \le j \le n : X_j \le X_i\}|$. So, R_i specifies the position of X_i when the observations $X_1, \ldots, X_n$ are reordered according to their size, so that $X_i = X_{R_i:n}$.

As both P and Q are assumed to be continuous, condition (8.15), the absence of ties, holds also in the present case. Hence, with probability 1, the rank statistics are unambiguously defined, and the mapping $i \to R_i$ forms a random permutation of the set $\{1, \ldots, n\}$. Figure 11.6 provides an illustration.

$$X_6 \; X_3 \quad X_5 \quad X_1 \qquad X_8 \; X_4 \qquad X_2 \qquad X_7$$

R_1	R_2	R_3	R_4	R_5	R_6	R_7	R_8
4	7	2	6	3	1	8	5

Figure 11.6. A realisation of $X_1, \ldots, X_8$ on $\mathbb{R}$ and the corresponding realisation of the rank statistics $R_1, \ldots, R_8$.

We will now distinguish between the sample $X_1, \ldots, X_k$ taken from the distribution P, and the sample $X_{k+1}, \ldots, X_{k+l}$ with common distribution Q. To this end, we introduce the rank sums of each of these samples, namely

$$W_P := R_1 + \cdots + R_k \quad \text{and} \quad W_Q = R_{k+1} + \cdots + R_{k+l}.$$

In other words, W_P is the sum of all ranks of the P-sample within the merged ensemble of *all* observations, and W_Q is the corresponding rank sum of the Q-sample.

An easy way of calculating the rank sums is as follows. The outcomes of both samples are marked on the real line (this corresponds to the process of determining the order statistics), with different marks for the outcomes from the two distinct samples. The rank sums W_P and W_Q are then simply obtained as the sums of the positions of the outcomes from the respective samples, see Figure 11.7.

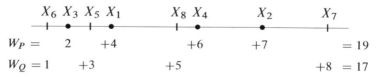

$$
\begin{array}{lccccc}
 & X_6 \; X_3 \; X_5 \; X_1 & & X_8 \; X_4 & X_2 & X_7 \\
W_P = & 2 \qquad +4 & & +6 & +7 & = 19 \\
W_Q = 1 & +3 & & +5 & & +8 \;\; = 17
\end{array}
$$

Figure 11.7. Calculation of the rank sums for $k = l = 4$.

The rank sums provide some overall information about the relative positioning of the two samples. Namely, if W_P is less than W_Q, the majority of outcomes in the P-sample is situated further to the left on the real axis than the outcomes of the Q-sample. It is then tempting to conclude that $P \prec Q$. But how much smaller than W_Q should W_P be for this conclusion to be justified?

To answer this question we note first that

$$W_P + W_Q = \sum_{i=1}^{n} R_i = \sum_{i=1}^{n} i = \frac{n(n+1)}{2}.$$

So it suffices to consider $W := W_P$ only. The following lemma relates the rank sums to the so-called U-statistic, which counts how often an observation in one of the samples exceeds an observation in the other sample.

(11.24) Lemma. Rank sums and U-statistics. *We have*

$$W = U + \frac{k(k+1)}{2} \quad for \ U = U_{k,l} := \sum_{i=1}^{k} \sum_{j=k+1}^{k+l} 1_{\{X_i > X_j\}},$$

and W_Q has an analogous description in terms of $U_Q = kl - U$.

Proof. Clearly, both W and U are invariant under permutations of the P-sample $X_1, \ldots, X_k$. Therefore, we only need to consider the case $X_1 < X_2 < \cdots < X_k$. The associated ranks $R_1 < R_2 < \cdots < R_k$ (in the entire sequence $X_1, \ldots, X_n$) then satisfy the equation $R_i = i + |\{j > k : X_j < X_i\}|$. Hence, the claim follows by summation over i. $\diamond$

As the lemma confirms, both W and U do serve as natural indicators for the relative positions of the individual outcomes of the two samples, and can thus be used to define a plausible decision rule as follows.

Definition. A test of the null hypothesis $H_0 : P = Q$ against the alternative $H_1 : P \prec Q$ with a rejection region of the form

$$\{U < c\} = \{W < c + k(k+1)/2\}$$

for $0 < c \leq kl$ is called a (one-sided) *Mann–Whitney U-test*, or *Wilcoxon two-sample rank sum test*.

The rank sum test based on W was proposed by Frank Wilcoxon (1945) and developed further using U by Henry B. Mann and his student Donald R. Whitney (1947).

As always, the threshold value c must be chosen to guarantee a given level α of significance. We thus need to know the distribution of U under the null hypothesis. This distribution is determined in the following theorem, which states, in particular, that the statistic U is in fact *distribution free*, in the sense that its distribution does not depend on the underlying P, as long as P is continuous and $Q = P$.

(11.25) Theorem. U-distribution under the null hypothesis. *For every continuous P and all $m = 0, \ldots, kl$,*

$$P^{\otimes n}(U = m) = N(m; k, l) / \binom{n}{k}.$$

Here, $N(m; k, l)$ denotes the number of all partitions $\sum_{i=1}^{k} m_i = m$ of m into k increasingly ordered numbers $m_1 \leq m_2 \leq \cdots \leq m_k$ from the set $\{0, \ldots, l\}$. In particular, this means that $P^{\otimes n}(U = m) = P^{\otimes n}(U = kl - m)$.

Proof. Under $P^{\otimes n}$, the random vector $(R_1, \ldots, R_n)$ is uniformly distributed on the set $\mathscr{S}_n$ of all permutations of $\{1, \ldots, n\}$. This is because

$$P^{\otimes n}\big((R_1, \ldots, R_n) = \pi^{-1}\big) = P^{\otimes n}(X_{\pi(1)} < \cdots < X_{\pi(n)}) = 1/n!$$

for each permutation $\pi \in \mathscr{S}_n$ and its inverse π^{-1}. The last equality follows from the absence of ties together with the fact that the product measure $P^{\otimes n}$ is invariant under permutations of the coordinates. (Indeed, the identity

(11.26) $$P^{\otimes n}\big((X_{\pi(1)}, \ldots, X_{\pi(n)}) \in A\big) = P^{\otimes n}(A)$$

is trivially true when $A = A_1 \times \cdots \times A_n$, and extends to arbitrary $A \in \mathscr{B}^n$ by the uniqueness theorem (1.12).)

Now, since the random permutation $(R_1, \ldots, R_n)$ has a uniform distribution, so does the random k-set $R_{[k]} := \{R_1, \ldots, R_k\}$, which takes values in the system $\mathscr{R}_{[k]}$ of all sets $r \subset \{1, \ldots, n\}$ with cardinality k. Each $r \in \mathscr{R}_{[k]}$ can be written in the form $r = \{r_1, \ldots, r_k\}$ with $r_1 < \cdots < r_k$, and the mapping

$$\vec{m} : r \to \vec{m}(r) := (r_1 - 1, \ldots, r_k - k)$$

is a bijection from $\mathscr{R}_{[k]}$ onto the set of all increasingly ordered k-tuples with entries from $\{0, \ldots, l\}$. The ith coordinate $m_i(r) = r_i - i = |\{1, \ldots, r_i\} \setminus r|$ of $\vec{m}(r)$ specifies how many elements of the complement r^c are smaller than r_i. This means that $U = \sum_{i=1}^{k} m_i(R_{[k]})$. Altogether, we therefore obtain

$$P^{\otimes n}(U = m) = \sum_{r \in \mathscr{R}_{[k]}: \sum_{i=1}^{k} m_i(r) = m} P^{\otimes n}(R_{[k]} = r).$$

By the above, the last probability is $1/|\mathscr{R}_{[k]}| = 1/\binom{n}{k}$, and the number of summands is $N(m; k, l)$. This yields the claim. $\diamond$

The partition numbers $N(m; k, l)$ can be calculated by elementary combinatorics, as follows. Each partition $\sum_{i=1}^{k} m_i = m$ with $0 \leq m_1 \leq m_2 \leq \cdots \leq m_k \leq l$ can be represented by a diagram as in Figure 11.8. By reflection at the diagonal, we obtain the symmetry $N(m; k, l) = N(m; l, k)$; interchanging ● and ○ and rotating the diagram by 180 degrees, we find that $N(m; k, l) = N(kl - m; k, l)$; finally, we can

$$
\begin{array}{llllllll}
m_1: & \circ & \circ & \circ & \circ & \circ \\
m_2: & \bullet & \circ & \circ & \circ & \circ \\
m_3: & \bullet & \bullet & \bullet & \circ & \circ \\
m_4: & \bullet & \bullet & \bullet & \circ & \circ
\end{array}
$$

Figure 11.8. Representation of the partition $(0, 1, 3, 3)$ of $m = 7$ in the case $k = 4, l = 5$.

remove the first column and distinguish the number j of $\bullet$ in this column to obtain the recursive formula

$$
N(m; k, l) = \sum_{j=0}^{k} N(m - j; j, l - 1),
$$

where $N(m; k, l) := 0$ for $m < 0$. This allows to calculate the partition numbers by induction, and therefore to compute the U-distribution. Some of the resulting quantiles are summarised in Table E in the appendix. The following example illustrates the use of the U-test.

(11.27) Example. *Age dependence of cholesterol values.* To determine whether the cholesterol level increases with age, the blood cholesterol values of men of different age groups are examined. Thus, 11 men of each of the age groups 20–30 and 40–50 are randomly chosen and their cholesterol values are measured. The null hypothesis is H_0: 'the cholesterol level is identically distributed in both age groups', and the alternative reads H_1: 'the cholesterol level increases with age'. This is a typical situation for the non-parametric U-test, because an assumption of normal distributions would be difficult to justify, whereas the assumption of a continuous distribution of the cholesterol values in each group seems unproblematic. For $k = l = 11$ and the level $\alpha = 0.05$, Table E gives the U-quantile $c = 35$. So, the U-test of level 0.05 has the rejection region $\{U < 35\}$. A classical study produced the data listed in Table 11.4. The rank sum W in the age group 20–30 takes the value 108, and therefore U evaluates to 42. So, at the chosen level, the measured data do not support the conjecture that the cholesterol level increases with age.

Table 11.4. Blood cholesterol values of men in two age groups, according to [5].

20–30 years:	Data	135	222	251	260	269	235	386	252	352	173	156
	Ranks	1	6	9	12	14	7	22	10	21	4	2
40–50 years:	Data	294	311	286	264	277	336	208	346	239	172	254
	Ranks	17	18	16	13	15	19	5	20	8	3	11

It is important to note that the U-test can only be applied when the two samples are independent. For matched pairs as in the Examples (8.6) and (11.21), this is not the case. A suitable non-parametric test then is the signed-rank test, which is described in Problem 11.18.

For large values of k and l, the calculation of the U-distribution can be avoided because a normal approximation is available. As the next theorem shows, the U-test with rejection region

$$\left\{ U < \frac{kl}{2} + \Phi^{-1}(\alpha) \sqrt{kl(n+1)/12} \right\}$$

has approximately size α when both k and l are sufficiently large.

(11.28) Theorem. *Wassily Hoeffding 1948. Let $X_1, X_2, \ldots$ be i.i.d. random variables, and suppose their common distribution P is continuous. For $k, l \geq 1$ let*

$$U_{k,l} := \sum_{i=1}^{k} \sum_{j=k+1}^{k+l} 1_{\{X_i > X_j\}}$$

and $v_{k,l} := kl(k+l+1)/12$. Then,

$$U_{k,l}^* := \frac{U_{k,l} - kl/2}{\sqrt{v_{k,l}}} \xrightarrow{d} \mathcal{N}_{0,1} \quad as\ k, l \to \infty.$$

Proof. By Theorem (11.25), the distribution of $U_{k,l}$ does not depend on the underlying P. Therefore, we can assume that $P = \mathcal{U}_{[0,1]}$, which means that the X_i are uniformly distributed on $[0, 1]$. Since the summands of $U_{k,l}$ are not independent of each other, the central limit theorem cannot be applied directly. The main idea of the proof is therefore to approximate $U_{k,l}$ by a sum of independent random variables. Namely, we will replace the centred summands $1_{\{X_i > X_j\}} - 1/2$ of $U_{k,l}$, which depend non-linearly on the observations, by the (linear) differences $X_i - X_j$. Hence, we define

$$Z_{k,l} = \sum_{i=1}^{k} \sum_{j=k+1}^{k+l} (X_i - X_j) = l \sum_{i=1}^{k} X_i - k \sum_{j=k+1}^{k+l} X_j$$

and $Z_{k,l}^* = Z_{k,l} / \sqrt{v_{k,l}}$. In Step 2 below we will show that

(11.29) $U_{k,l}^* - Z_{k,l}^* \to 0$ in probability,

and in Step 3 we will use the central limit theorem to prove that

(11.30) $Z_{k,l}^* \xrightarrow{d} \mathcal{N}_{0,1}$.

The theorem will then follow from Proposition (11.2b).

Step 1. To calculate the variance of $U_{k,l}$, we note first that $P(X_i > X_j) = 1/2$ for $i \neq j$; this follows from (8.15) and (11.26), the invariance under permutations.

Hence, $\mathbb{E}(U_{k,l}) = kl/2$. Further, if $i \neq j$ and $i' \neq j'$ then

$$\text{Cov}(1_{\{X_i > X_j\}}, 1_{\{X_{i'} > X_{j'}\}}) = P(X_i > X_j, X_{i'} > X_{j'}) - 1/4$$

$$= \begin{cases} 0 & \text{if } i \neq i', j \neq j', \\ 1/4 & \text{if } i = i', j = j', \\ 1/12 & \text{otherwise,} \end{cases}$$

and therefore

$$\mathbb{V}(U_{k,l}) = \sum_{1 \leq i, i' \leq k < j, j' \leq k+l} \text{Cov}(1_{\{X_i > X_j\}}, 1_{\{X_{i'} > X_{j'}\}}) = v_{k,l}.$$

In particular, $U_{k,l}^*$ is standardised.

  Step 2. We will now prove that $\mathbb{V}(U_{k,l} - Z_{k,l}) = kl/12$. First, a direct calculation shows that $\mathbb{V}(X_i) = \mathbb{V}(\mathcal{U}_{[0,1]}) = 1/12$. Together with Theorem (4.23), this implies that

$$\mathbb{V}(Z_{k,l}) = \sum_{i=1}^{k} \mathbb{V}(l X_i) + \sum_{j=k+1}^{k+l} \mathbb{V}(k X_j) = \frac{kl(k+l)}{12}.$$

On the other hand, using Example (3.30) and Corollary (4.13), we find for all $i < j$ that

$$\text{Cov}(1_{\{X_i > X_j\}}, X_i) = \int_0^1 dx_1 \int_0^1 dx_2 \, 1_{\{x_1 > x_2\}} x_1 - 1/4 = 1/12$$

and thus, using further that independent random variables are uncorrelated,

$$\text{Cov}(U_{k,l}, X_i) = \sum_{j=k+1}^{k+l} \text{Cov}(1_{\{X_i > X_j\}}, X_i) = l/12.$$

Since $\text{Cov}(1_{\{X_i > X_j\}}, X_j) = -\text{Cov}(1_{\{X_i > X_j\}}, X_i) = -1/12$, we finally obtain

$$\text{Cov}(U_{k,l}, Z_{k,l}) = l \sum_{i=1}^{k} \text{Cov}(U_{k,l}, X_i) - k \sum_{j=k+1}^{k+l} \text{Cov}(U_{k,l}, X_j)$$

$$= kl(k+l)/12 = \mathbb{V}(Z_{k,l})$$

and therefore

$$\mathbb{V}(U_{k,l} - Z_{k,l}) = \mathbb{V}(U_{k,l}) - 2\,\text{Cov}(U_{k,l}, Z_{k,l}) + \mathbb{V}(Z_{k,l})$$

$$= \mathbb{V}(U_{k,l}) - \mathbb{V}(Z_{k,l}) = kl/12,$$

as claimed. In particular, we see that

$$\mathbb{V}(U_{k,l}^* - Z_{k,l}^*) = \frac{1}{k+l+1} \xrightarrow[k,l \to \infty]{} 0,$$

which implies the claim (11.29) by Chebyshev's inequality (5.5).

Step 3. To prove (11.30) we write

$$Z^*_{k,l} = \sqrt{a_{k,l}}\, S^*_k + \sqrt{b_{k,l}}\, T^*_l$$

with $a_{k,l} = l/(k+l+1)$, $b_{k,l} = k/(k+l+1)$, and

$$S^*_k = \sum_{i=1}^{k} \frac{X_i - 1/2}{\sqrt{k/12}}, \quad T^*_l = \sum_{j=k+1}^{k+l} \frac{1/2 - X_j}{\sqrt{l/12}}.$$

We know from the central limit theorem (5.29) that

$$S^*_k \xrightarrow{d} S \quad \text{and} \quad T^*_l \xrightarrow{d} T \quad \text{as } k,l \to \infty,$$

where S and T are independent standard normal random variables on a suitable probability space. Next, consider an arbitrary sequence of pairs (k,l) with $k,l \to \infty$. We can choose a subsequence such that $a_{k,l} \to a \in [0,1]$ and so $b_{k,l} \to b = 1-a$ along this subsequence. By Proposition (11.2c), this implies further that

$$\sqrt{a_{k,l}}\, S^*_k \xrightarrow{d} \sqrt{a}\, S \quad \text{and} \quad \sqrt{b_{k,l}}\, T^*_l \xrightarrow{d} \sqrt{b}\, T$$

along the chosen subsequence. As in the second step in the proof of Theorem (11.5), we can then use Remark (11.1) and the independence of S^*_k and T^*_l to show that in fact

$$\left(\sqrt{a_{k,l}}\, S^*_k, \ \sqrt{b_{k,l}}\, T^*_l \right) \xrightarrow{d} \left(\sqrt{a}\, S, \ \sqrt{b}\, T \right),$$

again along the subsequence. Since addition is continuous, Proposition (11.2a) finally implies that

$$Z^*_{k,l} \xrightarrow{d} \sqrt{a}\, S + \sqrt{b}\, T,$$

still along the chosen subsequence. In view of Theorem (9.5), the latter random variable is $\mathcal{N}_{0,1}$-distributed. So we have shown that

$$Z^*_{k,l} \xrightarrow{d} \mathcal{N}_{0,1}, \quad \text{or equivalently,} \quad \| F_{Z^*_{k,l}} - \Phi \| \to 0$$

for a suitable subsequence of an arbitrary sequence of pairs (k,l) with $k,l \to \infty$. Taking a sequence of pairs (k,l) along which $\| F_{Z^*_{k,l}} - \Phi \|$ approaches its upper limit, we get the desired result (11.30). $\diamond$

So far, we have concentrated on the one-sided test problem $H_0 : P = Q$ against $H_1 : P \prec Q$. What can one say about the analogous two-sided test problem? Suppose, for instance, that we want to compare two medical treatments, none of which can a priori be expected to be more successful than the other. In such a situation, it would *not* make sense to search for a test of $H_0 : P = Q$ against the general alternative $H_1 : P \neq Q$. This is because the alternative is so large that no test can have a satisfactory power on the entire alternative. But if one takes for granted that the two treatments are in fact comparable ('one method must be better than the other'), one arrives at the two-sided test problem

$$H_0 : P = Q \quad \text{against} \quad H_1 : P \prec Q \text{ or } P \succ Q.$$

An efficient decision procedure for this problem is the two-sided U-test with a rejection region of the form $\{U_P < c\} \cup \{U_Q < c\}$. This is particularly suitable for testing the so-called *location problem*. Here, the class of all continuous probability measures P is reduced to the class $\{Q(\cdot - \vartheta) : \vartheta \in \mathbb{R}\}$ of all translates of a fixed continuous Q, so that the test problem boils down to $H_0' : \vartheta = 0$ against $H_1' : \vartheta \neq 0$.

Problems

11.1 *The 'portmanteau theorem' about convergence in distribution.* Let (E, d) be a metric space equipped with the σ-algebra $\mathscr{E}$ generated by the open sets, $(Y_n)_{n \geq 1}$ a sequence of random variables on a probability space $(\Omega, \mathscr{F}, P)$ taking values in $(E, \mathscr{E})$, and Q a probability measure on $(E, \mathscr{E})$. Prove the equivalence of the following four statements.

(a) $Y_n \xrightarrow{d} Q$, i.e., $\lim_{n \to \infty} P(Y_n \in A) = Q(A)$ for all $A \in \mathscr{E}$ with $Q(\partial A) = 0$.

(b) $\limsup_{n \to \infty} P(Y_n \in F) \leq Q(F)$ for all closed sets $F \subset E$.

(c) $\liminf_{n \to \infty} P(Y_n \in G) \geq Q(G)$ for all open sets $G \subset E$.

(d) $\lim_{n \to \infty} \mathbb{E}(f \circ Y_n) = \mathbb{E}_Q(f)$ for all bounded continuous functions $f : E \to \mathbb{R}$.

Hints: (a) $\Rightarrow$ (b): If $G^\varepsilon = \{x \in E : d(x, F) < \varepsilon\}$ is the open ε-neighbourhood of F, then $Q(\partial G^\varepsilon) = 0$ for all except at most countably many ε. (b), (c) $\Rightarrow$ (d): Suppose without loss of generality that $0 \leq f \leq 1$, and approximate f from above by functions of the form $\frac{1}{n} \sum_{k=0}^{n-1} 1_{\{f \geq k/n\}}$, and similarly from below.

11.2 Let $E = \{1, 2, 3\}$ and ρ be the density of the uniform distribution on E. Determine, for $c_1, c_2 \in \mathbb{R}$ and large n, a normal approximation of the multinomial probability

$$\mathcal{M}_{n,\rho}\big(\ell \in \mathbb{Z}_+^E : \ell(1) - \ell(2) \leq c_1 \sqrt{n/3}, \; \ell(1) + \ell(2) - 2\ell(3) \leq c_2 \sqrt{n/3}\big).$$

11.3 An algorithm for generating pseudo-random numbers is to be tested. So one performs an experiment, in which the algorithm produces, say, $n = 10\,000$ digits in $\{0, \dots, 9\}$. Suppose the following frequencies are observed:

Digit	0	1	2	3	4	5	6	7	8	9
Frequency	1007	987	928	986	1010	1029	987	1006	1034	1026

Choose an appropriate level and perform a χ^2-test of goodness of fit for a uniform distribution.

11.4^S In the setting of Section 11.2, let $c > 0$ be a fixed threshold, ρ a probability density on E, and $\{D_{n,\rho} \leq c\}$ the acceptance region of the associated χ^2-test of goodness of fit after n observations. Show that

$$\limsup_{n \to \infty} \frac{1}{n} \log P_\vartheta(D_{n,\rho} \leq c) \leq -H(\rho; \vartheta)$$

for all $\vartheta \neq \rho$, which means that the power tends to 1 at an exponential rate. *Hint:* Combine parts of the proofs of Proposition (11.10) and Theorem (10.4).

11.5 *Tendency towards the mean.* Teachers often face the suspicion that when awarding grades they tend to avoid the extremes. In a certain course, 17 students obtain the following average grades (on a scale from 1 to 6, where 1 is best):

$$
\begin{array}{cccccccc}
1.58 & 2.84 & 3.52 & 4.16 & 5.36 & 2.01 & 3.03 & 3.56 \\
4.19 & 2.35 & 3.16 & 3.75 & 4.60 & 2.64 & 3.40 & 3.99 & 4.75
\end{array}
$$

For simplicity, assume that these average grades are obtained from a large number of individual assessments, so that they can be regarded as continuously distributed. Use the χ^2-goodness-of-fit test with asymptotic size $\alpha = 0.1$ to verify whether the above data are $\mathcal{N}_{3.5,1}$-distributed. For this purpose, discretise the relative frequencies by subdividing them into the six groups

$$]-\infty, 1.5], \ ]1.5, 2.5], \ ]2.5, 3.5], \ ]3.5, 4.5], \ ]4.5, 5.5], \ ]5.5, \infty[.$$

11.6 [S] *Test against decreasing trend.* Consider the infinite product model for the χ^2-test of goodness of fit, and let ρ be the uniform distribution on $E = \{1, \ldots, s\}$. Suppose the null hypothesis $H_0 : \vartheta = \rho$ is not to be tested against the full $H_1 : \vartheta \neq \rho$, but only against $H_1' : \vartheta_1 > \vartheta_2 > \cdots > \vartheta_s$ ('decreasing trend'). The χ^2-test is then not particularly useful (why?). It is better to use the test statistic

$$
T_n = \frac{\sum_{i=1}^{s} i\, h_n(i) - n(s+1)/2}{\sqrt{n(s^2-1)/12}} \, .
$$

Calculate $\mathbb{E}_\vartheta(T_n)$ and $\mathbb{V}_\vartheta(T_n)$ and show that $T_n \xrightarrow{d} \mathcal{N}_{0,1}$. Use this result to develop a test of H_0 against H_1'. *Hint:* Represent T_n as a sum of independent random variables.

11.7 It is suspected that in a horse race on a circular racing track, the starting positions affect the chance of winning. The following table classifies 144 winners according to their starting positions (where the starting positions are numbered from the inside to the outside).

Starting position	1	2	3	4	5	6	7	8
Frequency	29	19	18	25	17	10	15	11

Test the null hypothesis 'same chance of winning' against the alternative 'decreasing chance of winning' for the level $\alpha = 0.01$ using (a) a χ^2-goodness-of-fit test, (b) the test developed in Problem 11.6.

11.8 The influence of vitamin C on the frequency of a cold is to be tested. Therefore, 200 test subjects are randomly split into two groups, of which one receives a certain dose of vitamin C and the other a placebo. The resulting data are the following:

Cold frequency	lower	higher	unchanged
Control group	39	21	40
Experimental group	51	20	29

For the level 0.05, test the null hypothesis that taking vitamins and the frequency of a cold do not depend on each other.

11.9 *Quadratic approximation of mutual information.* Consider the setting of Section 11.3 and establish the following convergence in probability result: For every $\alpha \otimes \beta \in \Theta_0$,

$$
n\, H(L_n; L_n^A \otimes L_n^B) - \widetilde{D}_n/2 \xrightarrow{P_{\alpha \otimes \beta}} 0
$$

in the limit as $n \to \infty$. *Hint:* Mimic the proof of Proposition (11.10) under the additional assumption that $L_n^A \otimes L_n^B \geq \frac{1}{2} \alpha \otimes \beta$, and show at the very end that this condition holds with a probability tending to 1.

11.10 *Pearson's theorem for 2×2 contingency tables.* Consider the situation of Theorem (11.18) for the case $a = b = 2$ and show the following.

(a) The squared deviations $(L_n(ij) - L_n^A(i)L_n^B(j))^2$ do not depend on the choice of $ij \in E$, so that $\widetilde{D}_n = Z_n^2$ for

$$Z_n = \sqrt{n} \left(L_n(11) - L_n^A(1)L_n^B(1) \right) \Big/ \sqrt{L_n^A(1)L_n^B(1)L_n^A(2)L_n^B(2)}.$$

(b) Let $X_k^A = 1$ if the A-coordinate of X_k is equal to 1, set $X_k^A = 0$ otherwise, and define X_k^B similarly. Then, on the one hand,

$$L_n(11) - L_n^A(1)L_n^B(1) = \frac{1}{n} \sum_{k=1}^{n} (X_k^A - L_n^A(1))(X_k^B - L_n^B(1)),$$

and on the other hand

$$\sum_{k=1}^{n} (X_k^A - \alpha(1))(X_k^B - \beta(1)) \Big/ \sqrt{n\, \alpha(1)\beta(1)\alpha(2)\beta(2)} \xrightarrow{d} \mathcal{N}_{0,1}$$

relative to each $P_{\alpha \otimes \beta}$ in the null hypothesis. (Note that the family $\{X_k^A, X_k^B : k \geq 1\}$ is independent under $P_{\alpha \otimes \beta}$.)

(c) Use Proposition (11.2) (in its one-dimensional version) to show that $Z_n \xrightarrow{d} \mathcal{N}_{0,1}$ with respect to $P_{\alpha \otimes \beta}$, and therefore $\widetilde{D}_n \xrightarrow{d} \chi_1^2$ under the null hypothesis.

11.11 [S] *Fisher's exact test of independence.* Consider the situation of Section 11.3 for $A = B = \{1, 2\}$. (For instance, A could represent two medical treatments and B indicate whether the patient has fully recovered or not.) Show the following:

(a) The null hypothesis $\vartheta = \vartheta^A \otimes \vartheta^B$ holds if and only if $\vartheta(11) = \vartheta^A(1)\, \vartheta^B(1)$.

(b) For all $n \in \mathbb{N}$, $k, n_A, n_B \in \mathbb{Z}_+$ and $\vartheta \in \Theta_0$,

$$P_\vartheta \left(h_n(11) = k \,\big|\, h_n^A(1) = n_A,\ h_n^B(1) = n_B \right)$$
$$= \mathcal{H}_{n_B; n_A, n - n_A}(\{k\}) = \mathcal{H}_{n_A; n_B, n - n_B}(\{k\}).$$

(c) How would you proceed to construct a level-α test of the hypothesis $H_0 : \vartheta = \vartheta^A \otimes \vartheta^B$ against the alternative $H_1 : \vartheta \neq \vartheta^A \otimes \vartheta^B$? And what would change if the alternative H_1 was replaced by the (smaller) alternative $H_1' : \vartheta(11) > \vartheta^A(1)\, \vartheta^B(1)$ ('therapy 1 has a higher success rate')?

11.12 [S] *Optimality of the randomised one-sided sign test.* Consider the non-parametric one-sided test problem $H_0 : \mu(Q) \leq 0$ against $H_1 : \mu(Q) > 0$ for the median, but (for simplicity) only within the class of probability measures Q on $\mathbb{R}$ with existing Lebesgue densities. Consider also the test statistic $S_n^- = \sum_{k=1}^{n} 1_{\{X_k \leq 0\}}$ which is based on n independent observations X_k. Finally, consider the binomial model $\{\mathcal{B}_{n, \vartheta} : 0 < \vartheta < 1\}$ and let φ be the uniformly most powerful level-α test for the right-sided test problem $H_0' : \vartheta \geq 1/2$ against $H_1' : \vartheta < 1/2$; recall Theorem (10.10). That is, $\varphi = 1_{\{0, \ldots, c-1\}} + \gamma 1_{\{c\}}$, where c and γ are such that $\mathbb{E}_{\mathcal{B}_{n, 1/2}}(\varphi) = \alpha$. Show the following.

(a) For each probability measure Q_1 with median $\mu(Q_1) > 0$ and density function ρ_1 there exists a probability measure Q_0 with $\mu(Q_0) = 0$ such that $\varphi \circ S_n^-$ is a Neyman–Pearson test of $Q_0^{\otimes n}$ against $Q_1^{\otimes n}$ of size $\mathbb{E}_0(\varphi \circ S_n^-) = \alpha$. In fact, one can take

$$Q_0 = \tfrac{1}{2} Q_1(\cdot \mid]-\infty, 0]) + \tfrac{1}{2} Q_1(\cdot \mid]-0, \infty[)\,.$$

(b) $\varphi \circ S_n^-$ is a uniformly most powerful level-α test of H_0 against H_1.

11.13 Consider the situation of (11.21) and the data set specified there. For the level $\alpha = 0.06$, perform a one-sided as well as a two-sided sign test for the median and interpret the seeming contradiction.

11.14[S] *Characterisation of stochastic domination.* Let Q_1, Q_2 be two probability measures on $(\mathbb{R}, \mathscr{B})$. Show the equivalence of the following statements.

(a) $Q_1 \preceq Q_2$.

(b) There exists a 'coupling', namely random variables X_1, X_2 on a suitable probability space $(\Omega, \mathscr{F}, P)$ such that $P \circ X_1^{-1} = Q_1$, $P \circ X_2^{-1} = Q_2$, and $P(X_1 \leq X_2) = 1$.

(c) For each bounded, increasing function $Y : \mathbb{R} \to \mathbb{R}$, $\mathbb{E}_{Q_1}(Y) \leq \mathbb{E}_{Q_2}(Y)$.

Hint: For (a) $\Rightarrow$ (b) use Proposition (1.30).

11.15 Let $G(P, Q) = P^{\otimes k} \otimes Q^{\otimes l}(U_{k,l} < c)$ be the power function of a one-sided U-test with threshold level c. Show that $G(P, Q)$ is decreasing in P and increasing in Q with respect to stochastic domination, i.e., $G(P, Q) \leq G(P', Q')$ when $P \succeq P'$ and $Q \preceq Q'$. *Hint:* Use Problem 11.14.

11.16 To investigate whether a certain medication increases the reaction time, 20 people take a reaction test, where 10 people receive the medication beforehand and the remaining 10 form a control group. The following reaction times are measured (in seconds):

Experimental group	.83	.66	.94	.78	.81	.60	.88	.90	.79	.86
Control group	.64	.70	.69	.80	.71	.82	.62	.91	.59	.63

Use a U-test of level $\alpha = 0.05$ to test the null hypothesis that the reaction time is not influenced against the alternative of a prolonged reaction time. Apply (a) the exact U-distribution, (b) the normal approximation.

11.17 *U-test as a t-test for rank statistics.* Consider the two-sample t-statistic T of Problem 10.24 and replace the total sample $X_1, \ldots, X_k, Y_1, \ldots, Y_l$ by the respective ranks $R_1, \ldots, R_{k+l}$. Show that the test statistic obtained in this way differs from the Wilcoxon statistic W only by a constant.

11.18[S] *Wilcoxon's signed-rank test.* Let $X_1, \ldots, X_n$ be independent real-valued random variables with identical distribution Q on $(\mathbb{R}, \mathscr{B})$. Assume Q is continuous and symmetric with respect to 0, i.e., $F_Q(-c) = 1 - F_Q(c)$ for all $c \in \mathbb{R}$. For each $1 \leq i \leq n$, let $Z_i = 1_{\{X_i > 0\}}$ and let R_i^+ be the rank of $|X_i|$ in the sequence $|X_1|, \ldots, |X_n|$ of absolute observations. Let $W^+ = \sum_{i=1}^n Z_i R_i^+$ be the corresponding signed-rank sum. Show the following:

(a) For each i, Z_i and $|X_i|$ are independent.

(b) The random vector $R^+ = (R_1^+, \ldots, R_n^+)$ is independent of the random set

$$Z = \{1 \leq i \leq n : Z_i = 1\}\,.$$

(c) Z is uniformly distributed on the power set $\mathscr{P}_n$ of $\{1, \ldots, n\}$, and R^+ is uniformly distributed on the permutation set $\mathscr{S}_n$.

(d) For each $0 \leq l \leq n(n+1)/2$, it is the case that $P(W^+ = l) = 2^{-n} N(l; n)$ for

$$N(l; n) = \left| \left\{ A \subset \{1, \ldots, n\} : \sum_{i \in A} i = l \right\} \right|.$$

(The $N(l; n)$ can be determined by combinatorial means, and an analogue of the central limit theorem (11.28) holds. Therefore, W^+ can be used as a test statistic for the null hypothesis $H_0 : Q$ is symmetric, which is suitable for the experimental design of matched pairs.)

11.19 Apply the signed-rank test of Problem 11.18 to the situation of Example (11.21) with the data from Table 11.3. Prescribe the level $\alpha = 0.025$. Compare your result with the outcome of the sign test of Theorem (11.22). *Hint:* For small l, the numbers $N(l; 10)$ can be determined by a simple count.

11.20 S *Kolmogorov–Smirnov test.* Let $X_1, \ldots, X_n$ be independent real random variables with continuous distribution function $F(c) = P(X_i \leq c)$. Consider the associated empirical distribution function

$$F_n(c) = \frac{1}{n} \sum_{i=1}^{n} 1_{\{X_i \leq c\}},$$

and let $\Delta_n = \sup_{c \in \mathbb{R}} |F_n(c) - F(c)|$. Show the following:

(a) $\Delta_n = \max_{1 \leq i \leq n} \max\left(\frac{i}{n} - F(X_{i:n}), F(X_{i:n}) - \frac{i-1}{n} \right)$.

(b) The distribution of Δ_n does not depend on F. *Hint:* Problem 1.18.

(Since this latter distribution can be calculated explicitly, one can verify the null hypothesis 'F is the true distribution function' by a test with rejection region $\{\Delta_n > c\}$; tables can be found e.g. in [46, 53].)

11.21 Consider the situation of Problem 11.5 about the tendency towards the mean when awarding grades, and the data given there.

(a) Plot the empirical distribution function.

(b) Use the Kolmogorov–Smirnov test of Problem 11.20 at level $\alpha = 0.1$ to decide whether or not the data are based on the normal distribution $\mathcal{N}_{3.5, 1}$. (You need to know that $P(\Delta_{17} \leq 0.286) = 0.9$.)

Chapter 12

Regression Models and Analysis of Variance

In many experiments, one can take for granted that the observations depend linearly on certain control parameters that can be adjusted arbitrarily. The problem is that this linear dependence is perturbed by random observational errors, and the true linear coefficients are unknown. How can these coefficients be recovered from the observations? This is the topic of linear regression, the method of determining an unknown linear dependence from a number of randomly perturbed observations. In the simplest case, one seeks the best approximation of a 'cloud' of data points in the plane by a straight line. A suitable theoretical framework is provided by the so-called linear model. If the error variables are independent and normally distributed, it is possible to specify the distribution of all relevant estimators and test statistics, and one arrives at various generalisations of the confidence intervals and tests developed in the Gaussian product model. An important special case is the analysis of variance, a method of determining the influence of experimental factors by comparing the data from different groups of samples.

12.1 Simple Linear Regression

As usually, we begin with an example for illustration.

(12.1) Example. *Thermal expansion of a metal.* Within a certain range, the length of a metal bar depends linearly on the temperature. To identify the expansion coefficient, one chooses n temperatures $t_1, \ldots, t_n$, of which at least two are different, and measures the length of the bar at each of theses temperatures. Because of random measurement errors, the length X_k at temperature t_k is random. Namely, X_k consists of a deterministic part, the true length of the bar, and a random error term. Accordingly, X_k is described by a *linear regression equation* of the form

$$(12.2) \qquad X_k = \gamma_0 + \gamma_1 t_k + \sqrt{v}\, \xi_k \,, \quad 1 \le k \le n \,;$$

here, $\gamma_0, \gamma_1 \in \mathbb{R}$ are two unknown coefficients, which are to be determined (γ_1 is the thermal expansion coefficient), $v > 0$ is an unknown dispersion parameter that governs the magnitude of the measurement errors, and $\xi_1, \ldots, \xi_n$ are suitable random variables that describe the random errors in the measurements. One assumes that the ξ_k are standardised, in that $\mathbb{E}(\xi_k) = 0$ and $\mathbb{V}(\xi_k) = 1$; this ensures that the parameters γ_0 and v are uniquely determined. It also means that the random errors of all measurements are supposed to have the same variance. The deterministic variable 'temperature' with the preselected values $t_1, \ldots, t_n$ is called the *regression variable* or the *regressor*, the measured variable 'bar length' with the random outcomes $X_1, \ldots, X_n$ is called the *dependent* or *response variable*. The pair $\gamma = (\gamma_0, \gamma_1)^\top$ is the so-called shift parameter, whereas v is a scaling parameter that determines the common dispersion of all measurements.

In the following, we write the regression equation (12.2) in the vector form

(12.3)
$$X = \gamma_0 \mathbf{1} + \gamma_1 t + \sqrt{v}\,\xi$$

with the (given) column vectors $\mathbf{1} = (1, \ldots, 1)^\top$ and $t = (t_1, \ldots, t_n)^\top$, the random observation vector $X = (X_1, \ldots, X_n)^\top$, and the random error vector $\xi = (\xi_1, \ldots, \xi_n)^\top$. Since, by assumption, not all the t_k are identical, the vectors $\mathbf{1}$ and t are linearly independent. So the parameters γ_0 and γ_1 are both relevant. If we denote by $P_{\gamma,v}$ the distribution of the random vector $\gamma_0 \mathbf{1} + \gamma_1 t + \sqrt{v}\,\xi$, then our modelling assumptions lead to the statistical model

$$\left(\mathbb{R}^n, \mathscr{B}^n, P_{\gamma,v} : (\gamma, v) \in \mathbb{R}^2 \times {]0, \infty[}\right).$$

In this model, $X_k : \mathbb{R}^n \to \mathbb{R}$ is simply the kth projection, and $X = \mathrm{Id}_{\mathbb{R}^n}$ is the identity mapping on $\mathbb{R}^n$.

How can one deduce the unknown parameters γ_0, γ_1 from the measurements $X_1, \ldots, X_n$? Since we assume a linear dependence of the bar length on the temperature, our task is to place a straight line 'through' the random data points $(t_1, X_1), \ldots, (t_n, X_n)$ in an intelligent way; see Figure 12.1. Such a straight line is called a *regression line*, or *line of best fit*. A practical procedure is provided by the following principle, which goes back to C. F. Gauss and A. M. Legendre (for the dispute over priority, see for instance [17]).

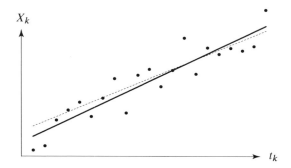

Figure 12.1. Scatter plot of data points (t_k, X_k) with equidistant t_k and random X_k that are generated by adding standard normal errors to the true (dashed) line. The corresponding regression line is bold.

The principle of least squares. Depending on the observation vector X, one determines $\hat{\gamma} = (\hat{\gamma}_0, \hat{\gamma}_1)^\top \in \mathbb{R}^2$ such that the mean squared error

$$E_\gamma := \frac{1}{n} \sum_{k=1}^n \left(X_k - (\gamma_0 + \gamma_1 t_k)\right)^2 = |X - \gamma_0 \mathbf{1} - \gamma_1 t|^2 / n$$

is minimal for $\gamma = \hat{\gamma}$.

Does such a minimiser $\hat{\gamma}$ exist, and how can we find it? Suppose the value of X is given. The function $\gamma \to E_\gamma$ then indeed has a global minimum. This is because $\lim_{|\gamma| \to \infty} E_\gamma = \infty$, so that E_γ only has to be considered on a sufficiently large compact set on which its global minimum is attained. Now, the gradient of E_γ vanishes at each such global minimiser γ; i.e., we have

$$0 = \frac{\partial}{\partial \gamma_0} E_\gamma = -\frac{2}{n} \sum_{k=1}^n (X_k - \gamma_0 - \gamma_1 t_k),$$

$$0 = \frac{\partial}{\partial \gamma_1} E_\gamma = -\frac{2}{n} \sum_{k=1}^n t_k (X_k - \gamma_0 - \gamma_1 t_k).$$

This is equivalent to the *normal equations*

$$(12.4) \qquad \gamma_0 + \gamma_1 M(t) = M(X), \quad \gamma_0 M(t) + \gamma_1 \frac{1}{n} \sum_{k=1}^n t_k^2 = \frac{1}{n} \sum_{k=1}^n t_k X_k,$$

where $M(x) = \frac{1}{n} \sum_{i=1}^n x_i$ again stands for the mean of a vector $x \in \mathbb{R}^n$. The second equation can be simplified. In terms of the variance

$$V(t) = \frac{1}{n} \sum_{k=1}^n t_k^2 - M(t)^2$$

and the covariance

$$c(t, X) := \frac{1}{n} \sum_{k=1}^n t_k X_k - M(t) M(X),$$

it can be written as

$$\gamma_0 M(t) + \gamma_1 \left(V(t) + M(t)^2\right) = c(t, X) + M(t) M(X).$$

Replacing $M(X)$ by $\gamma_0 + \gamma_1 M(t)$ according to the first normal equation, we find that $\gamma_1 V(t) = c(t, X)$. Since, by assumption, the t_k do not all agree, we know that $V(t) > 0$, and we arrive at the following result.

(12.5) Theorem. Regression line. *The statistics*

$$\hat{\gamma}_0 = M(X) - \frac{M(t)}{V(t)}\, c(t, X) \quad and \quad \hat{\gamma}_1 = \frac{c(t, X)}{V(t)}$$

are the uniquely determined least-squares estimators of γ_0 and γ_1. They are both unbiased.

Proof. It only remains to be shown that the estimators are unbiased. For all $\vartheta = (\gamma, v)$, we can use the linearity of $c(t, \cdot)$ to write

$$\mathbb{E}_\vartheta(\hat{\gamma}_1) = \mathbb{E}_\vartheta(c(t, X))/V(t) = c(t, \mathbb{E}_\vartheta(X))/V(t) \,;$$

here $\mathbb{E}_\vartheta(X) := (\mathbb{E}_\vartheta(X_1), \ldots, \mathbb{E}_\vartheta(X_n))^\top$. Using further the regression equation (12.3), the assumption that ξ is centred, and the rule (4.23a) for the covariance, we find that the last expression is equal to

$$c(t, \gamma_0 \mathbf{1} + \gamma_1 t)/V(t) = \gamma_1\, c(t, t)/V(t) = \gamma_1 \,.$$

Likewise, by the linearity of M,

$$\begin{aligned}
\mathbb{E}_\vartheta(\hat{\gamma}_0) &= \mathbb{E}_\vartheta\big(M(X) - \hat{\gamma}_1\, M(t)\big) = M(\mathbb{E}_\vartheta(X)) - \gamma_1\, M(t) \\
&= M(\gamma_0 \mathbf{1} + \gamma_1 t) - \gamma_1\, M(t) = \gamma_0 \,,
\end{aligned}$$

as claimed. $\diamond$

The results obtained so far can be summarised as follows. According to the principle of least squares, the regression line corresponding to the observation X has slope $\hat{\gamma}_1$ and axis intercept $\hat{\gamma}_0$. The first normal equation in (12.4) specifies that this straight line passes through the centre of gravity $(M(t), M(X))$ of the data points (t_k, X_k). Plainly, the regression line is uniquely determined by the regression vector $\hat{\gamma}_0 \mathbf{1} + \hat{\gamma}_1\, t$, which consists of its values at the points $t_1, \ldots, t_n$. This vector has the following geometric interpretation.

(12.6) Remark. *Regression line as a projection.* The regression vector $\hat{\gamma}_0 \mathbf{1} + \hat{\gamma}_1\, t$ is equal to the projection of the observation vector $X \in \mathbb{R}^n$ onto the subspace

$$L = L(\mathbf{1}, t) := \{\gamma_0 \mathbf{1} + \gamma_1 t : \gamma_0, \gamma_1 \in \mathbb{R}\} \,,$$

namely the span of the vectors $\mathbf{1}$ and t. Explicitly, writing Π_L for the projection of $\mathbb{R}^n$ onto L, we have that $\Pi_L X = \hat{\gamma}_0 \mathbf{1} + \hat{\gamma}_1 t$. Indeed, the principle of least squares states that $\hat{\gamma}_0 \mathbf{1} + \hat{\gamma}_1\, t$ is precisely the vector in L that has the least distance from X, and the normal equations express the equivalent fact that the difference vector $X - \hat{\gamma}_0 \mathbf{1} - \hat{\gamma}_1\, t$ is orthogonal to $\mathbf{1}$ and t and thus to L.

The regression line describes the dependence of the bar length on the temperature that is conjectured on the basis of the observation X. Therefore, one can make predictions about the bar length $\tau(\gamma) = \gamma_0 + \gamma_1 u$ at a temperature $u \notin \{t_1, \ldots, t_n\}$ for which no measurement was made. We will deal with this question later on in a general framework; see Theorem (12.15b) by Gauss. Here, we will continue with two additional remarks.

(12.7) Remark. *Random regressor.* In Example (12.1), we have assumed that the temperatures t_k can be adjusted deterministically and do not depend on chance. However, the method of least squares can also be used when the regression variable is random. Then it allows us to find an (approximate) linear relation between two random variables, as for instance air pressure and air humidity. Formally, one considers the pairs (X_k, T_k) of real random variables (which may correspond to the observations on different days), and replaces the deterministic vector t by the random vector $(T_1, \ldots, T_n)^\top$ in the above calculations.

(12.8) Remark. *Hidden factors.* Frequently, false conclusions are due to the fact that essential factors influencing the situation have been ignored. Suppose that the response variable X does not only depend on the regressor t that was observed, but was further influenced by an additional factor s. It may then happen that the regression line for X is increasing in t if the value of s is not taken into account, whereas the regression lines are decreasing if situations with fixed s are considered. In other words, if s is ignored, then t seems to have an enhancing character, whereas in fact the opposite is true. An example of this so-called *Simpson's paradox* follows in Problem 12.4.

12.2 The Linear Model

Abstractly, the linear regression equation (12.3) tells us that the observation vector X is obtained from the random error vector ξ by two operations: a scaling by $\sqrt{v}$ followed by a shift by a vector that depends linearly on the unknown parameter γ. We will now formulate this idea in a general setting.

Definition. Let $s, n \in \mathbb{N}$ be such that $s < n$. A *linear model* for n real-valued observations with unknown s-dimensional shift parameter $\gamma = (\gamma_1, \ldots, \gamma_s)^\top \in \mathbb{R}^s$ and unknown scaling parameter $v > 0$ consists of

▷ a real $n \times s$ matrix A with full rank s, the so-called *design matrix*, which consists of known control parameters, and

▷ a random vector $\xi = (\xi_1, \ldots, \xi_n)^\top$ of n standardised random variables $\xi_k \in \mathscr{L}^2$, the *additive error* or *noise terms*.

The n-dimensional observation vector $X = (X_1, \ldots, X_n)^\top$ is obtained from these quantities by the linear equation

(12.9) $X = A\gamma + \sqrt{v}\,\xi\,.$

(In particular, this means that all measurement errors are assumed to have the same variance.) A suitable statistical model is given by

$$(\mathcal{X}, \mathcal{F}, P_\vartheta : \vartheta \in \Theta) = \big(\mathbb{R}^n, \mathscr{B}^n, P_{\gamma,v} : \gamma \in \mathbb{R}^s, v > 0\big)\,,$$

where, for every $\vartheta = (\gamma, v) \in \Theta := \mathbb{R}^s \times]0, \infty[$, $P_\vartheta = P_{\gamma,v}$ is defined as the distribution of the random vector $A\gamma + \sqrt{v}\,\xi$. In this canonical model, X is simply the identity on $\mathbb{R}^n$ and X_k the kth projection. Since the ξ_k are centred, we have that $\mathbb{E}_{\gamma,v}(X) = A\gamma$.

Now, the statistician's task is the inference of the unknown parameters γ and v from a given realisation of X. So we need to find estimators and hypothesis tests for these parameters. A special case, which will turn out to be particularly accessible, is the *Gaussian linear model*. In this model, ξ has the n-dimensional standard normal distribution $\mathcal{N}_n(0, E)$, so that $P_{\gamma,v} = \mathcal{N}_n(A\gamma, vE)$ by Theorem (9.2). In the next section we shall deal with the Gaussian case in more detail, but first we will discuss various specific choices of the design matrix A.

(12.10) Example. *Gaussian product model.* Let $s = 1$, $A = \mathbf{1}$, and $\gamma = m \in \mathbb{R}$, and suppose ξ has the multivariate standard normal distribution $\mathcal{N}_n(0, E)$. Then $P_{\gamma,v} = \mathcal{N}_n(m\mathbf{1}, vE) = \mathcal{N}_{m,v}^{\otimes n}$. We are thus back to the Gaussian product model considered previously, which describes n independent, normally distributed observations each with the same (unknown) mean m and the same (unknown) variance v.

(12.11) Example. *Simple linear regression.* Let $s = 2$, $A = (\mathbf{1}\,t)$ for a regressor vector $t = (t_1, \ldots, t_n)^\top \in \mathbb{R}^n$ with at least two distinct coordinates, and $\gamma = (\gamma_0, \gamma_1)^\top$. Then, the equation (12.9) for the linear model is identical to the linear regression equation (12.3).

(12.12) Example. *Polynomial regression.* Let $t = (t_1, \ldots, t_n)^\top$ be a non-constant vector of known regressor values. If, in contrast to Example (12.1), a linear dependence between regression variable and response variable cannot be taken for granted, it is natural to generalise the linear regression equation (12.2) to the polynomial regression equation

$$X_k = \gamma_0 + \gamma_1\,t_k + \gamma_2\,t_k^2 + \cdots + \gamma_d\,t_k^d + \sqrt{v}\,\xi_k\,, \quad 1 \le k \le n\,.$$

Here, $d \in \mathbb{N}$ is the maximal degree that is taken into account. This situation is also a special case of (12.9), provided we set $s = d + 1$, $\gamma = (\gamma_0, \dots, \gamma_d)^\top$ and

$$
A = \begin{pmatrix}
1 & t_1 & t_1^2 & \cdots & t_1^d \\
\vdots & \vdots & \vdots & & \vdots \\
1 & t_n & t_n^2 & \cdots & t_n^d
\end{pmatrix}.
$$

(12.13) Example. *Multiple linear regression.* In the case of simple linear regression and polynomial regression, it is assumed that the observed quantities are only influenced by a single variable. However, in many cases it is necessary to consider the influence of several different factors. If the influence of each of these factors can be assumed to be linear, we arrive at the multiple linear regression equation

$$
X_k = \gamma_0 + \gamma_1 t_{k,1} + \cdots + \gamma_d t_{k,d} + \sqrt{v}\, \xi_k , \quad 1 \le k \le n .
$$

Here, d is the number of relevant regression variables, and $t_{k,i}$ the value of the ith variable used for the kth observation. Clearly, to allow for rational conclusions about the unknown coefficients $\gamma_0, \dots, \gamma_d$, one must make sure that $n > d$ and the matrix

$$
A = \begin{pmatrix}
1 & t_{1,1} & \cdots & t_{1,d} \\
\vdots & \vdots & & \vdots \\
1 & t_{n,1} & \cdots & t_{n,d}
\end{pmatrix}
$$

has full rank $s = d + 1$. Setting $\gamma = (\gamma_0, \dots, \gamma_d)^\top$, we once again arrive at the fundamental equation (12.9) of the linear model.

As in Remark (12.6) we will now consider the linear subspace

$$
L = L(A) := \{ A\gamma : \gamma \in \mathbb{R}^s \} \subset \mathbb{R}^n ,
$$

which is spanned by the s column vectors of A. According to the principle of least squares, one tries to find, for each observed vector $x \in \mathbb{R}^n$, the respective element of L that has the least distance from x. This can be obtained from x using the orthogonal projection $\Pi_L : \mathbb{R}^n \to L$ onto L. Π_L is characterised by either of the following properties:

(a) $\Pi_L x \in L$, and $|x - \Pi_L x| = \min_{u \in L} |x - u|$ for all $x \in \mathbb{R}^n$;

(b) $\Pi_L x \in L$, and $x - \Pi_L x \perp L$ for all $x \in \mathbb{R}^n$.

The orthogonality statement in (b) corresponds to the *normal equations* (12.4) and is equivalent to the property that $(x - \Pi_L x) \cdot u = 0$ for all u in a basis of L, so it is equivalent to the equation $A^\top (x - \Pi_L x) = 0$ for all $x \in \mathbb{R}^n$.

(12.14) Remark. *Representation of the projection matrix.* The $s \times s$ matrix $\mathsf{A}^\top \mathsf{A}$ is invertible, and the projection Π_L of $\mathbb{R}^n$ onto $L = L(\mathsf{A})$ is given by the $n \times n$ matrix

$$\Pi_L = \mathsf{A}(\mathsf{A}^\top \mathsf{A})^{-1} \mathsf{A}^\top .$$

In particular, $\hat{\gamma} := (\mathsf{A}^\top \mathsf{A})^{-1} \mathsf{A}^\top X$ is the unique solution of the equation $\Pi_L X = \mathsf{A}\hat{\gamma}$.

Proof. Suppose $\mathsf{A}^\top \mathsf{A}$ were not invertible. Then $\mathsf{A}^\top \mathsf{A} c = 0$ for some $0 \neq c \in \mathbb{R}^s$, and therefore $|\mathsf{A}c|^2 = c^\top \mathsf{A}^\top \mathsf{A} c = 0$. Hence $\mathsf{A}c = 0$, in contradiction to the assumption that A has full rank. Next, it is clear that $\mathsf{A}(\mathsf{A}^\top \mathsf{A})^{-1} \mathsf{A}^\top x \in L$ for every $x \in \mathbb{R}^n$, while, on the other hand,

$$\mathsf{A}^\top \big(x - \mathsf{A}(\mathsf{A}^\top \mathsf{A})^{-1} \mathsf{A}^\top x\big) = \mathsf{A}^\top x - \mathsf{A}^\top x = 0 .$$

Property (b) above thus implies that $\Pi_L x = \mathsf{A}(\mathsf{A}^\top \mathsf{A})^{-1} \mathsf{A}^\top x$. $\diamond$

The following theorem deals with the estimation of the parameters γ and v, and of arbitrary linear functions of γ.

(12.15) Theorem. *Estimators in the linear model. In the linear model with uncorrelated error terms $\xi_1, \ldots, \xi_n$, the following statements hold.*

(a) *The least-squares estimator $\hat{\gamma} := (\mathsf{A}^\top \mathsf{A})^{-1} \mathsf{A}^\top X$ is an unbiased estimator of γ.*

(b) *Gauss' theorem. Let $\tau : \mathbb{R}^s \to \mathbb{R}$ be a linear characteristic of γ that is to be estimated. That is, $\tau(\gamma) = c \cdot \gamma$ for some $c \in \mathbb{R}^s$ and every $\gamma \in \mathbb{R}^s$. Then $T := c \cdot \hat{\gamma}$ is an unbiased estimator of τ, its variance is minimal among all linear unbiased estimators, and there is no other linear estimator of τ with these properties.*

(c) *The sample variance*

$$V^* := \frac{|X - \Pi_L X|^2}{n-s} = \frac{|X|^2 - |\Pi_L X|^2}{n-s} = \frac{|X - \mathsf{A}\hat{\gamma}|^2}{n-s}$$

is an unbiased estimator of v.

An estimator T as in statement (b) is called a *best linear unbiased estimator*, with the acronym BLUE. In the literature, statement (b) is commonly known as the Gauss–Markov theorem. But the attribution to Markov seems to be incorrect, see [28]. Statement (c) generalises Theorem (7.13). Since we need to estimate the s-dimensional parameter γ (rather than only the mean m), we loose s degrees of freedom. This explains the division by $n-s$.

Proof. (a) For all $\vartheta = (\gamma, v)$, it follows from the linearity of the expectation that

$$\mathbb{E}_\vartheta(\hat{\gamma}) = (\mathsf{A}^\top \mathsf{A})^{-1} \mathsf{A}^\top \, \mathbb{E}_\vartheta(X) = (\mathsf{A}^\top \mathsf{A})^{-1} \mathsf{A}^\top \mathsf{A}\gamma = \gamma .$$

Here, the expectation of a random vector is again defined coordinate-wise as the vector of the expectations of the coordinate variables.

(b) Statement (a) implies that T is unbiased. We will now show that T has minimal variance among all linear unbiased estimators of τ. To this end, we consider the vector $a = \mathsf{A}(\mathsf{A}^{\mathsf{T}}\mathsf{A})^{-1}c \in L$. Since $\Pi_L a = a$ and Π_L is symmetric, we know that $a^{\mathsf{T}}\Pi_L = a^{\mathsf{T}}$. On the other hand, since $\mathsf{A}^{\mathsf{T}}a = c$ by definition, we find that $c^{\mathsf{T}} = a^{\mathsf{T}}\mathsf{A}$ and therefore $\tau(\gamma) = c^{\mathsf{T}}\gamma = a^{\mathsf{T}}\mathsf{A}\gamma$. Altogether, we conclude that $T = c^{\mathsf{T}}\hat{\gamma} = a^{\mathsf{T}}\Pi_L X = a^{\mathsf{T}}X$. In particular, this shows that T is linear.

Now, let $S : \mathbb{R}^n \to \mathbb{R}$ be an arbitrary linear unbiased estimator of τ. By the linearity of S, there exists a vector $b \in \mathbb{R}^n$ such that $S = b \cdot X$. Exploiting that S is unbiased, we further find that

$$b \cdot \mathsf{A}\gamma = \mathbb{E}_{\vartheta}(b \cdot X) = \mathbb{E}_{\vartheta}(S) = \tau(\gamma) = a \cdot \mathsf{A}\gamma$$

for all $\vartheta = (\gamma, v)$; this means that $b \cdot u = a \cdot u$ for all $u \in L$. Hence, $b - a$ is orthogonal to L. It follows that $a = \Pi_L b$, and so, in particular, $|a| \leq |b|$. We now write

$$\begin{aligned}
\mathbb{V}_{\vartheta}(S) - \mathbb{V}_{\vartheta}(T) &= \mathbb{E}_{\vartheta}\left([b^{\mathsf{T}}(X - \mathsf{A}\gamma)]^2 - [a^{\mathsf{T}}(X - \mathsf{A}\gamma)]^2\right) \\
&= v\,\mathbb{E}\left(b^{\mathsf{T}}\xi\,\xi^{\mathsf{T}}b - a^{\mathsf{T}}\xi\,\xi^{\mathsf{T}}a\right) \\
&= v\left(b^{\mathsf{T}}\,\mathbb{E}(\xi\,\xi^{\mathsf{T}})b - a^{\mathsf{T}}\,\mathbb{E}(\xi\,\xi^{\mathsf{T}})a\right).
\end{aligned}$$

Here, we use the vector notation $\xi\,\xi^{\mathsf{T}}$ for the matrix $(\xi_k\xi_l)_{1 \leq k,l \leq n}$, and the expectation $\mathbb{E}(\xi\,\xi^{\mathsf{T}})$ is defined elementwise. Since the ξ_k are uncorrelated, we obtain that $\mathbb{E}(\xi\,\xi^{\mathsf{T}}) = \mathsf{E}$, the unit matrix. It follows that

$$\mathbb{V}_{\vartheta}(S) - \mathbb{V}_{\vartheta}(T) = v\,(|b|^2 - |a|^2) \geq 0,$$

and this proves the optimality and uniqueness of T.

(c) The second expression for V^* follows from Pythagoras' theorem, see Figure 12.2 below, and the third from Remark (12.14). Now, let $u_1, \ldots, u_n$ be an orthonormal basis of $\mathbb{R}^n$ such that $L = \mathrm{span}(u_1, \ldots, u_s)$, and let O be the orthogonal matrix with column vectors $u_1, \ldots, u_n$, which maps the linear subspace $H = \{x \in \mathbb{R}^n : x_{s+1} = \cdots = x_n = 0\}$ onto L. The projection onto H is described by the diagonal matrix E_s that consists of ones in the diagonal entries $1, \ldots, s$ and zeroes everywhere else. This shows that $\Pi_L = \mathsf{O}\mathsf{E}_s\mathsf{O}^{\mathsf{T}}$.

We now set $\eta := \mathsf{O}^{\mathsf{T}}\xi$ and write $\eta_k = \sum_{j=1}^{n} \mathsf{O}_{jk}\xi_j$ for the kth coordinate of η. Then we obtain

(12.16) $$(n-s)\,V^* = v\,|\xi - \Pi_L\xi|^2 = v\,|\eta - \mathsf{E}_s\eta|^2 = v \sum_{k=s+1}^{n} \eta_k^2\,.$$

In the above, the first equality follows from (12.9) and the fact that $\Pi_L \mathsf{A}\gamma = \mathsf{A}\gamma \in L$, and the second from the rotational invariance of the Euclidean norm. Finally, since the ξ_j are assumed to be uncorrelated and standardised, we can conclude that

$$\mathbb{E}(\eta_k^2) = \mathbb{E}\left(\sum_{1 \le i,j \le n} \mathsf{O}_{ik}\mathsf{O}_{jk}\xi_i\xi_j\right) = \sum_{i=1}^{n} \mathsf{O}_{ik}^2 = 1$$

for each k. This and (12.16) jointly imply that V^* is unbiased. $\diamond$

Finding estimates for γ and v is only the first step. To get more reliable conclusions, one needs to specify confidence regions and test hypotheses for these parameters. The calculation of error probabilities then requires more detailed information about the underlying distributions. Complete results can be obtained under the conventional assumption that the errors ξ_k are independent and standard normally distributed. This case is the subject of the next section.

12.3   The Gaussian Linear Model

Throughout this section, we assume that the error vector ξ has the multivariate standard normal distribution $\mathcal{N}_n(0, \mathsf{E})$, so that, by equation (12.9) and Theorem (9.2), $P_\vartheta = \mathcal{N}_n(\mathsf{A}\gamma, v\mathsf{E})$ for all $\vartheta = (\gamma, v) \in \mathbb{R}^s \times]0, \infty[$. In particular, this means that the observations $X_1, \dots, X_n$ are independent under every P_ϑ. The corresponding linear model is called the *normally distributed* or *Gaussian linear model*. The following theorem determines the distributions of the estimators $\hat{\gamma}$ and V^* considered above, and of some further statistics that can be used for constructing confidence regions and tests.

(12.17) Theorem. Generalised Student's theorem. *Consider the Gaussian linear model, and take an arbitrary parameter (γ, v). Then the following statements hold under $P_{\gamma,v}$.*

(a) *The least-squares estimator $\hat{\gamma}$ has the distribution $\mathcal{N}_s(\gamma, v(\mathsf{A}^\top\mathsf{A})^{-1})$.*

(b) *The (rescaled) variance estimator $\frac{n-s}{v} V^*$ has the distribution χ_{n-s}^2.*

(c) *The statistic $|\mathsf{A}(\hat{\gamma} - \gamma)|^2/v = |\Pi_L X - \mathbb{E}_{\gamma,v}(X)|^2/v$ is χ_s^2-distributed and independent of V^*. Therefore, $|\mathsf{A}(\hat{\gamma} - \gamma)|^2/(sV^*)$ has the F-distribution $\mathcal{F}_{s,n-s}$.*

(d) *Suppose $H \subset L$ is a linear subspace such that $\dim H = r < s$ and $\mathsf{A}\gamma \in H$. Then $|\Pi_L X - \Pi_H X|^2/v$ has the distribution χ_{s-r}^2 and is independent of V^*. The Fisher statistic*

$$(12.18) \qquad F_{H,L} := \frac{n-s}{s-r} \frac{|\Pi_L X - \Pi_H X|^2}{|X - \Pi_L X|^2} = \frac{|\mathsf{A}\hat{\gamma} - \Pi_H X|^2}{(s-r) V^*}$$

thus has the F-distribution $\mathcal{F}_{s-r,n-s}$.

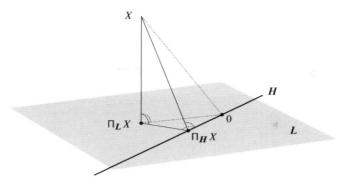

Figure 12.2. The faces of the tetrahedron formed by the projection rays are right-angled triangles.

By Pythagoras' theorem, the Fisher statistic can also be written as

$$F_{H,L} = \frac{n-s}{s-r}\,\frac{|\Pi_L X|^2 - |\Pi_H X|^2}{|X|^2 - |\Pi_L X|^2} = \frac{n-s}{s-r}\,\frac{|X - \Pi_H X|^2 - |X - \Pi_L X|^2}{|X - \Pi_L X|^2}\,,$$

see Figure 12.2.

Proof. (a) Theorems (9.5) and (12.15a) show that $\hat{\gamma} = (\mathsf{A}^\top\mathsf{A})^{-1}\mathsf{A}^\top X$ has the multivariate normal distribution with mean vector γ and covariance matrix

$$(\mathsf{A}^\top\mathsf{A})^{-1}\mathsf{A}^\top(v\mathsf{E})\mathsf{A}(\mathsf{A}^\top\mathsf{A})^{-1} = v\,(\mathsf{A}^\top\mathsf{A})^{-1}\,.$$

(b)–(d) Let $H \subset L$ and $u_1,\ldots,u_n$ be an orthonormal basis of $\mathbb{R}^n$ such that

$$\mathrm{span}(u_1,\ldots,u_r) = H\,,\quad \mathrm{span}(u_1,\ldots,u_s) = L\,.$$

Let O be the orthogonal matrix with columns $u_1,\ldots,u_n$. Consider first the $\mathcal{N}_n(0,\mathsf{E})$-distributed error vector ξ. By Corollary (9.4), the vector $\eta := \mathsf{O}^\top\xi$ still has the distribution $\mathcal{N}_n(0,\mathsf{E})$, so its coordinates $\eta_1,\ldots,\eta_n$ are independent and $\mathcal{N}_{0,1}$-distributed. Hence, statement (b) follows immediately from equation (12.16) and Theorem (9.10). Moreover, we can write

$$|\Pi_L\xi - \Pi_H\xi|^2 = |(\mathsf{E}_s - \mathsf{E}_r)\eta|^2 = \sum_{k=r+1}^{s}\eta_k^2\,.$$

Therefore, $|\Pi_L\xi - \Pi_H\xi|^2$ is χ_{s-r}^2-distributed and, by (12.16) and Theorem (3.24), it is also independent of V^*. For $H = \{0\}$ and $r = 0$, this implies that $|\Pi_L\xi|^2$ is χ_s^2-distributed and independent of V^*.

Passing now to the observation vector X for the parameter (γ, v), we find from (12.9) that

$$\mathsf{A}(\hat{\gamma} - \gamma) = \Pi_L(X - \mathsf{A}\gamma) = \sqrt{v}\,\Pi_L\xi$$

and, if $A\gamma \in H$,

$$\Pi_L X - \Pi_H X = \Pi_L(X - A\gamma) - \Pi_H(X - A\gamma) = \sqrt{v}\,(\Pi_L\xi - \Pi_H\xi)\,.$$

Together with the above and Theorem (9.12) we obtain (c) and (d). ◇

Theorem (12.17) allows us to construct confidence intervals and tests for the un-known parameters γ and v.

(12.19) Corollary. Confidence intervals in the Gaussian linear model. *For any given error level* $0 < \alpha < 1$, *the following holds.*

(a) Confidence region for γ. If $f_{s,n-s;1-\alpha}$ is the α-fractile of $\mathcal{F}_{s,n-s}$, then the ran-dom ellipsoid

$$C(\cdot) = \{\gamma \in \mathbb{R}^s : |A(\gamma - \hat{\gamma})|^2 < s\, f_{s,n-s;1-\alpha}\, V^*\}$$

is a confidence region for γ with error level α.

(b) Confidence interval for a linear characteristic $\tau(\gamma) = c \cdot \gamma$. Let $c \in \mathbb{R}^s$, $t_{n-s;1-\alpha/2}$ be the $\alpha/2$-fractile of t_{n-s}, and $\delta = t_{n-s;1-\alpha/2}\sqrt{c^{\mathsf{T}}(A^{\mathsf{T}}A)^{-1}c}$. Then

$$C(\cdot) = \,]c \cdot \hat{\gamma} - \delta\sqrt{V^*},\; c \cdot \hat{\gamma} + \delta\sqrt{V^*}\,[$$

is a confidence interval for $\tau(\gamma) = c \cdot \gamma$ with error level α.

(c) Confidence interval for the variance. Let $q_- = \chi^2_{n-s;\alpha/2}$ resp. $q_+ = \chi^2_{n-s;1-\alpha/2}$ be the $\alpha/2$-quantile resp. the $\alpha/2$-fractile of χ^2_{n-s}. Then

$$C(\cdot) = \,](n-s)\, V^*/q_+,\; (n-s)\, V^*/q_-\,[$$

is a confidence interval for v with error level α.

Proof. Statement (a) follows directly from Theorem (12.17c). Turning to the proof of (b), we conclude from Theorems (12.17a) and (9.5) that $Z := c \cdot \hat{\gamma}$ is normally distributed under $P_{\gamma,v}$ with mean $c \cdot \gamma$ and variance $v\, c^{\mathsf{T}}(A^{\mathsf{T}}A)^{-1}c$. Hence,

$$Z^* := \frac{Z - c \cdot \gamma}{\sqrt{v c^{\mathsf{T}}(A^{\mathsf{T}}A)^{-1}c}}$$

is $\mathcal{N}_{0,1}$-distributed. From the proof of Theorem (12.15b) we know that Z^* is a func-tion of $A\hat{\gamma}$ and thus, by Theorem (12.17c), independent of $(n-s)V^*/v$, which is itself χ^2_{n-s}-distributed. Corollary (9.15) thus shows that the statistic $T = Z^*\sqrt{v/V^*}$ is t_{n-s}-distributed. So we can write

$$P_{\gamma,v}\big(C(\cdot) \ni \tau(\vartheta)\big) = P_{\gamma,v}\big(|T| \le t_{n-s;1-\alpha/2}\big) = 1 - \alpha\,,$$

which proves claim (b). Finally, statement (c) follows from Theorem (12.17b). ◇

(12.20) Corollary. Tests in the Gaussian linear model. *For every given significance level $0 < \alpha < 1$, the following statements hold.*

(a) *t-test of the null hypothesis $c \cdot \gamma = m_0$. Take any $c \in \mathbb{R}^s$ and $m_0 \in \mathbb{R}$. With $t_{n-s;1-\alpha/2}$ the $\alpha/2$-fractile of t_{n-s}, the rejection region*

$$\left\{ |c \cdot \hat{\gamma} - m_0| > t_{n-s;1-\alpha/2} \sqrt{c^{\mathsf{T}}(\mathsf{A}^{\mathsf{T}}\mathsf{A})^{-1} c \; V^*} \right\}$$

defines a level-α test for the two-sided test problem $H_0 : c \cdot \gamma = m_0$ against $H_1 : c \cdot \gamma \neq m_0$. Level-$\alpha$ tests for the one-sided null hypotheses $c \cdot \gamma \leq m_0$ resp. $c \cdot \gamma \geq m_0$ can be constructed analogously.

(b) *F-test of the linear hypothesis $\mathsf{A}\gamma \in \mathbf{H}$. Let $\mathbf{H} \subset \mathbf{L}$ be a linear space of dimension $\dim \mathbf{H} =: r < s$ and $f_{s-r,n-s;1-\alpha}$ the α-fractile of $F_{s-r,n-s}$. For $F_{\mathbf{H},\mathbf{L}}$ as in (12.18), the rejection region*

$$\left\{ F_{\mathbf{H},\mathbf{L}} > f_{s-r,n-s;1-\alpha} \right\}$$

defines a level-α test for the test problem $H_0 : \mathsf{A}\gamma \in \mathbf{H}$ against $H_1 : \mathsf{A}\gamma \notin \mathbf{H}$.

(c) *χ^2-test for the variance. For $v_0 > 0$, the rejection region*

$$\left\{ (n-s)\, V^* > v_0\, \chi^2_{n-s;1-\alpha} \right\}$$

determines a test of level α for the left-sided test problem $H_0 : v \leq v_0$ against $H_1 : v > v_0$. Here, $\chi^2_{n-s;1-\alpha}$ is the α-fractile of the χ^2_{n-s}-distribution. The right-sided test problem is similar, and the two-sided case can be treated as in Problem 10.18.

Proof. Statement (a) is analogous to Corollary (12.19b) and can be proved in the same way. Claims (b) and (c) follow immediately from statements (12.17d) and (12.17b), respectively. ◇

Note that the null hypotheses of the t-tests in part (a) and the F-tests in part (b) above are sharp, so that Remark (10.24) applies, and the 'true' test problem is determined by the associated power function. To calculate the power function of an F-test, one needs to know the distribution of $F_{\mathbf{H},\mathbf{L}}$ in the case $\mathsf{A}\gamma \notin \mathbf{H}$ as well. As the proof of Theorem (12.17) and Problem 9.9 show, the test statistic $|\Pi_{\mathbf{L}} X - \Pi_{\mathbf{H}} X|^2/v$ then has the non-central χ^2_{r-s}-distribution with non-centrality parameter $\lambda = |\mathsf{A}\gamma - \Pi_{\mathbf{H}}\mathsf{A}\gamma|^2/v$. Likewise, $F_{\mathbf{H},\mathbf{L}}$ (whose numerator and denominator are still independent) then has the non-central $\mathcal{F}_{s-r,n-s}$-distribution in the sense of Problem 9.14 for this λ. This means that the power function of an F-test can be viewed as a function of λ; cf. Figure 12.3. (Note that $\sqrt{\lambda}$ is simply the distance of the mean vector $\mathsf{A}\gamma$ from $\mathbf{H}$, scaled by $\sqrt{v}$.) The power function of a t-test as in part (a) above can be determined in the same way by means of the non-central t_{n-s}-distributions as defined in Problem 9.15. Finally, the family of non-central $\mathcal{F}_{s-r,n-s}$-distributions with varying non-centrality parameter can be shown to have increasing likelihood ratios. From this one can deduce (in analogy with Theorem (10.10)) that the F-test is

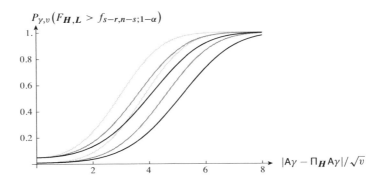

Figure 12.3. Power functions of the F-tests for $s = 20$, $n = 50$, $r = \dim H = 5, 10, 15$ (from black to light grey), and size $\alpha = 0.05$ and 0.01, respectively.

a uniformly most powerful test within the class of all tests that are invariant under all linear transformations of $\mathbb{R}^n$ that map the subspaces L and H onto themselves. More details can be found, for instance, in [20, 42].

We will now apply the previous corollaries to the special choices of A discussed in Section 12.2.

(12.21) Example. *Gaussian product model, see* (12.10). Let $s = 1$, $A = 1$, and $\gamma = m \in \mathbb{R}$, so that $P_{m,v} = \mathcal{N}_{m,v}{}^{\otimes n}$. Then, $A^\top A = n$. For $c = 1$, the t-test in Corollary (12.20a) reduces to the two-sided t-test of the null hypothesis $H_0 : m = m_0$, which was considered in Theorem (10.21).

(12.22) Example. *Simple linear regression, see* (12.11). Let $s = 2$ and $A = (1\,t)$ for a regression vector $t = (t_1, \ldots, t_n)^\top \in \mathbb{R}^n$ with $V(t) > 0$. Then we obtain

$$A^\top A = \begin{pmatrix} n & \sum_{k=1}^n t_k \\ \sum_{k=1}^n t_k & \sum_{k=1}^n t_k^2 \end{pmatrix},$$

so $\det A^\top A = n^2 V(t)$ and thus

$$(A^\top A)^{-1} = \frac{1}{nV(t)} \begin{pmatrix} \sum_{k=1}^n t_k^2/n & -M(t) \\ -M(t) & 1 \end{pmatrix}.$$

This in turn implies that

$$\hat{\gamma} = (A^\top A)^{-1} A^\top X = (A^\top A)^{-1} \begin{pmatrix} \sum_{k=1}^n X_k \\ \sum_{k=1}^n t_k X_k \end{pmatrix}$$

$$= \frac{1}{V(t)} \begin{pmatrix} M(X) \sum_{k=1}^n t_k^2/n - M(t) \sum_{k=1}^n t_k X_k/n \\ -M(X)M(t) + \sum_{k=1}^n t_k X_k/n \end{pmatrix}$$

$$= \begin{pmatrix} M(X) - c(t, X)\, M(t)/V(t) \\ c(t, X)/V(t) \end{pmatrix} = \begin{pmatrix} \hat{\gamma}_0 \\ \hat{\gamma}_1 \end{pmatrix},$$

in accordance with Theorem (12.5). In particular,

$$V^* = |X - \hat{\gamma}_0 \mathbf{1} - \hat{\gamma}_1 t|^2/(n-2).$$

If we substitute these results into the Corollaries (12.19b) and (12.20a), we obtain confidence intervals and tests for linear characteristics of γ, such as the slope parameter γ_1 or an interpolated value $\gamma_0 + \gamma_1 u$, namely, the value of the response variable at a point u. (In Example (12.1), γ_1 is the thermal expansion coefficient and $\gamma_0 + \gamma_1 u$ the length of the metal bar at the temperature u. Alternatively, one can think of the performance of a motor in dependence on the number of revolutions per minute; then $\gamma_0 + \gamma_1 u$ is the engine power when the number of revolutions is equal to some norm value u.)

▷ *The slope parameter.* For $c = (0, 1)^\top$, it follows from (12.19b) and (12.20a) that the random interval

$$] \hat{\gamma}_1 - t_{n-2;1-\alpha/2} \sqrt{V^*/nV(t)}, \; \hat{\gamma}_1 + t_{n-2;1-\alpha/2} \sqrt{V^*/nV(t)} [$$

is a confidence interval for γ_1 with error level α, and

$$\left\{ |\hat{\gamma}_1 - m_0| > t_{n-2;1-\alpha/2} \sqrt{V^*/nV(t)} \right\}$$

is the rejection region of a level-α test of the null hypothesis $H_0 : \gamma_1 = m_0$.

▷ *Interpolated values.* For $u \in \mathbb{R}$ we find, setting $c = (1, u)^\top$,

$$\begin{aligned} c^\top (A^\top A)^{-1} c &= \left(\sum_{k=1}^n t_k^2/n - 2u M(t) + u^2 \right)\big/ nV(t) \\ &= \left(V(t) + M(t)^2 - 2u M(t) + u^2 \right)\big/ nV(t) \\ &= \frac{1}{n} + \frac{(M(t) - u)^2}{nV(t)}. \end{aligned}$$

Hence,

$$\left\{ |\hat{\gamma}_0 + \hat{\gamma}_1 u - m_0| > t_{n-2;1-\alpha/2} \sqrt{V^*} \sqrt{\frac{1}{n} + \frac{(M(t)-u)^2}{nV(t)}} \right\}$$

is the rejection region of a level-α test of the null hypothesis $H_0 : \gamma_0 + \gamma_1 u = m_0$.

(12.23) Example. *Multiple linear and polynomial regression, see* (12.12) *and* (12.13). In the case of polynomial regression, it is natural to ask whether the data can actually be described by a regression polynomial of a degree ℓ that is less than the maximal degree d considered a priori. Similarly, in the case of multiple linear regression, one might wonder whether, in fact, some of the influencing factors (say those with an index $i > \ell$) can be ignored. These questions lead to the null hypothesis

$$H_0 : \gamma_{\ell+1} = \gamma_{\ell+2} = \cdots = \gamma_d = 0$$

for $\ell < d$. This hypothesis can be tested by an F-test. Namely, let

$$\boldsymbol{H} = \{\mathsf{A}\gamma : \gamma = (\gamma_0, \ldots, \gamma_\ell, 0, \ldots, 0)^\top, \gamma_0, \ldots, \gamma_\ell \in \mathbb{R}\}$$

be the span of the first $\ell + 1$ columns of A. Then the null hypothesis takes the form $H_0 : \mathsf{A}\gamma \in \boldsymbol{H}$. Writing

$$\mathsf{B} = \begin{pmatrix} 1 & t_{1,1} & \cdots & t_{1,\ell} \\ \vdots & \vdots & & \vdots \\ 1 & t_{n,1} & \cdots & t_{n,\ell} \end{pmatrix}$$

for the 'left part' of A, we see that $\boldsymbol{H} = \{\mathsf{B}\beta : \beta \in \mathbb{R}^{\ell+1}\}$, and therefore $\Pi_{\boldsymbol{H}} X = \mathsf{B}(\mathsf{B}^\top \mathsf{B})^{-1}\mathsf{B}^\top X$. This expression can be substituted into the definition (12.18) of $F_{\boldsymbol{H},\boldsymbol{L}}$, and one can apply the test described in Corollary (12.20b).

We conclude this section with an application of polynomial regression.

(12.24) Example. *Climate time series.* In many places of the world, the average monthly temperatures have been recorded for a long time. Is it possible to recognise any trends in these data sets? One possibility is to test the properties of the corresponding regression polynomials. Let us prescribe a significance level $\alpha = 0.05$ and let us consider, for instance, the average August temperatures in Karlsruhe (a city in the German south-west) for the years 1800 until 2006, which can be found under `www.klimadiagramme.de/Europa/special01.htm`. Since there are no data for the years 1854 and 1945, the vector $x = (x_1, \ldots, x_n)$ of observed temperatures has the length $n = 205$. We will measure time in centuries and start in the year 1800; hence the regression vector is $t = (0, 0.01, \ldots, \widehat{0.54}, \ldots, \widehat{1.45}, \ldots, 2.06)^\top$, from which the values 0.54 and 1.45 have been excluded. Figure 12.4 shows the data points (t_k, x_k) for $k = 1$ to n and the corresponding regression polynomials p_d of degrees $d = 1$ to 4, which can be conveniently calculated with the help of appropriate software. In `Mathematica`, for instance, one can obtain p_3 with the command `Fit[list,{1,t,t^2,t^3},t]`, where `list` represents the list of the n points. For example, $\mathsf{p}_1(\tau) \approx 18.7 + 0.1\,\tau$ and $\mathsf{p}_3(\tau) \approx 19.5 - 0.5\,\tau - 2.0\,\tau^2 + 1.3\,\tau^3$.

What can we infer from these results? First of all, it is plausible that the random fluctuations of the average August temperatures in different years are mutually independent. Also, the assumption of normal distributions seems to be justifiable, at least in a first approximation. We are then dealing with a Gaussian regression model. On the other hand, there is hardly any evidence of the appropriate degree of the regression polynomial (apart from the requirement that the degree should be as small as possible, so that long-term trends can be recognised). First of all, let us consider

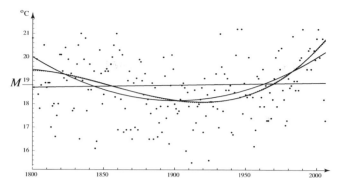

Figure 12.4. Scatter plot of the average August temperatures in Karlsruhe between 1800 and 2006, overall average M, and the regression polynomials p_d of degrees $d = 1$ to 4. p_4 is dotted (and almost indistinguishable from p_3).

the regression line p_1. Can we deduce from its slight increase that the temperatures are increasing on average? If we denote the slope parameter by γ_1, then this question leads to the one-sided test problem $H_0 : \gamma_1 \leq 0$ against $H_1 : \gamma_1 > 0$. By Corollary (12.20a) and Example (12.22), a suitable test is the one-sided t-test with rejection region $\{\hat{\gamma}_1 > t_{n-2,1-\alpha}\sqrt{V^*/nV(t)}\}$. Table C in the appendix yields the value $t_{n-2,1-\alpha} = 1.65$, and one can calculate that $1.65\sqrt{V^*(x)/nV(t)} = 0.29$. This is larger than the estimated value $\hat{\gamma}_1(x) = 0.1$, the leading coefficient of p_1. Therefore, the null hypothesis $H_0 : \gamma_1 \leq 0$ can *not* be rejected.

This result is not really surprising: The scatter plot in Figure 12.4 shows a temperature decrease in the 19th century followed by an increase, which is a shape that cannot be described by a straight line. But since p_4 is hardly distinguishable from p_3, one can assume that a polynomial of fourth (or even third) degree should be sufficient to capture the properties of the scatter plot. So let us set $d = 4$ and let us first consider the null hypothesis $H_0' : \gamma_2 = \gamma_3 = \gamma_4 = 0$ that the scatter plot is based on a linear increase. According to Example (12.23), H_0' can be checked by an F-test. Writing L_ℓ for the subspace of $\mathbb{R}^n$ generated by the vectors $\mathbf{1}, t, (t_k^2), \dots, (t_k^\ell)$, we find

$$F_{L_1,L_4}(x) = \frac{n-5}{5-2}\frac{\sum_{k=1}^n(\mathsf{p}_4(t_k)-\mathsf{p}_1(t_k))^2}{\sum_{k=1}^n(x_k-\mathsf{p}_4(t_k))^2} = 14.26 > 2.65 = f_{3,200;0.95}.$$

(The last value is taken from Table D in the appendix. Note also that, for each ℓ, the vector $(\mathsf{p}_\ell(t_k))_{1 \leq k \leq n}$ coincides with the projection $\Pi_{L_\ell}x$.) This leads to a clear rejection of the null hypothesis H_0', and confirms that a regression *line* is indeed insufficient. However, $F_{L_3,L_4}(x) = 0.03 < 3.89 = f_{1,200;0.95}$, so that the null hypothesis $\gamma_4 = 0$ is accepted. In the following, we therefore ignore the fourth power and apply polynomial regression of degree $d = 3$. Then, $F_{L_2,L_3}(x) = 4.78 > 3.89 = f_{1,201;0.95}$, so that the null hypothesis $\gamma_3 = 0$ is rejected. This means that the cubic part in the regression polynomial is significant for an adequate description of the scatter plot. In other words, p_3 is significantly better than p_2. But p_3 gives a clear indication that the recent increase of temperature is considerably stronger than the decrease in the 19th century; see also Problem 12.10. As a final comment, however, it must be emphasised that the above discussion is purely phenomenological. Moreover, it is only based on a single time series. It is the task of climatologists to put this analysis into context and to identify the possible causes. More about the recent climate change discussion can be found, for instance, under `www.ipcc.ch`.

12.4 Analysis of Variance

What is the influence of a certain causal factor on a random phenomenon? This is the central question of the analysis of variance, a technique of comparing the effects of various experimental specifications. We will explain this method in the context of a classical example.

(12.25) Example. *Influence of fertilisers on the crop yield.* How much does a certain fertiliser influence the crop yield when compared with other fertilisation methods? Let G be the finite set of different types of fertilisers that are to be compared with each other, say $G = \{1, \dots, s\}$. To obtain statistically relevant statements, each fertiliser $i \in G$ is applied to $n_i \geq 2$ different lots $L_{i1}, \dots, L_{in_i}$ of farm land. One assumes that the (random) yield X_{ik} of lot L_{ik} is given by the formula

$$X_{ik} = m_i + \sqrt{v}\, \xi_{ik}, \quad i \in G,\ 1 \leq k \leq n_i .$$

That is, X_{ik} consists of a deterministic part m_i, the unknown average yield when the fertiliser $i \in G$ is used, and a random perturbation $\sqrt{v}\, \xi_{ik}$ caused by the weather and other factors.

The abstract scheme is the following. For the factor 'fertilisation', one distinguishes $s = |G|$ different levels. At level $i \in G$, the factor has a certain effect $m_i \in \mathbb{R}$, which is, however, perturbed by random noise. In order to determine this effect, one makes $n_i \geq 2$ observations X_{ik} with expectation m_i. These observations form the ith observation group, or sample. The resulting *s-sample problem* can be summarised as follows:

Group	Observations	Expectation
1	$X_{11}, \dots, X_{1n_1}$	m_1
2	$X_{21}, \dots, X_{2n_2}$	m_2
$\vdots$	$\vdots$	$\vdots$
s	$X_{s1}, \dots, X_{sn_s}$	m_s

The full observation vector is thus

$$X = (X_{ik})_{ik \in B} \quad \text{with } B = \{ik : i \in G, 1 \leq k \leq n_i\},$$

which we can think of as being arranged as

$$X = (X_{11}, \ldots, X_{1n_1}, X_{21}, \ldots, X_{2n_2}, \ldots)^\top,$$

so that it is a random column vector in $\mathbb{R}^n$, where $n = \sum_{i \in G} n_i$.

The unknown parameters are the vector $\gamma = (m_i)_{i \in G} \in \mathbb{R}^s$ of group expectations, and the error variance $v > 0$ (which, by assumption, does not depend on the group $i \in G$). The expectation vector

$$(12.26) \qquad \mathbb{E}(X) = (\underbrace{m_1, \ldots, m_1}_{n_1}, \underbrace{m_2, \ldots, m_2}_{n_2}, \ldots, \underbrace{m_s, \ldots, m_s}_{n_s})^\top$$

of X can be written as $\mathbb{E}(X) = A\gamma$ for short, with the $n \times s$ matrix

$$A = (\delta_{ij})_{ik \in B,\, j \in G},$$

which reads explicitly

$$(12.27) \qquad A = \begin{pmatrix} \begin{matrix} 1 \\ \vdots \\ 1 \end{matrix} & & & \\ & \begin{matrix} 1 \\ \vdots \\ 1 \end{matrix} & & \\ & & \ddots & \\ & & & \begin{matrix} 1 \\ \vdots \\ 1 \end{matrix} \end{pmatrix} \begin{matrix} \left.\vphantom{\begin{matrix}1\\1\\1\end{matrix}}\right\} n_1 \\ \left.\vphantom{\begin{matrix}1\\1\\1\end{matrix}}\right\} n_2 \\ \vdots \\ \left.\vphantom{\begin{matrix}1\\1\\1\end{matrix}}\right\} n_s \end{matrix}$$

with zeroes everywhere else. So, our model for finding the effects of different fertilisers turns out to be a linear model with a specific design matrix A.

Definition. The *model for the analysis of variance* consists of

▷ a finite set G of $s := |G| \geq 2$ different observation groups,

▷ a number $n_i \geq 2$ of observations for each group $i \in G$, and the corresponding index set $B = \{ik : i \in G, 1 \leq k \leq n_i\}$ with cardinality $n = |B| = \sum_{i \in G} n_i$,

▷ an unknown vector $\gamma = (m_i)_{i \in G}$ of mean values m_i in group $i \in G$, an unknown scaling parameter $v > 0$ (which does not depend on the group), and pairwise uncorrelated standardised noise terms ξ_{ik}, $ik \in B$, and finally

▷ the observation variables

$$X_{ik} = m_i + \sqrt{v}\, \xi_{ik}, \quad ik \in B.$$

It is given by the linear model with the $n \times s$ design matrix A defined in (12.27). A common acronym for 'analysis of variance' is ANOVA.

To apply the general results obtained for the linear model, we need to analyse the design matrix A in (12.27). Let us collect a number of facts.

(a) The columns of A are pairwise orthogonal, and their span $L = L(A)$ is

$$L = \{x = (x_{ij})_{ij \in B} \in \mathbb{R}^n : x_{ij} = x_{i1} \text{ for all } ij \in B\}.$$

(b) One has

$$A^\top A = \begin{pmatrix} n_1 & & \\ & \ddots & \\ & & n_s \end{pmatrix},$$

where all the off-diagonal entries are zero.

(c) The observation vector $X = (X_{ik})_{ik \in B}$ satisfies

$$A^\top X = (n_1 M_1, \dots, n_s M_s)^\top,$$

where

$$M_i = \frac{1}{n_i} \sum_{k=1}^{n_i} X_{ik}$$

is the *sample mean of group i*.

(d) The unbiased estimator $\hat{\gamma}$ of $\gamma = (m_i)_{i \in G}$ is given by

$$\hat{\gamma} = (A^\top A)^{-1} A^\top X = \begin{pmatrix} 1/n_1 & & \\ & \ddots & \\ & & 1/n_s \end{pmatrix} \begin{pmatrix} n_1 M_1 \\ \vdots \\ n_s M_s \end{pmatrix} = \begin{pmatrix} M_1 \\ \vdots \\ M_s \end{pmatrix},$$

viz. the vector of the sample means within the groups. This is hardly surprising!

(e) For the unbiased variance estimator V^*, we find that

$$V^* = \frac{1}{n-s} |X - \Pi_L X|^2 = \frac{1}{n-s} |X - A(M_1, \dots, M_s)^\top|^2$$

$$= \frac{1}{n-s} |X - (\underbrace{M_1, \dots, M_1}_{n_1}, \underbrace{M_2, \dots, M_2}_{n_2}, \dots)^\top|^2$$

$$= \frac{1}{n-s} \sum_{ik \in B} (X_{ik} - M_i)^2.$$

Letting

$$V_i^* = \frac{1}{n_i - 1} \sum_{k=1}^{n_i} (X_{ik} - M_i)^2$$

denote the unbiased variance estimator within group i, we can rewrite the preceding formula as

(12.28) $$V^* = V_{\mathrm{wg}}^* := \frac{1}{n-s} \sum_{i \in G} (n_i - 1) V_i^*.$$

V_{wg}^* is called the *average sample variance within the groups*.

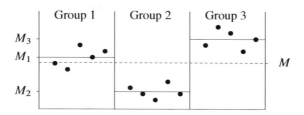

Figure 12.5. Comparison of the group means M_i and the total mean M.

(f) V_{wg}^* must be distinguished from the *total sample variance*

$$V_{tot}^* = \frac{1}{n-1} \sum_{ik \in B} (X_{ik} - M)^2 = \frac{|X - M\,\mathbf{1}|^2}{n-1},$$

where $M = \frac{1}{n} \sum_{ik \in B} X_{ik} = \frac{1}{n} \sum_{i \in G} n_i M_i$ denotes the *total sample mean*. By Pythagoras' theorem, we obtain the equation

$$|X - M\,\mathbf{1}|^2 = |X - \Pi_L X|^2 + |\Pi_L X - M\,\mathbf{1}|^2,$$

cf. Figure 12.2. This is equivalent to the *partitioning of variation*

$$(n-1)\,V_{tot}^* = (n-s)\,V_{wg}^* + (s-1)\,V_{bg}^*.$$

Here,

(12.29) $$V_{bg}^* = \frac{1}{s-1} \sum_{i \in G} n_i\,(M_i - M)^2 = \frac{|\Pi_L X - M\,\mathbf{1}|^2}{s-1}$$

is the *sample variance between the groups*, namely the empirical variance of the group means. (Note the weighting by the number of observations n_i in group $i \in G$.) Figure 12.5 illustrates the meaning of the different parts of V_{tot}^*. If the true group expectations $m_1, \ldots, m_s$ do not all agree (as in the figure), then the variation of the different group means adds up to the 'natural' fluctuations caused by the noise ξ_{ik}. Correspondingly, V_{tot}^* is not an unbiased estimator of v. Explicitly, we obtain the following.

(12.30) Remark. *Expected total sample variance.* For all $\gamma = (m_i)_{i \in G}$ and $v > 0$,

$$\mathbb{E}_{\gamma,v}(V_{tot}^*) = v + \frac{1}{n-1} \sum_{i \in G} n_i\,(m_i - \overline{m})^2,$$

where $\overline{m} = \frac{1}{n} \sum_{i \in G} n_i\,m_i$. Consequently, $\mathbb{E}_{\gamma,v}(V_{tot}^*) = v$ if and only if all group expectations m_i coincide.

Proof. Let $H = \{x \in \mathbb{R}^n : x_1 = \cdots = x_n\}$ be the subspace of $\mathbb{R}^n$ generated by the diagonal vector $\mathbf{1}$, and $H^\perp$ its orthogonal complement. Then $M\mathbf{1} = \Pi_H X$. Using (12.9), we can then write

$$
(n-1)\, V_{\mathrm{tot}}^* = |X - \Pi_H X|^2 = |\Pi_{H^\perp} X|^2
$$
$$
= |\Pi_{H^\perp} A\gamma|^2 + v\, |\Pi_{H^\perp} \xi|^2 + 2\sqrt{v}\, \gamma^\mathsf{T} A^\mathsf{T} \Pi_{H^\perp} \xi .
$$

Applying Theorem (12.15c) to H in place of L and using $\mathbb{E}(\xi) = 0$, we can conclude that

$$
(n-1)\, \mathbb{E}_{\gamma,v}(V_{\mathrm{tot}}^*) = |\Pi_{H^\perp} A\gamma|^2 + v(n-1) .
$$

Since $A\gamma$ coincides with the vector (12.26), the claim follows. ◇

Now, the preceding list of facts allows us to apply the results obtained for the linear model. We restrict our attention to the case that the noise terms ξ_{ik} are independent and standard normal. So we are in the situation of the *Gaussian linear model*, and we can apply Corollaries (12.19) and (12.20). Instead of repeating all statements, we rather present a couple of typical special cases as examples.

(12.31) Example. *Confidence ellipsoid for the vector of group expectations.* Since $A\gamma$ is given by (12.26) and $A\hat{\gamma}$ by the analogous vector of empirical group means, we find that

$$
|A(\hat{\gamma} - \gamma)|^2 = \sum_{i \in G} n_i\, (M_i - m_i)^2 .
$$

We also know that the variance estimator V^* coincides with V_{wg}^*, the sample variance within the groups defined in (12.28). Applying Corollary (12.19a), we thus conclude that the random ellipsoid

$$
C(\cdot) = \left\{ (m_i)_{i \in G} \in \mathbb{R}^s : \tfrac{1}{s} \sum_{i \in G} n_i (m_i - M_i)^2 < f_{s,n-s;1-\alpha}\, V_{\mathrm{wg}}^* \right\}
$$

is a confidence region for $\gamma = (m_i)_{i \in G}$ with error level α. Note that this ellipsoid is centred at the vector of empirical group means, and its halfaxis in direction $i \in G$ is proportional to $1/\sqrt{n_i}$. This illustrates again the 'square root effect' of the number of observations on the precision of the result.

(12.32) Example. *t-test for the two-sample problem.* Suppose we want to decide whether two fertilisers perform equally well. Then the null hypothesis $H_0 : m_1 = m_2$ must be tested against the alternative $H_1 : m_1 \neq m_2$. So we have $s = 2$. In terms of the vector $c = (1, -1)^\mathsf{T}$, the null hypothesis takes the form $H_0 : c \cdot \gamma = 0$. Corollary (12.20a) therefore applies, showing that the rejection region

$$
\left\{ |M_1 - M_2| > t_{n-2;1-\alpha/2}\, \sqrt{\left(\tfrac{1}{n_1} + \tfrac{1}{n_2}\right) V_{\mathrm{wg}}^*} \right\}
$$

defines a suitable level-α test; you might remember this from Problem 10.24.

(12.33) Example. *F-test in the multi-sample problem.* Suppose now that more than two, say s, types of fertilisers shall be compared. Then it is not advisable to perform the $\binom{s}{2}$ tests for each of the null hypotheses $H_0^{ii'} : m_i = m_{i'}$ with $i \neq i'$. This is because the error probabilities would then add up; if α were chosen small enough to compensate for this, the power would become too small. Instead, one should consider the linear subspace $\boldsymbol{H} = \{m\mathbf{1} : m \in \mathbb{R}\}$ of $\mathbb{R}^n$. Fact (a) above shows that $\boldsymbol{H}$ is a subspace of $\boldsymbol{L}$, and the null hypothesis $H_0 : m_1 = \cdots = m_s$ is equivalent to $H_0 : A\gamma \in \boldsymbol{H}$. According to fact (f), the associated Fisher statistic $F_{\boldsymbol{H},\boldsymbol{L}}$ from (12.18) coincides with the ratio $V_{\mathrm{bg}}^* / V_{\mathrm{wg}}^*$. Corollary (12.20b) therefore implies that the test of the null hypothesis $H_0 : m_1 = \cdots = m_s$ with rejection region

$$\{V_{\mathrm{bg}}^* > f_{s-1,n-s;1-\alpha}\, V_{\mathrm{wg}}^*\}$$

has size α. When applying this procedure to concrete data, it is convenient to summarise all relevant quantities in a so-called ANOVA table as in Table 12.1.

Table 12.1. ANOVA table. 'df' stands for 'degrees of freedom'. By partitioning of variation, one has $S_{\mathrm{bg}} + S_{\mathrm{wg}} = S_{\mathrm{tot}}$, and likewise for the associated degrees of freedom.

	df	sums of squares	variances	F-ratio
between	$s-1$	$S_{\mathrm{bg}} = \sum_{i \in G} n_i\,(M_i - M)^2$	$V_{\mathrm{bg}}^* = S_{\mathrm{bg}}/(s-1)$	
within	$n-s$	$S_{\mathrm{wg}} = \sum_{i \in G} (n_i - 1)\,V_i^*$	$V_{\mathrm{wg}}^* = S_{\mathrm{wg}}/(n-s)$	$V_{\mathrm{bg}}^*/V_{\mathrm{wg}}^*$
total	$n-1$	$S_{\mathrm{tot}} = \sum_{ik \in B} (X_{ik} - M)^2$	$V_{\mathrm{tot}}^* = S_{\mathrm{tot}}/(n-1)$	

So far in this section, we have considered the influence of a single factor on a random phenomenon; this case is known as the *one-factor ANOVA*, or the *one-way layout*. But, of course, one can also ask whether and how the crop yield depends on the influence of several factors, as for instance fertilisation, sowing date, and the humidity of the soil. This question leads to the *multi-factor analysis of variance*, which does in fact fit into the framework considered so far. For simplicity, we confine ourselves to the case of two factors only.

(12.34) Example. *Two-factor analysis of variance, or the two-way layout.* Consider two factors, say 1 and 2. Each factor is divided into finitely many levels, which form two sets G_1 and G_2 with cardinalities $|G_1| = s_1$ and $|G_2| = s_2$. For example, G_1 is the set of fertilisation methods considered, and G_2 the set of the calender weeks in which part of the seeds are sown. The Cartesian product

$$G = G_1 \times G_2 = \{ij : i \in G_1, j \in G_2\}$$

is then the set of different observation groups; its cardinality is $s = |G| = s_1 s_2$. With G defined in this way, one can work exactly as before. However, we need to confine ourselves to the 'balanced' case, in which the number of observations is the same for all 'cells' $ij \in G$. (For the problems that arise in the unbalanced case see [1], for example.) So, let us assume we make $\ell \geq 2$ observations for each cell $ij \in G$. This means that the ensemble of all observations is indexed by the set $B = \{ijk : ij \in G, 1 \leq k \leq \ell\}$ of cardinality $n := |B| = \ell s$. The complete observation vector is thus $X = (X_{ijk})_{ijk \in B}$.

The point is now that the product structure of G leads to new kinds of test hypotheses, which concern the individual effects and the interaction of the two factors. Specifically, we set

$$\overline{m} = \frac{1}{s} \sum_{ij \in G} m_{ij}, \quad \overline{m}_{i\bullet} = \frac{1}{s_2} \sum_{j \in G_2} m_{ij}, \quad \overline{m}_{\bullet j} = \frac{1}{s_1} \sum_{i \in G_1} m_{ij}.$$

The difference $\alpha_i := \overline{m}_{i\bullet} - \overline{m}$ then describes the ith row effect, namely the influence that factor 1 has on the crop yield if it is in state $i \in G_1$. Likewise, $\beta_j := \overline{m}_{\bullet j} - \overline{m}$ is the jth column effect, i.e., the influence of factor 2 if it is in state $j \in G_2$. Finally,

$$\gamma_{ij} := (m_{ij} - \overline{m}) - \alpha_i - \beta_j = m_{ij} - \overline{m}_{i\bullet} - \overline{m}_{\bullet j} + \overline{m}$$

is the interactive effect between the two factors in the joint state ij. These three effects add up to the total effect of the factors, since

$$m_{ij} = \overline{m} + \alpha_i + \beta_j + \gamma_{ij}$$

for all $ij \in G$. Now, the following null hypotheses are of interest:

$$H_0^1 : \alpha_i = 0 \quad \text{for all } i \in G_1,$$

which asserts that factor 1 does, in fact, not influence the crop yield; the analogous hypothesis for factor 2; and, in particular, the hypothesis

$$\widetilde{H}_0 : \gamma_{ij} = 0 \quad \text{for all } ij \in G,$$

which states that the effects of the two factors simply add up, so that there is no interaction between the two factors. How can one test these hypotheses?

Consider the vector $\vec{m} = (m_{ij})_{ij \in G}$. The null hypothesis H_0^1 then takes the form $A\vec{m} \in H$ for a subspace H of L of dimension $\dim H = s - s_1 + 1$. (Note that one of the equations in H_0^1 is redundant because $\sum_{i \in G_1} \alpha_i = 0$; so, $\vec{m} \in \mathbb{R}^s$ is restricted by $s_1 - 1$ equations.) We thus need to determine the corresponding F-statistic $F_{H,L}$. As before, $\Pi_L X = (M_{ij})_{ijk \in B}$ for

$$M_{ij} = \frac{1}{\ell} \sum_{k=1}^{\ell} X_{ijk}.$$

On the other hand, one can easily check that $\Pi_H X = (M_{ij} - M_{i\bullet} + M)_{ijk \in B}$ for

$$(12.35) \qquad M_{i\bullet} = \frac{1}{s_2} \sum_{j \in G_2} M_{ij} = \frac{1}{\ell s_2} \sum_{j \in G_2, 1 \le k \le \ell} X_{ijk} \,.$$

Consequently, we obtain that

$$|\Pi_L X - \Pi_H X|^2 = (s_1 - 1)\, V^*_{\mathrm{bg1}} := \ell s_2 \sum_{i \in G_1} (M_{i\bullet} - M)^2 \,.$$

In analogy to (12.29), V^*_{bg1} is the *empirical variance between the groups of factor 1.* The empirical variance within the groups, or cells, is, as in (12.28), given by

$$V^*_{\mathrm{wg}} = \frac{1}{n-s} \sum_{ijk \in B} (X_{ijk} - M_{ij})^2 \,,$$

where in fact $n - s = (\ell - 1) s_1 s_2$. By (12.18), it now follows that $F_{H,L} = V^*_{\mathrm{bg1}} / V^*_{\mathrm{wg}}$. Under the Gaussian assumption, we also know that $F_{H,L}$ has the Fisher distribution $\mathcal{F}_{s_1 - 1, (\ell - 1) s_1 s_2}$. Hence, the rejection region

$$\{ V^*_{\mathrm{bg1}} > f_{s_1 - 1, (\ell - 1) s_1 s_2; 1 - \alpha}\, V^*_{\mathrm{wg}} \}$$

defines a level-α test of the hypothesis H^1_0 : *'The crop yield depends on factor 2 only'.*

The null hypothesis $\widetilde{H}_0$ of no interaction can also be described by a linear subspace $\widetilde{H}$ of L; its dimension is $\dim \widetilde{H} = s - (s_1 - 1)(s_2 - 1) = s_1 + s_2 - 1$. (Indeed, since $\gamma_{i\bullet} = \gamma_{\bullet j} = 0$ for all $ij \in G$, $\widetilde{H}_0$ merely requires the validity of the $(s_1 - 1)(s_2 - 1)$ equations $\gamma_{ij} = 0$ with $ij \in \{2, \ldots, s_1\} \times \{2, \ldots, s_2\}$.) The corresponding F-statistic (12.18) is given by $F_{\widetilde{H},L} = V^*_{\mathrm{bg1\text{-}2}} / V^*_{\mathrm{wg}}$, where

$$V^*_{\mathrm{bg1\text{-}2}} = \frac{\ell}{(s_1 - 1)(s_2 - 1)} \sum_{ij \in G} (M_{ij} - M_{i\bullet} - M_{\bullet j} + M)^2$$

and $M_{\bullet j}$ is defined in obvious analogy to (12.35). Under the Gaussian assumption for the error terms, we therefore see that

$$\{ V^*_{\mathrm{bg1\text{-}2}} > f_{(s_1 - 1)(s_2 - 1), (\ell - 1) s_1 s_2; 1 - \alpha}\, V^*_{\mathrm{wg}} \}$$

is the rejection region of a level-α test of the null hypothesis $\widetilde{H}_0$: *'Regarding the crop yield, the factors 1 and 2 do not influence each other'.*

We conclude this section with an application of the two-factor analysis of variance.

(12.36) Example. *The combined influence of a drug and alcohol on the reaction time.* For a certain drug, it is to be determined whether it influences the fitness to drive, either alone or in combination with alcohol. Thus, 24 candidates are divided into groups of four, and classified according to the levels of factor 1, the drug (medication yes or no), and factor 2, the blood alcohol content. Finally they take a reaction test, say with the results listed in Table 12.2. What can one deduce from this outcome? To get a first idea, one calculates the group averages and factor averages; these are given in Table 12.3.

The last column, which contains the group means of factor 1, indicates a noticeable influence of the drug, and the last row a clear influence of alcohol. However, which of these differences are significant? To find out, one calculates the characteristics of the two-factor analysis of variance from Example (12.34) and collects the results in an ANOVA table analogous to Table 12.1. For the given data, one obtains Table 12.4. In practice, such a table is calculated using suitable statistics software such as R, S-Plus, SPSS, or XPloRe.

Comparing the last two columns of this ANOVA table, we arrive at the following conclusions at the significance level 0.05 (provided we accept that the measurement errors are normally distributed with equal variances, which is fairly plausible in this situation). Since $2.35 < 4.41$, the data do not allow to reject the null hypothesis that the drug alone does not influence the reaction time. However, the influence of alcohol is highly significant, and there is also a significant interaction between the drug and alcohol. Table 12.3 shows that the two factors enhance each other.

Table 12.2. Reaction times (in hundredth seconds) for 24 test persons, classified according to medication and blood alcohol content (BAC).

medication	BAC per mil		
	0.0	0.5	1.0
no	23, 21, 20, 19	22, 25, 24, 25	24, 25, 22, 26
yes	22, 19, 18, 20	23, 22, 24, 28	26, 28, 32, 29

Table 12.3. Group and factor averages of the data in Table 12.2.

medication	BAC per mil			$M_{i\bullet}$
	0.0	0.5	1.0	
no	20.75	24	24.25	23
yes	19.75	24.25	28.75	24.25
$M_{\bullet j}$	20.25	24.125	26.5	$M = 23.625$

Table 12.4. Two-factor ANOVA table for the data from Table 12.2.

	df	sums S	variances V^*	F-values	5% F-fractiles
bg1	1	9.375	9.375	2.35	4.41
bg2	2	159.25	79.625	19.98	3.55
bg1-2	2	33.25	16.625	4.17	3.55
wg	18	71.75	3.986		

Problems

12.1 How does the mortality caused by malignant melanomas depend on the sunshine intensity? To answer this question, the mortality was recorded for each state of the USA (in deaths per 10 millions of the white population between 1950 and 1969), together with the latitude. The following table contains the data for seven states:

State	Delaware	Iowa	Michigan	New Hampshire	Oklahoma	Texas	Wyoming
Mortality	200	128	117	129	182	229	134
Latitude	39	42	44	44	35	31	43

Determine the corresponding regression line. Which mortality would you expect in Ohio (latitude $40°$)?

12.2 Consider the simple linear regression model $X_k = \gamma_0 + \gamma_1 t_k + \sqrt{v}\,\xi_k$, $k = 1,\ldots,n$, and assume that the variance $v > 0$ is known. Show the following.

(a) If the error vector ξ has a multivariate standard normal distribution, then the least squares estimator $\hat{\gamma} = (\hat{\gamma}_0, \hat{\gamma}_1)$ is also a maximum likelihood estimator of $\gamma = (\gamma_0, \gamma_1)$.

(b) Defining the residual and regression variances by

$$V^*_{\text{resid}} = V^* = \frac{1}{n-2} \sum_{k=1}^{n} (X_k - \hat{\gamma}_0 - \hat{\gamma}_1 t_k)^2 \quad \text{and} \quad V^*_{\text{regr}} = \sum_{k=1}^{n} (\hat{\gamma}_0 + \hat{\gamma}_1 t_k - M)^2 ,$$

one obtains the *partitioning of variation*

$$(n-1)\, V^*_{\text{tot}} := \sum_{k=1}^{n} (X_k - M)^2 = (n-2)\, V^*_{\text{resid}} + V^*_{\text{regr}} .$$

(c) The statistic $T = \hat{\gamma}_1 \sqrt{nV(t)/V^*_{\text{resid}}}$, which can be used to test the null hypothesis $H_0 : \gamma_1 = 0$, can be written in the form $T^2 = V^*_{\text{regr}}/V^*_{\text{resid}}$.

12.3 [S] *Autoregressive model.* To describe a time evolution with deterministic growth and random errors, one often uses the following autoregressive model (of order 1):

$$X_k = \gamma\, X_{k-1} + \sqrt{v}\,\xi_k , \quad 1 \le k \le n .$$

Here, $\gamma \in \mathbb{R}$ and $v > 0$ are unknown parameters, $X_0, \ldots, X_n$ the observations, and $\xi_1, \ldots, \xi_n$ are independent random errors with $\mathbb{E}(\xi_k) = 0$, $\mathbb{V}(\xi_k) = 1$.

(a) Introduce an appropriate expression for the mean squared error and find the least squares estimator of γ.

(b) Let $X_0 = 0$ and suppose the error variables ξ_k are standard normally distributed. Show that the likelihood function of the model is given by

$$\rho_{\gamma,v} = (2\pi v)^{-n/2} \exp\left[-\sum_{k=1}^{n} (X_k - \gamma X_{k-1})^2/2v \right].$$

(c) Consider the test problem $H_0 : \gamma = 0$ ('no dependence') against $H_1 : \gamma \neq 0$. Show that, under the assumptions of part (b), the likelihood ratio is a monotone function of the modulus of the sample correlation coefficient

$$\hat{r} = \sum_{k=1}^{n} X_k X_{k-1} \Big/ \sqrt{\sum_{k=1}^{n} X_k^2} \sqrt{\sum_{k=1}^{n} X_{k-1}^2} \,.$$

12.4 *Simpson's paradox.* The long-term student Andy has contacted 8 of his former fellow students (with the same final grade) and has collected the duration of study (in semesters) and the starting salary (in € 1000):

Duration of study	10	9	11	9	11	12	10	11
Starting salary	35	35	34	36	41	39	40	38

He draws a regression line for the starting salary against the length of study and announces triumphantly: 'Studying longer leads to a higher starting salary!' His friend Beth doubts this and notices that the first four students in the table chose a different area of specialisation than the other four. Then she draws a regression line for each of the two groups and notices: 'Length of study and starting salary are negatively correlated!' Determine and draw the respective regression lines.

12.5[S] A spatially homogeneous force field with known direction and unknown strength f is to be studied. Therefore, a test object of mass 1 is placed into the field at time 0, and its spatial coordinates (in direction of the field) are measured at the times $0 < t_1 < t_2 < \cdots < t_n$. Introduce a suitable mean squared error and determine the least squares estimator of f in the following two cases: (a) starting position and speed of the test object are known, so without loss of generality equal to zero, (b) both are unknown.

12.6[S] *Autoregressive errors.* Consider the model $X_k = m + \sqrt{v}\,\xi_k$, $k = 1, \ldots, n$, for n real-valued observations with unknown expectation $m \in \mathbb{R}$ and unknown $v > 0$. For the errors, suppose that $\xi_k = \gamma \xi_{k-1} + \eta_k$; here $\gamma > 0$ is supposed to be known, $\xi_0 = 0$, and $\eta_1, \ldots, \eta_n$ are uncorrelated and standardised random variables in $\mathcal{L}^2$. Show the following:

(a) Besides the sample mean M,

$$S := \Big(X_1 + (1-\gamma)\sum_{k=2}^{n}(X_k - \gamma X_{k-1})\Big)\Big/\big(1 + (n-1)(1-\gamma)^2\big)$$

is another unbiased estimator of m. Note that $S \neq M$ because $\gamma \neq 0$.

(b) For all m and v we have $\mathbb{V}_{m,v}(S) < \mathbb{V}_{m,v}(M)$.

Hint: Write $B\xi = \eta$ for a suitable matrix B, and set up a linear model for $Y := BX$.

12.7 *F-test as a likelihood ratio test.* Consider the Gaussian linear model and let H be an r-dimensional hypothesis space. Show that the likelihood ratio

$$R = \sup_{\gamma,v:\, A\gamma \notin H} \phi_{A\gamma,v\mathsf{E}} \Big/ \sup_{\gamma,v:\, A\gamma \in H} \phi_{A\gamma,v\mathsf{E}}$$

is an increasing function of the statistic $F_{H,L}$ from (12.18), namely $R = (1 + \frac{s-r}{n-s} F_{H,L})^{n/2}$. Conclude that Fisher's F-test is a likelihood ratio test.

12.8 The content of silicon (in percent) was measured in 7 basalt rock samples taken from the moon and 7 from the Pacific Ocean. The following sample means and standard deviations were obtained:

	Moon	Pacific
M	19.6571	23.0429
$\sqrt{V^*}$	1.0861	1.4775

Suppose the measurements are normally distributed, and perform the following statistical procedure. Design first a variance ratio test of level 0.1 to verify the null hypothesis $v_{\text{moon}} = v_{\text{Pacific}}$. Then test the null hypothesis $m_{\text{moon}} = m_{\text{Pacific}}$ for the level 0.05.

12.9 In $n = 8$ measurements, a direct current motor produces the following values for its power [kW] at various rotational speeds [1000 rpm]:

t_k	0.8	1.5	2.5	3.5	4.2	4.7	5.0	5.5
X_k	8.8	14.7	22.8	29.4	38.2	44.1	47.8	51.5

Assume we are in the case of the Gaussian linear model.

(a) Calculate the regression line and hence deduce the interpolated (estimated) value for the missing power at 4000 rpm.

(b) Perform a test of $H_0 : \gamma_1 \geq 7.5$ of level $\alpha = 0.05$.

(c) Determine a confidence interval for γ_1 with error level $\alpha = 0.05$.

12.10 Perform polynomial regression of third degree for the temperature data of Example (12.24). Choose a suitable error level α, determine a confidence interval for the leading coefficient γ_3 of the regression polynomial, and test the one-sided null hypothesis $H_0 : \gamma_3 \leq 0$ against $H_1 : \gamma_3 > 0$. Use appropriate software to perform the calculations.

12.11 It is to be tested how water resistant the cladding panels of two different manufacturers are. From earlier measurements, it is known that the logarithmic water resistance is approximately normally distributed and that it varies on a similar scale for both manufacturers. The measurements yield

Manufacturer A	1.845	1.790	2.042
Manufacturer B	1.583	1.627	1.282

For the level $\alpha = 0.01$, test whether the water resistance of the cladding panels is the same for both manufacturers.

12.12[S] *Generalised linear models.* Let $(Q_\lambda)_{\lambda \in \Lambda}$ be an exponential family with respect to the statistic $T = \text{id}$ and a function $a : \Lambda \to \mathbb{R}$, see p. 211. (By assumption, a is invertible.) Also, let $n \in \mathbb{N}$, $\Theta = \Lambda^n$, $P_\vartheta = \bigotimes_{i=1}^n Q_{\vartheta_i}$ for $\vartheta = (\vartheta_1, \dots, \vartheta_n)^\top \in \Theta$, and ρ_ϑ be the density of P_ϑ. It is assumed that the quantities $a(\vartheta_i)$ appearing in ρ_ϑ depend linearly on an unknown parameter $\gamma \in \mathbb{R}^s$. That is, there exists a real $n \times s$ matrix A (determined by the design of the experiment) such that $a(\vartheta_i) = A_i \gamma$; here A_i is the ith row vector of A. The vector $a(\vartheta) := (a(\vartheta_i))_{1 \leq i \leq n}$ is thus given by $a(\vartheta) = A\gamma$, and conversely we have $\vartheta = \vartheta(\gamma) = a^{-1}(A\gamma) := (a^{-1}(A_i \gamma))_{1 \leq i \leq n}$. Let $X = (X_1, \dots, X_n)^\top$ be the identity map on $\mathbb{R}^n$. Show the following.

(a) Writing $\tau(\lambda) := \mathbb{E}(Q_\lambda)$ and $\tau \circ a^{-1}(A\gamma) := (\tau \circ a^{-1}(A_i \gamma))_{1 \leq i \leq n}$, one has

$$\text{grad}_\gamma \log \rho_{\vartheta(\gamma)} = A^\top \left(X - \tau \circ a^{-1}(A\gamma) \right).$$

Therefore, a maximum likelihood estimator $\hat{\gamma}$ of γ is a zero of the right-hand side and can be determined numerically, for example by Newton's method.

(b) The Gaussian linear model fits into this framework, and the maximum likelihood estimator is equal to the estimator $\hat{\gamma}$ from Theorem (12.15a).

12.13 [S] *Non-linear regression.* (a) The efficiency of a new medication in treating a certain disease is to be tested on n patients. One is interested in the chance of a full recovery within a certain time span as a function of the type of administration and the dose of the medication as well as certain characteristics of the patient such as weight, age, etc. Let $X_i = 1$ or 0 depending on whether the ith patient recovers or not, and let γ_j be the parameter that determines the effect of the jth factor on the chance of recovery. Specialise Problem 12.12a to this situation. (Note that $f := a^{-1}$ satisfies the logistic differential equation $f' = f(1 - f)$ of limited growth; one therefore speaks of *logistic regression.*)

(b) For comparison, consider also the case that the observed quantities X_i are independent random variables that can be assumed to be Poisson distributed for a parameter ϑ_i the logarithm of which is a linear function of an unknown $\gamma \in \mathbb{R}^s$. As an example, imagine a study of the antibacterial properties of some material, in which one observes the number of bacteria left after a certain time in dependence on their type and various properties of the material. In such cases, one speaks of *Poisson regression.*

12.14 A farmer is growing potatoes. He subdivides his field into 18 (more or less similar) lots and fertilises each of them with one of the three fertilisers Elite, Elite-plus, and Elite-extra. The following table shows the logarithmic crop yield for each lot:

Fertiliser	Yield					
Elite	2.89	2.81	2.78	2.89	2.77	
Elite-plus	2.73	2.88	2.98	2.82	2.90	2.85
Elite-extra	2.84	2.81	2.80	2.66	2.83	2.58	2.80

For this data set, work out the corresponding ANOVA table and, under the Gaussian assumption, test the null hypothesis that all three fertilisers have the same influence on the crop yield for the level 0.1.

12.15 [S] *Cuckoo eggs.* Each year cuckoos return to their home territory and lay their eggs in the nests of a certain host species. This way regional subspecies are created that can adapt to the population of their step-parents. In a study by O. M. Latter (1902), the sizes of 120 cuckoo eggs were measured for 6 different types of host birds. Find the data under `http://lib.stat.cmu.edu/DASL/Datafiles/cuckoodat.html` and test the null hypothesis 'the size of the cuckoo eggs does not depend on the host species' against the alternative 'the egg size is adapted to the host species'. Choose a suitable level, and use the Gaussian assumption.

12.16 *ANOVA with equal weights.* Consider the model of the analysis of variance for s groups with respective expectations m_i and sample sizes n_i. Let $\widetilde{m} = \frac{1}{s} \sum_{i=1}^{s} m_i$ be the averaged effect of all groups, and $\alpha_i = m_i - \widetilde{m}$ the additional effect of the ith group. (In the unbalanced case when n_i does depend on i, $\widetilde{m}$ differs from the overall mean $\overline{m}$ introduced in (12.30).) Show the following:

(a) The estimator $\widetilde{M} = \frac{1}{s} \sum_{i=1}^{s} M_i$ (which is, in general, different from M) is a best linear unbiased estimator of $\widetilde{m}$, and $\hat{\alpha}_i = M_i - \widetilde{M}$ is a best linear unbiased estimator of α_i.

(b) For every parameter $(m, v) \in \mathbb{R}^s \times]0, \infty[$, the variances are given by

$$\mathbb{V}_{m,v}(\widetilde{M}) = \frac{v}{s^2} \sum_{i=1}^{s} \frac{1}{n_i} \quad \text{and} \quad \mathbb{V}_{m,v}(\hat{\alpha}_i) = \frac{v}{s^2}\left(\frac{(s-1)^2}{n_i} + \sum_{j \neq i} \frac{1}{n_j}\right).$$

(c) If $k \in \mathbb{N}$ and $n = sk$, then $\mathbb{V}_{m,v}(\widetilde{M})$ is minimal for the sample sizes $n_1 = \cdots = n_s = k$. On the other hand, if $n = 2(s-1)k$, then $\mathbb{V}_{m,v}(\hat{\alpha}_i)$ is minimal whenever $n_i = (s-1)k$ and $n_j = k$ for $j \neq i$.

12.17 ^{S} *Two-factor ANOVA, or two-way layout, with one observation per cell.* Consider the situation of Example (12.34) for the case that $\ell = 1$. Then $V^*_{\mathrm{wg}} = 0$, and the results from (12.34) are no longer applicable. Therefore, this experimental design only makes sense when it is a priori clear that the factors do not interact, so that the 'additive two-factor model'

$$X_{ij} = \mu + \alpha_i + \beta_j + \sqrt{v}\,\xi_{ij}\,, \quad ij \in G,$$

applies; here μ, α_i, β_j are unknown parameters that satisfy $\sum_{i \in G_1} \alpha_i = 0$ and $\sum_{j \in G_2} \beta_j = 0$. Characterise the linear space L of all vectors of the form $(\mu + \alpha_i + \beta_j)_{ij \in G}$ by a system of equations, determine the projection of the observation vector X onto L, and design an F-test of the null hypothesis H_0 : 'factor 1 has no influence'.

12.18 *Analysis of covariance (ANCOVA).* The model of the one-factor analysis of covariance with d observation groups of sizes $n_1, \ldots, n_d$ is given by

$$X_{ik} = m_i + \beta\, t_{ik} + \sqrt{v}\,\xi_{ik}, \quad k = 1, \ldots, n_i, \ i = 1, \ldots, d$$

for unknown group expectations $m_1, \ldots, m_d$ and an unknown regression coefficient β that indicates the dependence on a regressor vector $t = (t_{ik})$. (For example, this model can be used to study the effect of different treatment methods when the age of the respective patient is also taken into account.)

(a) Determine the defining parameters of the corresponding linear model and find a (necessary and sufficient) condition on t that ensures that the design matrix A has full rank.

(b) Determine the least squares estimator $\hat{\gamma} = (\hat{m}_1, \ldots, \hat{m}_d, \hat{\beta})$.

(c) Verify the partitioning of variation

$$(n-1)\, V^*_{\mathrm{tot}} = \sum_{i=1}^{d} (n_i - 1)\, V_i^*\big(X - \hat{\beta}t\big)$$

$$+ \sum_{i=1}^{d} n_i \big(M_i(X) - M(X)\big)^2 + \hat{\beta}^2 \sum_{i=1}^{d} (n_i - 1)\, V_i^*(t)$$

into a residual variance within the groups, a sample variance between the groups, and a regression variance. Then find the Fisher statistic for a test of the null hypothesis $H_0 : m_1 = \cdots = m_d$.

12.19 ^{S} *Non-parametric analysis of variance.* Let $G = \{1, \ldots, s\}$ and $B = \{ik : i \in G, 1 \leq k \leq n_i\}$ for certain $n_i \geq 2$. Also, let $n = |B|$ and $X = (X_{ik})_{ik \in B}$ be a vector of independent real-valued observations X_{ik} with unknown continuous distributions Q_i. Furthermore, let R_{ik} be the rank of X_{ik} in X and $R = (R_{ik})_{ik \in B}$. Test the null hypothesis $H_0 : Q_1 = \cdots = Q_s =: Q$ against the alternative $H_1 : Q_i \prec Q_j$ for a pair $(i, j) \in G^2$ by applying the F-test from Example (12.33) to R instead of X. With this in mind, verify the following:

(a) $M(R) = (n+1)/2$ and $V^*_{\text{tot}}(R) = n(n+1)/12$.

(b) The F-statistic $V^*_{\text{bg}}(R)/V^*_{\text{wg}}(R)$ is an increasing function of the *Kruskal–Wallis test* statistic

$$T = \frac{12}{n(n+1)} \sum_{i \in G} n_i \left(M_i(R) - \frac{n+1}{2} \right)^2 .$$

(c) In the case $s = 2$, T is given by

$$T = \frac{12}{n_1 n_2 (n+1)} \left(U_{n_1,n_2} - \frac{n_1 n_2}{2} \right)^2 ,$$

where U_{n_1,n_2} is the U-statistic from Lemma (11.24). Hence, a Kruskal–Wallis test with rejection region $\{T > c\}$ of H_0 against H_1 is equivalent to a two-sided Mann–Whitney U-test.

(d) Under the null hypothesis H_0, T has the expectation $\mathbb{E}(T) = s-1$, and the distribution of T does not depend on Q. *Hint:* Recall the proofs of Theorems (11.25) and (11.28).

(e) Convince yourself that $T \xrightarrow{d} \chi^2_{s-1}$ under H_0, in the limit as $n_i \to \infty$ for all $i \in G$.

Solutions to Marked Problems

Chapter 1

1.3 Each box $\prod_{i=1}^{n} [a_i, b_i]$ is an element of $\mathscr{B}^{\otimes n}$, for it is the intersection of the projection preimages $X_i^{-1}[a_i, b_i]$ that belong to $\mathscr{B}^{\otimes n}$ by (1.9). As $\mathscr{B}^n$ is the smallest σ-algebra containing all such boxes, it follows that $\mathscr{B}^n \subset \mathscr{B}^{\otimes n}$. To prove the converse inclusion one may observe that the sets $X_i^{-1}[a_i, b_i]$ are closed and thus belong to $\mathscr{B}^n$ by (1.8b). Using (1.25), one concludes that $X_i^{-1} A_i \in \mathscr{B}^n$ for all $A_i \in \mathscr{B}$. This means that $\mathscr{B}^n$ includes a generator of $\mathscr{B}^{\otimes n}$ and therefore also the whole σ-algebra $\mathscr{B}^{\otimes n}$.

1.6 The system $\mathscr{A} = \{A \subset \Omega_1 \times \Omega_2 : A_{\omega_1} \in \mathscr{F}_2 \text{ for all } \omega_1 \in \Omega_1\}$ can easily be seen to be a σ-algebra that contains all product sets $A_1 \times A_2$ with $A_i \in \mathscr{F}_i$. Since these product sets constitute a generator of $\mathscr{F}_1 \otimes \mathscr{F}_2$, it follows that $\mathscr{A} \supset \mathscr{F}_1 \otimes \mathscr{F}_2$. This implies the first claim. Since $\{f(\omega_1, \cdot) \leq c\} = \{f \leq c\}_{\omega_1}$, the second claim then follows by (1.26).

1.7 (a) The sets B_J are pairwise disjoint, and $\bigcap_{k \in K} A_k = \bigcup_{K \subset J \subset I} B_J$ for all K. Now use that P is additive.

(b) Inserting the result of (a) into the right-hand side of the proposed equation and interchanging summations, one finds

$$\sum_{K: K \supset J} (-1)^{|K \setminus J|} \sum_{L: L \supset K} P(B_L) = \sum_{L: L \supset J} P(B_L) \sum_{K: J \subset K \subset L} (-1)^{|K \setminus J|} = P(B_J).$$

Indeed, the last sum (over all K between J and L) is equal to 1 for $J = L$ and 0 otherwise. In the case $J = \varnothing$ one notices that $B_\varnothing$ is the complement of $\bigcup_{j \in I} A_j$, so that

$$P\left(\bigcup_{i \in I} A_i\right) = 1 - P(B_\varnothing) = \sum_{\varnothing \neq K \subset I} (-1)^{|K|+1} P\left(\bigcap_{k \in K} A_k\right).$$

1.11 Bob should insist on splitting the € 10 at the ratio $1 : 3$.

1.13 Let $N = 52$, $I = \{1, \ldots, N\}$, Ω the set of all permutations of I, and $P = \mathcal{U}_\Omega$. Each $\omega \in \Omega$ describes the relative permutation of the cards of the second pile relative to those of the first. For $i \in I$ let $A_i = \{\omega \in \Omega : \omega(i) = i\}$. Then $A = \bigcup_{i \in I} A_i$ is the event that Bob wins. For each $K \subset I$, the probability that the cards at positions in K agree for both piles is $P(\bigcap_{k \in K} A_k) = (N - |K|)!/N!$. Applying Problem 1.7b one obtains

$$P(A) = -\sum_{\varnothing \neq K \subset I} (-1)^{|K|} \frac{(N - |K|)!}{N!} = -\sum_{k=1}^{N} (-1)^k \binom{N}{k} \frac{(N-k)!}{N!} = 1 - \sum_{k=0}^{N} (-1)^k \frac{1}{k!}.$$

Hence $P(A) \approx 1 - e^{-1} > 1/2$ (with an approximation error of at most $1/N!$).

1.15   (a) Apply (1.26): For a monotone X, all sets of the form $\{X \leq c\}$ are intervals (open, closed, or halfopen) and therefore Borel sets. If X is only piecewise monotone, the set $\{X \leq c\}$ is an at most countable union of intervals and thus also Borelian.

   (b) Apply (1.27) and Problem 1.14c: X' is the limit of the continuous difference ratios $X_n : \omega \to n(X(\omega + 1/n) - X(\omega))$ as $n \to \infty$.

1.18   We need to show that $P \circ X^{-1} \circ F^{-1} = \mathcal{U}_{]0,1[}$ if F is continuous. Since $P \circ X^{-1}$ is uniquely determined by its distribution function F, one can assume that $P = \mathcal{U}_{]0,1[}$ and X is defined as in (1.30). Then $X(u) \leq c \Leftrightarrow u \leq F(c)$ for $u \in]0, 1[$ and $c \in \mathbb{R}$. Setting $c = X(u)$ one concludes that $F(X(u)) \geq u$. Conversely, if $c < X(u)$ then $u > F(c)$, and letting $c \uparrow X(u)$ one finds that $F(X(u)) \leq u$ by the continuity of F. This shows that $F(X)$ is the identity mapping, and the claim $P \circ F(X)^{-1} = \mathcal{U}_{]0,1[}$ follows. Suppose next F exhibits a jump at some $c \in \mathbb{R}$, in that $F(c-) := \lim_{x \uparrow c} F(x) < F(c)$. Then

$$P\big(F(X) = F(c)\big) \geq P(X = c) = F(c) - F(c-) > 0 = \mathcal{U}_{]0,1[}\big(\{F(c)\}\big)$$

and therefore $P \circ F(X)^{-1} \neq \mathcal{U}_{]0,1[}$.

Chapter 2

2.2   Arguing as in (2.2) and Section 2.3.2 one finds that the probability in question is equal to

$$\binom{n+N-1}{n}^{-1} \prod_{a \in E} \binom{k_a + N_a - 1}{k_a}, \quad \text{where } \sum_{a \in E} k_a = n \text{ and } \sum_{a \in E} N_a = N.$$

2.3   In case (a), the distribution density of X on $[0, r[$ is $\rho_1(x) = 2x/r^2$. In case (b) one obtains $\rho_2(x) = 2/\big(\pi \sqrt{r^2 - x^2}\big)$, $x \in [0, r[$.

2.7   We take the model $\Omega = \{-1, 1\}^{2N}$ with the uniform distribution $P = \mathcal{U}_\Omega$, which corresponds to taking ordered samples with replacement from an urn with two balls of colour A and B, respectively. X_i stands for the projection onto the ith coordinate. The event G_n depends only on the variables $X_1, \ldots, X_{2n}$ and corresponds to the set of all broken lines from $(0, 0)$ to $(2n, 0)$ (with vertices in $\mathbb{Z}^2$) which hit the horizontal axis at their endpoints only. By symmetry, their total number is twice the number of all paths from $(1, 1)$ to $(2n - 1, 1)$ which never hit the horizontal axis. The total number of *all* paths from $(1, 1)$ to $(2n - 1, 1)$ (either hitting the axis or not) is $\binom{2n-2}{n-1}$. This is because each such path is uniquely determined by the position of its $n - 1$ 'upward steps'. This number has to be diminished by the number of paths that hit the axis. But each of these can be bijectively mapped to a path from $(1, -1)$ to $(2n - 1, 1)$ by reflecting its initial piece between $(1, 1)$ and its first zero in the horizontal axis. In the same way as above, one finds that the total number of these bijection images is $\binom{2n-2}{n}$. This gives the first identity of (a). The second identity follows by direct computation. For the proof of (b) it is then sufficient to notice that $G_{>n}^c = \bigcup_{k=1}^n G_k$ and $u_0 = 1$.

2.10   A direct computation is possible, but it is more instructive to exploit the derivation of $\mathcal{M}_{n,\rho}$ in (2.9). Observe that the statement in question depends only on the distribution of X. One can therefore assume that X coincides with the random variable S in (2.7) defined on E^n with the underlying probability measure $P = \rho^{\otimes n}$, the urn model with replacement. Next, imagine that the balls are drawn by a colour-blind person which can only distinguish the prescribed colour $a \in E$. This person observes the outcome of the random variable

$Z : E^n \to \{0, 1\}^n$ with $Z = (Z_i)_{1 \leq i \leq n}$, $Z_i(\omega) := 1_{\{a\}}(\omega_i)$ for $\omega \in E^n$. You can then easily verify that Z has the distribution $\sigma^{\otimes n}$, where $\sigma(1) := \rho(a)$ and $\sigma(0) := 1 - \rho(a)$. Now, as $X = S$, we have $X_a = \sum_{i=1}^n Z_i$. But (2.9) implies that the sum of the coordinates of a $\sigma^{\otimes n}$-distributed random variable has distribution $\mathcal{B}_{n,\sigma(1)}$.

2.11 Applying Problem 1.7b in a way similar to the solution of Problem 1.13 one obtains for each $k \in \mathbb{Z}_+$

$$P(X = k) = \sum_{J \subset I : |J| = k} P(B_J) = \sum_{J \subset I : |J| = k} \sum_{L : L \supset J} (-1)^{|L \setminus J|} \frac{(N - |L|)!}{N!}$$

$$= \sum_{J \subset I : |J| = k} \sum_{l=k}^N \binom{N-k}{l-k} (-1)^{l-k} \frac{(N-l)!}{N!} = \frac{1}{k!} \sum_{m=0}^{N-k} \frac{(-1)^m}{m!}.$$

The last term tends to $e^{-1} \frac{1}{k!} = \mathcal{P}_1(\{k\})$ as $N \to \infty$. That is, the number of fixed points of a random permutation is asymptotically Poisson distributed with parameter 1. Alternatively, this Poisson asymptotics can be derived directly from Problem 1.13 using a recursion over N as follows (the respective value of N is indicated by an index). Namely, for all $N \geq 2$ and $1 \leq k \leq N$ one has by symmetry

$$k\, P_N(X_N = k) = \sum_{i=1}^N P_N\big(i \text{ is fixed and there exist exactly } k - 1 \text{ further fixed points} \big)$$

$$= \sum_{i=1}^N \frac{|\{X_{N-1} = k-1)\}|}{N!} = P_{N-1}(X_{N-1} = k-1)$$

and thus $k!\, P_N(X_N = k) = P_{N-k}(X_{N-k} = 0) \to e^{-1}$ as $N \to \infty$.

2.14 This is a short computation using the identity in the line before (2.21). Intuitively, this result means the following. If n particles are thrown randomly into the interval $[0, s_n]$ and s_n varies as indicated, then the position of the rth smallest particle has asymptotically the same distribution as the rth smallest point in the Poisson model. The parameter α^{-1} represents the asymptotic particle density.

Chapter 3

3.4 Replacing all terms on the right-hand side of the claim by their definitions and integrating over p one obtains the expression $\binom{n}{l} B(a, b)^{-1} B(a+l, b+n-l)$. In view of (2.23) and the symmetry of the beta function, this equals

$$\binom{n}{l} \prod_{0 \leq i \leq l-1} \frac{a+i}{a+i+b+n-l} \prod_{0 \leq j \leq n-l-1} \frac{b+j}{a+b+j} = \frac{\binom{-a}{l}\binom{-b}{n-l}}{\binom{-a-b}{n}}.$$

The equality above is obtained by expanding all fractions with -1.

3.8 To prove the 'only if' direction, conclude from the self-independence that $F_X(s) = F_X(s)^2$, and thus $F_X(s) \in \{0, 1\}$, for all $s \in \mathbb{R}$. Then use (1.29) to conclude that F_X jumps from 0 to 1 at a unique point c.

3.12 (a) Define $\rho(n) = P(X = n)$ and $c_n = P(X + Y = n)/(n+1)$. The condition then implies that $\rho(k)\,\rho(l) = c_{k+l}$ for all $k, l \in \mathbb{Z}_+$. In particular, if $\rho(m) > 0$ then $\rho(k) > 0$ for all $k \leq 2m$. Hence, either $\rho(0) = 1$, and thus $P \circ X^{-1} = \delta_0$, or $\rho(k) > 0$ for all $k \in \mathbb{Z}_+$. In the second case, one can further conclude that $\rho(n)\,\rho(1) = \rho(n+1)\,\rho(0)$ for all $n \in \mathbb{Z}_+$. Setting $q = \rho(1)/\rho(0)$, it follows by induction that $\rho(n) = \rho(0)\,q^n$ for all $n \in \mathbb{Z}_+$. As ρ is a probability density, one has $\rho(0) = 1 - q$. Consequently, X is geometrically distributed with parameter $p = \rho(0) \in\,]0, 1[$.

(b) In this case, X is either degenerate with constant value 0 or has a Poisson distribution.

3.14 (a) Writing X_i for the number shown by dice i, one finds $P(X_1 > X_2) = 7/12$.

(b) Possible labellings for D_3 are 4 4 4 4 1 1 or 4 4 4 4 4 1.

3.16 The partial fraction decomposition is verified by direct computation. To prove the second hint one can assume that $x \geq 0$. Then

$$\int_{x-n}^{x+n} z\, c_a(z)\, dz = \int_{x-n}^{n-x} z\, c_a(z)\, dz + \int_{n-x}^{n+x} z\, c_a(z)\, dz$$

for all n. The first integral on the right-hand side vanishes by symmetry. For $n > x$, the second integral is not larger than $2x\,(n+x)\,c_a(n-x)$, which tends to zero as $n \to \infty$. In view of (3.31b), the main result will follow once it is shown that

$$\int_{-n}^{n} c_a(y)\, c_b(x-y)/c_{a+b}(x)\, dy \xrightarrow[n\to\infty]{} 1\,.$$

This is done by inserting the partial fraction decomposition and noting that, by the second hint, the integrals $\int_{-n}^{n} y\, c_a(y)\, dy$ and $\int_{-n}^{n} (x-y)\, c_b(x-y)\, dy$ vanish as $n \to \infty$. Since c_a and c_b are probability densities, the sum of the remaining integrals necessarily tends to 1.

3.20 (a) In view of (3.21a), each event of the form $\{L_j = \ell_j$ for $1 \leq j \leq m\}$ with $m \in \mathbb{N}$ and $\ell_j \in \mathbb{Z}_+$ must be shown to have probability $\prod_{j=1}^{m} \mathcal{G}_p(\{\ell_j\})$. Let $n_k = \sum_{j=1}^{k}(\ell_j + 1)$. The event above then means that $X_{n_k} = 1$ for all $1 \leq k \leq m$, while $X_n = 0$ for all other $n \leq n_m$. As the X_n are assumed to be Bernoulli, this has the probability

$$p^m (1-p)^{\sum_{j=1}^{m} \ell_j} = \prod_{j=1}^{m} \mathcal{G}_p(\{\ell_j\})\,.$$

Claim (b) follows similarly.

3.23 (a) Let $k \in \mathbb{Z}_+$ be fixed. It is known from Section 2.5.2 (or Problem 3.15a) that T_k is distributed according to $\Gamma_{\alpha,k}$, and $\tilde{T}_1$ according to $\mathcal{E}_{\tilde\alpha}$. Moreover, T_k and $\tilde{T}_1$ are independent. Applying (3.30), setting $q := \alpha/(\alpha + \tilde\alpha)$ and noting that $\gamma_{\alpha,k}(s)\,e^{-\tilde\alpha s} = q^k\,\gamma_{\alpha+\tilde\alpha,k}(s)$ one thus obtains

$$P(N_{\tilde{T}_1} \geq k) = P(T_k < \tilde{T}_1) = \int_0^\infty ds\,\gamma_{\alpha,k}(s) \int_s^\infty dt\,\tilde\alpha e^{-\tilde\alpha t} = q^k\,.$$

This means that $N_{\tilde{T}_1}$ is geometrically distributed with parameter $p := \tilde\alpha/(\alpha + \tilde\alpha)$.

(b) Pick any $l \in \mathbb{N}$ and $n_1, \ldots, n_l \in \mathbb{Z}_+$. The event $\{N_{\tilde{T}_k} - N_{\tilde{T}_{k-1}} = n_k$ for $1 \leq k \leq l\}$ can then be expressed in terms of the independent, exponentially distributed random variables $\tilde{L}_i$ and L_j; cf. (3.34). In view of (3.30), its probability can be written as an

$(l + \sum_{k=1}^{l} n_k + 1)$-fold Lebesgue integral. Together with Fubini's theorem, one thus finds that

$$P\left(N_{\tilde{T}_k} - N_{\tilde{T}_{k-1}} = n_k \text{ for } 1 \leq k \leq l\right)$$

$$= \int_0^\infty \cdots \int_0^\infty ds_1 \ldots ds_l \, \tilde{\alpha}^l \, e^{-\tilde{\alpha}(s_1 + \cdots + s_l)} \, P\left(N_{t_k} - N_{t_{k-1}} = n_k \text{ for } 1 \leq k \leq l\right),$$

where $t_k := s_1 + \cdots + s_k$. Using (3.34), one finally checks that the last expression is equal to $\prod_{k=1}^{l} a(n_k)$, where

$$a(n) = \int_0^\infty ds \, \tilde{\alpha} e^{-\tilde{\alpha} s} \, \mathcal{P}_{\alpha s}(\{n\}).$$

Since $a(n) = pq^n$ by (a), this implies the result.

(c) Interchanging the roles of the two Poisson processes and applying (b) and (3.24) one finds that the X_n are independent. As $N_{\tilde{T}_1}$ is the first n with $X_{n+1} = 1$, Section 2.5.1 therefore implies that its distribution is geometric.

3.25 For brevity we write λ for the d-dimensional Lebesgue measure λ^d. Define $\rho(j) = \lambda(B_j)/\lambda(\Lambda)$ for $1 \leq j \leq n$. Theorem (2.9) then implies that, for each k, the random vector $\left(\sum_{i=1}^{k} 1_{B_j}(X_i)\right)_{1 \leq j \leq n}$ has the multinomial distribution $\mathcal{M}_{k,\rho}$. Picking any $k_1, \ldots, k_n \in \mathbb{Z}_+$ and setting $k = \sum_{j=1}^{n} k_j$ one thus finds, using that N_Λ and the X_i are independent,

$$P(N_{B_j} = k_j \text{ for } 1 \leq j \leq n) = P(N_\Lambda = k) \, P\left(\sum_{i=1}^{k} 1_{B_j}(X_i) = k_j \text{ for } 1 \leq j \leq n\right)$$

$$= e^{-\alpha\lambda(\Lambda)} \frac{(\alpha\lambda(\Lambda))^k}{k!} \frac{k!}{k_1! \cdot \ldots \cdot k_n!} \prod_{j=1}^{n} \rho(j)^{k_j} = \prod_{j=1}^{n} \mathcal{P}_{\alpha\lambda(B_j)}(\{k_j\}).$$

This shows that the N_{B_j} are independent and Poisson distributed with parameter $\alpha \, \lambda(B_j)$.

3.27 The differential equation for F is $F' = r(1 - F)$, or $(\log(1 - F))' = -r$. By definition, $\rho = F'$.

3.29 A_1 and A_4 are tail events. This can be proved in complete analogy to (3.48) because, for each k, the convergence of a series and the lim sup of a sequence do not depend on the first k terms. In general, A_2 and A_3 are no tail events (except, for instance, if all Y_k are constant). For example, let $\Omega = [0, \infty[^{\mathbb{N}}$ and Y_k the kth projection. If $A_2 \in \mathscr{T}(Y_k : k \geq 1)$, then there was a set $B \neq \varnothing$ such that $A_2 = [0, \infty[\times B$. But this is impossible because $A_2 \cap \{Y_1 = 2\} = \varnothing$ but $([0, \infty[\times B) \cap \{Y_1 = 2\} = \{2\} \times B \neq \varnothing$.

3.31 For fixed $k \in \mathbb{N}$, the events $A_n = \{|S_{nk+k} - S_{nk}| \geq k\}$ are independent by (3.24) and have the same probability $2 \cdot 2^{-k} > 0$. The first claim thus follows from (3.50b). Next, one finds that $P(|S_n| \leq m \text{ for all } n) \leq P(|S_{n+2m+1} - S_n| \leq 2m \text{ for all } n) = 0$ for all m. In the limit $m \to \infty$, this gives $P(\sup_n |S_n| < \infty) = 0$ and thus

$$P\left(\{\sup_n S_n = \infty\} \cup \{\inf_n S_n = -\infty\}\right) = 1.$$

Now, the two events $\{\sup S_n = \infty\}$ and $\{\inf S_n = -\infty\}$ both belong to the tail σ-algebra $\mathscr{T}(X_i : i \geq 1)$ and, by symmetry, have the same probability. Together with (3.49), this implies the last assertion. It follows that, with probability one, the sequence (S_n) oscillates back and forth and thus visits every point of $\mathbb{Z}$ infinitely often. For an alternative proof of the last statement see (6.31).

Chapter 4

4.5 (a) As all terms in the double sum $\sum_{k,l\geq 1: l\geq k} P(X = l)$ are positive, the order of summation is irrelevant. Summing either first over k and then over l, or vice versa, one obtains either the left or the right side of the identity in question. (b) For each n one finds

$$\int_{1/n}^{\infty} P(X \geq s)\, ds \leq \mathbb{E}(X_{(n)}) = \frac{1}{n} \sum_{k\geq 1} P(X \geq k/n) \leq \int_{0}^{\infty} P(X \geq s)\, ds\,.$$

Here, the equality in the middle is justified by applying (a) to the random variable $nX_{(n)}$. To verify the two inequalities, one should observe that the function $s \to P(X \geq s)$ is decreasing. In view of Problem 1.15, this implies that this function is measurable and thus (properly or improperly) integrable. On the other hand, comparing the kth term of the series with either the integral over $[k/n, (k+1)/n[$, or over $[(k-1)/n, k/n[$, one obtains the two inequalities. Finally, the claim follows by letting $n \to \infty$.

4.8 The required identity is easily checked if $f = 1_{A_1 \times A_2}$. Next, observe that the product sets $A_1 \times A_2$ form an intersection-stable generator of $\mathscr{E}_1 \otimes \mathscr{E}_2$. Consider the system $\mathscr{D}$ of all events $A \in \mathscr{E}_1 \otimes \mathscr{E}_2$ for which $(1_A)_1$ is a random variable. Using Theorem (4.11c) and Problem 1.14c one then finds that $\mathscr{D}$ is a Dynkin system. Hence $\mathscr{D} = \mathscr{E}_1 \otimes \mathscr{E}_2$ by (1.13). A further application of (4.11c) shows that both $\mathbb{E}(1_A(X_1, X_2))$ and $\mathbb{E}((1_A)_1(X_1))$ depend σ-additively on A, and thus coincide by the uniqueness theorem (1.12). Finally, let f be an arbitrary bounded random variable. Its $1/n$-discretisation $f_{(n)}$ attains only finitely many values and thus is a linear combination of finitely many indicator functions. The required identity thus also holds for $f_{(n)}$. Letting $n \to \infty$ and applying Problem 4.7b one finally gets the required identity for f.

4.10 Suppose first that all X_i are non-negative. Applying Theorem (4.11c) and Problem 4.5 and interchanging the order of summations one then finds

$$\mathbb{E}(S_\tau) = \sum_{n,i\geq 1: i\leq n} \mathbb{E}(1_{\{\tau=n\}} X_i) = \sum_{i\geq 1} P(\tau \geq i)\, \mathbb{E}(X_i) = \mathbb{E}(\tau)\, \mathbb{E}(X_1)\,.$$

In the general case, it follows that $\mathbb{E}\big(\sum_{i=1}^{\tau} |X_i|\big) = \mathbb{E}(\tau)\, \mathbb{E}(|X_1|) < \infty$. Hence, both series in the display above are absolutely convergent, so that one can argue as before.

4.12 Let $\varphi(s) = (s - K)_+$ and $Y_{N+1} = e^{2\sigma Z_{N+1} - \mu}$. Then $\mathbb{E}^*(Y_{N+1}) = 1$ by the choice of p^*, and $X_{N+1} = X_N Y_{N+1}$ by definition. Moreover, X_N and Y_{N+1} are independent. Problem 4.8 thus gives $\Pi(N+1) = \mathbb{E}^*(\varphi(X_N Y_{N+1})) = \mathbb{E}^*(f_1(X_N))$, where $f_1(x) = \mathbb{E}^*(\varphi(x Y_{N+1}))$ for $x > 0$. As the function $y \to \varphi(xy)$ is convex, Problem 4.4 shows that $f_1(x) \geq \varphi(x\, \mathbb{E}^*(Y_{N+1})) = \varphi(x)$. Hence $\Pi(N+1) \geq \mathbb{E}^*(\varphi(X_N)) = \Pi(N)$.

A similar argument shows that the optimal hedging strategy $\alpha\beta$ with maturity $N+1$ is also a hedging strategy with maturity N. Indeed, for each $\omega \in \Omega$ one can write

$$\begin{aligned}
W_N^{\alpha\beta}(\omega_{\leq N}) &= \mathbb{E}^*\big(\varphi\big(X_N(\omega_{\leq N})\, Y_{N+1}\big)\big) \\
&\geq \varphi\big(X_N(\omega_{\leq N})\, \mathbb{E}^*(Y_{N+1})\big) = \big(X_N(\omega_{\leq N}) - K\big)_+\,.
\end{aligned}$$

Together with the definition of the Black–Scholes price, this yields an alternative proof of the claim.

4.15 (a) Expanding the squares one finds $\mathbb{E}((X-a)^2) = \mathbb{V}(X) + (a-m)^2$. (b) One can assume that $a < \mu$; otherwise replace X by $-X$. The identity in the hint then follows by distinguishing the cases $X \le a$, $a < X < \mu$ and $X \ge \mu$. Taking expectations, one obtains

$$\mathbb{E}(|X-a|) - \mathbb{E}(|X-\mu|) = (\mu-a)\left(2\,P(X \ge \mu) - 1\right) + 2\,\mathbb{E}\left((X-a)\,1_{\{a<X<\mu\}}\right).$$

As μ is a median and $(X-a)\,1_{\{a<X<\mu\}} \ge 0$, the right-hand side above is non-negative. It vanishes precisely if $P(X \ge \mu) = 1/2$ and $\mathbb{E}\left((X-a)\,1_{\{a<X<\mu\}}\right) = 0$. By Problem 4.1b, this implies that $P(a < X < \mu) = 0$, and therefore $P(X \le a) = P(X < \mu) = 1 - P(X \ge \mu) = 1/2$ and $P(X \ge a) \ge P(X \ge \mu) = 1/2$. So, the right-hand side of the display vanishes only if a is a median. Conversely, if a is a median, the roles of a and μ can be interchanged to obtain the desired identity.

4.18 Let $A_i = \{\omega \in \Omega : \omega(i) = i\}$ be the set of all permutations with fixed point i. Then $X = \sum_{1 \le i \le n} 1_{A_i}$, $P(A_i) = (n-1)!/n! = 1/n$, $\mathbb{V}(1_{A_i}) = 1/n - (1/n)^2$, and

$$\mathrm{Cov}(1_{A_i}, 1_{A_j}) = P(A_i \cap A_j) - \left(\tfrac{1}{n}\right)^2 = \tfrac{1}{n(n-1)} - \tfrac{1}{n^2} = \tfrac{1}{n^2(n-1)} \qquad \text{for } i \ne j\,.$$

Hence $\mathbb{E}(X) = \sum_{1 \le i \le n} 1/n = 1$ and

$$\mathbb{V}(X) = \sum_{1 \le i \le n} \mathbb{V}(1_{A_i}) + \sum_{i \ne j} \mathrm{Cov}(1_{A_i}, 1_{A_j}) = 1\,.$$

Since the equality of expectation and variance is a characteristic feature of the Poisson distribution, this result fits nicely with that of Problem 2.11.

4.20 (a) Let $A = \{T_{r+1} = n+d,\ X_{n+d} = i\}$ and $B = \{D_j = d_j \text{ for } 1 \le j \le r,\ \Xi_n = I\}$. For $i \in I$ one has $A \cap B = \emptyset$ and thus $P(A|B) = 0$. In the case $i \notin I$ let

$$A' = \{X_{n+k} \in I \text{ for } 1 \le k < d,\ X_{n+d} = i\}\,.$$

Then $A \cap B = A' \cap B$, and the events A' and B are independent by (3.24). So one can conclude that $P(A|B) = P(A'|B) = P(A') = (r/N)^{d-1}/N$.
  (b) Summing in (a) over $i \in I^c$ and applying (3.3a) one obtains

$$P\left(D_{r+1} = d \,\middle|\, D_j = d_j \text{ for } 1 \le j \le r\right) = \left(\tfrac{r}{N}\right)^{d-1} \tfrac{N-r}{N}\,.$$

On the other hand, $D_1 \equiv 1$, so that $D_1 - 1$ has the degenerate geometric distribution with parameter 1. The claim then follows by (3.7) and (3.21a).
  (c) Since $\mathbb{E}(D_r) = N/(N-r+1)$ and $\mathbb{V}(D_r) = N(r-1)/(N-r+1)^2$ by (4.6) and (4.35), one finds using (4.23cd)

$$\mathbb{E}(T_N) = N \sum_{k=1}^{N} \frac{1}{k} \underset{N \to \infty}{\sim} N \log N\,,$$

$$\mathbb{V}(T_N) = N \sum_{k=1}^{N-1} \frac{N-k}{k^2} \underset{N \to \infty}{\sim} \frac{N^2 \pi^2}{6}\,.$$

Setting $N = 20$ yields $\mathbb{E}(T_{20}) \approx 72$ and $\mathbb{V}(T_{20}) \approx 567$.

4.26   Problem 2.7 implies that $\varphi_\tau(s) = \sum_{n\geq 1} s^{2n}(u_{n-1} - u_n)$. Moreover, $u_n = (-1)^n \binom{-1/2}{n}$, as is easily seen by cancelling $n!$ against the even terms of $(2n)!$. The general binomial theorem thus implies that $\varphi_\tau(s) = 1 - \sqrt{1-s^2}$. It follows that $\mathbb{E}(\tau) = \varphi_\tau'(1) = \infty$; see (6.36) for a 'Markov proof' of this result.

Chapter 5

5.3   (a) Observe that $D_n^2 = \left(\sum_{i=1}^n \cos \Psi_i\right)^2 + \left(\sum_{i=1}^n \sin \Psi_i\right)^2$ and thus

$$D_n^2 = n + 2 \sum_{1\leq i<j\leq n} (\cos \Psi_i \cos \Psi_j + \sin \Psi_i \sin \Psi_j).$$

By the independence of the Ψ_i, it follows that

$$\mathbb{E}(D_n^2) = n + 2 \sum_{1\leq i<j\leq n} \left(\mathbb{E}(\cos \Psi_i)\,\mathbb{E}(\cos \Psi_j) + \mathbb{E}(\sin \Psi_i)\,\mathbb{E}(\sin \Psi_j)\right).$$

As $\mathbb{E}(\cos \Psi_i) = \frac{1}{2\pi}\int_0^{2\pi} \cos x \, dx = 0$ and likewise for the sine, one obtains $\mathbb{E}(D_n^2) = n$.
   (b) To follow the hint, apply (5.22) to see that

$$P\left(\sum_{i=1}^{30} Z_i > 15\right) = \mathcal{B}_{30,p}(\{16,\dots,30\}) \approx 1 - \Phi\left(\frac{15.5-30p}{\sqrt{30p(1-p)}}\right) = \Phi\left(\frac{30p-15.5}{\sqrt{30p(1-p)}}\right).$$

The last expression is at least 0.9 if $p \gtrsim p_0 := 0.63$ because $\Phi(1.28) \approx 0.9$. Next, writing $D_{n,i}$ for the distance of the ith particle from the origin after n steps, one observes that the random variables $Z_i := 1_{\{D_{n,i}\leq r_n\}}$ form a Bernoulli sequence with parameter $p :=$ $P(D_n \leq r_n)$. In view of (5.4) and part (a), one has $1 - p \leq n/r_n^2$. Hence $p \geq p_0$ as soon as $r_n = 1.65\sqrt{n}$.

5.5   Pick any $0 < p_1 < p_3 < 1$ and $0 < \alpha < 1$ and define $p_2 = \alpha p_1 + (1-\alpha)p_3$, so that $\alpha = (p_3-p_2)/(p_3-p_1)$. Also, let $\alpha_Z = (Z_3-Z_2)/(Z_3-Z_1)$ if $Z_3 > Z_1$, and $\alpha_Z = \alpha$ otherwise, so that $Z_2 = \alpha_Z Z_1 + (1-\alpha_Z) Z_3$. Then one can write, using the convexity of f,

$$f_n(p_2) = \mathbb{E}\big(f(Z_2/n)\big) \leq \mathbb{E}\big(\alpha_Z f(Z_1/n) + (1-\alpha_Z) f(Z_3/n)\big)$$
$$= \sum_{0\leq k\leq l\leq n} P(B_{k,l})\Big[\mathbb{E}(\alpha_Z|B_{k,l}) f(k/n) + \big(1-\mathbb{E}(\alpha_Z|B_{k,l})\big) f(l/n)\Big],$$

where $B_{k,l} := \{Z_1 = k,\, Z_3 = l\}$ and $\mathbb{E}(\cdot\,|B_{k,l})$ stands for the expectation relative to the conditional probability $P(\cdot\,|B_{k,l})$. Next, Theorem (2.9) implies that the random vector

$$\big(Z_1, Z_2-Z_1, Z_3-Z_2, n-Z_3\big) = \sum_{i=1}^n \big(1_{[0,p_1]}, 1_{]p_1,p_2]}, 1_{]p_2,p_3]}, 1_{]p_3,1]}\big) \circ U_i$$

has the multinomial distribution $\mathcal{M}_{n,\rho}$ with $\rho = (p_1, p_2-p_1, p_3-p_2, 1-p_3)$. So, if $k < l$ and $m \leq l-k$ then a short computation gives $P(Z_3-Z_2 = m\,|B_{k,l}) = \mathcal{B}_{l-k,\alpha}(\{m\})$, and therefore $\mathbb{E}(\alpha_Z|B_{k,l}) = \alpha$. By definition of α_Z, the last identity also holds when $k = l$. Inserting this into the display above one obtains $f_n(p_2) \leq \alpha f_n(p_1) + (1-\alpha) f_n(p_3)$, which proves the convexity of f_n.

5.7 For $k \in \mathbb{N}$ let $T_k = \sum_{i=1}^{k} L_i$. If $\mathbb{E}(L_1) = 0$ then $T_k = 0$ almost surely for all k, so that $N_t = \infty$ for all $t > 0$. So, in this case the result is trivial. Suppose therefore that $\mathbb{E}(L_1) > 0$, possibly $= \infty$. The event $A := \{\lim_{k \to \infty} T_k/k = \mathbb{E}(L_1)\}$ then has probability one, by (5.16) resp. Problem 5.6c. On the set A, one knows that $T_k \to \infty$, hence $N_t < \infty$ for all $t > 0$ and further $N_t \to \infty$ for $t \to \infty$. Now, by the definition of N_t, $T_{N_t} \le t < T_{N_t+1}$, and therefore $T_{N_t}/N_t \le t/N_t < T_{N_t+1}/N_t$ for all $t > 0$. This shows that $t/N_t \to \mathbb{E}(L_1)$ on A as $t \to \infty$, and the result follows. In the case of the Poisson process of intensity α, one has $\mathbb{E}(L_1) = 1/\alpha$ by (4.14), so that $N_t/t \to \alpha$ almost surely.

5.8 Define T_k as in the solution to Problem 5.7. As the function $s \to f(L_{(s)})$ is non-negative and constant on each interval $[T_{k-1}, T_k[$, one deduces for each $t > 0$ the sandwich inequality

$$\frac{N_t - 1}{t} \frac{1}{N_t - 1} \sum_{i=1}^{N_t-1} L_i \, f(L_i) \le \frac{1}{t} \int_0^t f(L_{(s)}) \, ds \le \frac{N_t}{t} \frac{1}{N_t} \sum_{i=1}^{N_t} L_i \, f(L_i).$$

In the limit as $t \to \infty$, N_t/t tends to $1/\mathbb{E}(L_1)$ almost surely, by Problem 5.7. In particular, $N_t \to \infty$ almost surely. Combining this with (5.16) one finds that both sides of the sandwich inequality converge to $\mathbb{E}(L_1 f(L_1))/\mathbb{E}(L_1)$ almost surely, and the result follows. If the L_i are exponentially distributed with parameter α, the size biased distribution has the expectation $\mathbb{E}(L_1^2)/\mathbb{E}(L_1) = 2/\alpha$, in agreement with (4.16). Finally, in order to prove that the random probability measure $\frac{1}{t} \int_0^t \delta_{L_{(s)}} \, ds$ converges in distribution towards Q almost surely, one can exploit the fact that, by monotonicity, it is sufficient to show that $\frac{1}{t} \int_0^t 1_{[0,c]}(L_{(s)}) \, ds \to F_Q(c)$ almost surely for all c in a countable dense subset of $\mathbb{R}$.

5.11 (a) Let P_p be the Bernoulli measure with parameter p; cf. (3.29). Then, by Problem 3.4,

$$P(S_n/n \le c) = \int_0^1 dp \, \beta_{a,b}(p) \, P_p(S_n/n \le c)$$

for all $0 < c < 1$. Theorem (5.6) shows that, in the limit $n \to \infty$, $P_p(S_n/n \le c)$ tends to 0 for $c < p$, and to 1 for $c > p$. Hence $P(S_n/n \le c) \to \beta_{a,b}([0,c])$ by dominated convergence, see Problem 4.7b.

(b) As can be seen from Figure 2.7, the fraction of the 'red' population behaves as follows when n is large. It concentrates around $a/(a+b)$ in case (i) ('coexistence'), around 1 in case (ii) ('suppression of the minority'), around either 0 or 1 in case (iii) ('one species wins, but both have a chance'), and nowhere in case (iv) ('complete randomness').

5.14 Fix any $c > 0$, and consider any $k \in \mathbb{Z}_+$ with $|x_\lambda(k)| \le c$. Then, as $\lambda \to \infty$, $k \sim \lambda$ and thus by (5.1)

$$\mathcal{P}_\lambda(\{k\}) \sim \exp[-\lambda \, \psi(k/\lambda)]/\sqrt{2\pi\lambda},$$

where $\psi(s) = 1 - s + s \log s$. Since $\psi(1) = \psi'(1) = 0$ and $\psi''(1) = 1$, Taylor's formula gives

$$\psi(k/\lambda) = \psi\big(1 + x_\lambda(k)/\sqrt{\lambda}\big) = x_\lambda(k)^2/2\lambda + O(\lambda^{-3/2}),$$

which implies the result. (See also Figure 11.3.)

5.17 By definition, $G_N = a \min(S_N, S) - b \max(S_N - S, 0)$. The recursion in the hint can be verified by distinguishing the cases $S_N < S$ and $S_N \ge S$. Since S_N and X_{N+1} are independent, it follows that

$$\mathbb{E}(G_{N+1}) - \mathbb{E}(G_N) = \big((a+b)P(S_N < S) - b\big) \, p,$$

which implies the second hint. Hence, $\mathbb{E}(G_N)$ is maximal for the smallest N satisfying $P(S_N < S) < r := b/(a+b)$. In terms of the normal approximation (5.22), this condition can be rewritten in the form $(S-0.5-Np)/\sqrt{Np(1-p)} \lesssim \Phi^{-1}(r)$. For the specific numbers given, this leads to $N = 130$ in case (a), and $N = 576$ in case (b).

5.22   In view of (5.24) or (5.29), $\sqrt{\varepsilon/(vt)} \, X_{t/\varepsilon}$ converges in distribution to $\mathcal{N}_{0,1}$ as $\varepsilon \to 0$. Problem 2.15 thus shows that $B_t^{(\varepsilon)} \xrightarrow{d} \mathcal{N}_{0,vt}$, which means that $\rho_t = \phi_{0,vt}$. The heat equation (with $D = v$) follows by differentiation.

5.25   (a) The event $\{L_{2N} = 2n\}$ requires that both (i) $S_{2n} = 0$ and (ii) $S_{2j} - S_{2n} \neq 0$ for all $j \in \{n+1, \ldots, N\}$. By (3.24) and the independence of the X_i, the events in (i) and (ii) are independent. The probability of the former is $\mathcal{B}_{2n,1/2}(\{n\}) = u_n$, and that of the latter is u_{N-n} by Problem 2.7b.

   (b) Part (a) and (5.2) imply that

$$P(L_{2N} = 2n) \underset{N \to \infty}{\sim} \frac{1}{N\pi \sqrt{(n/N)(N-n)/N}}$$

uniformly for all n with $a \leq n/N \leq b$. Now take the sum over these n and observe that the result on the right-hand side is simply a Riemann sum for the integral in the problem. The name arcsine law refers to the fact that the integrand has the primitive $(2/\pi) \arcsin \sqrt{x}$.

Chapter 6

6.2   (a) For example, let X_n be the deterministic Markov chain that runs cyclically through $E = \{1, 2, 3\}$ and starts with the uniform distribution. That is, $\alpha = \mathcal{U}_E$ and $\Pi(1,2) = \Pi(2,3) = \Pi(3,1) = 1$. Also, let $F = \{a, b\}$ and define $f : E \to F$ by $f(1) = f(2) = a$, $f(3) = b$. Finally, let $Y_n = f \circ X_n$. Then one can easily check that

$$P(Y_2 = a \mid Y_1 = a, Y_0 = a) = 0 \neq 1 = P(Y_2 = a \mid Y_1 = a, Y_0 = b).$$

   (b) For the sequence $Y_n := f \circ X_n$ to be a Markov chain, it is necessary and sufficient that there exist an initial distribution $\hat{\alpha}$ and a stochastic matrix $\hat{\Pi}$ on F such that

$$P(Y_0 = a_0, \ldots, Y_n = a_n) = \hat{\alpha}(a_0) \prod_{1 \leq i \leq n} \hat{\Pi}(a_{i-1}, a_i)$$

for all $n \geq 0$ and $a_i \in F$. By definition, the term on the left-hand side coincides with

$$P\big(X_0 \in f^{-1}\{a_0\}, \ldots, X_n \in f^{-1}\{a_n\}\big) = \sum_{x_0 \in f^{-1}\{a_0\}} \alpha(x_0) \prod_{1 \leq i \leq n} \sum_{x_i \in f^{-1}\{a_i\}} \Pi(x_{i-1}, x_i).$$

The last term will have the desired product form as soon as, for all $b \in F$ and $x \in E$, the expression $\sum_{y \in f^{-1}\{b\}} \Pi(x, y)$ only depends on b and $f(x)$, rather than b and x itself. Explicitly, this condition means that

$$\sum_{y \in f^{-1}\{b\}} \Pi(x, y) = \hat{\Pi}(f(x), b)$$

for a suitable matrix $\hat{\Pi}$. In that case let $\hat{\alpha}(a) = \sum_{x \in f^{-1}\{a\}} \alpha(x)$.

6.5 (a) One needs to verify the induction step $\rho_{n,\vartheta} \Pi = \rho_{n+1,\vartheta}$ for arbitrary n. Let $x, y \in E$ be such that $N(x) = n$, $N(y) = n+1$ and $\Pi(x, y) > 0$. Then

$$\frac{\rho_{n,\vartheta}(x)}{\rho_{n+1,\vartheta}(y)} = \begin{cases} \frac{\vartheta+n}{n+1}\frac{y_1}{\vartheta} & \text{if } y = x + (1,0,0,\dots), \\ \frac{\vartheta+n}{n+1}\frac{j+1}{j}\frac{y_{j+1}}{y_j+1} & \text{if } y = x + (0,\dots,0,-1,1,0,\dots) \text{ with } -1 \text{ at position } j, \end{cases}$$

and thus

$$\sum_{x \in E : N(x)=n} \rho_{n,\vartheta}(x)\, \Pi(x, y) = \rho_{n+1,\vartheta}(y)\left[\frac{y_1}{n+1} + \sum_{j=1}^{n}\frac{(j+1)y_{j+1}}{n+1}\right] = \rho_{n+1,\vartheta}(y).$$

(b) Since $\rho_{0,\vartheta} = \delta_0$, part (a) implies that all $\rho_{n,\vartheta}$ are probability densities. Next, by the Poisson assumption on Y, there exists a constant $c_{n,\vartheta}$ so that $\rho_{n,\vartheta}(x) = c_{n,\vartheta}\, P(Y = x)$ for all x with $N(x) = n$. Summing over all these x one finds that $P(N(Y) = n) = 1/c_{n,\vartheta}$. Together with the previous equation, this gives the result.

6.7 (a) Let $I := \{1,\dots,N\}$ be the set of individuals of each generation, and $(\psi_n)_{n\geq1}$ the sequence of random mappings describing the selection of the parental gene. By assumption, the ψ_n are independent and uniformly distributed on I^I. The time evolution of the set Ξ_n of A-individuals at time n is then described by the inverse mappings $\varphi_n := \psi_n^{-1}$ that map $\mathscr{P}(I)$ into itself. These random mappings are still i.i.d., and for any initial set $\Xi_0 \subset I$ one has the recursion $\Xi_n := \varphi_n(\Xi_{n-1})$, $n \geq 1$. According to (6.2d), the sequence $(\Xi_n)_{n\geq1}$ is thus a Markov chain on $\mathscr{P}(I)$ with transition matrix

$$\Pi(\xi, \eta) = P(\varphi_1(\xi) = \eta)$$
$$= P(\psi_1(i) \in \xi \Leftrightarrow i \in \eta) = \left(\frac{|\xi|}{N}\right)^{|\eta|}\left(\frac{N-|\xi|}{N}\right)^{N-|\eta|}, \quad \xi, \eta \in \mathscr{P}(I).$$

(b) Consider the mapping $f : \xi \to |\xi|$. Then

$$\sum_{\eta \in f^{-1}\{y\}} \Pi(\xi, \eta) = \binom{N}{y}\left(\frac{|\xi|}{N}\right)^{y}\left(\frac{N-|\xi|}{N}\right)^{N-y} = \mathcal{B}_{N,f(\xi)/N}(\{y\})$$

for all $\xi \in \mathscr{P}(I)$ and $y \in \{0,\dots,N\}$. In view of Problem 6.2b, this means that the sequence $(X_n)_{n\geq0}$ is Markovian with transition matrix

$$\hat{\Pi}(x, y) = \mathcal{B}_{N,x/N}(\{y\}), \quad x, y \in \{0,\dots,N\}.$$

(c) The states 0 and N are absorbing for $\hat{\Pi}$. So, by (6.9), the limit

$$h_N(x) := \lim_{N\to\infty} P^x(X_n = N)$$

exists and satisfies $h_N(N) = 1$, $h_N(0) = 0$ and $\hat{\Pi}h_N = h_N$. The function $h(x) := x/N$ has the same properties; indeed, the equation $\hat{\Pi}h = h$ follows directly from (4.5). To show that both functions agree consider their difference $g := h_N - h$ and let $m = \max_{0\leq y\leq N} g(y)$. Suppose that $m > 0$. Then there is a state $0 < x < N$ with $g(x) = m$. Hence

$$0 = g(x) - (\hat{\Pi}g)(x) = \sum_{y=0}^{N} \hat{\Pi}(x, y)\,(m-g(y))$$

and therefore $g(y) = m$ for all y. But this is false for $y = 0$. It follows that $m = 0$, and the same argument for $-g$ gives the result $g \equiv 0$.

6.9 Only the case $p \neq 1/2$ is to be considered. As in (6.10) one then finds the winning probability $h_0(m) = p^m (p^{N-m} - q^{N-m})/(p^N - q^N)$, where $q = 1 - p$.

6.12 (a) The state space is $E = \mathbb{Z}_+^N$, where an N-tuple $x = (x_n)_{1 \leq n \leq N} \in E$ means that x_n particles are located at n. For $1 \leq n \leq N$ let $e_n = (\delta_{n,m})_{1 \leq m \leq N} \in E$ be the state with a single living particle at position n. Also, let $\mathbf{0} = (0, \ldots, 0) \in E$ be the null state and $e_0 = e_{N+1} := \mathbf{0}$. The assumptions then imply that

$$\Pi(e_n, \cdot) = \frac{1}{2} \sum_{k \in \mathbb{Z}_+} \rho(k) \left[\delta_{ke_{n-1}} + \delta_{ke_{n+1}} \right] \quad \text{for } 1 \leq n \leq N.$$

To deal with the general case of an arbitrary number of independent particles, the right concept is that of convolution. Its definition for probability densities on E is the same as in the case of only one dimension, provided the addition on E is defined coordinatewise. Then one can write

$$\Pi(x, \cdot) = \mathop{\bigstar}_{n=1}^{N} \Pi(e_n, \cdot)^{\star x_n}, \quad x \in E.$$

In explicit terms, this means that

$$\Pi(x, y) = \sum_{y_{n,j} \in E : (n,j) \in J(x)}^{y} \prod_{(n,j) \in J(x)} \Pi(e_n, y_{n,j}),$$

where $J(x) = \{(n, j) : 1 \leq n \leq N, \ 1 \leq j \leq x_n\}$ and the sum extends over all $(y_{n,j}) \in E^{J(x)}$ with $\sum_{(n,j) \in J(x)} y_{n,j} = y$.

(b) By definition, $q(n) = h_0(e_n)$. Theorem (6.9) thus shows that

$$q(n) = \Pi h_0(e_n) = \frac{1}{2} \sum_{k \in \mathbb{Z}_+} \rho(k) \left[h_0(ke_{n-1}) + h_0(ke_{n+1}) \right].$$

Arguing as in (6.11) one finds that $P^{ke_n}(X_l = \mathbf{0}) = P^{e_n}(X_l = \mathbf{0})^k$ for all n, k, l, and therefore $h_0(ke_n) = q(n)^k$. Hence the proposed equation.

(c) Let $q := \min_{1 \leq n \leq N} q(n)$. Part (b) and the monotonicity of φ imply that $q \geq \varphi(q)$. If $\rho(0) > 0$, the hypothesis $\varphi'(1) \leq 1$ implies as in (6.11) that necessarily $q = 1$. In the case $\rho(0) = 0$, the assumption $\varphi'(1) \leq 1$ can only be satisfied if $\rho(1) = 1$ and therefore $\varphi(s) = s$ for all s. By (b), this implies that $q(\cdot)$ is linear, and therefore identically equal to its value 1 at the boundary.

(d) In the case $N = 2$, symmetry implies that $q(1) = q(2)$ and therefore $q(1) = f(q(1))$, where $f = (1 + \varphi)/2$. Since $f'(1) = 3/4 < 1$, one can conclude that $q(1) = 1$. Next, let $N = 3$. Then $q(1) = q(3)$ and thus $q(2) = \varphi(q(1))$ by (b). Applying (b) once again, one finds that $q(1)$ is the smallest fixed point of $g = (1 + \varphi \circ \varphi)/2$; cf. (6.11). As $g'(1) = 9/8 > 1$, one has $q(1) < 1$. Furthermore, $q(2) = \varphi(q(1)) < q(1)$ because otherwise $g(q(1)) > q(1)$.

6.18 (a) The transition matrix $\tilde{\Pi}$ is given by $\tilde{\Pi}(x, y) = P(L_1 = y + 1)$ for $x = 0$ and $\tilde{\Pi}(x, x - 1) = 1$ for $x > 0$. Then $\tilde{\Pi}^N(x, y) > 0$ for all $x, y \in E$ because it is possible to move in x steps from x to 0, stay there for $N - 1 - x$ steps and then jump to y. So, the ergodic theorem applies and gives the renewal theorem as in (6.29).

(b) The equation $\alpha \Pi = \alpha$ implies that $\alpha(y)/\alpha(y - 1) = P(L_1 > y)/P(L_1 > y - 1)$ for all $y > 0$, and therefore $\alpha(y) = \alpha(0) P(L_1 > y)$. In view of Problem 4.5, this shows that $\alpha(y) = P(L_1 > y)/\mathbb{E}(L_1)$. It is then easy to check that also $\alpha \tilde{\Pi} = \alpha$.

(c) One observes that $\alpha(x)\Pi(x, y) = \alpha(y)\tilde{\Pi}(y, x)$ for all $x, y \in E$. This implies that, under the initial distribution α, the age process $(X_n)_{0 \le n \le T}$ up to some T has the same distribution as the time-reversed process $(Y_{T-n})_{0 \le n \le T}$ of remaining life spans.

6.21 (a) Pick any $x, y \in E_{\text{rec}}$ such that $\Pi^k(x, y) > 0$ for some k. As $F_\infty(x, x) = 1$, an application of Theorem (6.7) shows that

$$0 < \Pi^k(x, y) = P^x(X_k = y, \ X_n = x \text{ for infinitely many } n > k)$$
$$\le \Pi^k(x, y) \sum_{n \ge 1} \Pi^n(y, x).$$

Hence, $\Pi^n(y, x) > 0$ for some n, which proves the symmetry of the relation "$\to$". Reflexivity and transitivity are obvious.

(b) Let $C \subset E_{\text{rec}}$ be an irreducible class. Suppose further that $x \in C$ is positive recurrent and $x \to y$, so that $y \in C$. By the proof of (6.34), $\alpha(z) = \mathbb{E}^x\left(\sum_{n=1}^{\tau_x} 1_{\{X_n = z\}}\right)/\mathbb{E}^x(\tau_x)$ is a stationary distribution. Since obviously $\alpha(C^c) = 0$, one has $\alpha(C) = 1$. As the restriction $\Pi^C = (\Pi(x, y))_{x, y \in C}$ of Π to C is irreducible, Theorem (6.27) implies that α is the only such stationary distribution and is given by $\alpha(y) = 1/\mathbb{E}^y(\tau_y) > 0$. It follows that y is positive recurrent, and the two expressions for α imply the proposed identity.

6.23 Suppose a stationary distribution α exists. Using the equation $\alpha\Pi = \alpha$, one then obtains by induction on n that $\alpha(n)\Pi(n, n+1) = \alpha(n+1)\Pi(n+1, n)$ for all $n \in \mathbb{Z}_+$. Hence $\alpha(n) = \alpha(0)\prod_{k=1}^n \Pi(k-1, k)/\Pi(k, k-1)$. Summation over n then yields the necessary condition

$$z := \sum_{n \in \mathbb{Z}_+} \prod_{k=1}^n \frac{\Pi(k-1, k)}{\Pi(k, k-1)} < \infty.$$

Reversing this reasoning one finds that the condition $z < \infty$ is also sufficient. Indeed, setting $\alpha(0) := 1/z$ and defining $\alpha(n)$ for $n > 0$ as above one obtains a probability density α which satisfies the detailed balance equation. A fortiori, this α is stationary.

6.28 (a) The inequality $|G^n(x, y)| \le (2a)^n$ follows by induction on n. The series defining e^{tG} is therefore absolutely convergent. In particular, this series can be differentiated term by term, which gives the last conclusion. Next, for any $x, y \in E$ and $t \ge 0$ one finds that

$$P^x(X_t = y) = \sum_{k \ge 0} P^x(N_t = k, \ Z_k = y) = e^{-at} \sum_{k \ge 0} \frac{(at)^k}{k!} \Pi^k(x, y).$$

As $a\Pi = aE + G$ for the identity matrix E, one has $a^k \Pi^k = \sum_{l=0}^k \binom{k}{l} a^l G^{k-l}$ for all k. Inserting this into the above and interchanging the order of summations one arrives at the first conclusion.

(b) Similarly as in the last display, the event under consideration can be decomposed in subevents with a fixed number of Poisson points in each subinterval $]t_{k-1}, t_k]$. Then apply (3.34), (6.7) and (a).

(c) By definition, (Z_k^*) is the embedded jump chain associated to (Z_n) as introduced in Problem 6.3. Its transition matrix is $\Pi^*(x, y) = \Pi(x, y)/(1 - \Pi(x, x)) = G(x, y)/a(x)$ for $y \ne x$, and $\Pi^*(x, x) = 0$. Next, let (τ_k) be the sequence of jump times of (Z_n) and $\delta_k = \tau_k - \tau_{k-1}$ the kth holding time of (Z_n). By Problem 6.3, the δ_k are conditionally independent given (Z_k^*), and the conditional distribution of $\delta_{k+1} - 1$ is geometric with parameter

$1 - \Pi(Z_k^*, Z_k^*) = a(Z_k^*)/a$. Further, let (T_i) be the sequence of jump times of the Poisson process (N_t). By definition, the differences $L_i = T_i - T_{i-1}$ are i.i.d. with distribution $\mathcal{E}_a$. By hypothesis, the L_i are also independent of (Z_k^*). Note that $T_k^* = T_{\tau_k}$ by construction. Observe then the following: If $\delta - 1$ is geometrically distributed with parameter p and independent of the L_i, then $\sum_{i=1}^{\delta} L_i$ is exponentially distributed with parameter pa. This can be seen, e.g., by combining Problems 3.20b and 3.18 (or by direct computation). Hence, choosing any $N \in \mathbb{N}$, $c_n > 0$ and $x_n \in E$ and letting $x_0 = x$, one finally finds that

$$P^x\big(T_{n+1}^* - T_n^* \le c_n \text{ for } 0 \le n < N \,\big|\, Z_n^* = x_n \text{ for } 0 \le n < N\big)$$

$$= P^x\Big(\sum_{i=\tau_n+1}^{\tau_{n+1}} L_i \le c_n \text{ for } 0 \le n < N \,\Big|\, Z_n^* = x_n \text{ for } 0 \le n < N \Big)$$

$$= \prod_{n=0}^{N-1} \sum_{l_n \ge 1} \frac{a(x_n)}{a}\Big(1 - \frac{a(x_n)}{a}\Big)^{l_n-1} P\Big(\sum_{i=1}^{l_n} L_i \le c_n \Big) = \prod_{n=0}^{N-1} \mathcal{E}_{a(x_n)}\big([0, c_n]\big).$$

6.30 (a) Fix any $x, y \in E$ and pick a smallest k satisfying the irreducibility condition. Then $G^k(x, y) > 0$, $G^l(x, y) = 0$ for $l < k$, and thus $\lim_{t \to 0} \Pi_t(x, y)/t^k = G^k(x, y)/k! > 0$. As E is finite, it follows that all entries of Π_t are strictly positive when t is small enough. In view of the identity $\Pi_s \Pi_t = \Pi_{s+t}$, this property then carries over to all $t > 0$.

(b) By (a) and (6.13), the limit $\alpha = \lim_{n \to \infty} \Pi_n(x, \cdot) > 0$ (along the integers $n \in \mathbb{Z}_+$) exists for all $x \in E$ and does not depend on x. If $t = n + \delta$ with $n \in \mathbb{Z}_+$ and $\delta > 0$, one can write

$$\|\Pi_t(x, \cdot) - \alpha\| \le \sum_{z \in E} \Pi_\delta(x, z)\, \|\Pi_n(z, \cdot) - \alpha\| \le \max_{z \in E} \|\Pi_n(z, \cdot) - \alpha\|.$$

The limit $\alpha = \lim_{t \to \infty} \Pi_t(x, \cdot)$ thus even exists along the real numbers. In particular, $\alpha \Pi_s = \lim_{t \to \infty} \Pi_{t+s}(x, \cdot) = \alpha$ for all $s > 0$. This in turn implies that $\alpha G = \frac{d}{ds} \alpha \Pi_s|_{s=0} = 0$. Conversely, the last property implies that $\alpha \Pi_s = \alpha e^{sG} = \alpha(sG)^0 + 0 = \alpha$ for every $s > 0$.

Chapter 7

7.3 (a) A natural statistical model is $(\mathcal{X}, \mathcal{F}, P_\vartheta : \vartheta \in \Theta)$, where $\mathcal{X} = \mathbb{N}^n$, $\mathcal{F} = \mathcal{P}(\mathcal{X})$, $\Theta = \mathbb{N}$ and P_ϑ is the uniform distribution on $\{1, \dots, \vartheta\}^n$. The likelihood function for the outcome $x \in \mathcal{X}$ is $\rho_x(\vartheta) = \vartheta^{-n} 1_{\{x_1, \dots, x_n \le \vartheta\}}$, which becomes maximal for $\vartheta = \max\{x_1, \dots, x_n\}$. Hence, the unique MLE is $T(x) = \max\{x_1, \dots, x_n\}$. Since clearly $P_\vartheta(T \le \vartheta) = 1$ and $P_\vartheta(T < \vartheta) > 0$, Problem 4.1b shows that $\mathbb{E}_\vartheta(\vartheta - T) > 0$. This means that T is biased.

(b) Using Problem 4.5a one can write

$$\frac{\mathbb{E}_N(T)}{N} = \frac{1}{N} \sum_{k=1}^{N} P_N(T \ge k) = 1 - \frac{1}{N} \sum_{k=1}^{N} \Big(\frac{k-1}{N}\Big)^n \xrightarrow{N \to \infty} 1 - \int_0^1 x^n\, dx = \frac{n}{n+1}.$$

(Note the analogy to (7.8) and (7.3).)

7.6 The formulation of the problem suggests the statistical model $(\mathbb{Z}_+, \mathcal{P}(\mathbb{Z}_+), P_\vartheta : \vartheta > 0)$ with $P_\vartheta = \mathcal{P}_{\mu\vartheta} \star \mathcal{P}_{\mu\vartheta} = \mathcal{P}_{2\mu\vartheta}$. Here, ϑ is the unknown age of V, and $\mu\vartheta$ is the expected number of mutations along each of the two ancestral lines. The likelihood function is $\rho_x(\vartheta) = e^{-2\mu\vartheta}(2\mu\vartheta)^x/x!$. Differentiating its logarithm, one finds that its maximum is attained precisely at $\vartheta = T(x) := x/(2\mu)$.

7.8 The model is given by $\mathcal{X} = \{0, \ldots, K\}$, $\Theta = \{0, \ldots, N\}$ and $P_\vartheta = \mathcal{H}_{\vartheta; K, N-K}$. Arguing as in (7.4) one finds the MLE $T(x) = \lfloor (N+1)x/K \rfloor$, with the possible alternative $T(x) = (N+1)(x/K) - 1$ when $(N+1)x/K \in \mathbb{N}$.

7.14 If S, T are minimum variance estimators then $\mathbb{V}_\vartheta(S) = \mathbb{V}_\vartheta(T) \leq \mathbb{V}_\vartheta((S+T)/2)$. With the help of (4.23c), it follows that $\text{Cov}_\vartheta(S, T) \geq \mathbb{V}_\vartheta(S)$ and thus $\mathbb{V}_\vartheta(T - S) \leq 0$. Together with Problem 4.1b or Corollary (5.5), this gives the result.

7.17 Let $0 < \vartheta < 1$ be the unknown probability of affirming the delicate question A. The probability of a person replying 'Yes' is then $p_\vartheta = (\vartheta + p_B)/2$. As n persons are interviewed, the model is thus given by $\mathcal{B}_{n, p_\vartheta}$, $\vartheta \in \,]0, 1[$. Proceeding as in Example (7.25) one can check that this is an exponential model with underlying statistic $T(x) = (2x/n) - p_B$. Further, $\mathbb{E}_\vartheta(T) = \vartheta$. By (7.24), it follows that T is a best estimator of ϑ. One computes $\mathbb{V}_\vartheta(T) = 4p_\vartheta(1 - p_\vartheta)/n$.

7.18 By definition, $P_{1/2}(\min_i X_i \geq s, \, \max_i X_i \leq t) = (t - s)^n$ for all $0 < s < t < 1$. Differentiation with respect to both s and t thus shows that, under $P_{1/2}$, $\min_i X_i$ and $\max_i X_i$ have the joint distribution density $\rho(s, t) = n(n-1)(t-s)^{n-2} 1_{\{0 < s < t < 1\}}$. This implies that for any $0 < c < 1/2$

$$P_{1/2}(T \leq c) = \int_0^1 ds \int_0^1 dt \, \rho(s, t) \, 1_{\{s + t \leq 2c\}} = 2^{n-1} c^n .$$

Hence, T has the distribution density $\tau(x) = n \, 2^{n-1} x^{n-1}$ on $[0, 1/2]$ relative to $P_{1/2}$. The distribution density on $[1/2, 1]$ then follows by reflection at the point $1/2$. Together with (2.21) and (2.23), one concludes that

$$\mathbb{V}_\vartheta(T) = \mathbb{V}_{1/2}(T) = 2 \int_0^{1/2} dx \, \tau(x) \left(\frac{1}{2} - x \right)^2 = \frac{n \, B(3, n)}{4} = \frac{1}{2(n+1)(n+2)} .$$

On the other hand, $\mathbb{V}_\vartheta(M) = \mathbb{V}_{1/2}(X_1)/n = 1/(12n)$; see (4.29) for $a = b = 1$. As $T = M$ for $n = 2$, it is clear that $\mathbb{V}_\vartheta(T) = \mathbb{V}_\vartheta(M)$ in this case. However, already for $n = 3$ one has $\mathbb{V}_\vartheta(T) < \mathbb{V}_\vartheta(M)$, and the larger n the more advantageous is T.

7.19 (a) As T is sufficient, one finds

$$\mathbb{E}_\vartheta(g_S \circ T) = \sum_{s \in \Sigma} P_\vartheta(T = s) \, g_S(s) = \sum_{s \in \Sigma} P_\vartheta(T = s) \, \mathbb{E}_{Q_s}(S) = \mathbb{E}_\vartheta(S) = \tau(\vartheta)$$

for all $\vartheta \in \Theta$. That is, $g_S \circ T$ is unbiased. Next, consider the difference $U = S - g_S \circ T$. Then $\mathbb{E}_\vartheta(U) = 0$ and also $\mathbb{E}_{Q_s}(U) = 0$ for all $s \in \Sigma$, so that

$$\mathbb{E}_\vartheta(U \, g_S \circ T) = \sum_{s \in \Sigma} P_\vartheta(T = s) \, g_S(s) \, \mathbb{E}_{Q_s}(U) = 0 .$$

Hence, $\text{Cov}_\vartheta(U, g_S \circ T) = 0$. By (4.23c), this means that $\mathbb{V}_\vartheta(S) = \mathbb{V}_\vartheta(U) + \mathbb{V}_\vartheta(g_S \circ T)$, which gives the result.

(b) Let S be an arbitrary unbiased estimator and g_S as in (a). Then $\mathbb{E}_\vartheta(g_S \circ T) = \mathbb{E}_\vartheta(g \circ T)$ for all ϑ. As T is complete, this implies that $g = g_S$. Hence, by (a), one has $\mathbb{V}_\vartheta(g \circ T) = \mathbb{V}_\vartheta(g_S \circ T) \leq \mathbb{V}_\vartheta(S)$ for all ϑ.

7.23   (a) Without loss of generality, $t_- < t_+$, so that T is non-constant. By (7.23b), the function $\tau(\vartheta) = \mathbb{E}_\vartheta(T)$ has the derivative $\tau'(\vartheta) = -\mathbb{V}_\vartheta(T) < 0$. In particular, it is continuous and strictly decreasing. Moreover, $\tau(\vartheta) \to t_-$ in the limit $\vartheta \to \infty$. Indeed, one has $\tau(\vartheta) - t_- = \sum_{x \in \mathcal{X}} [T(x) - t_-] \rho_\vartheta(x)$ and

$$\rho_\vartheta(x) = \frac{e^{-\vartheta[T(x)-t_-]}}{\sum_{y \in \mathcal{X}} e^{-\vartheta[T(y)-t_-]}} \xrightarrow[\vartheta \to \infty]{} 0 \quad \text{for } T(x) > t_-.$$

Likewise, $\tau(\vartheta) \to t_+$ as $\vartheta \to -\infty$. The result thus follows from the intermediate value theorem.

(b) For any P with $\mathbb{E}_P(T) = t$ one obtains straight from the definitions

$$0 \leq H(P; P_\vartheta) = -H(P) + \vartheta t + \log Z(\vartheta).$$

Equality holds precisely for $P = P_\vartheta$. Hence the result. (Compare this with Problem 9.5.)

7.24   (a) The probability densities $\rho_{n,\vartheta}$ can be written in the form (7.21) with $T = K_n$, $a(\vartheta) = \log \vartheta$ and $b(\vartheta) = \log \vartheta^{(n)}$. In view of Remark (7.23a), it follows that $\mathbb{E}_{n,\vartheta}(K_n) = b'(\vartheta)/a'(\vartheta) = \tau_n(\vartheta)$, and (7.24) implies that K_n is a best unbiased estimator of τ_n.

(b) Differentiation with respect to ϑ shows that the function $\vartheta \to \rho_{n,\vartheta}(x)$ is increasing in ϑ as long as $\tau_n(\vartheta) \leq K_n(x)$, and decreasing for larger values of ϑ.

(c) The sandwich estimate

$$\vartheta \log \frac{n+\vartheta}{\vartheta} = \vartheta \int_\vartheta^{n+\vartheta} \frac{1}{x} \, dx \leq \tau_n(\vartheta) \leq 1 + \vartheta \int_\vartheta^{n+\vartheta-1} \frac{1}{x} \, dx \leq 1 + \vartheta \log \frac{n+\vartheta}{\vartheta}$$

implies that $\tau_n(\vartheta) = \vartheta \log n + O(1)$ as $n \to \infty$. The estimator $K_n^* := K_n / \log n$ of ϑ thus has a bias of order $\mathbb{E}_{n,\vartheta}(K_n^*) - \vartheta = O(1/\log n)$. So, K_n^* is asymptotically unbiased. Next, one concludes from (7.23b) that K_n has the variance

$$\mathbb{V}_{n,\vartheta}(K_n) = \tau_n'(\vartheta)/a'(\vartheta) = \sum_{i=1}^{n-1} \frac{\vartheta i}{(\vartheta+i)^2} =: v_n(\vartheta),$$

which has the asymptotics $v_n(\vartheta) \sim \tau_n(\vartheta) \sim \vartheta \log n$ as $n \to \infty$ because $1 \geq i/(\vartheta+i) \geq 1-\varepsilon$ for $i \geq \vartheta/\varepsilon$. The estimator K_n^* thus has the mean squared error $\mathbb{E}_{n,\vartheta}((K_n^*-\vartheta)^2) = \mathbb{V}_\vartheta(K_n^*) + O(1/\log^2 n) \sim \vartheta/\log n$. By (5.4), it follows that the sequence (K_n^*) is consistent.

7.26   As the log-likelihood function is concave, one can apply (7.30). (When combined with Problem 7.10, this gives a special case of Problem 8.19.)

7.29   Example (7.39) states that $\theta_{n,x}$ is normally distributed with mean $T_n(x) = \frac{vm+nuM_n(x)}{v+nu}$ and variance $u_n = \frac{uv}{v+nu}$. By Problem 2.15 this means that $\sqrt{n/v}\,(\theta_{n,x} - M_n(x))$ is normally distributed with mean

$$T_n^*(x) = \sqrt{\frac{n}{v}} \left(\frac{vm+nuM_n(x)}{v+nu} - M_n(x) \right) = \frac{\sqrt{nv}}{v+nu} (m - M_n(x))$$

and variance $u_n^* = nu_n/v = \frac{nu}{v+nu}$. In the limit as $n \to \infty$, one finds that $u_n^* \to 1$ and $T_n^*(x) \to 0$; the latter relies on the assumption that the sequence $M_n(x)$ stays bounded. This gives the result.

Chapter 8

8.4 (a) Arguing as in Section 2.5.3 one finds that all T_ϑ have the distribution Q with density $\rho(x) = 1_{]1,2[}(x)\, n(x-1)^{n-1}$.

(b) As ρ is increasing, the shortest interval I satisfying $Q(I) = 1-\alpha$ is $I =]c, 2[$ with $c = 1 + \alpha^{1/n}$. Hence, $C(\cdot) =]\max_{1 \le i \le n} X_i/2, \max_{1 \le i \le n} X_i/c[$.

8.9 (a) For fixed $x \in \mathbb{R}^n$ and any $\vartheta \in \mathbb{Z}$, (7.10) gives

$$\log \rho(x, \vartheta) = -\tfrac{n}{2}\log(2\pi v) - \tfrac{n}{2v} V(x) - \tfrac{n}{2v}(\vartheta - M(x))^2.$$

So, the likelihood function is maximal for the integer ϑ minimising $|\vartheta - M(x)|$, which is $\vartheta = \widetilde{M}(x)$.

(b) In view of (3.32), M has the distribution $\mathcal{N}_{\vartheta, v/n}$ under $P_\vartheta = \mathcal{N}_{\vartheta,v}{}^{\otimes n}$. The distribution of $\widetilde{M} - \vartheta = \operatorname{ni}(M - \vartheta)$ under P_ϑ thus does not depend on ϑ. So one can assume that $\vartheta = 0$. Further, $P_0(\widetilde{M} = l) = \mathcal{N}_{0, v/n}([l - 1/2, l + 1/2[)$ for each $l \in \mathbb{Z}$. This gives the first claim and also shows that $\mathbb{E}_0(|\widetilde{M}|) < \infty$. In view of the symmetry $P_0(\widetilde{M} = l) = P_0(\widetilde{M} = -l)$, it follows that $\mathbb{E}_0(\widetilde{M}) = 0$, which means that $\widetilde{M}$ is unbiased.

(c) Since $P_0(\widetilde{M} = 0) = 2\Phi(0.5\sqrt{n/v}) - 1$, the condition $P_0(\widetilde{M} = 0) \ge 1-\alpha$ requires that $n \ge 4v\, \Phi^{-1}(1 - \tfrac{\alpha}{2})^2$.

8.11 Method 1 of Section 8.2 leads to the condition $n \ge 1/(4\alpha\varepsilon^2) = 2000$, while Method 2 gives $2\varepsilon\sqrt{n} \ge \Phi^{-1}(1 - \tfrac{\alpha}{2})$, i.e., $n \ge 385$.

8.14 Suppose first that q' is a further α-quantile. Then $\alpha \le Q(]-\infty, q]) \le Q(]-\infty, q'[) \le \alpha$, and thus $Q(]-\infty, q]) = \alpha$ and $Q(]q, q'[) = 0$. Conversely, the last two identities imply that $Q(]-\infty, q'[) = Q(]-\infty, q]) = \alpha$ and $Q(]-\infty, q']) \ge Q(]-\infty, q]) = \alpha$, which means that q' is an α-quantile.

8.15 (a) It is sufficient to show the following: If P is a probability measure on $\mathbb{R}$ and μ a median of P then $f(\mu)$ is a median of $P \circ f^{-1}$. Suppose f is increasing. Then $\{f \le f(\mu)\} \supset]-\infty, \mu]$ and therefore $P(f \le f(\mu)) \ge 1/2$. Similarly, $P(f \ge f(\mu)) \ge 1/2$. By way of contrast, the expectation value is only preserved if f is affine, in that $f(x) = ax + b$ for some constants $a, b \in \mathbb{R}$.

(b) Let $n = 2k + 1$, so that $T = X_{k+1:n}$. By Section 2.5.3, T has under $\mathcal{U}_{]0,1[}{}^{\otimes n}$ the beta distribution $\beta_{k+1, k+1}$. By symmetry, the latter has the unique median $1/2$. Here is another, direct argument that the distribution of T is symmetric relative to $1/2$: Consider the reflection $\tau : (x_i)_{1 \le i \le n} \to (1 - x_i)_{1 \le i \le n}$ on $]0, 1[^n$. The uniform distribution $\mathcal{U}_{]0,1[}{}^{\otimes n}$ is invariant under τ, and the sample median T satisfies $T \circ \tau = 1 - T$. Hence,

$$\mathcal{U}_{]0,1[}{}^{\otimes n} \circ T^{-1} = \mathcal{U}_{]0,1[}{}^{\otimes n} \circ \tau^{-1} \circ T^{-1} = \mathcal{U}_{]0,1[}{}^{\otimes n} \circ (1-T)^{-1},$$

which gives the desired symmetry.

(c) Suppose Q is continuous and $f = f_Q$ the associated quantile transformation as in (1.30). As f is increasing, f preserves the order of the order statistics. By (b) and the proof of (a), $f(1/2)$ is a median of Q. (If the median of Q is not unique, $f(1/2)$ can be set equal to a given median of Q. f is then still increasing, and since $\mathcal{U}_{]0,1[}(\{1/2\}) = 0$ it is still true that $\mathcal{U}_{]0,1[} \circ f^{-1} = Q$.) Applying (a) one gets the result. *Comment:* In the even case $n = 2k$ one faces the problem that there is no natural definition of the sample median. The linear definition (8.22) does allow the symmetry argument in the proof of (b), but is lost in general when f is applied. A suitable definition of the sample median would be $f([F(X_{k:n}) + F(X_{k+1:n})]/2)$, where $F = F_Q$. But this is not practical.

8.18 Let F be the distribution function of the X_i. The hypothesis implies that F is continuous and strictly increasing on $\mathcal{X}$. In view of (3.24) and Problem 1.18, it follows that the random variables $U_i = F(X_i)$ are independent with uniform distribution $\mathcal{U}_{]0,1[}$. By Section 2.5.3, the associated order statistics $U_{i:n}$ have distribution $\beta_{i,n-i+1}$. Also, $U_{i:n} = F(X_{i:n})$. Hence, for each $c \in \mathcal{X}$ the substitution $x = F^{-1}(u)$ gives

$$P(X_{i:n} \le c) = P(U_{i:n} \le F(c)) = \int_0^{F(c)} \beta_{i,n-i+1}(u)\, du = \int_{-\infty}^c \beta_{i,n-i+1}(F(x))\, \rho(x)\, dx \,.$$

This shows that $X_{i:n}$ has the distribution density $\rho_{i:n}(x) = \beta_{i,n-i+1}(F(x))\, \rho(x)$ on $\mathcal{X}$. Alternatively, the key identity $P(X_{i:n} \le c) = \beta_{i,n-i+1}([0, F(c)])$ follows directly from (8.17) and (8.8b).

8.19 Let q be the α-quantile of Q, $\varepsilon > 0$, and $0 < \delta < F(q+\varepsilon) - \alpha$. Then, for n large enough, one has $j_n/n < F(q+\varepsilon) - \delta$ and thus by (8.17)

$$P\left(X_{j_n:n} > q+\varepsilon\right) \le P\left(\frac{1}{n} \sum_{i=1}^n 1_{\{X_i \le q+\varepsilon\}} < F(q+\varepsilon) - \delta\right).$$

By (5.6), the term on the right tends to 0 as $n \to \infty$. A similar argument shows that also $P(X_{j_n:n} \le q-\varepsilon) \to 0$.

8.20 Let α and q be as required and pick any $c \in \mathbb{R}$. Then, by (8.17),

$$P\left(\sqrt{n}\,(X_{j_n:n} - q) \le c\right) = P\left(S_{c,n} \ge j_n\right),$$

where $S_{c,n} = \sum_{i=1}^n 1_{\{X_i \le q_{c,n}\}}$ and $q_{c,n} = q + c/\sqrt{n}$. With the quantities $\alpha_{c,n} = F(q_{c,n})$, $v_{c,n} = n\,\alpha_{c,n}(1 - \alpha_{c,n})$, $\gamma_{c,n} = (n\alpha_{c,n} - j_n)/\sqrt{v_{c,n}}$, and $S_{c,n}^* = (S_{c,n} - n\alpha_{c,n})/\sqrt{v_{c,n}}$, the last probability can be rewritten as $P(S_{c,n}^* \ge -\gamma_{c,n})$. Next, the mean value theorem shows that $\alpha_{c,n} = \alpha + \rho(q_{c,n}^*)\, c/\sqrt{n}$ for some $q_{c,n}^*$ between q and $q_{c,n}$. Hence,

$$\gamma_{c,n} \sim \sqrt{n}\, \rho(q_{c,n}^*)\, c/\sqrt{v_{c,n}} \to c/\sqrt{v} \,.$$

Together with the hint, one thus ends up with

$$P\left(\sqrt{n}\,(X_{j_n:n} - q) \le c\right) = P\left(S_{c,n}^* \ge -\gamma_{c,n}\right) \to 1 - \Phi(-c/\sqrt{v}) = \Phi(c/\sqrt{v}) \,.$$

The hint is verified by a careful examination of the proofs in Section 5.2.

Chapter 9

9.2 (a) Suppose $X_1, \ldots, X_n$ are uncorrelated. Their covariance matrix C is then a diagonal matrix, so that their joint distribution density (9.3) is a product density. By (3.28) and (3.30), this implies that $X_1, \ldots, X_n$ are independent. The converse holds in general, recall (4.23d).

(b) Without loss of generality, one can assume that all X_i are centred. Then $a_0 = 0$. For $\mathbf{a} = (a_1, \ldots, a_{n-1})$ let $X_{\mathbf{a}} = \sum_{j=1}^{n-1} a_j X_j$. We are looking for an $\mathbf{a}$ for which $X_{\mathbf{a}} - X_n$ is independent of $X_1, \ldots, X_{n-1}$. Now, the random vector $(X_1, \ldots, X_{n-1}, X_{\mathbf{a}} - X_n)^\top$ is the image of $(X_1, \ldots, X_n)$ under a linear transformation. So, by (9.5), its distribution is multivariate normal. In view of (a), it is therefore sufficient to find an $\mathbf{a}$ for which the components of this vector are uncorrelated. Equivalently, the vector $X_{\mathbf{a}} - X_n$ is required to be orthogonal to the subspace of $\mathscr{L}^2$ spanned by $X_1, \ldots, X_{n-1}$. Hence, $X_{\mathbf{a}}$ should be the projection of X_n onto this subspace. The existence of such an $X_{\mathbf{a}}$ is known from general Hilbert space theory. For a direct proof, consider the function $d(\mathbf{a}) = \mathbb{E}((X_{\mathbf{a}} - X_n)^2)$. Expanding the square, one finds that $d(\cdot)$ is continuously differentiable and satisfies $d(\mathbf{a}) \to \infty$ as $|\mathbf{a}| \to \infty$. Hence, $d(\cdot)$ attains its global minimum at some point $\mathbf{a}$. As its gradient vanishes there, it follows that $0 = \frac{\partial}{\partial a_j} d(\mathbf{a}) = 2 \operatorname{Cov}(X_{\mathbf{a}} - X_n, X_j)$ for each $1 \le j < n$.

9.5   For $P \in \mathscr{W}_{\mathsf{C}}$ one has $H(P) + \mathbb{E}_P(\log \phi_{0,\mathsf{C}}) = -H(P; \mathcal{N}_n(0, \mathsf{C})) \le 0$ with equality if and only if $P = \mathcal{N}_n(0, \mathsf{C})$. Also, writing $X = (X_1, \ldots, X_n)^{\top}$,

$$\mathbb{E}_P(\log \phi_{0,\mathsf{C}}) = -\frac{n}{2} \log(2\pi) - \frac{1}{2} \log \det \mathsf{C} - \frac{1}{2} \mathbb{E}_P(X^{\top} \mathsf{C}^{-1} X).$$

Finally, as the X_i are centred and C is symmetric, one obtains

$$\mathbb{E}_P(X^{\top} \mathsf{C}^{-1} X) = \sum_{i,j=1}^{n} (\mathsf{C}^{-1})_{ij} \, \mathbb{E}_P(X_i X_j) = \sum_{i,j=1}^{n} (\mathsf{C}^{-1})_{ij} \mathsf{C}_{ji} = \operatorname{Tr}(\mathsf{C}^{-1} \mathsf{C}) = \operatorname{Tr}(\mathsf{E}) = n,$$

where $\operatorname{Tr} \mathsf{A}$ denotes the trace of a matrix A. Hence the result.

9.6   For $n \in \mathbb{N}$, $m \in \mathbb{R}^n$ and $r > 0$ let $p_n(m, r) = \mathcal{N}_n(0, \mathsf{E})(|X - m| < r)$. The claim then reads $p_n(m, r) \le p_n(0, r)$. One possible proof is by induction on n, as follows. For $n = 1$, observe that $\frac{d}{dm} p_1(m, r) = \phi_{0,1}(m + r) - \phi_{0,1}(m - r)$ is negative when $m > 0$ and positive for $m < 0$. Hence, $p_1(\cdot, r)$ is maximal at the point $m = 0$. In the case $n > 1$, pick any $k, l \in \mathbb{N}$ with $k + l = n$. Let $X_{\flat}$ be the projection from $\mathbb{R}^n$ onto the first k coordinates and $X_{\sharp}$ the projection onto the remaining l coordinates. For $m \in \mathbb{R}^n$ let $m_{\flat} = X_{\flat}(m)$ and $m_{\sharp} = X_{\sharp}(m)$. As the random variables $X_{\flat}$ and $X_{\sharp}$ are independent relative to $\mathcal{N}_n(0, \mathsf{E})$, one can apply Problem 4.8 to obtain with $r(X_{\flat}) = \sqrt{r^2 - |X_{\flat} - m_{\flat}|^2}$

$$p_n(m, r) = \mathcal{N}_n(0, \mathsf{E})\big(|X_{\sharp} - m_{\sharp}|^2 < r(X_{\flat})^2\big) = \mathbb{E}\big(1_{\{|X_{\flat} - m_{\flat}| < r\}} \, p_l(m_{\sharp}, r(X_{\flat}))\big).$$

By induction hypothesis, $p_l(m_{\sharp}, r_{\flat}(X)) \le p_l(0_{\sharp}, r_{\flat}(X))$. Hence $p_n(m, r) \le p_n(m_{\flat} 0_{\sharp}, r)$. Interchanging the roles of $X_{\flat}$ and $X_{\sharp}$, one further obtains that $p_n(m_{\flat} 0_{\sharp}, r) \le p_n(0, r)$, which proves the claim. Alternatively, this can be derived directly from Problem 9.10c.

9.10   (a) Differentiating the distribution function $P(Y \le c) = \Phi(\sqrt{c} - a) - \Phi(-\sqrt{c} - a)$ for $c > 0$, one finds by (1.31) the distribution density

$$\rho_a(y) = \big(\phi_{0,1}(\sqrt{y} - a) + \phi_{0,1}(-\sqrt{y} - a)\big)/\big(2\sqrt{y}\big), \quad y > 0,$$

which by elementary calculations coincides with the proposed expression.

   (b) Using the identity $(2k)! \, 2^{-2k} = k! \, \Gamma(k + \frac{1}{2})/\Gamma(\frac{1}{2})$ and the series expansion of $\cosh$, one finds the formula

$$\rho_a(y) = \sum_{k \ge 0} \mathcal{P}_{a^2/2}(\{k\}) \, \chi_{2k+1}^2(y)$$

for $y > 0$, which is a rewriting of the claim.

(c) The proposed identity is equivalent to the convolution identity

$$\overset{n}{\underset{i=1}{\bigstar}} \, \rho_{a_i} = \sum_{k \geq 0} \mathcal{P}_{\lambda/2}(\{k\}) \, \chi^2_{2k+n} \,.$$

In view of (9.9) and (3.36), the probability density on the right is equal to

$$\sum_{k_1,\ldots,k_n \geq 0} \Big(\prod_{i=1}^{n} \mathcal{P}_{a_i^2/2}(\{k_i\}) \Big) \Big(\overset{n}{\underset{i=1}{\bigstar}} \, \chi^2_{2k_i+1} \Big) \,.$$

Moreover, (3.31b) shows that the convolution map $(\sigma, \tau) \to \sigma \star \tau$ is bilinear. The last expression thus coincides with the convolution of the densities ρ_{a_i} as identified in part (b).

9.12 Let $(X_i)_{i \geq 1}$ be independent standard normal random variables and $S_n = \sum_{i=1}^{n} X_i^2$. By (9.10), S_n has the χ_n^2-distribution. Note further that the X_i^2 have expectation $m = 1$ and (by (7.27b)) variance $v = 2$. Applying Problem 5.21 to the function $f(x) = \sqrt{x}$, one can therefore conclude that

$$S_n^{\sharp} := \sqrt{2S_n} - \sqrt{2n} = \frac{\sqrt{n/v}}{f'(m)} \Big(f\big(\tfrac{1}{n} \sum_{i=1}^{n} X_i^2\big) - f(m) \Big) \xrightarrow{d} \mathcal{N}_{0,1} \,.$$

Next, let $0 < \alpha < 1$ and $0 < \varepsilon < \min(\alpha, 1-\alpha)$. Then $P(S_n^{\sharp} \leq \Phi^{-1}(\alpha \pm \varepsilon)) \to \alpha \pm \varepsilon$. Hence, if n is sufficiently large then

$$P\big(S_n^{\sharp} \leq \Phi^{-1}(\alpha - \varepsilon)\big) < \alpha = P\big(S_n^{\sharp} \leq \sqrt{2\chi^2_{n;\alpha}} - \sqrt{2n}\big) < P\big(S_n^{\sharp} \leq \Phi^{-1}(\alpha + \varepsilon)\big)$$

and therefore $\Phi^{-1}(\alpha - \varepsilon) < \sqrt{2\chi^2_{n;\alpha}} - \sqrt{2n} < \Phi^{-1}(\alpha + \varepsilon)$. As Φ^{-1} is continuous, the second claim follows by letting $\varepsilon \to 0$. Comparing the numerical values, one finds the following: Fisher's approximation of $\chi^2_{n;\alpha}$ improves that of Problem 9.11 when α is close to 0 or 1; for α close to $1/2$, both approximations are equivalent; for other values of α, Fisher's approximation is worse.

9.17 Let $M = \frac{1}{n} \sum_{i=1}^{n} X_i$ and $M' = \frac{1}{n} \sum_{j=1}^{n} Y_j$. In view of (3.24) and (9.17b), the random variables $(M - m)\sqrt{n/v}$ and $(M' - m')\sqrt{n/v}$ are independent with standard normal distribution. Hence, the sum of their squares has distribution χ_2^2. This leads to the confidence circle

$$C(\cdot) = \big\{ (m, m') \in \mathbb{R}^2 : (m - M)^2 + (m' - M')^2 < \chi_{2;1-\alpha} \, v/n \big\} \,.$$

9.19 By (9.17), M_n and V_n^* are independent. This implies that, for each $k > n$, $M_k = \frac{n}{k} M_n + \frac{1}{k} \sum_{i=n+1}^{k} X_i$ is independent of V_n^* and thus also of N. Theorem (9.17) shows further that all $(M_k - m)\sqrt{k/v}$ have the standard normal distribution under $P_{m,v}$. This gives for arbitrary $a, b \in \mathbb{R}$

$$\begin{aligned}
P_{m,v}\big((M_N - m)\sqrt{N/v} &\leq a, \, V_n^* \leq b\big) \\
&= \sum_{k \geq n} P_{m,v}\big((M_k - m)\sqrt{k/v} \leq a, \, V_n^* \leq b, \, N = k\big) \\
&= \Phi(a) \sum_{k \geq n} P_{m,v}\big(V_n^* \leq b, \, N = k\big) = \Phi(a) \, P_{m,v}(V_n^* \leq b) \,,
\end{aligned}$$

proving (a). Part (b) now follows directly from (9.15).

Chapter 10

10.4 The most powerful test of level α in case (a) is $\varphi = 1_{[2,3[} + 2\alpha 1_{]1,2[}$. In case (b) one finds $\varphi = 1_{]1-2\alpha,1]}$ when $\alpha \in]0, 1/4]$, while $\varphi = 1_{]1/2,1[} + (2\alpha - \frac{1}{2}) 1_{]1,2[}$ for $\alpha \in [1/4, 1/2[$.

10.7 (a) It is clear that G^* is increasing. To prove its concavity let $0 < \alpha < \alpha' < 1$, $0 < s < 1$ and ψ, ψ' any tests of level α resp. α'. Then, $\psi_s = s\psi + (1-s)\psi'$ is a test of level $\alpha_s = s\alpha + (1-s)\alpha'$, so that $G^*(\alpha_s) \geq \mathbb{E}_1(\psi_s) = s\mathbb{E}_1(\psi) + (1-s)\mathbb{E}_1(\psi')$. Taking the sup over all ψ, ψ', one finds that $G^*(\alpha_s) \geq sG^*(\alpha) + (1-s)G^*(\alpha')$, as desired.

(b) Let φ be a Neyman–Pearson test with threshold value $c > 0$ and size $\alpha = \mathbb{E}_0(\varphi)$. Since φ is most powerful at level α, one has $G^*(\alpha) = \mathbb{E}_1(\varphi)$. Further, pick any $\alpha' \in]0, 1[$ and construct a Neyman–Pearson test ψ of size $\mathbb{E}_0(\psi) = \alpha'$ as in (10.3a). Then $G^*(\alpha') = \mathbb{E}_1(\psi)$. As in the proof of (10.3b), it follows that $\mathbb{E}_1(\varphi) - \mathbb{E}_1(\psi) \geq c\,(\alpha - \alpha')$. Hence

$$G^*(\alpha') \leq G^*(\alpha) + c\,(\alpha' - \alpha) \quad \text{for all } \alpha' \in]0, 1[.$$

This shows that c is the slope of a tangent of G^* at α.

10.9 If $P_1(\rho_0 = 0) = 1$, the test $\varphi = 1_{\{\rho_1 > 0\}}$ satisfies $\mathbb{E}_0(\varphi) = \mathbb{E}_1(1-\varphi) = 0$ and is therefore a minimax test. So let $P_1(\rho_0 = 0) < 1$. Consider the likelihood ratio R (defined as in Section 10.2) and the probability measure $P = (P_0 + P_1)/2$. The hint suggests to construct a Neyman–Pearson test φ satisfying $\mathbb{E}_P(\varphi) = 1/2$. Contrary to the proof of (10.3a), it is now possible that $P(R = \infty) > 0$. However, $P(R = \infty) = P(\rho_0 = 0) = P_1(\rho_0 = 0)/2 < 1/2$, i.e., $P \circ R^{-1}$ has a finite median. So, one can construct φ verbatim as in the proof of (10.3a). For this φ, the error probabilities of types I and II coincide and are equal to $\alpha := \mathbb{E}_0(\varphi) = \mathbb{E}_1(1-\varphi)$. So, if ψ is any other test then either $\mathbb{E}_0(\psi) > \alpha$ or, by (10.3b), $\mathbb{E}_1(1-\psi) \geq \alpha$. This shows that indeed the maximum of the type I and type II error probabilities for ψ is at least as large as the corresponding maximum for φ.

10.12 Let c_n be the threshold value of the Neyman–Pearson test φ_n of size $\mathbb{E}_0(\varphi_n) = \alpha$ after n observations. Further, for $|\eta| < \min(\alpha, 1-\alpha)$ let

$$a_{n,\eta} = -n\,H(Q_0; Q_1) + \sqrt{nv_0}\,\Phi^{-1}(1-\alpha+\eta).$$

Recall from the proof of (10.4) that $\log R_n = -\sum_{i=1}^n h(X_i)$ and $\mathbb{E}_0(h) = H(Q_0; Q_1)$. Hence, applying (5.29) to the standardised sum $S_n^* = \sum_{i=1}^n (h(X_i) - \mathbb{E}_0(h))/\sqrt{nv_0}$ one finds that

$$P_0\big(\log R_n \geq a_{n,\eta}\big) = P_0\big(S_n^* \leq \Phi^{-1}(\alpha - \eta)\big) \xrightarrow[n\to\infty]{} \alpha - \eta.$$

On the other hand,

$$P_0\big(\log R_n > \log c_n\big) \leq P_0\big(\varphi_n = 1\big) \leq \alpha \leq P_0\big(\varphi_n > 0\big) \leq P_0\big(\log R_n \geq \log c_n\big).$$

In view of (1.11c), both displays together imply that, for each $\delta > 0$, $\log c_n \in [a_{n,-\delta}, a_{n,\delta}]$ for sufficiently large n. But this is an explicit way of stating the desired result.

10.14 In view of (10.10), φ has the form $\varphi = 1_{\{T > c\}} + \gamma 1_{\{T = c\}}$, where c and γ are such that $\mathbb{E}_{\vartheta_0}(\varphi) = \alpha$. Consider now the reparametrised model $(\mathcal{X}, \mathcal{F}, P'_{\vartheta'} : \vartheta' \in \Theta')$ with $\Theta' = -\Theta$ and $P'_{\vartheta'} = P_{-\vartheta'}$ for $\vartheta' \in \Theta'$. This model has increasing likelihood ratios relative to $-T$. The test $\varphi' = 1 - \varphi$ can be written in the form $\varphi' = 1_{\{-T > -c\}} + (1 - \gamma) 1_{\{-T = -c\}}$ and satisfies $\mathbb{E}'_{-\vartheta_0}(\varphi') = 1 - \alpha =: \alpha'$. According to the proof of (10.10), the power of φ' exceeds that of every other test ψ' with $\mathbb{E}'_{-\vartheta_0}(\psi') \leq \alpha'$. That is, $\mathbb{E}'_{\vartheta'}(\varphi') \geq \mathbb{E}'_{\vartheta'}(\psi')$ for all $\vartheta' > -\vartheta_0$. Now let $\psi' = 1 - \psi$ and translate the preceding inequality back into the original model.

10.18   Assume that $v_0 = 1$; otherwise replace v by v/v_0. (a) By (9.17b), the power function of the test under consideration reads

$$G(v) = \mathcal{N}_{m,v}^{\otimes n}\big((n-1)V^*/v \notin [c_1/v, c_2/v]\big) = 1 - \int_{c_1/v}^{c_2/v} \chi_{n-1}^2(x)\, dx$$

and does not depend on m. In particular, $G'(v) = -\chi_{n-1}^2(c_1/v)\, c_1/v^2 + \chi_{n-1}^2(c_2/v)\, c_2/v^2$. Now insert (9.11) to get the result.

(b) The naive choice is $c_1 = \chi_{n-1;\alpha/2}^2$ and $c_2 = \chi_{n-1;1-\alpha/2}^2$. For $\alpha = 0.02$ and $n = 3$, Table B gives $G'(1) > 0$ by (a). Hence $G(v) < G(1) = \alpha$ when $v \gtrless 1$.

(c) In view of (a), G has a unique local extremum, and this is a global minimum. Consequently, for the test to be unbiased of level α, it is necessary and sufficient that $v_0 = 1$ is the argument of this minimum, and $G(1) = \alpha$ its value. With the notation $f(c) = c^{n-1}e^{-c}$, this means that $f(c_1) = f(c_2)$ and $\chi_{n-1}^2([c_1, c_2]) = 1 - \alpha$. As the function f is unimodal with mode $n-1$, for each $s \in {]0, f(n-1)[}$ there exist precisely two solutions $c_1(s) < c_2(s)$ of the equation $f(c) = s$. Also, $c_1(\cdot)$ increases from 0 to $n-1$, and $c_2(\cdot)$ decreases from ∞ to $n-1$. This implies that $\chi_{n-1}^2([c_1(\cdot), c_2(\cdot)])$ decreases from 1 to 0, so that by the intermediate value theorem there exists a unique s with $\chi_{n-1}^2([c_1(s), c_2(s)]) = 1 - \alpha$.

(d) A computation similar to that before (10.16) shows that $R = a\, g((n-1)V^*)^{-1/2}$ with the constant $a = (n/e)^{n/2}$ and the function $g(c) = c^n e^{-c} = c\, f(c)$. A likelihood ratio test thus has an acceptance region of the form $\{g((n-1)V^*) \geq s\}$, and is therefore quite similar to the unbiased test described in part (c), except that the condition $f(c_1) = f(c_2)$ is to be replaced by the condition $g(c_1) = g(c_2)$.

10.20   By construction, $G_\varphi(m_0, v) = \alpha$ for all $v > 0$. So, the unbiasedness of φ will follow once it is shown that the power function is increasing in m. To check this let $t = t_{n-1;1-\alpha}$ and $P_{m,v} = \mathcal{N}_{m,v}^{\otimes n}$. Consider the shift mapping $S_m : (x_i)_{1 \leq i \leq n} \to (x_i + m)_{1 \leq i \leq n}$ on $\mathbb{R}^n$. Then $P_{m,v} = P_{0,v} \circ S_m^{-1}$ by Problem 2.15. Hence

$$G_\varphi(m, v) := P_{0,v}\big(M \circ S_m > m_0 + t\sqrt{V^* \circ S_m/n}\,\big)$$
$$= P_{0,v}\big(M + m > m_0 + t\sqrt{V^*/n}\,\big),$$

and (1.11c) shows that the last expression is increasing in m. Compare Figure 10.5.

10.22   Let M_n and V_n^* be as in (7.29), and for $\vartheta = (m, v)$ let $P_\vartheta = \mathcal{N}_{m,v}^{\otimes \mathbb{N}}$. The t-test φ_n is determined by its rejection region $\{M_n > -u_n\sqrt{V_n^*/n}\}$, where $u_n = t_{n-1;\alpha} = -t_{n-1;1-\alpha}$. Now let $\vartheta = (m, v)$ be fixed and consider the standardised mean $M_n^* = (M_n - m)\sqrt{n/v}$, which by (9.17b) is standard normal. With the abbreviation $b_n = m\sqrt{n/v}$, the power function can then be written as

$$G_\varphi(\vartheta) = P_\vartheta\big(M_n^* > -u_n\sqrt{V^*/v} - b_n\big).$$

Theorem (7.29) ensures that $V_n^* \xrightarrow{P_\vartheta} v$, and Problem 9.13 implies that $u_n \to u := \Phi^{-1}(\alpha)$. So, for each $\varepsilon > 0$ one can conclude that $P_\vartheta(|u_n\sqrt{V^*/v} - u| \geq \varepsilon) \leq \varepsilon$ for sufficiently large n. For these n, it follows that

$$G_\varphi(\vartheta) \leq P_\vartheta\big(M_n^* > -u - \varepsilon - b_n\big) + \varepsilon = \Phi(u + \varepsilon + b_n) + \varepsilon \leq \Phi(u + b_n) + 2\varepsilon$$

because $\Phi' \leq 1$. Likewise, $G_\varphi(\vartheta) \geq \Phi(u + b_n) - 2\varepsilon$, and the proof is complete.

10.24   One can proceed in a similar way as before Theorem (10.18). Let $n = k + l$, $X = (X_1, \ldots, X_k)^\top$, X' be the analogous l-vector, and

$$\tilde{V}_{m,m'} = |X - m\mathbf{1}|^2/n + |X' - m'\mathbf{1}|^2/n = \frac{n-2}{n} V^* + \frac{k}{n}(M - m)^2 + \frac{l}{n}(M' - m')^2 \,;$$

the last equality comes from (7.10). By hypothesis, the likelihood function is given by

$$\rho_{m,m';v} = (2\pi v)^{-n/2} \exp\left[-\frac{n}{2v} \tilde{V}_{m,m'}\right],$$

and its maximum over v is attained at $v = \tilde{V}_{m,m'}$. Hence, $\sup_{v>0} \rho_{m,m';v} = (2\pi e \tilde{V}_{m,m'})^{-n/2}$. The generalised likelihood ratio R in (10.14) thus satisfies the identity

$$R^{2/n} = \inf_{m,m': m \leq m'} \tilde{V}_{m,m'} \Big/ \inf_{m,m': m > m'} \tilde{V}_{m,m'} \,.$$

In the case $M \leq M'$, the infimum in the numerator is attained for $m = M$ and $m' = M'$; so, in this case the numerator equals $\tilde{V}_{M,M'} = \frac{n-2}{n} V^*$. In the alternative case $M > M'$, the infimum in the numerator can only be attained when $m = m'$, and solving for $\frac{d}{dm} \tilde{V}_{m,m} = 0$ one then finds that necessarily

$$m = m' = \frac{k}{n} M + \frac{l}{n} M',$$

so that the minimal value is $\frac{n-2}{n} V^* + \frac{kl}{n^2}(M - M')^2$. Together with a similar computation for the denominator, one thus arrives at the identity

$$R^{2/n} = \begin{cases} 1 + T^2 & \text{for } T \geq 0, \\ 1/(1 + T^2) & \text{for } T \leq 0, \end{cases}$$

where $T = (M - M')/\sqrt{(\frac{1}{k} + \frac{1}{l})V^*}$. Hence, R is an increasing function of T, and this gives the result. *Note:* In (12.20) it will be shown that T has the t_{n-2}-distribution.

10.25   (a) Relative to P_ϑ with $\vartheta \in \Theta_0$, $F \circ T$ is uniformly distributed on $]0, 1[$; see Problem 1.18. By symmetry, it follows that also $p(\cdot)$ has the uniform distribution.

(b) Writing α for the size of the test with rejection region $\{T > c\}$, we have by hypothesis that $c = F^{-1}(1 - \alpha)$. Hence $\{T > c\} = \{F \circ T > 1 - \alpha\} = \{p(\cdot) < \alpha\}$.

(c) Let $\vartheta \in \Theta_0$. Under P_ϑ, the random variables $-2 \log p_i(\cdot)$ are independent and, by (a) and (3.40), $\mathcal{E}_{1/2}$-distributed. Due to (3.36) or (9.7), S thus has the distribution $\Gamma_{1/2,n} = \chi^2_{2n}$. The rest is evident.

Chapter 11

11.4   Consider the likelihood ratio

$$R_{\vartheta,n} = \prod_{k=1}^n \frac{\vartheta(X_k)}{\rho(X_k)} = \prod_{i \in E} \left(\frac{\vartheta(i)}{\rho(i)}\right)^{h_n(i)} = \exp\big(n\, H(L_n; \rho) - n\, H(L_n; \vartheta)\big)$$

of P_ϑ relative to P_ρ after n observations. Changing the measure as in the proof of Theorem (10.4) one obtains

$$P_\vartheta(D_{n,\rho} \leq c) = \mathbb{E}_\rho\big(R_{\vartheta,n} \mathbf{1}_{\{D_{n,\rho} \leq c\}}\big).$$

Pick any $\varepsilon > 0$ and follow the proof of Proposition (11.10) to deduce the following: If n is sufficiently large then, on the one hand, $n\,H(L_n;\rho) \le c$ on the set $\{D_{n,\rho} \le c\}$. On the other hand, L_n is then so close to ρ that $H(L_n;\vartheta) \ge H(\rho;\vartheta) - \varepsilon$, by the continuity of $H(\cdot\,;\vartheta)$. Both observations together yield an upper estimate of $R_{\vartheta,n}$ on the set $\{D_{n,\rho} \le c\}$, and thus of $P_\vartheta(D_{n,\rho} \le c)$, which immediately gives the result.

11.6 Since $h_n(i) = \sum_{k=1}^n 1_{\{X_k = i\}}$, one can write

$$T_n = \frac{1}{\sqrt{n}} \sum_{k=1}^n X_k^* \quad \text{with} \quad X_k^* = \frac{X_k - (s+1)/2}{\sqrt{(s^2-1)/12}}.$$

Further, under the null hypothesis that each X_k is uniformly distributed, one observes that $\mathbb{E}_\rho(X_k) = \frac{1}{s}\sum_{i=1}^s i = (s+1)/2$ and $\mathbb{E}_\rho(X_k^2) = \frac{1}{s}\sum_{i=1}^s i^2 = (s+1)(2s+1)/6$, so that $\mathbb{V}_\rho(X_k) = (s^2-1)/12$. That is, X_k^* is standardised. The first claim thus follows from (5.29). A natural candidate for a test of H_0 against H_1' of asymptotic size α is therefore the test with rejection region $\{T_n < \Phi^{-1}(\alpha)\}$. On the restricted alternative H_0', this test has a larger power than the χ^2-test because the χ^2-statistic $D_{n,\rho}$ takes no advantage of the particular structure of H_0' and is therefore less sensitive to deviations of L_n from ρ towards H_0'.

11.11 Part (a) is elementary; note that, e.g., $\vartheta(12) = \vartheta^A(1) - \vartheta(11)$.
  (b) For all n_A and n_B in $\{0,\ldots,n\}$ and all $0 \le k \le \min(n_A, n_B)$ one can write

$$P_\vartheta\big(h_n(11) = k,\, h_n^A(1) = n_A,\, h_n^B(1) = n_B\big)$$
$$= \mathcal{M}_{n,\vartheta}\big(\{(k, n_A-k, n_B-k, n-n_A-n_B+k)\}\big)$$
$$= \binom{n}{k,\, n_A-k,\, n_B-k,\, n-n_A-n_B+k}$$
$$\qquad \cdot\, \vartheta(11)^k\, \vartheta(12)^{n_A-k}\, \vartheta(21)^{n_B-k}\, \vartheta(22)^{n-n_A-n_B+k}.$$

The multinomial coefficient in the last expression differs from $\mathcal{H}_{n_B;n_A,n-n_A}(\{k\})$ by a factor which depends on n, n_A, n_B but not on k. Further, for $\vartheta \in \Theta_0$, the ϑ-dependent term above coincides with $\vartheta^A(1)^{n_A}\, \vartheta^B(1)^{n_B}\, \vartheta^A(2)^{n-n_A}\, \vartheta^B(2)^{n-n_B}$, which also does not depend on k. This implies the first equation, and the second follows by symmetry.
  (c) To obtain a non-randomised test of level α one can proceed as follows. For any two frequencies $n_A, n_B \in \{0,\ldots,n\}$ determine two numbers $c_\pm = c_\pm(n_A, n_B) \in \{0,\ldots,n\}$ such that $c_- \le n_A n_B/n \le c_+$ and $\mathcal{H}_{n_B;n_A,n-n_A}(\{c_-,\ldots,c_+\}) \ge 1-\alpha$. Then let φ be the test with acceptance region

$$\Big\{c_-\big(h_n^A(1), h_n^B(1)\big) \le h_n(11) \le c_+\big(h_n^A(1), h_n^B(1)\big)\Big\}.$$

In view of (b), this φ has level α. In the one-sided case one should simply take $c_- = 0$.

11.12 (a) Define $p_\pm = Q_1(\mathbb{R}_\pm)$. As $\mu(Q_1) > 0$, one has $p_- < p_+$. By definition, Q_0 has the density

$$\rho_0 = 1_{]-\infty,0]}\,\frac{\rho_1}{2p_-} + 1_{]0,\infty[}\,\frac{\rho_1}{2p_+}.$$

Hence, $\rho_1/\rho_0 = 2p_-\,1_{]-\infty,0]} + 2p_+\,1_{]0,\infty[} = 2p_+\, r^{1_{]-\infty,0]}}$, where $r = p_-/p_+$. The likelihood ratio after n observations is therefore $R = \rho_1^{\otimes n}/\rho_0^{\otimes n} = (2p_+)^n\, r^{S_n^-}$. In particular, R is a strictly decreasing function of S_n^-. It follows that $\varphi \circ S_n^-$ is a Neyman–Pearson test of $Q_0^{\otimes n}$ against $Q_1^{\otimes n}$. Moreover, $\mathbb{E}_0(\varphi \circ S_n^-) = \alpha$ because, under $Q_0^{\otimes n}$, S_n^- is binomially distributed with parameter $Q_0(]-\infty, 0]) = 1/2$.

(b) For each Q in the null hypothesis H_0 one has $\vartheta(Q) := Q(]-\infty, 0]) \geq 1/2$. Hence, by the choice of φ, $\mathbb{E}_Q(\varphi \circ S_n^-) = \mathbb{E}_{\mathcal{B}_{n,\vartheta(Q)}}(\varphi) \leq \alpha$. That is, $\varphi \circ S_n^-$ has level α. For Q_1 and Q_0 as in (a), Theorem (10.3) implies that $\varphi \circ S_n^-$ is at least as powerful at Q_1 as every other test of level α. This holds for every Q_1 in the alternative, so $\varphi \circ S_n^-$ is UMP.

11.14 (a) $\Rightarrow$ (b): For $i = 1, 2$ let F_i be the distribution function of Q_i, and X_i the associated quantile transformation as in (1.30). By (a), $F_1 \geq F_2$ on $\mathbb{R}$, and thus $X_1 \leq X_2$ on $]0, 1[$. So, by (1.30), (b) holds with $P = \mathcal{U}_{]0,1[}$. (b) $\Rightarrow$ (c): Since $\mathbb{E}_{Q_i}(f) = \mathbb{E}_P(f \circ X_i)$, one can apply (4.11a). (c) $\Rightarrow$ (a): Consider $f = 1_{]c,\infty[}$.

11.18 (a) Note that $P(Z_i = 1, |X_i| \leq c) = P(0 < X_i \leq c) = P(|X_i| \leq c)/2$ for all $c > 0$, where the second equality follows from the symmetry assumption. Now apply (3.19).

(b) By (a) and the independence of the X_i, the family $Z_1, |X_1|, \ldots, Z_n, |X_n|$ is independent. Together with (3.24), this gives the claim.

(c) By the above, $\vec{Z} = (Z_1, \ldots, Z_n)$ is a Bernoulli sequence with parameter $1/2$, and therefore uniformly distributed on $\{0, 1\}^n$. But Z arises from $\vec{Z}$ by the natural identification of $\{0, 1\}^n$ and $\mathscr{P}_n$. The uniform distribution of R^+ follows as in the first part of the proof of (11.25).

(d) For $A \in \mathscr{P}_n$ let $R^+(A) = \{R_i^+ : i \in A\}$. The random permutation R^+ can thus be viewed as a bijection from $\mathscr{P}_n$ onto itself. Define $\mathscr{A}_{l,n} = \{A \in \mathscr{P}_n : \sum_{i \in A} i = l\}$. Then by (b) and (c)

$$P(W^+ = l) = \sum_{A \in \mathscr{P}_n} P(Z = A) \, P\left(\sum_{i \in A} R_i^+ = l\right) = 2^{-n} \sum_{A \in \mathscr{P}_n} P\left(R^+(A) \in \mathscr{A}_{l,n}\right)$$

$$= 2^{-n} \, \mathbb{E}\left(|(R^+)^{-1}\mathscr{A}_{l,n}|\right) = 2^{-n} \, \mathbb{E}\left(|\mathscr{A}_{l,n}|\right) = 2^{-n} \, |\mathscr{A}_{l,n}| = 2^{-n} N(l; n) \,.$$

11.20 (a) The empirical distribution function F_n is piecewise constant with jumps of size $1/n$ at the points $X_{i:n}$, and F is increasing. Hence, the absolute difference $|F_n - F|$ attains its maximum at one of the points $X_{i:n}$.

(b) By Problem 1.18, the random variables $U_i = F(X_i)$ are uniform on $]0, 1[$. By the monotonicity of F, one has $U_{i:n} = F(X_{i:n})$. In view of (a), it follows that the distribution of Δ_n only depends on the joint distribution of the U_i, which is $\mathcal{U}_{]0,1[}^n$.

Chapter 12

12.3 (a) The mean squared error and the least-squares estimator are

$$E_\gamma = \frac{1}{n} \sum_{k=1}^n \left(X_k - \gamma X_{k-1}\right)^2 \quad \text{and} \quad \hat{\gamma} = \sum_{k=1}^n X_k X_{k-1} \Big/ \sum_{k=1}^n X_{k-1}^2 \,.$$

(b) For fixed parameters γ, v let B be the matrix with entries $B_{ij} = \delta_{ij} - \gamma \delta_{i-1,j}$. In particular, $\det B = 1$. The observation vector $X = (X_1, \ldots, X_n)^\top$ and the error vector $\xi = (\xi_1, \ldots, \xi_n)^\top$ are then related by $BX = \sqrt{v}\, \xi$ resp. $X = \sqrt{v}\, B^{-1}\xi$. Since by hypothesis ξ is multivariate standard normal, (9.2) shows that X has distribution $\mathcal{N}_n(0, v(B^\top B)^{-1})$. But $\phi_{0, v(B^\top B)^{-1}}(x) = (2\pi v)^{-n/2} \exp[-|Bx|^2/2v] = \rho_{\gamma,v}(x)$ for $x \in \mathbb{R}^n$.

(c) By continuity, the sup of $\rho_{\gamma,v}$ over the alternative coincides with its maximum over *all* γ, v. Hence, by (a) and (b), one has $\sup_{\gamma,v} \rho_{\gamma,v} = \sup_{v>0} \rho_{\hat{\gamma},v} = (2\pi e E_{\hat{\gamma}})^{-n/2}$. Likewise,

$\sup_{v>0} \rho_{0,v} = (2\pi e E_0)^{-n/2}$. A short computation thus shows that the likelihood ratio equals $R = (1 - \hat{r}^2)^{-n/2}$.

12.5   As the test object has mass 1, f coincides with the object's acceleration. So, case (a) is described by the linear model $X_k = f\, t_k^2 + \sqrt{v}\, \xi_k$. Solving the normal equation for f one finds the least-squares estimator

$$\hat{f} = \sum_{k=1}^{n} t_k^2 X_k \Big/ \sum_{k=1}^{n} t_k^4 \,.$$

Case (b) corresponds to the quadratic regression model $X_k = a + b\, t_k + f t_k^2 + \sqrt{v}\, \xi_k$. Reasoning as in Section 12.1 one obtains three normal equations for the unknowns a, b, f. In terms of the notations introduced there, one ends up with the least-squares estimator

$$\hat{f} = \frac{c(t^2, X) - c(t, t^2)\, c(t, X)}{V(t)\, V(t^2) - c(t, t^2)^2} \,,$$

where $t = (t_1, \ldots, t_n)^\top$ and $t^2 = (t_1^2, \ldots, t_n^2)^\top$. As these two vectors are linearly independent, the Cauchy–Schwarz inequality shows that the denominator above is strictly positive.

12.6   Fix any $\gamma > 0$ and let B be as in the solution to Problem 12.3. Then $B\xi = \eta$. The vector $Y = BX$ satisfies the equation $Y = mb + \sqrt{v}\, \eta$, where $b = B1$ is the vector with coordinates $b_i = 1 - \gamma(1 - \delta_{i1})$. Theorem (12.15a) thus shows that $\hat{m} = (b^\top b)^{-1} b^\top BX$ is an unbiased estimator of m. But $\hat{m} = S$, as is easily checked. Next, (12.15b) states that S is strictly better than every other linear unbiased estimator of m in the model $Y = mb + \sqrt{v}\, \eta$. In particular, it is strictly better than its competitor $M(X) = M(B^{-1}Y)$.

12.12   (a) By hypothesis, the likelihood function reads

$$\rho(X, \vartheta(\gamma)) = \exp\left[(A\gamma)^\top X - \sum_{i=1}^{n} b \circ a^{-1}(A_i\, \gamma) \right] \prod_{i=1}^{n} h(X_i)$$

with b and h as in (7.21). Next, by (7.23a), $(b \circ a^{-1})' = b' \circ a^{-1} / a' \circ a^{-1} = \tau \circ a^{-1}$, and thus $(\partial/\partial\gamma_j)\, b \circ a^{-1}(A_i\, \gamma) = \tau \circ a^{-1}(A_i\, \gamma)\, A_{ij}$. The claim is now evident.

(b) By definition, the Gaussian linear model with design matrix A and fixed variance $v > 0$ has the expectation vector $\vartheta = A\gamma \in \mathbb{R}^n$. In view of (7.27a), $a(\lambda) = \lambda/v$ for $\lambda \in \mathbb{R}$, so that $a(\vartheta) = A_v \gamma$ with $A_v = A/v$. That is, we are in the setting of (a) with A_v in place of A. Also, $\tau(\lambda) = \mathbb{E}(\mathcal{N}_{\lambda,v}) = \lambda$. The equation in (a) for the maximum likelihood estimator is therefore equivalent to the equation $A^\top(X - A\gamma) = 0$, which is solved by the least-squares estimator $\hat{\gamma}$ of Theorem (12.15a).

12.13   (a) In this case, $\Lambda = \,]0, 1[$, $Q_\lambda = \mathcal{B}_{1,\lambda}$ and, by Example (7.25), $a(\lambda) = \log \frac{\lambda}{1-\lambda}$ with the inverse $f(t) = a^{-1}(t) = (1 + e^{-t})^{-1}$. Also, $\tau(\lambda) = \lambda$. The matrix A can be chosen as in (12.13).

(b) Here, $\Lambda = \,]0, \infty[$, $Q_\lambda = \mathcal{P}_\lambda$ and, by Example (7.26), $a(\lambda) = \log \lambda$ and $\tau(\lambda) = \lambda$. The maximum likelihood equation of Problem 12.12a thus takes the form $A^\top(X - \exp(A\gamma)) = 0$, where the exponential function is to be applied coordinate-wise.

12.15   The data hold the length (in mm) of $n = 120$ cuckoo eggs in the nests of $s = 6$ different host species. The relevant values are the following.

host species i	n_i	M_i	V_i^*	
meadow pipit	45	22.30	0.85	$M = 22.46$
tree pipit	15	23.09	0.81	$V_{bg}^* = 8.59$
hedge sparrow	14	23.12	1.14	
robin	16	22.57	0.47	$V_{wg}^* = 0.83$
pied wagtail	15	22.90	1.14	$F = 10.39$
wren	15	21.13	0.55	

The F-ratio is significantly larger than $f_{5,114;0.99} = 3.18$, so that the null hypothesis can be clearly rejected at level 0.01. In fact, the p-value is as minute as $\approx 3 \cdot 10^{-8}$. That is, the size of the cuckoo eggs is adapted to the host species.

12.17 In contrast to (12.34), the number of observations per cell is now $\ell = 1$, so that $B = G = G_1 \times G_2$ and $n = s_1 s_2$. A vector $x \in \mathbb{R}^G$ belongs to L if and only if $x_{ij} = \overline{x}_{i\bullet} + \overline{x}_{\bullet j} - \overline{x}$ for all $ij \in G$ (using the notation of (12.34)). Consequently, $\Pi_L X = (\overline{X}_{i\bullet} + \overline{X}_{\bullet j} - \overline{X})_{ij \in G}$, as is verified by a short computation. The null hypothesis H_0 is described by the space H of all $x \in L$ which are such that $\overline{x}_{i\bullet} = \overline{x}$ for all $i \in G_1$. Equivalently, H is the space of all $x \in \mathbb{R}^G$ with $x_{ij} = \overline{x}_{\bullet j}$ for all $ij \in G$. In particular, $\Pi_H X = (\overline{X}_{\bullet j})_{ij \in G}$. Moreover, $\dim L = s_1 + s_2 - 1$ and $\dim H = s_2$. The associated Fisher statistic is therefore given by

$$F_{H,L} := \frac{(s_1-1)(s_2-1)}{s_1-1} \frac{|\Pi_L X - \Pi_H X|^2}{|X - \Pi_L X|^2} = \frac{s_2(s_2-1) \sum_{i \in G_1} (\overline{X}_{i\bullet} - \overline{X})^2}{\sum_{ij \in G} (X_{ij} - \overline{X}_{i\bullet} - \overline{X}_{\bullet j} + \overline{X})^2},$$

which by (12.17) and the Gaussian assumption has distribution $\mathcal{F}_{s_1-1,(s_1-1)(s_2-1)}$. This leads to an F-test according to (12.20b).

12.19 (a) As R is a random permutation of $\{1, \ldots, n\}$, one finds

$$\sum_{ik \in B} R_{ik} = \sum_{j=1}^n j = \frac{n(n+1)}{2} \quad \text{and} \quad (n-1)\, V_{tot}^*(R) = \sum_{j=1}^n \left(j - \frac{n+1}{2}\right)^2.$$

The rest is straightforward.

(b) In view of (a) and the definition of T, one has $T = (s-1)\, V_{bg}^*(R)/V_{tot}^*(R)$. Also, by the partitioning of variation, $(n-s)\, V_{wg}^* = (n-1)\, V_{tot}^* - (s-1)\, V_{bg}^*$. Hence,

$$\frac{V_{bg}^*(R)}{V_{wg}^*(R)} = \frac{n-s}{s-1} \frac{T}{n-1-T}.$$

In particular, the denominator is always non-negative.

(c) Lemma (11.24) shows that $M_1(R) - (n+1)/2 = (U_{n_1,n_2} - n_1 n_2/2)/n_1$ and likewise $M_2(R) - (n+1)/2 = (n_1 n_2/2 - U_{n_1,n_2})/n_2$. Insert this into the definition of T.

(d) Let $B_i = \{ik : 1 \le k \le n_i\}$ and $U_i = \sum_{ik \in B_i} \sum_{jl \in B_i^c} 1_{\{X_{ik} > X_{jl}\}}$. Lemma (11.24) then shows that $n_i M_i(R) = U_i + n_i(n_i + 1)/2$. Hence, under the null hypothesis, one has $\mathbb{E}_Q(M_i(R)) = (n+1)/2$. Together with Step 1 of the proof of (11.28), this gives

$$\mathbb{E}_Q\big((M_i(R) - (n+1)/2)^2\big) = \mathbb{V}_Q(U_i)/n_i^2 = (n-n_i)(n+1)/(12 n_i)$$

and therefore $\mathbb{E}_Q(T) = s - 1$. Next, as shown before (11.26), R is uniformly distributed on $\mathscr{S}_n$. So, if $R_i = \{R_{ik} : ik \in B_i\}$ then $[R] := (R_i)_{1 \le i \le s}$ is uniformly distributed on the set of all ordered partitions of $\{1, \ldots, n\}$ into s subsets of cardinalities $n_1, \ldots, n_s$. From the proof of (11.25) it is also known that $U_i = \sum_{k=1}^{n_i} m_k(R_i)$. One can therefore conclude that the joint distribution of the U_i only depends on the distribution of $[R]$, which does not depend on Q. This fact carries over to the joint distribution of the $M_i(R)$, and thus to the distribution of T.

(e) Define $v_i = n_i n(n+1)/12$, $U_i^* = (U_i - n_i(n-n_i)/2)/\sqrt{v_i}$ and $U^* = (U_1^*, \ldots, U_s^*)^\top$. Then $T = |U^*|^2$ by the proof of (d) above. In view of statement (d), one can assume that the X_{ik} are uniformly distributed on $]0, 1[$, and then proceed as in the proof of (11.28) by replacing the indicator functions $1_{\{X_{ik} > X_{jl}\}}$ with the differences $X_{ik} - X_{jl}$. Consider, therefore, the random vector Z with coordinates $Z_i = n S_i - n_i \sum_j S_j$, where $S_i = \sum_{ik \in B_i}(X_{ik} - \frac{1}{2})$. Consider further the standardised quantities

$$S_i^* = S_i / \sqrt{n_i/12} \quad \text{and} \quad Z_i^* = Z_i / \sqrt{v_i}$$

as well as the random vectors $S^* = (S_1^*, \ldots, S_s^*)^\top$ and $Z^* = (Z_1^*, \ldots, Z_s^*)^\top$. Then one can write $Z^* = A S^*$, where A is the matrix with entries

$$\mathsf{A}_{ij} = \sqrt{n/(n+1)}\,\delta_{ij} - \sqrt{n_i n_j / n(n+1)}\,.$$

Now one can pass to the limit $\min_i n_i \to \infty$. Step 2 of the proof of (11.28) shows that then $U^* - Z^* \to 0$ in probability. Using (5.29) and the independence of the S_i^*, one also finds that $S^* \xrightarrow{d} \mathcal{N}_s(0, \mathsf{E})$. Moreover, a subsequence trick as in Step 3 of the proof of (11.28) allows to assume that $n_i/n \to a_i$ for some $a_i \in [0, 1]$. Let u be the column vector with coordinates $u_i = \sqrt{a_i}$. Then $|u| = 1$ and $\mathsf{A} \to \mathsf{E} - uu^\top$. This limiting matrix satisfies $\mathsf{O}(\mathsf{E} - uu^\top) = \mathsf{E}_{s-1}\mathsf{O}$ for any orthogonal matrix O with last row $u^\top$. Hence

$$\mathsf{O}Z^* = \mathsf{O}\mathsf{A}S^* \xrightarrow{d} \mathcal{N}_s(0, \mathsf{E}_{s-1})$$

by (11.2c) and (9.4). Finally, (11.2a) and (9.10) yield $|Z^*|^2 = |\mathsf{O}Z^*|^2 \xrightarrow{d} \chi_{s-1}^2$, and (11.2b) then implies the claim $T = |U^*|^2 \xrightarrow{d} \chi_{s-1}^2$.

Tables

A Normal Distribution

Cumulative distribution function $\Phi(c) = \mathcal{N}_{0,1}(]-\infty, c]) = 1 - \Phi(-c)$ of the standard normal distribution. Its value for $c = 1.16$, say, can be found in row 1.1 and column .06: $\Phi(1.16) = 0.8770$. The α-quantile of $\mathcal{N}_{0,1}$ is determined by localising α in the table and adding the values of the associated row and column: $\Phi^{-1}(0.975) = 1.96$; some quantiles can also be found in Table C. For large values of c see Problem 5.15.

c	.00	.01	.02	.03	.04	.05	.06	.07	.08	.09
0.0	.5000	.5040	.5080	.5120	.5160	.5199	.5239	.5279	.5319	.5359
0.1	.5398	.5438	.5478	.5517	.5557	.5596	.5636	.5675	.5714	.5753
0.2	.5793	.5832	.5871	.5910	.5948	.5987	.6026	.6064	.6103	.6141
0.3	.6179	.6217	.6255	.6293	.6331	.6368	.6406	.6443	.6480	.6517
0.4	.6554	.6591	.6628	.6664	.6700	.6736	.6772	.6808	.6844	.6879
0.5	.6915	.6950	.6985	.7019	.7054	.7088	.7123	.7157	.7190	.7224
0.6	.7257	.7291	.7324	.7357	.7389	.7422	.7454	.7486	.7517	.7549
0.7	.7580	.7611	.7642	.7673	.7704	.7734	.7764	.7794	.7823	.7852
0.8	.7881	.7910	.7939	.7967	.7995	.8023	.8051	.8078	.8106	.8133
0.9	.8159	.8186	.8212	.8238	.8264	.8289	.8315	.8340	.8365	.8389
1.0	.8413	.8438	.8461	.8485	.8508	.8531	.8554	.8577	.8599	.8621
1.1	.8643	.8665	.8686	.8708	.8729	.8749	.8770	.8790	.8810	.8830
1.2	.8849	.8869	.8888	.8907	.8925	.8944	.8962	.8980	.8997	.9015
1.3	.9032	.9049	.9066	.9082	.9099	.9115	.9131	.9147	.9162	.9177
1.4	.9192	.9207	.9222	.9236	.9251	.9265	.9279	.9292	.9306	.9319
1.5	.9332	.9345	.9357	.9370	.9382	.9394	.9406	.9418	.9429	.9441
1.6	.9452	.9463	.9474	.9484	.9495	.9505	.9515	.9525	.9535	.9545
1.7	.9554	.9564	.9573	.9582	.9591	.9599	.9608	.9616	.9625	.9633
1.8	.9641	.9649	.9656	.9664	.9671	.9678	.9686	.9693	.9699	.9706
1.9	.9713	.9719	.9726	.9732	.9738	.9744	.9750	.9756	.9761	.9767
2.0	.9772	.9778	.9783	.9788	.9793	.9798	.9803	.9808	.9812	.9817
2.1	.9821	.9826	.9830	.9834	.9838	.9842	.9846	.9850	.9854	.9857
2.2	.9861	.9864	.9868	.9871	.9875	.9878	.9881	.9884	.9887	.9890
2.3	.9893	.9896	.9898	.9901	.9904	.9906	.9909	.9911	.9913	.9916
2.4	.9918	.9920	.9922	.9925	.9927	.9929	.9931	.9932	.9934	.9936
2.5	.9938	.9940	.9941	.9943	.9945	.9946	.9948	.9949	.9951	.9952
2.6	.9953	.9955	.9956	.9957	.9959	.9960	.9961	.9962	.9963	.9964
2.7	.9965	.9966	.9967	.9968	.9969	.9970	.9971	.9972	.9973	.9974
2.8	.9974	.9975	.9976	.9977	.9977	.9978	.9979	.9979	.9980	.9981
2.9	.9981	.9982	.9982	.9983	.9984	.9984	.9985	.9985	.9986	.9986
3.0	.9987	.9987	.9987	.9988	.9988	.9989	.9989	.9989	.9990	.9990

B   Chi-square and Gamma Distributions

Quantiles $\chi^2_{n;\alpha}$ of the chi-square distributions $\chi^2_n = \Gamma_{1/2,n/2}$ with n degrees of freedom. $\chi^2_{n;\alpha}$ is defined as the value $c > 0$ satisfying $\chi^2_n([0,c]) = \alpha$. By a change of scale, one obtains the quantiles of the gamma distributions $\Gamma_{\lambda,r}$ with $\lambda > 0$ and $2r \in \mathbb{N}$. For large n one can use the approximations in Problems 9.11 and 9.12. Notation: $^{-5}3.9 = 3.9 \cdot 10^{-5}$.

$\alpha =$	0.005	0.01	0.02	0.05	0.1	0.9	0.95	0.98	0.99	0.995
$n = 1$	$^{-5}3.9$	$^{-4}1.6$	$^{-4}6.3$	$^{-3}3.9$	.0158	2.706	3.841	5.412	6.635	7.879
2	.0100	.0201	.0404	.1026	.2107	4.605	5.991	7.824	9.210	10.60
3	.0717	.1148	.1848	.3518	.5844	6.251	7.815	9.837	11.34	12.84
4	.2070	.2971	.4294	.7107	1.064	7.779	9.488	11.67	13.28	14.86
5	.4117	.5543	.7519	1.145	1.610	9.236	11.07	13.39	15.09	16.75
6	.6757	.8721	1.134	1.635	2.204	10.64	12.59	15.03	16.81	18.55
7	.9893	1.239	1.564	2.167	2.833	12.02	14.07	16.62	18.48	20.28
8	1.344	1.646	2.032	2.733	3.490	13.36	15.51	18.17	20.09	21.95
9	1.735	2.088	2.532	3.325	4.168	14.68	16.92	19.68	21.67	23.59
10	2.156	2.558	3.059	3.940	4.865	15.99	18.31	21.16	23.21	25.19
11	2.603	3.053	3.609	4.575	5.578	17.28	19.68	22.62	24.72	26.76
12	3.074	3.571	4.178	5.226	6.304	18.55	21.03	24.05	26.22	28.30
13	3.565	4.107	4.765	5.892	7.042	19.81	22.36	25.47	27.69	29.82
14	4.075	4.660	5.368	6.571	7.790	21.06	23.68	26.87	29.14	31.32
15	4.601	5.229	5.985	7.261	8.547	22.31	25.00	28.26	30.58	32.80
16	5.142	5.812	6.614	7.962	9.312	23.54	26.30	29.63	32.00	34.27
17	5.697	6.408	7.255	8.672	10.09	24.77	27.59	31.00	33.41	35.72
18	6.265	7.015	7.906	9.390	10.86	25.99	28.87	32.35	34.81	37.16
19	6.844	7.633	8.567	10.12	11.65	27.20	30.14	33.69	36.19	38.58
20	7.434	8.260	9.237	10.85	12.44	28.41	31.41	35.02	37.57	40.00
21	8.034	8.897	9.915	11.59	13.24	29.62	32.67	36.34	38.93	41.40
22	8.643	9.542	10.60	12.34	14.04	30.81	33.92	37.66	40.29	42.80
23	9.260	10.20	11.29	13.09	14.85	32.01	35.17	38.97	41.64	44.18
24	9.886	10.86	11.99	13.85	15.66	33.20	36.42	40.27	42.98	45.56
25	10.52	11.52	12.70	14.61	16.47	34.38	37.65	41.57	44.31	46.93
26	11.16	12.20	13.41	15.38	17.29	35.56	38.89	42.86	45.64	48.29
27	11.81	12.88	14.13	16.15	18.11	36.74	40.11	44.14	46.96	49.64
28	12.46	13.56	14.85	16.93	18.94	37.92	41.34	45.42	48.28	50.99
29	13.12	14.26	15.57	17.71	19.77	39.09	42.56	46.69	49.59	52.34
30	13.79	14.95	16.31	18.49	20.60	40.26	43.77	47.96	50.89	53.67
35	17.19	18.51	20.03	22.47	24.80	46.06	49.80	54.24	57.34	60.27
40	20.71	22.16	23.84	26.51	29.05	51.81	55.76	60.44	63.69	66.77
45	24.31	25.90	27.72	30.61	33.35	57.51	61.66	66.56	69.96	73.17
50	27.99	29.71	31.66	34.76	37.69	63.17	67.50	72.61	76.15	79.49
55	31.73	33.57	35.66	38.96	42.06	68.80	73.31	78.62	82.29	85.75
60	35.53	37.48	39.70	43.19	46.46	74.40	79.08	84.58	88.38	91.95
70	43.28	45.44	47.89	51.74	55.33	85.53	90.53	96.39	100.4	104.2
80	51.17	53.54	56.21	60.39	64.28	96.58	101.9	108.1	112.3	116.3
90	59.20	61.75	64.63	69.13	73.29	107.6	113.1	119.6	124.1	128.3
100	67.33	70.06	73.14	77.93	82.36	118.5	124.3	131.1	135.8	140.2

C   Student's t-Distributions

Quantiles $t_{n;\alpha}$ of the t-distributions t_n with n degrees of freedom. $t_{n;\alpha}$ is the value $c > 0$ such that $t_n(]-\infty, c]) = \alpha$. The row $n = \infty$ contains the quantiles $\lim_{n\to\infty} t_{n;\alpha} = \Phi^{-1}(\alpha)$ of the standard normal distribution, see Problem 9.13.

$\alpha =$	0.9	0.95	0.96	0.975	0.98	0.99	0.995
$n = 1$	3.078	6.314	7.916	12.71	15.89	31.82	63.66
2	1.886	2.920	3.320	4.303	4.849	6.965	9.925
3	1.638	2.353	2.605	3.182	3.482	4.541	5.841
4	1.533	2.132	2.333	2.776	2.999	3.747	4.604
5	1.476	2.015	2.191	2.571	2.757	3.365	4.032
6	1.440	1.943	2.104	2.447	2.612	3.143	3.707
7	1.415	1.895	2.046	2.365	2.517	2.998	3.499
8	1.397	1.860	2.004	2.306	2.449	2.896	3.355
9	1.383	1.833	1.973	2.262	2.398	2.821	3.250
10	1.372	1.812	1.948	2.228	2.359	2.764	3.169
11	1.363	1.796	1.928	2.201	2.328	2.718	3.106
12	1.356	1.782	1.912	2.179	2.303	2.681	3.055
13	1.350	1.771	1.899	2.160	2.282	2.650	3.012
14	1.345	1.761	1.887	2.145	2.264	2.624	2.977
15	1.341	1.753	1.878	2.131	2.249	2.602	2.947
16	1.337	1.746	1.869	2.120	2.235	2.583	2.921
17	1.333	1.740	1.862	2.110	2.224	2.567	2.898
18	1.330	1.734	1.855	2.101	2.214	2.552	2.878
19	1.328	1.729	1.850	2.093	2.205	2.539	2.861
20	1.325	1.725	1.844	2.086	2.197	2.528	2.845
21	1.323	1.721	1.840	2.080	2.189	2.518	2.831
22	1.321	1.717	1.835	2.074	2.183	2.508	2.819
23	1.319	1.714	1.832	2.069	2.177	2.500	2.807
24	1.318	1.711	1.828	2.064	2.172	2.492	2.797
25	1.316	1.708	1.825	2.060	2.167	2.485	2.787
29	1.311	1.699	1.814	2.045	2.150	2.462	2.756
34	1.307	1.691	1.805	2.032	2.136	2.441	2.728
39	1.304	1.685	1.798	2.023	2.125	2.426	2.708
49	1.299	1.677	1.788	2.010	2.110	2.405	2.680
59	1.296	1.671	1.781	2.001	2.100	2.391	2.662
69	1.294	1.667	1.777	1.995	2.093	2.382	2.649
79	1.292	1.664	1.773	1.990	2.088	2.374	2.640
89	1.291	1.662	1.771	1.987	2.084	2.369	2.632
99	1.290	1.660	1.769	1.984	2.081	2.365	2.626
149	1.287	1.655	1.763	1.976	2.072	2.352	2.609
199	1.286	1.653	1.760	1.972	2.067	2.345	2.601
299	1.284	1.650	1.757	1.968	2.063	2.339	2.592
∞	1.282	1.645	1.751	1.960	2.054	2.326	2.576

D   Fisher and Beta Distributions

Quantiles $f_{m,n;\alpha}$ of the $\mathcal{F}_{m,n}$-distributions with m and n degrees of freedom in the numerator and denominator, respectively. $f_{m,n;\alpha}$ is the value $c > 0$ satisfying $\mathcal{F}_{m,n}([0, c]) = \alpha$. The associated quantiles of the beta distributions can be derived from Remark (9.14). The row $n = \infty$ contains the limiting values $\lim_{n\to\infty} f_{m,n;\alpha} = \chi^2_{m;\alpha}/m$, cf. Problem 9.13.

95%-Quantiles $f_{m,n;0.95}$

$m =$	1	2	3	4	5	6	7	8	9	10
$n = 1$	161.	199.	216.	225.	230.	234.	237.	239.	241.	242.
2	18.5	19.0	19.2	19.2	19.3	19.3	19.4	19.4	19.4	19.4
3	10.1	9.55	9.28	9.12	9.01	8.94	8.89	8.85	8.81	8.79
4	7.71	6.94	6.59	6.39	6.26	6.16	6.09	6.04	6.00	5.96
5	6.61	5.79	5.41	5.19	5.05	4.95	4.88	4.82	4.77	4.74
6	5.99	5.14	4.76	4.53	4.39	4.28	4.21	4.15	4.10	4.06
7	5.59	4.74	4.35	4.12	3.97	3.87	3.79	3.73	3.68	3.64
8	5.32	4.46	4.07	3.84	3.69	3.58	3.50	3.44	3.39	3.35
9	5.12	4.26	3.86	3.63	3.48	3.37	3.29	3.23	3.18	3.14
10	4.96	4.10	3.71	3.48	3.33	3.22	3.14	3.07	3.02	2.98
11	4.84	3.98	3.59	3.36	3.20	3.09	3.01	2.95	2.90	2.85
12	4.75	3.89	3.49	3.26	3.11	3.00	2.91	2.85	2.80	2.75
13	4.67	3.81	3.41	3.18	3.03	2.92	2.83	2.77	2.71	2.67
14	4.60	3.74	3.34	3.11	2.96	2.85	2.76	2.70	2.65	2.60
15	4.54	3.68	3.29	3.06	2.90	2.79	2.71	2.64	2.59	2.54
16	4.49	3.63	3.24	3.01	2.85	2.74	2.66	2.59	2.54	2.49
17	4.45	3.59	3.20	2.96	2.81	2.70	2.61	2.55	2.49	2.45
18	4.41	3.55	3.16	2.93	2.77	2.66	2.58	2.51	2.46	2.41
19	4.38	3.52	3.13	2.90	2.74	2.63	2.54	2.48	2.42	2.38
20	4.35	3.49	3.10	2.87	2.71	2.60	2.51	2.45	2.39	2.35
21	4.32	3.47	3.07	2.84	2.68	2.57	2.49	2.42	2.37	2.32
22	4.30	3.44	3.05	2.82	2.66	2.55	2.46	2.40	2.34	2.30
23	4.28	3.42	3.03	2.80	2.64	2.53	2.44	2.37	2.32	2.27
24	4.26	3.40	3.01	2.78	2.62	2.51	2.42	2.36	2.30	2.25
25	4.24	3.39	2.99	2.76	2.60	2.49	2.40	2.34	2.28	2.24
26	4.23	3.37	2.98	2.74	2.59	2.47	2.39	2.32	2.27	2.22
27	4.21	3.35	2.96	2.73	2.57	2.46	2.37	2.31	2.25	2.20
28	4.20	3.34	2.95	2.71	2.56	2.45	2.36	2.29	2.24	2.19
29	4.18	3.33	2.93	2.70	2.55	2.43	2.35	2.28	2.22	2.18
30	4.17	3.32	2.92	2.69	2.53	2.42	2.33	2.27	2.21	2.16
35	4.12	3.27	2.87	2.64	2.49	2.37	2.29	2.22	2.16	2.11
40	4.08	3.23	2.84	2.61	2.45	2.34	2.25	2.18	2.12	2.08
45	4.06	3.20	2.81	2.58	2.42	2.31	2.22	2.15	2.10	2.05
50	4.03	3.18	2.79	2.56	2.40	2.29	2.20	2.13	2.07	2.03
60	4.00	3.15	2.76	2.53	2.37	2.25	2.17	2.10	2.04	1.99
70	3.98	3.13	2.74	2.50	2.35	2.23	2.14	2.07	2.02	1.97
80	3.96	3.11	2.72	2.49	2.33	2.21	2.13	2.06	2.00	1.95
90	3.95	3.10	2.71	2.47	2.32	2.20	2.11	2.04	1.99	1.94
100	3.94	3.09	2.70	2.46	2.31	2.19	2.10	2.03	1.97	1.93
150	3.90	3.06	2.66	2.43	2.27	2.16	2.07	2.00	1.94	1.89
200	3.89	3.04	2.65	2.42	2.26	2.14	2.06	1.98	1.93	1.88
∞	3.84	3.00	2.60	2.37	2.21	2.10	2.01	1.94	1.88	1.83

99%-Quantiles $f_{m,n;0.99}$

$m =$	1	2	3	4	5	6	7	8	9	10
$n = 6$	13.7	10.9	9.78	9.15	8.75	8.47	8.26	8.10	7.98	7.87
7	12.2	9.55	8.45	7.85	7.46	7.19	6.99	6.84	6.72	6.62
8	11.3	8.65	7.59	7.01	6.63	6.37	6.18	6.03	5.91	5.81
9	10.6	8.02	6.99	6.42	6.06	5.80	5.61	5.47	5.35	5.26
10	10.0	7.56	6.55	5.99	5.64	5.39	5.20	5.06	4.94	4.85
11	9.65	7.21	6.22	5.67	5.32	5.07	4.89	4.74	4.63	4.54
12	9.33	6.93	5.95	5.41	5.06	4.82	4.64	4.50	4.39	4.30
13	9.07	6.70	5.74	5.21	4.86	4.62	4.44	4.30	4.19	4.10
14	8.86	6.51	5.56	5.04	4.69	4.46	4.28	4.14	4.03	3.94
15	8.68	6.36	5.42	4.89	4.56	4.32	4.14	4.00	3.89	3.80
16	8.53	6.23	5.29	4.77	4.44	4.20	4.03	3.89	3.78	3.69
17	8.40	6.11	5.18	4.67	4.34	4.10	3.93	3.79	3.68	3.59
18	8.29	6.01	5.09	4.58	4.25	4.01	3.84	3.71	3.60	3.51
19	8.18	5.93	5.01	4.50	4.17	3.94	3.77	3.63	3.52	3.43
20	8.10	5.85	4.94	4.43	4.10	3.87	3.70	3.56	3.46	3.37
21	8.02	5.78	4.87	4.37	4.04	3.81	3.64	3.51	3.40	3.31
22	7.95	5.72	4.82	4.31	3.99	3.76	3.59	3.45	3.35	3.26
23	7.88	5.66	4.76	4.26	3.94	3.71	3.54	3.41	3.30	3.21
24	7.82	5.61	4.72	4.22	3.90	3.67	3.50	3.36	3.26	3.17
25	7.77	5.57	4.68	4.18	3.85	3.63	3.46	3.32	3.22	3.13
26	7.72	5.53	4.64	4.14	3.82	3.59	3.42	3.29	3.18	3.09
27	7.68	5.49	4.06	4.11	3.78	3.56	3.39	3.26	3.15	3.06
28	7.64	5.45	4.57	4.07	3.75	3.53	3.36	3.23	3.12	3.03
29	7.60	5.42	4.54	4.04	3.73	3.50	3.33	3.20	3.09	3.00
30	7.56	5.39	4.51	4.02	3.70	3.47	3.30	3.17	3.07	2.98
31	7.53	5.36	4.48	3.99	3.67	3.45	3.28	3.15	3.04	2.96
32	7.50	5.34	4.46	3.97	3.65	3.43	3.26	3.13	3.02	2.93
33	7.47	5.31	4.44	3.95	3.63	3.41	3.24	3.11	3.00	2.91
34	7.44	5.29	4.42	3.93	3.61	3.39	3.22	3.09	2.98	2.89
35	7.42	5.27	4.40	3.91	3.59	3.37	3.20	3.07	2.96	2.88
40	7.42	5.27	4.40	3.91	3.59	3.37	3.20	3.07	2.96	2.88
45	7.31	5.18	4.31	3.83	3.51	3.29	3.12	2.99	2.89	2.80
50	7.23	5.11	4.25	3.77	3.45	3.23	3.07	2.94	2.83	2.74
55	7.17	5.06	4.20	3.72	3.41	3.19	3.02	2.89	2.78	2.70
60	7.08	4.98	4.13	3.65	3.34	3.12	2.95	2.82	2.72	2.63
70	7.01	4.92	4.07	3.60	3.29	3.07	2.91	2.78	2.67	2.59
80	6.96	4.88	4.04	3.56	3.26	3.04	2.87	2.74	2.64	2.55
90	6.93	4.85	4.01	3.53	3.23	3.01	2.84	2.72	2.61	2.52
100	6.90	4.82	3.98	3.51	3.21	2.99	2.82	2.69	2.59	2.50
120	6.85	4.79	3.95	3.48	3.17	2.96	2.79	2.66	2.56	2.47
150	6.81	4.75	3.91	3.45	3.14	2.92	2.76	2.63	2.53	2.44
200	6.76	4.71	3.88	3.41	3.11	2.89	2.73	2.60	2.50	2.41
300	6.72	4.68	3.85	3.38	3.08	2.86	2.70	2.57	2.47	2.38
400	6.70	4.66	3.83	3.37	3.06	2.85	2.68	2.56	2.45	2.37
500	6.69	4.65	3.82	3.36	3.05	2.84	2.68	2.55	2.44	2.36
∞	6.63	4.61	3.78	3.32	3.02	2.80	2.64	2.51	2.41	2.32

E   Wilcoxon–Mann–Whitney U-Distributions

Quantiles $u_{k;\alpha}$ of the distribution of the U-statistics $U_{k,k}$ under the null hypothesis. $u_{k;\alpha}$ is the largest integer c such that $P^{\otimes 2k}(U_{k,k} < c) \leq \alpha$. The actual probabilities $p_{k-;\alpha} = P^{\otimes 2k}(U_{k,k} < u_{k;\alpha})$ and $p_{k+;\alpha} = P^{\otimes 2k}(U_{k,k} \leq u_{k;\alpha})$ are listed for comparison. In view of the symmetry of the $U_{k,k}$-distribution, one obtains the α-fractiles by the equation $u_{k;1-\alpha} = k^2 - u_{k;\alpha}$. For large k one can use Theorem (11.28).

5%-Quantiles

k	4	5	6	7	8	9	10	11	12
$u_{k;0.05}$	2	5	8	12	16	21	28	35	43
$p_{k-;0.05}$	.0286	.0476	.0465	.0487	.0415	.0470	.0446	.0440	.0444
$p_{k+;0.05}$	.0571	.0754	.0660	.0641	.0525	.0568	.0526	.0507	.0567

2.5%-Quantiles

k	4	5	6	7	8	9	10	11	12
$u_{k;0.025}$	1	3	6	9	14	18	24	31	38
$p_{k-;0.025}$	.0143	.0159	.0206	.0189	.0249	.0200	.0216	.0236	.0224
$p_{k+;0.025}$	.0286	.0278	.0325	.0265	.0325	.0252	.0262	.0278	.0259

1%-Quantiles

k	5	6	7	8	9	10	11	12	13
$u_{k;0.01}$	2	4	7	10	15	20	26	32	40
$p_{k-;0.01}$	.0079	.0076	.0087	.0074	.0094	.0093	.0096	.0086	.0095
$p_{k+;0.01}$	.0159	.0130	.0131	.0103	.0122	.0116	.0117	.0102	.0111

References

Besides the literature cited above, the list below offers a selection of textbooks which complement this text or can be used for a deeper study of stochastics.

[1] S. F. Arnold. *The Theory of Linear Models and Multivariate Analysis.* J. Wiley & Sons, New York etc., 1981.

[2] R. B. Ash. *Basic Probability Theory.* J. Wiley & Sons, Chichester, 1970. Republished by Dover Publications, 2008.

[3] H. Bauer. *Probability Theory.* Walter de Gruyter, Berlin – New York, 1996.

[4] H. Bauer. *Measure and Integration Theory.* Walter de Gruyter, Berlin – New York, 2001.

[5] P. J. Bickel and K. J. Doksum. *Mathematical Statistics, Basic Ideas and Selected Topics, Vol. I.* Prentice Hall, 2nd ed., updated printing, 2006.

[6] P. Billingsley. *Probability and Measure.* J. Wiley & Sons, anniversary ed., 2012.

[7] L. Breiman. *Probability and Stochastic Processes: With a View Toward Applications.* Course Technology, 2nd reprint ed., 1986.

[8] L. Breiman. *Statistics: With a View Toward Applications.* Houghton Mifflin, 1973.

[9] L. Breiman. *Probability.* SIAM: Society for Industrial and Applied Mathematics, reprint ed. 1992.

[10] P. Brémaud. *An Introduction to Probabilistic Modeling.* Springer, New York, 2nd printing 1994.

[11] K. L. Chung and F. AitSahlia. *Elementary Probability Theory.* Springer, New York etc., 4th ed., 2003.

[12] D. L. Cohn. *Measure Theory.* Birkhäuser, Boston, 1980, reprinted 1997.

[13] I. Csiszár and J. Körner. *Information Theory: Coding Theorems for Discrete Memoryless Systems.* Cambridge University Press, 2nd ed., 2011.

[14] F. M. Dekking, C. Kraaikamp, H. P. Lopuhaä, and L. E. Meester. *A Modern Introduction to Probability and Statistics.* Springer, London, 2005, reprint 2010.

[15] R. M. Dudley. *Real Analysis and Probability.* Cambridge University Press, 2nd ed., 2002.

[16] R. Durrett. *Probability: Theory and Examples.* Cambridge University Press, 4th ed., 2010.

[17] J. Dutka. On Gauss' Priority in the Discovery of the Method of Least Squares. *Arch. Hist. Exact. Sci.* **49** (1996), 355–370.

[18]   W. Feller. *An Introduction to Probability Theory and its Applications*, Vol. I. J. Wiley & Sons, Chichester, 3rd ed., 1968.

[19]   W. Feller. *An Introduction to Probability Theory and its Applications*, Vol. II. J. Wiley & Sons, Chichester, 2nd ed., 1971.

[20]   T. S. Ferguson. *Mathematical Statistics: A Decision Theoretic Approach.* Academic Press, New York – London, 1967.

[21]   G. Gallavotti. Ergodicity, Ensembles, Irreversibility in Boltzmann and Beyond. *J. Statist. Phys.* **78** (1995), 1571–1589.

[22]   G. Gigerenzer, *Calculated Risks: How to Know When Numbers Deceive You.* Simon & Schuster, New York, 2002.

[23]   G. Gigerenzer, Z. Swijtink, Th. Porter, L. Daston, J. Beatty, and L. Krüger. *The Empire of Chance: How Probability Changed Science and Everyday Life.* Cambridge University Press, 1990.

[24]   G. R. Grimmett and D. R. Stirzaker. *Probability and Random Processes.* Oxford University Press, 3rd ed., 2001.

[25]   G. R. Grimmett and D. R. Stirzaker. *One Thousand Exercises in Probability.* Oxford University Press, 2nd ed., 2001.

[26]   C. M. Grinstead and J. L. Snell. *Introduction to Probability.* American Mathematical Society, Providence, 2nd ed., 1997. Also available under `www.math.dartmouth.edu/ ~prob/prob/prob.pdf`

[27]   O. Häggström. *Finite Markov Chains and Algorithmic Applications.* Cambridge University Press, 2002.

[28]   A. Hald. *A History of Mathematical Statistics from 1750 to 1930.* Wiley, New York, 1998.

[29]   K. Jacobs. *Discrete Stochastics.* Birkhäuser, Basel, 1992.

[30]   J. Jacod and P. Protter. *Probability Essentials.* Springer, Berlin etc., 2nd ed., 2003, corr. 2nd printing 2004.

[31]   Th. Jech. *Set Theory.* Springer, Berlin etc., 3rd millenium ed., 4th printing, 2006.

[32]   O. Kallenberg. *Foundations of Modern Probability.* Springer, New York, 2nd ed. 2002, reprint 2010.

[33]   N. Keyfitz. *Introduction to the Mathematics of Population.* Addison-Wesley, Reading, Mass., rev. print., 1977.

[34]   J. C. Kiefer. *Introduction to Statistical Inference.* Springer, New York, 1987.

[35]   A. Klenke. *Probability Theory. A Comprehensive Course.* Springer, Berlin etc., 2008.

[36]   L. B. Koralov and Y. G. Sinai. *Theory of Probability and Random Processes.* Springer, Berlin etc., 2007.

[37]   D. E. Knuth. *The Art of Computer Programming*, Vol. 2 / Seminumerical Algorithms. Addison Wesley, Reading, Mass., 3rd ed., 1997.

[38] U. Krengel. *Einführung in die Wahrscheinlichkeitstheorie und Statistik.* Vieweg, Braunschweig, 8th ed., 2005.

[39] K. Krickeberg and H. Ziezold. *Stochastische Methoden.* Springer, Berlin etc., 4th ed., 1995.

[40] S. Lang. *Linear Algebra.* Springer, Berlin etc., 3rd ed. 1987, reprint 2010.

[41] E. L. Lehmann and G. Casella. *Theory of Point Estimation.* Springer, New York, 2nd ed. 1998, 4th printing 2003.

[42] E. L. Lehmann and J. P. Romano. *Testing Statistical Hypotheses.* Springer, New York, 3rd ed. 2005, reprint 2010.

[43] J. Lehn and H. Wegmann. *Einführung in die Statistik.* B. G. Teubner, Stuttgart etc., 5th ed., 2006.

[44] D. A. Levin, Y. Peres, and E. L. Wilmer. *Markov Chains and Mixing Times.* American Mathematical Society, Providence, 2009. Also available under http://pages.uoregon.edu/dlevin/MARKOV/markovmixing.pdf

[45] B. W. Lindgren. *Statistical Theory.* Chapman and Hall / CRC, London – New York, 4th ed., 1993.

[46] D. V. Lindley and W. F. Scott. *New Cambridge Statistical Tables.* Cambridge University Press, 2nd ed., 1995.

[47] J. K. Lindsey. *Parametric Statistical Inference.* Oxford University Press, 1996.

[48] R. Meester and R. Roy. *Continuum Percolation.* Cambridge University Press, 1996.

[49] J. D. Miller, E. C. Scott, and S. Okamoto. Public Acceptance of Evolution, *Science* **313** (2006), 765–766.

[50] J. P. Morgan, N. R. Chaganty, R. C. Dahiya, and M. J. Doviak. Let's Make a Deal: The Player's Dilemma. *Amer. Statist.* **45** (1991), 284–287.

[51] W. R. Pestman. *Mathematical Statistics.* Walter de Gruyter, Berlin – New York, 2nd ed., 2009.

[52] J. Pitman. *Probability.* Springer, Berlin etc., 7th printing, 1999.

[53] L. Rade and B. Westergren. *Mathematics Handbook for Science and Engineering.* Springer, Berlin etc., 5th ed. 2004, reprint 2010.

[54] J. Rosenhouse. *The Monty Hall Problem: The Remarkable Story of Math's Most Contentious Brain Teaser.* Oxford University Press, New York etc., 2009.

[55] S. M. Ross. *Introduction to Probability Models.* Academic Press, Amsterdam etc., 10th ed., 2010.

[56] H. L. Royden and P. M. Fitzpatrick. *Real Analysis.* Prentice Hall, Boston, 4th ed., 2009.

[57] W. Rudin. *Principles of Mathematical Analysis.* McGraw-Hill, New York, 3rd ed., 1976.

[58] W. Rudin. *Real and Complex Analysis.* McGraw-Hill, New York, 3rd ed., 1987.

[59] A. N. Shiryayev. *Probability.* Springer, New York, 2nd ed., 1996.

[60] W. A. Stahel. *Statistische Datenanalyse.* Eine Einführung für Naturwissenschaftler. Vieweg, Braunschweig, 5th ed., 2007.

[61] D. Stoyan, W. S. Kendall, and J. Mecke. *Stochastic Geometry and its Applications.* J. Wiley & Sons, Chichester, 2nd ed. 1995, softcover 2008.

[62] J. M. Stoyanov. *Counterexamples in Probability.* J. Wiley & Sons, Chichester, 2nd ed., 1997.

[63] H. M. Taylor and S. Karlin. *An Introduction to Stochastic Modelling.* Academic Press, San Diego etc., 3rd ed., 1998.

[64] F. Topsoe. *Spontaneous Phenomena: A Mathematical Analysis.* Academic Press, Boston, 1990.

[65] D. Williams. *Weighing the Odds. A course in Probability and Statistics.* Cambridge University Press, 2001.

List of Notation

General Notations

$:=$	defining equality
$\mathbb{N} := \{1, 2, \ldots\}$	set of natural numbers
$\mathbb{Z} := \{\ldots, -1, 0, 1, \ldots\}$	set of integers
$\mathbb{Z}_+ := \{0, 1, 2, \ldots\}$	set of non-negative integers
$\mathbb{Q}$	set of rational numbers
$\mathbb{R}$	real line
$[a, b]$	closed interval
$]a, b[$	open interval
$[a, b[, \,]a, b]$	half-open intervals
$\mathscr{P}(\Omega) := \{A : A \subset \Omega\}$	power set (i.e., set of all subsets) of Ω
$\varnothing$	empty set
$A \subset B$	A is a (not necessarily proper!) subset of B
A^c	complement of a set A
$\bar{A}, A^o, \partial A$	closure, interior, boundary of a set $A \subset \mathbb{R}^n$
$|A|$	cardinality of a finite set A
$TA = \{T(\omega) : \omega \in A\}$	image of A under the mapping T
$\{X \in A\} = X^{-1}A$	preimage of A under the mapping X
$A \times B, \prod_{i \in I} A_i$	Cartesian product of sets
$A_n \uparrow A$	$A_1 \subset A_2 \subset \cdots$ and $A = \bigcup_{n=1}^{\infty} A_n$
$A_n \downarrow A$	$A_1 \supset A_2 \supset \cdots$ and $A = \bigcap_{n=1}^{\infty} A_n$
$\lfloor x \rfloor$ for $x \in \mathbb{R}$	greatest integer not greater than x
$\lceil x \rceil$ for $x \in \mathbb{R}$	least integer not less than x
$|x|$	modulus of $x \in \mathbb{R}$, Euclidean norm of $x \in \mathbb{R}^n$
x_i	ith coordinate of an n-tuple $x \in E^n$
$x \cdot y = \sum_{i=1}^{n} x_i y_i$	Euclidean scalar product of $x, y \in \mathbb{R}^n$
$x \perp y$	$x \cdot y = 0$, i.e., $x, y \in \mathbb{R}^n$ are orthogonal
$\|f\|$	supremum norm of a real-valued function f
$\log$	natural logarithm

δ_{ij}	Kronecker's delta, i.e., $\delta_{ij} = 1$ if $i = j$, and 0 otherwise
$\mathsf{E} = (\delta_{ij})_{1 \le i, j \le n}$	identity matrix (of appropriate dimension n)
M^{T}	transpose of a matrix (or a vector) M
$\mathbf{1} = (1, \dots, 1)^{\mathsf{T}}$	diagonal vector in $\mathbb{R}^n$
$\mathrm{span}(u_1, \dots, u_s)$	linear span of $u_1, \dots, u_s \in \mathbb{R}^n$
$a(k) \sim b(k)$	asymptotic equivalence: $a(k)/b(k) \to 1$ for $k \to \infty$
$O(\cdot)$	Landau symbol, 132

σ-Algebras, Probability Measures, Random Variables, Statistics

$\mathscr{B}, \mathscr{B}^n$	Borel σ-algebra on $\mathbb{R}$ resp. $\mathbb{R}^n$, 11
$\mathscr{B}_\Omega, \mathscr{B}_\Omega^n$	restriction of $\mathscr{B}$ resp. $\mathscr{B}^n$ to Ω, 12
$\bigotimes_{i \in I} \mathscr{E}_i, \mathscr{E}^{\otimes I}$	product σ-algebra, 12
$P \otimes Q$	product of two probability measures, 72
$\rho^{\otimes n}, P^{\otimes n}$	n-fold product of probability measures, 32, 72
$P \star Q, P^{\star n}$	convolution, n-fold convolution of probability measures, 74
$P \prec Q$	stochastic domination, 312
$P \circ X^{-1}$	distribution of a random variable X, 22
P^α, P^x	canonical Markov distributions, 152
F_X, F_P	(cumulative) distribution function of X resp. P, 23
$\mathbb{E}(X)$	expectation of a real random variable X, 92, 97
$\mathbb{E}^\alpha, \mathbb{E}^x, \mathbb{E}_\vartheta$	expectations relative to P^α, P^x, resp. P_ϑ, 169, 194
$\mathbb{E}(P)$	expectation of a probability measure P on $\mathbb{R}$, 109
$\mathbb{V}(X), \mathbb{V}(P)$	variance, 107, 109
$\mathrm{Cov}(X, Y)$	covariance, 107
$\mathscr{L}^m, \mathscr{L}^m(P)$	space of real random variables with existing mth moment, 107
$Y_n \uparrow Y$	$Y_1 \le Y_2 \le \cdots$ and $Y_n \to Y$ (pointwise)
$Y_n \xrightarrow{P} Y$	convergence in probability relative to P, 120
$Y_n \xrightarrow{d} Y$ resp. Q	convergence in distribution, 138, 289
1_A	indicator function of a set A, 17
Id_E	identity map $x \to x$ on E
L	empirical distribution, 240, 297
M	sample mean, 204
V	sample variance, 204
V^*	corrected (unbiased) sample variance, 205, 332
$X_{k:n}$	kth order statistics, 44, 240
$G_\varphi(\vartheta) = \mathbb{E}_\vartheta(\varphi)$	power function of a test φ, 263

Special Distributions and their Densities

$\mathcal{B}_{n,p}$	binomial distribution, 33
$\overline{\mathcal{B}}_{r,p}$	negative binomial distribution, 41
$\boldsymbol{\beta}_{a,b}$	beta distribution with density $\beta_{a,b}$, 45
χ_n^2	chi-square distribution with density χ_n^2, 251
δ_ξ	Dirac distribution at ξ, 13
$\mathcal{D}_\rho$	Dirichlet distribution, 226
$\mathcal{E}_\alpha$	exponential distribution, 43
$\mathcal{F}_{m,n}$	Fisher distribution with density $f_{m,n}$, 253
$\Gamma_{\alpha,r}$	gamma distribution with density $\gamma_{\alpha,r}$, 43
$\mathcal{G}_p$	geometric distribution, 41
$\mathcal{H}_{n,\vec{N}}, \mathcal{H}_{n;N_1,N_0}$	hypergeometric distribution, 36
$\mathcal{M}_{n,\rho}$	multinomial distribution, 33
$\mathcal{N}_{m,v}$	normal distribution with density $\phi_{m,v}$, 48
ϕ, Φ	density and distribution function of $\mathcal{N}_{0,1}$, 134
$\mathcal{N}_n(m, C)$	multivariate normal distribution with density $\phi_{m,C}$, 248
$\mathcal{P}_\alpha$	Poisson distribution, 40
t_n	Student's distribution with density t_n, 253
$\mathcal{U}_\Omega$	(discrete or continuous) uniform distribution on Ω, 27, 29

The Greek Alphabet

α	A	alpha	ι	I	iota	ρ	P	rho
β	B	beta	κ	K	kappa	σ, ς	Σ	sigma
γ	Γ	gamma	λ	Λ	lambda	τ	T	tau
δ	Δ	delta	μ	M	mu	υ	Υ	upsilon
ε	E	epsilon	ν	N	nu	φ, ϕ	Φ	phi
ζ	Z	zeta	ξ	Ξ	xi	χ	X	chi
η	H	eta	o	O	omicron	ψ	Ψ	psi
ϑ	Θ	theta	π	Π	pi	ω	Ω	omega

Index

Printed by Amazon Italia Logistica S.r.l.
Torrazza Piemonte (TO), Italy

11236996R00243